Intelligence
is not Artificial
(Expanded Edition)

Piero Scaruffi

ISBN 978-1984101457

To the machines that will never become intelligent enough to understand the books that humans write about them and why we write them

Table of Contents

4

5

What this Book is About – 2019 Edition

(I apologize for the long preface, but, paraphrasing Mark Twain, "I didn't have time to write a short one, so I wrote a long one instead").

This book contains a boring history of Artificial Intelligence but also something more interesting. This book, basically, deals with three fascinating sociological and anthropological phenomena: 1. an old-fashioned apocalyptic religion, according to which the robots are coming and will kill us all (or, at least, steal our jobs and make us irrelevant); 2. an old-fashioned prophetic religion, according to which a superhuman entity (the singularity) is coming and will save us all (i.e. will make us immortal); and 3. a new kind of society, the "vast algorithmic bureaucracy", in which everything that is not forbidden is mandatory (which, i feel, is the real news, not just an irrational belief like the other two).

Since the publication of the first edition of this book (the "philosophical" edition), it has become difficult to say something intelligent about Artificial Intelligence. The market is flooded with books that alternatively predict the apocalypse or the panacea to all problems.

Luckily, an increasing number of scholars are coming out to speak against the hype, the unreasonable expectations and the exaggerated fears that have been created by the not-too-well-informed media and by not-too-honest writers and publishers.

I originally wrote the book "Intelligence is not Artificial" in 2013 when the media started reporting stunning progress in Artificial Intelligence that in my opinion was wildly exaggerated and when discussions about the Singularity were becoming, in my opinion, a bit ridiculous. My 2013 book was more philosophical than technical.

Now the term "Artificial Intelligence" has become so popular that i literally don't know anymore what we are talking about: just about everything is being tagged "Artificial Intelligence". The devices of the so-called "Internet of Things" are regularly marketed as "Artificial Intelligence" as are all of data science and most of statistics. This fad was almost singlehandedly created by one corporation's press releases, hailing humble experiments in neural networks (often based on very old theory) as steps towards a technological, social and economic revolution. Countless firms are rushing to reprint and restyle their marketing material to include popular terms such as "machine learning" and "bot" (short for "robot", typically a software robot). The term "Artificial Intelligence" is so abused that i wonder why a light switch is not called "Artificial Intelligence": after all it does something that is short of miraculous, it turns a dark room into a bright room. When i asked a startup founder why he was calling his device "Artificial Intelligence" but not his TV set, he couldn't come up with a good explanation. A TV set uses sophisticated algorithms to "learn" what the original image was, and the "app" is pretty spectacular: i press a button and i see someone who is in another city. I can keep pressing buttons and see people in different cities. It looks like a pretty amazing application to me, certainly more amazing than that startup's wearable device that checks some bodily data and displays a warning if they are too high or too low.

Much has changed since i first published this book in 2013. The main change is not in technological progress, but in the definition of Artificial Intelligence. What a departure from the 1990s, when the expression "Artificial Intelligence" was ridiculed. "Artificial Intelligence" is rapidly becoming synonymous with "automation". All automation is now "Artificial Intelligence". For example, the A.I. community never considered a factory robot that simply repeats the same movement all the time as "intelligent", but now it is. Some of these robots replace the dumbest of human activities, but they are now routinely classified into the "Artificial Intelligence" category to the point that old-school A.I. researchers literally don't know what "automation" means anymore: is there anything that is "automation" but not "Artificial Intelligence"?

The railway or airplane seat booking systems are sophisticated computerized systems that we never called A.I. but it looks like today they would be called A.I. They certainly do something much more useful than playing a board game, and they serve billions of passengers. If they weren't decades old, the various weather forecast systems (some of the most challenging simulation programs in the world) would also be classified as A.I. And what about the various "malware" that infiltrate millions of computers worldwide? Aren't those A.I.? In the 1960s, A.I. scientists had some dignity and never claimed that the Apollo mission guidance system was A.I. but today more trivial guidance systems for drones are routinely called A.I.

John McCarthy is credited with saying: "As soon as it works, no one calls it A.I. anymore" (incidentally, i have never been able to prove that he really said it). Today we are rapidly moving towards the opposite bias: "If it works, everybody calls it A.I."

We have seen this movie before. In the 1980s the most reputable names in business studies were counting billions of dollars of investment in A.I. simply because everything was being tagged A.I. It was popular to be an A.I. company, group, researcher (i was one of them, the founding director of Olivetti's Artificial Intelligence Center in California). In the 2010s, we are witnessing a similar craze. A few months after Bloomberg estimated the total 2015 investment in A.I. startups at $128 million (a decline by 50% over the previous year), VentureScanner estimated $2.2 billion, 20 times more. How could Bloomberg be so wrong? It all depends on what you count as A.I. What happened in those few months between one and the other study was that just about every startup rebranded itself as being an A.I. company or at least having an A.I. component. A similar phenomenon is spreading through the corporate world, rebranding old projects and products as A.I.-based. Artificial Intelligence will soon encompass every software on your mobile phone.

I have been using a messaging application since 2014. Recently i noticed that they changed the top line of their website. It now boasts "Voice Calls: Secure, Crystal-Clear, AI-Powered", but it is the exact same app of 2014. The camera feature that Canon dubbed "image stabilization" is now routinely marketed as an "intelligent" feature. Canon introduced it in 1995 with the EF 75-300/4-5.6 IS zoom lens, and the other camera manufacturers followed suit (Nikon called it "vibration reduction" but now the acronym VR is being monopolized by virtual reality). It is based on simple optical formulas, and in 1995 nobody would have dreamed of relating it to Artificial Intelligence. In 1992 Mattel released a talking Barbie doll which spoke a few sentences such as "Wanna have a pizza party?" Today this would probably be hailed as another feat of A.I. (In a prelude to all the controversies that would arise in the age of chatbots, this talking doll was parodied on the TV show "The Simpsons" and some dolls were later recalled because accused of being sexist by the American Association of University Women). When (in 1999) Tim Westergren and Will Glaser wrote the algorithm (the Music Genome Project) to classify musical compositions based on a few hundred features, and later (in 2000) launched the application Pandora that picks music for you based on your taste, they didn't call it "Artificial Intelligence". But that's what it is now. Now it would be silly not to call it A.I., given that much simpler algorithms are marketed as A.I. I suspect that today Akihiro Yokoi's Tamagotchi pets, released in 1996, whose life story depends on the actions of the owner, would be marketed as A.I. And certainly Eyepet, the Sony PlayStation3 game of 2009 developed by Playlogic in the Netherlands (that was a top-selling game) should qualify as A.I.: this virtual pet (augmented reality before it became fashionable) reacts to objects and people. In 2017 Huawei introduced the smartphone Mate 10 Pro equipped with a "neural processing unit" (NPU) that reportedly accelerates Microsoft's translation software: it is just a faster processor. In Hangzhou i was told that they are building an "A.I. hotel". I asked what is an "A.I. hotel" and they told me it's a hotel where guests use a card to enter the front door and to register themselves at booths. There is no reception. I used similar hotels twice, in Sweden and France, but back then nobody thought of calling them "A.I. hotels". Your dishwasher will soon be called A.I. In fact, you don't know it, but your house is already full of A.I.

machines: the manufacturers are changing the brochures while you are using them. The "smartphone" somehow contributed to the misunderstanding. Ericsson was the first brand to coin the phrase "smartphone", with the release of its GS88 in 1997 and the term "smartphone" took off around 1999-2002, especially after the launch of Research in Motion's first BlackBerry phone (the 5810) in 2002. By accident, Apple launched the iPhone in 2007 just when "deep learning" was invented. The two events had nothing in common, but, one being called "smart" and the other one being called "intelligent", some confusion was inevitable. And, by accident, 2012 (the annus mirabilis of deep learning) was the year when the world went mobile: smartphone sales skyrocketed to 680 million units, up 30% yearly (according to Gartner Group), a feat repeated the following year and never repeated again. The peak of the smartphone frenzy was the fourth quarter of 2012: 208 million smartphones were sold, a 38% increase over the fourth quarter of the previous year. By coincidence, that's exactly when the A.I. world was shaken by deep learning's spectacular success in the Large Scale Visual Recognition Challenge (ILSVRC), whose competition results were announced in December. "Smart" and "intelligent" became mandatory adjectives for just about anything. The word "intelligent" is being applied to all sorts of features in all sorts of appliances, gadgets and devices, but the founders of Artificial Intelligence would turn in their graves if they were told what features now qualify as "intelligent".

The prediction that "A.I. will be pervasive" is becoming a self-fulfilling prophecy: if we call everything "A.I.", then, yes, A.I. will be pervasive. Just like if we called everything Nonsense, then Nonsense would be pervasive.

My feeling today is that 99% of the research in A.I. is not used for practical commercial applications, and that 99% of the commercial applications that are being marketed as A.I. have little or nothing in common with research in A.I. The market is coming up with a definition of A.I. that the founders and assorted philosophers would never endorse.

Technically speaking, renaming all Computer Science as Artificial Intelligence is not completely wrong because the very first computers were publicized as the "electronic brains", hence every software ever written is the by-product of an Artificial Intelligence program that started with the very first computer. The border between A.I. and plain Computer Science has always been blurred.

Whatever the current definition, it is important to understand that A.I. is not magic: the border between A.I. and magic is NOT blurred! A.I. is just computational mathematics applied to automation.

Therefore this book tries to explain what A.I. scientists do. Whatever your theoretical definition of A.I. is, and whatever your theoretical definition of "intelligence" is, there is a history of working on very interesting mathematics. Artificial Intelligence practitioners are more like artisans than scientists: the artisan doesn't care what the scientists proved, the artisan keeps doing what he thinks can be done.

While the hype was growing, i had the opposite problem. I kept wondering why algorithms are so stupid. We are increasingly surrounded by incredibly stupid algorithms that want to turn us into dumb robots. I got tempted to write a book subtitled "A manual on how to cope with the age of hyper-stupid machines". I am not a dumb algorithm but i increasingly have no way to tell the dumb algorithm that i am not a dumb algorithm: the dumb algorithm insists in treating me like a dumb algorithm. I feel like shouting "I am intelligent!" to a crowd of incredibly stupid algorithms that are closing in on me.

As we lower the degree of intelligence that is expected from humans, it becomes natural to see machines as "intelligent". The lower your intelligence, the more intelligent the machines around you will look. How can we tell if machines are getting more intelligent or we are getting less intelligent? Everything is relative, as Albert Einstein said, and you see the train as moving forward but someone on the train sees you as moving backwards.

I have a vision of a world increasingly dominated by "vast algorithmic bureaucracies". That's the real dystopia. The algorithm is a consequence, not a cause. Those bureaucracies are created by humans to get society organized. Initially the algorithm is performed by a human being. You can still see human algorithms when you order a sandwich at one of those fast-food chains: you pick the kind of sandwich, the kind of bread, the kind of vegetables, and so on, and then the kids follow a number of repetitive steps to prepare your sandwich; you move down the line and pay at the register. Once you turn a service into an algorithm, it is trivial to replace the human being with a computer. But it is important to realize that we are the ones replacing human interaction with algorithms. A "smart city" is a city where everything has been turned into an efficient algorithm connected to all other algorithms. The problem, of course, is that cities are not just buildings, streets and cars. There are also people. The term "smart city" makes you think of an intelligent city working for its citizens when in fact a "smart city" is simply a high-tech concentration camp where citizens are treated like numbers.

Humanity is not at risk because very intelligent machines will take over. Humanity is at risk because it is increasingly forced to coexist with very stupid machines in these vast algorithmic bureaucracies. The risk is that we will end up creating not superhuman technology but subhuman societies.

This book now is many books in one. It is an introduction to the methods of Artificial Intelligence, and probably one of the most extensive histories of the field ever published. It is also a book on the risk of declining human intelligence when human minds are constantly surrounded by incredibly stupid machines. It is a book about these "vast algorithmic bureaucracies". And it is the original book, a book on the emerging religion of the 21st century, a religion that replaces even the God of monotheistic religions with an algorithm. The ultimate thesis of this book is perhaps more sociological than technological.

Alas, A.I. has not solved the mystery of the mind at all, and is not even remotely close to doing so. We understand so little of how our brain works.

In fact, what we understand is not enough to understand why we understand it.

A Preface that was Originally Omitted

I published my first two books in the late 1980s: the first one was a book on Artificial Intelligence and the second one, almost at the same time, was a history of rock music. Since that year, i have published six books on A.I. and seven on rock music. A friend once asked me what A.I. and rock music have in common. It took me almost 30 years to find the answer: the potential and the hype.

Both rock music and A.I. had and have a huge potential to have an impact on the world (respectively, on music and on science). But both rock music and A.I. suffer from poor historiography and even worse commentary. There is a popular website that "aggregates" reviews of popular music: metacritic.com. Judging from the average rating on that website, popular music is blessed with an incredible number of Beethovens who are constantly and effortlessly producing scores of masterpieces every year. The truth? There are few albums of popular music that are worth your time, let alone your money.

At the same time, there are scores of articles published in all sorts of magazines and newspapers, describing the amazing feats of "intelligent" machines. Every year it looks like machines are about to get smarter than the smartest humans, and that we are doomed to becoming their slaves if not extinct. The truth? Most machines "beep". That is the best that they can do. The most intelligent of them are incapable of doing what even the least intelligent of all creatures do every single day: survive.

Alas, my detractors think that i too fit the pattern of A.I. and rock music: i have become famous for my skepticism, and my detractors think that the word "hype" applies to me too.

Coincidentally, A.I. and rock music were born in the same year. (Coincidentally, that's also the year when i was born).

The Preface to the Original Edition

Writers, inventors and entrepreneurs, impressed by progress in several scientific fields and notably in Artificial Intelligence, are debating whether we may be heading for a "singularity" in which machines with super-human intelligence will arise and multiply. At the same time, enthusiastic coverage in the media has widely publicized machines performing sophisticated tasks, from beating masters of go/weiqi to driving a car, from recognizing cats in videos to outperforming human experts on TV quiz shows. These stories have re-ignited interest in Artificial Intelligence, whose goal is to create machines that are as intelligent as humans, and generated fears in the public that these intelligent machines might cause harm to humans, if not steal their jobs.

First of all, this book provides a "reality check" of sorts on Artificial Intelligence. I show that, in a society driven by news media that desperately need sensational news to make money and in an academic world increasingly driven by the desire to translate academic research into Silicon Valley start-ups, technological progress in general, and progress in computer science in particular, is often overrated. I wanted to dispel some notions and misconceptions, and my version of the facts may sound controversial until you read my explanations. I think that (real) progress in (real) Artificial Intelligence, since its founding, has been negligible, and one reason is, ironically, that computers have become so much more (computationally) powerful.

In general, we tend to exaggerate the uniqueness of our age, just as previous generations had done. The very premise of the singularity theory is that progress is accelerating like never before. I argue that there have been other eras of accelerating progress, and it is debatable if ours is truly so special. The less you know about the past the more you are amazed by the present.

There is certainly a lot of change in our era. But change is not necessarily progress, or, at least, it is not necessarily progress for everybody. Disruptive innovation is frequently more about disruption than about innovation because disruption creates huge new markets for the consumer electronics industry. This has nothing to do with machine intelligence, and sometimes not even much to do with innovation.

There is also an exaggerated metaphysical notion that human intelligence is some sort of an evolutionary climax. Maybe so, but it is worth cautioning that non-human intelligence is already among us, and is multiplying rapidly, but it is not a machine: countless animals are capable of feats that elude the smartest humans. For a long time we have also had machines capable of performing "superhuman" tasks. Think of the clock, invented almost 1,000 years ago, that can do something that no human can do: it can tell how many hours, minutes and seconds elapse between two events.

Once we realize that non-human intelligence has always been around, and that we were already building super-human machines centuries ago, the discussion about super-intelligent machines can be reframed in more historically and biologically meaningful terms.

The last generation or two missed out on the debates of the previous decades (the "Turing test", the "ghost in the machine", the "Chinese room", etc). Therefore it is much easier for new A.I. practitioners to impress the younger generations. I have summarized the various philosophical arguments in favor of and against the feasibility of machine intelligence in my book "Thinking about Thought" and i won't repeat them here. I will, however, at least caution the new generations that "grew up" (as far as cognitive science goes) at a time when the term "intelligence" was not "cool" at all: it was held to be too vague, too unscientific, too abused in popular literature to lend itself to scientific investigation. It is regrettable that it is being abused again, and, just like back then, without a proper definition of what we mean by "intelligence". Ask one hundred psychologists and you will get one hundred different definitions. Ask philosophers and you will get thick tomes written in a cryptic language. Ask neurobiologists and they may simply ignore you.

This is the mother of all problems in the debate on the "singularity": "singularity" and "superhuman intelligence" are non-scientific terms based on non-scientific coffee-house chatting.

The term "artificial intelligence" is even more confusing, a veritable moving target. In this book i capitalize Artificial Intelligence when i am referring to the discipline, while using lowercase "artificial intelligence" to refer to an intelligent machine or an intelligent software. A.I. practitioners also use the term "Artificial General Intelligence" (A.G.I.) to refer to a machine that will exhibit human-level intelligence, not just one intelligent skill.

I also feel that any discussion on machine intelligence should be complemented with an important (more important?) discussion about the changes in human intelligence due to the increased "intelligence" of machines. This change in human intelligence may have a stronger impact on the future of human civilization than the improvements in machine intelligence. To wit: the program of turning machines into humans is not very successful yet, but the program of turning humans into machines (via an endless repertory of rules and regulations) is very successful.

My perspective is a little different from the perspective of the many writers who have written, or are writing, books on Artificial Intelligence: i am a historian, not a futurist. I may not know the future, but at least i know the past.

I am intrigued by another sociological/anthropological aspect of this discussion: humans seem to have a genetic propensity to believe in higher forms of intelligence (gods, saints, UFOs, ...) and the Singularity (capitalized "S") could simply be the latest manifestation of this propensity in our post-religious 21st century.

However, most people don't really care for how we call it: they are afraid not of some electromechanical monster that will kill the human race, but, quite simply, of losing their job to smarter and smarter machines. This too seems to me a wild exaggeration. New machines have always created new jobs, that are also better-paid. I fail to see why this time should be different. Remove all the sensational hyperboles, and it should be obvious that smarter machines will create more jobs, and better-paid jobs.

All of this explains why i am not afraid of Artificial Intelligence: 1. A reality check shows that most of its achievements are not that impressive; 2. Most of the "intelligence" displayed by machines is actually due to the structured environment that humans build for them; 3. The accelerating progress that we perceive is not unique in history; 4. We have always been surrounded by super-human (or, better, non-human) intelligence; 5. I am more concerned about the future of human intelligence than about the future of machine intelligence.

We actually need intelligent machines. Technological progress has solved many problems, but there are still people dying of diseases and dangerous jobs, and we will soon have an ageing society that will need even more help from technology. I am not afraid that "intelligent" machines are coming. I am afraid that they will not come soon enough.

piero scaruffi

P.S.: Yes, i don't like to capitalize the first person pronoun "i".
P.P.S. I don't like footnotes. All the footnotes are chapters in this book. I also dislike listing the technical papers at the end of the book. I list them where they are referenced.

Sociological Background

Historians, scientists, philosophers and poets alike have written that the human being strives for the infinite. In the old days this meant (more or less) that s/he strives to become one with the god who created and rules the world. As atheism began to make strides in Western civilization, Arthur Schopenhauer rephrased the concept as a "will to power". Friedrich Nietzsche confirmed that the god of the Bible is dead, and the Western search for the "infinite" became a mathematical and scientific program instead of a mystical one. About a century ago, European mathematicians such as Bertrand Russell and David Hilbert launched a logical program that basically aimed at making it easy to prove and discover everything that can be. The perspective therefore changed: instead of something that humans have to attain, the infinite has become something that humans will build.

One of the consequences of this line of research was the creation of the digital electronic computer, the physical implementation of a thought experiment by the British mathematician Alan Turing. He also wrote a pioneering paper about machine intelligence ("Computing Machinery and Intelligence", 1950) and a few years later the term "artificial intelligence" was already popular among both scientists and philosophers. The first conference on Artificial Intelligence was held in 1956 at Dartmouth College in New Hampshire, organized by John McCarthy with help from Harvard University scientist Marvin Minsky and others. That was just ten years after the introduction of the first general-purpose computer, the ENIAC. From the very beginning, the electronic computer had been dubbed by the media "the electronic brain".

The first book ever published on electronic computing had been Edmund Berkeley's "Giant Brains or Machines that Think" (1949) that at one point says "we shall now consider how we can design a very simple machine that will think". In 1950, announcing the construction of the ACE by Alan Turing, a British newspaper screamed: "Electronic brain to be made at Teddington". In 1953, a Canadian newspaper wondered: "Does the Univac 120 really think?" In 1957 the magazine Radio & Television News ran an article on computers by Frank Leary titled "Behind the Giant Brains". A documentary released in 1958 was titled "Horizons of Science: Thinking Machines". (It was about IBM's programmer and chess player Alex Bernstein playing one of the first computer chess games on an IBM 704).

The idea behind the Singularity, a concept popularized by Ray Kurzweil's "The Age of Intelligent Machines" (1990) and by his subsequent, highly-successful, public-relations campaign, is that we are about to witness the advent of machines that are more intelligent than humans, so intelligent that humans can neither control them nor understand them.

Admittedly, the discipline of Artificial Intelligence, that had largely languished in the 1990s and 2000s, has staged a revival of sorts, both in the eyes of the public and in the eyes of big business. Achievements in the field of A.I. are often hailed by the mainstream media as steps towards machine domination, and investment in A.I. startups has multiplied several times over to reach record levels.

In the age that has seen the end of human space exploration, the retirement of the only commercial supersonic airplane, the decline of nuclear power, and the commercialization of the Internet (an event that basically turned a powerful scientific tool into a marketing tool and a form of light entertainment), machine intelligence seems to bring some kind of collective reassurance that we are not, after all, entering a new Dark Age; on the contrary, we are witnessing the dawn of a superhuman era. Of course, decades of science fiction books and movies helped create the ideal audience for this kind of scenario.

However, the tone and the (very weak) arguments in favor of the Singularity do remind one of religious prophecies, except that this time the coming messiah will be a product made by us instead of being sent by an external divinity. In a sense, this is a religion according to which we are creating the divinity.

The idea of the Singularity is fascinating because it plays the history of religion backwards. Religion traditionally is meant to explain the mystery and complexity of the universe, the miracle of life, the purpose of consciousness and so on. Even some of today's eminent scientists subscribe to the "creationist" notion that a superhuman intelligence was required to create the world. This theory is frequently called "intelligent design" but it would be more appropriate to call it "super-intelligent design" because "intelligent" only refers to human intelligence. The whole point of religion was precisely to posit the existence of something that human intelligence could never possibly build. The hidden assumption of religion is that all the laws of nature that humans can possibly discover will never be enough to explain the mysteries of the universe, of life, of the soul. Whatever can be explained by those mathematical laws can also be implemented by humans, and therefore does not require the existence of supernatural forces. God, instead, is a singularity, the singularity that preceded human intelligence and created it, and is infinitely superior to it.

Luckily for us, this supreme deity is also capable of, and somewhat interested in, granting us immortality, which, at the end of the day, is what believers hope to obtain from believing.

Today's hypothesis of a coming singularity due to super-intelligent machines provides a mirror image of this story. The original singularity (God) was needed to explain the inexplicable. The new singularity (the machine) will be unexplainable. Human intelligence could not, in the past, understand the nature of a God who created human intelligence; nor can it, in the present, understand the workings of the super-intelligent machine of the future that human intelligence will have created.

The Singularity movement is split into two camps: the optimists and the pessimists. The optimists think that machines will make us immortal. The pessimists think that machines will kill us all. I still haven't heard anyone take the kind of intermediary position that most religions take: good people will go to paradise, bad people will go to hell. Apparently, the Singularity will not distinguish between good and bad people: either it kills everybody or it makes everybody immortal. (Money may be more important than good deeds for one to become immortal because, if i understand correctly, immortality will be a service available for sale or for rent just like cloud computing is today).

"This is the whole point of technology. It creates an appetite for immortality on the one hand. It threatens universal extinction on the other. Technology is lust removed from nature.... It's what we invented to conceal the terrible secret of our decaying bodies". (Don DeLillo, "White Noise")

It is sometimes difficult to argue with the Singularity crowd because they often seem unaware that some of the topics they discuss have been discussed for a long time, with pros and cons, by philosophers and scientists. In its worst manifestation the Singularity movement is becoming the religion of high-tech nerds who did not study history, philosophy, or science, not even computer science. At its best, however, it helps acquaint the general public with a society of (software and hardware) robots that is inevitably coming, although its imminence might be wildly exaggerated.

It may not be a coincidence that the boom of interest in the Singularity originated in the USA (a country well acquainted with apocalyptic evangelism, conspiracy theories, UFO sightings and cryptic prophets like Nostradamus) and that it originated after the year 2,000, a year that had three zeroes– according to the calendar introduced by Pope Gregory in 1582– because of which it was thought by many to herald a major discontinuity in history if not the end of the world itself. For a while the world was shaken almost yearly by catastrophic predictions, most famously (in the USA) by Harold Camping's Biblical calculations that the end of the world was coming on the 21st of October of 2011 and the theory that the end of the Mayan calendar (December 21, 2012) marked the end of the world. Luckily, they were all proven wrong, but maybe they created a public opinion ready to be fascinated by a technological version of the same general story (the end of the human race).

Irony aside, it is fascinating to see how religion is being reinvented on completely different foundations in Silicon Valley.

A Brief History of Artificial Intelligence/ Prequel

One can start way back in the past with the ancient Greek and Chinese automata of two thousand years ago, or with the first electromechanical machines of a century ago, but to me a history of machine intelligence begins in earnest with the "universal machine", originally conceived in 1936 by the British mathematician Alan Turing. He did not personally build it, but Turing realized that one could create the perfect mathematician by simulating the way logical problems are solved: by manipulating symbols. The first computers were not Universal Turing Machines (UTM), but most computers built since the ENIAC (1946), including all the laptops and smartphones that are available today, are UTMs. Because it was founded on predicate logic, which only admits two values ("true" and "false"), the computer at the heart of any "intelligent" machine relies on binary logic (ones and zeroes).

Cybernetics (that can be dated back to the 1943 paper "Behavior, Purpose and Teleology" co-written by MIT mathematician Norbert Wiener, physiologist Arturo Rosenblueth and engineer Julian Bigelow) did much to show the relationship between machines and living organisms. One can argue that machines are a form of life or, vice versa, that living organisms are forms of machinery.

However, "intelligence" is commonly considered one or many steps above the merely "alive": humans are generally considered intelligent (by fellow humans), whereas worms are not.

Using digital electronic computers to mimic the brain is particularly tempting because Santiago Ramon y Cajal had discovered in approximately 1891 that neurons work like on/off switches. They "fire" when the cumulative signal that they receive from other neurons exceeds a certain threshold value, otherwise they don't. Plenty of mathematicians felt vindicated by that discovery: binary logic, invented in 1854 by the British philosopher George Boole in a book titled "The Laws of Thought", does seem to lie at the very foundation of human thinking. In 1943 an unlikely pair at University of Chicago wed the digital electronic computer and the neuron: the psychiatrist and part-time poet Warren McCulloch and the young homeless runaway and math prodigy Walter Pitts described mathematically an "artificial" neuron that can only be in one of two possible states, and how to connect a population of such artificial binary neurons in a very intricate network to mimic the way the brain works. When signals are sent into the network, they spread to its neurons according to the simple rule that any neuron receiving enough positive signals from other neurons sends a signal to other neurons. It gets better: their seminal paper "A Logical Calculus of the Ideas Immanent in Nervous Activity" (1943) proved that such a network of binary neurons is equivalent to a Universal Turing Machine. As Von Neumann said (in a speech at the Hixon Symposium of 1948), the McCulloch-Pitts theorem proved that "anything that can be completely and unambiguously put into words is ipso facto realizable by a suitable finite neural network". McCulloch, influenced by the mathematical logic of Leibniz that Pitts had described to him, and by the physiologist Arturo Rosenblueth (befriended at a Macy conference), was looking for the logic of the nervous system, and Pitts, who was studying (unofficially) with the influential philosopher of logic Rudolf Carnap, found a way to apply Boolean algebra to the behavior of the brain's neurons. The result was that the brain seemed to work according to the logical calculus discovered by Russell and Whitehead in the "Principia Mathematica". (Except that, of course, things are much more complicated in real neurons). Their paper only cited three works: Carnap's "Logical Syntax of Language" next to Bertrand Russell's "Principia Mathematica" and David Hilbert's "Foundations of Theoretical Logic". Carnap had moved from Europe to University of Chicago, and Carnap's hostility towards metaphysics may have helped Pitts to think in terms of a new logical language, so much so that Pitts' "logical calculus of ideas" feels like an application of Carnap's logical calculus to the functioning of the nervous system. Note that the McCulloch-Pitts networks do not learn. There was no learning rule to automatically change the weights of the connections.

When in 1945 John Von Neumann published the lengthy and highly influential "First Draft of a Report on the EDVAC" about the "stored-program architecture" (which is still the architecture of today's programmable computers), he cited only one paper, the McCulloch-Pitts paper. He didn't even mention the colleagues of University of Pennsylvania's Moore School of Electrical Engineering, notably John Mauchly and Presper Eckert, the designers of ENIAC, who had provided Von Neumann with the idea.

Upon opening a panel titled "The Design of Machines to Simulate the Behavior of the Human Brain" during the Institute of Radio Engineers' Offsite Link Convention, held in New York in 1955, McCulloch would confidently state that "we need not ask, theoretically, whether machines can be built to do what brains can do" because, in creating our own brain, Nature already showed us that it is possible. (Of course, the McCulloch-

Pitts theorem still fails because of Goedel's theorem, but that's a detail). Pitts, instead, moved to MIT and worked with Norbert Wiener, and the two had the prescient intuition of turning the deterministic laws of neurons into the statistical laws of behavior, but it was too early for the world to appreciate it. McCulloch and Von Neumann were working in a world that still had not seen the first digital electronic computer for sale (that would happen in 1951).

The intellectual motivation was not "computer science" but cybernetics. The founders of cybernetics regularly convened from 1946 until 1953 at the Macy Conference on Cybernetics, organized by the Macy Foundation of New York when Willard Rappleye was its president. These conferences, that sometimes occurred twice a year, were truly interdisciplinary. Speakers at the first conference were: John Von Neumann (computer science), Norbert Wiener (mathematics), Walter Pitts (mathematics), Arturo Rosenblueth (physiology), Rafael Lorente de No (neurophysiology), Ralph Gerard (neurophysiology), Warren McCulloch (neuropsychiatry), Gregory Bateson (anthropology), Margaret Mead (anthropology), Heinrich Kluever (psychology), Molly Harrower (psychology), Lawrence Kubie (psychoanalysis), Filmer Northrop (philosophy), Lawrence Frank (sociology), and Paul Lazarsfeld (sociology). It was at the Second Cybernetic Conference in 1947 that Pitts announced that he was writing his doctoral dissertation on probabilistic three-dimensional neural networks. Unfortunately, he burned his unfinished doctoral dissertation.

In January 1951 in Paris a cybernetic conference titled "Calculating Machines and Human Thought" was organized by Joseph Pérès with help from Louis Couffignal (who one year later published the book "The Thinking Machines"). The participants included Norbert Wiener from MIT, Howard Aiken from Harvard University, McCulloch from University of Illinois, Maurice Wilkes from University of Cambridge (the project leader of the EDSAC), Ross Ashby from Barnwood House Hospital (who in 1948 had designed the first homeostat), Pete Uttley from the Telecommunications Research Establishment, William Grey-Walter from the Burden Neurological Institute (who in 1949 had built electronic tortoises,), the French physicist Louis de Broglie (the discoverer of the dualism of waves and matter), the Polish-born mathematician Benoit Mandelbrot (the future inventor of the science of fractals), Frederick Williams from the University in Manchester (who in 1947 had developed the tube for Random Access Memory), Tom Kilburn from University of Manchester (the engineer who in 1948 had run the first computer program ever), plus representatives from Ferranti (that was about to introduce the first commercial computer) and from IBM. At this conference Wiener played against the electromechanical chess-playing automaton of Leonardo Torres y Quevedo, a Spanish builder of cableways, first demonstrated in 1914. (Trivia: in 1951 France food was still rationed, and three months later France, West Germany, Italy, Belgium, Holland and Luxembourg, meeting in Paris, would create the European Coal and Steel Community, the forerunner of the European Union).

The impact of cybernetics extended beyond biology and engineering. For example, the Czech social scientist Karl Deutsch at Harvard applied cybernetic methods to sociopolitical problems in a series of articles beginning with "Mechanism, Organism, and Society" (1951), later collected in the book "The Nerves of Government" (1963) in which he tried to reduce even politics to neural networks.

The idea that computers were "giant brains" wasn't just a myth invented by the media. Some psychologists enthusiastically signed on to this metaphor. George Miller was a psychologist at Harvard University who in 1950 visited the Institute for Advanced Study in Princeton, one of the pioneering centers in computer science. The following year he was hired by MIT to lead the psychology group at the newly formed Lincoln Laboratories (a hotbet of military technology for the Cold War) and published an influential book titled "Language and Communication" in which he launched the program of studying the human mind using the information theory just developed by Claude Shannon at Bell Labs in his article "A Mathematical Theory of Communication" (1948).

The "Turing Test", introduced by Alan Turing in his paper "Computing Machinery and Intelligence" (1950), has often been presented as the kind of validation that a machine has to pass in order to be considered "intelligent": if a human observer, asking all sorts of questions, cannot tell whether the agent providing the answers is human or mechanical, then the machine has become intelligent (or, better, as intelligent as the human being).

To start with, Claude Shannon at Bell Labs, shortly after writing his groundbreaking "A Mathematical Theory of Communication", delivered a lecture titled "Programming a Computer for Playing Chess" at the national IRE (Institute of Radio Engineers) convention of March 1949 in New York. According to Shannon's biographer Rob Goodman, later in life Shannon (a jazz player and a juggling unicyclist himself) planned a memorial parade for his own funeral, featuring a jazz combo, a 417-instrument marching band, acrobats, a chess-playing computer and juggling robots.

A Brief History of Artificial Intelligence/ Part 1

The first influential conference for designers of intelligent machines took place in 1955 in Los Angeles: the Western Joint Computer Conference. At this conference Allen Newell and Herbert Simon presented the "Logic Theory Machine", Newell also presented his "Chess Machine", Walter Pitts' housemate Oliver Selfridge gave a talk on "Pattern Recognition and Modern Computers", and Wesley Clark and Belmont Farley reported the first computer simulations of a neural network ("Generalization of Pattern Recognition in a Self-organizing System").

Then in 1956 John McCarthy organized the conference at Dartmouth College that gave Artificial intelligence its name (called "Dartmouth Summer Research Project on Artificial Intelligence"). The 1955 proposal was also signed by his friend Marvin Minsky (then at Harvard University), Claude Shannon (the Bell Labs scientist who had published "A Mathematical Theory of Communication" in 1948 and trained both McCarthy and Minsky in 1953), and Nathaniel Rochester (the IBM scientist who had designed the IBM 701 of 1952, considered the first mass-produced computer, and who was toying with neural networks on the new IBM 704). Minsky, McCarthy and Ray Solomonoff where the only scientists who attended the whole summer workshop of two months (June 18 to August 17). Ray Solomonoff was president of a sci-fi club at University of Chicago. Others who attended were: Claude Shannon, Julian Bigelow, Warren McCulloch, Oliver Selfridge, Ross Ashby, Arthur Samuel, Herbert Simon, Allen Newell, and nine more. Wiener was not there because the workshop was meant as a reaction against cybernetics: this group, mainly driven by Newell and Simon, wanted to found a new discipline of symbolic information processing. Turing had committed suicide two years earlier (1954) and didn't live to witness the use of the term "artificial intelligence".

Note that there was no inherent need to invent a new term for the discipline that Wiener (also at the MIT at the same time) had already coined "Cybernetics". McCarthy was trying to make a point that his "Artificial Intelligence", unlike Wiener's Cybernetics, was to be based on mathematical logic, not on engineering. But today the term "Artificial Intelligence" has stuck, and we use it also for projects (such as neural networks) that perhaps, historically speaking, should be more appropriately referred to as Cybernetics.

That seems to have been a particularly fertile time for young people with revolutionary ideas. The plan for that conference matured in August 1955, a few days after the London premiere of Samuel Beckett's play "Waiting for Godot" (originally written in French), a month after Chuck Berry started the vogue for rock'n'roll with his single "Maybellene" (July 1955), a month after Miles Davis performed with Thelonious Monk at the

Newport Jazz Festival and became the new rising star of jazz music, a few weeks after RCA Victor released a box-set featuring four records of electronic music, "The Sounds and Music of the RCA Electronic Music Synthesizer" (the synthesizer had been built by Harry Olson and Herbert Belar at RCA's Princeton labs), two months after Billy Wilder's film "The Seven Year Itch" launched Marilyn Monroe as the ultimate sex symbol, and five months after Richard Brooks' film "Blackboard Jungle" depicted the lives of juvenile delinquents (and the soundtrack's "Rock Around the Clock", performed by Bill Haley & His Comets, was becoming an anthem for young rebels). One year earlier, in 1954, Edgar Varese had inaugurated electronic tape music with "Deserts", (1954), the Russian-born illustrator Boris Artzybasheff had published "As I See", a book of visionary drawings about the post-human society (humanized machines and/or cyborgs), and Gyorgy Kepes was working on his book "The New Landscape in Art and Science", mixing pictures of modern art with images obtained from x-ray machines, electron microscopes, sonars, radars, telescopes and infrared sensors. It was a time of upheaval: one month later William Shockley would found the first startup of Silicon Valley, three months later the "beat" poet Allen Ginsberg would recite his poem "Howl" in San Francisco, and four months later Martin Luther King would start a civil-rights movement (sparked by the arrest of a black woman, Rosa Parks, who refused to give up her bus seat to a white passenger). An era ended when Charlie Parker died in March 1955, Albert Einstein died in April, Winston Churchill retired in April, and Thomas Mann died in August.

In 1956 McCarthy also co-edited with Claude Shannon a volume on "Automata Studies" that included papers continuing the mission of the McCulloch-Pitts neuron. by Ross Ashby, Pete Uttley, Minsky, Von Neumann, besides Shannon and McCarthy themselves. Notably, the volume contains "Some Uneconomical Robots" by the philosopher James Culbertson from the RAND Corporation (who had already published mathematical neural-network models in "A Mechanism for Optic Nerve Conduction and Form Perception" of 1948) and in "Hypothetical robots" of 1952) and a revised version of Stephen Kleene's "Representation of Events in Nerve Nets and Finite Automata" (originally published in 1951 also at RAND). Ross Ashby's paper "Design for an Intelligence-amplifier" even discussed how to generate creativity in machines.

The Symbolic School (Knowledge-based Systems)

The practitioners of Artificial Intelligence quickly split into two fields. One, pioneered by Herbert Simon and his student Allen Newell at Carnegie Institute of Technology with their "Logic Theorist" (1956), basically understood intelligence as the pinnacle of mathematical logic, and focused on symbolic processing. Logic Theorist (written by Clifford Shaw, who implemented Newell's Information Processing Language or IPL) effortlessly proved 38 of the first 52 theorems in chapter 2 of Bertrand Russell's "Principia Mathematica." A proof that differed from Russell's original was submitted to the Journal of Symbolic Logic, the first case of a paper co-authored by a computer program. In 1955 Allen had published a paper titled "The Chess Machine" and in 1958 the trio presented a program to play chess, NSS (the initials of the three).

In 1955 Arthur Samuel at IBM in New York wrote not only the first computer program that could play checkers but the first self-learning program. That program implemented the alpha-beta search algorithm that would dominate A.I. for the next 20 years. In February 1956 a television channel broadcast a checkers game played by an IBM 701 computer running Samuel's program against a human expert. A few years later Samuel devised another learning method, which he called "learning by generalization", and that was an embryonic version of the "temporal-difference learning" method ("Some Studies in Machine Learning Using the Game of Checkers", 1959).

The first breakthrough in this branch of A.I. (the "symbolic branch") was probably John McCarthy's article "Programs with Common Sense" (1958): McCarthy (then at Stanford University) understood that someday machines would easily be better than humans at many repetitive and computational tasks, but "common sense" is what really makes someone "intelligent" and common sense comes from knowledge of the world. That article spawned the discipline of "knowledge representation": how can a machine learn about the world and use that knowledge to make inferences. McCarthy's approach relied on symbolic logic, notably first-order predicate logic (true/false predicates), to describe human knowledge in a formal language that the computer (a binary-logic machine) can process.

This approach was somehow "justified" by the idea, introduced by MIT linguist Noam Chomsky in "Syntactic Structures" (1957), that language competence is due to some grammatical rules that express which sentences are correct in a language. The grammatical rules express "knowledge" of how a language works, and, once you have that knowledge (and a vocabulary), you can produce any sentence in that language, including sentences you have never heard or read before. Chomsky had studied with Zellig Harris and wed Harris' "rules of transformation" (discussed in 1952 in two seminal papers, "Culture and Style" and "Discourse Analysis") to the production systems invented by Emil Post in 1943 at City College of New York ("Formal Reductions of the General Combinatorial Decision Problem", 1943). Post was a mathematician who had done his dissertation on Bertrand Russell's "Principia Mathematica" (and who had almost invented the Turing Machine before Turing). In fact, the first English parser, running on a Univac 1, was completed in 1959 at University of Pennsylvania within the Transformations and Discourse Analysis Project (TDAP) directed by Zellig Harris.

AI was helped by the Sputnik, the first artificial satellite, launched by the Soviet Union in October 1957. The US government panicked that the Soviet Union was technologically ahead, and the new-born discipline of Artificial Intelligence was ready to accept the military funding that started flocking onto any technology remotely promising.

At the peak of the Cold War, McCarthy himself wrote a chess program that challenged a Soviet chess program, the one originally developed in 1963 by Alexander Kronrod's team at the Moscow Institute of Theoretical and Experimental Physics (ITEP). The Soviets won that match.

The first compendium of research in Artificial Intelligence was compiled by two young UC Berkeley researchers, Edward Feigenbaum and Julian Feldman, both former students of Herbert Simon at the Carnegie Institute of Technology. The compendium was called "Computers and Thought" (1963). It included articles by Minsky, Simon and Newell (the Logic Theorist), Samuel, Selfridge (the Pandemonium), Shaw, and Turing himself, plus Earl Hunt's as well as Carl Hovland's model of concept learning (CLS) and Feigenbaum's own Elementary Perceiver and Memorizer (EPAM), both early experiments with decision trees.

The rapid development of computer programming helped this field take off, as computers were getting better and better at processing symbols: knowledge was represented in symbolic structures and "reasoning" was reduced to a matter of processing symbolic expressions. This line of research led to "knowledge-based systems" (or "expert systems"), such as Ed Feigenbaum's Dendral (1965) at Stanford, that consisted of an "inference engine" (the repertory of legitimate reasoning techniques recognized by the mathematicians of the world) and a "knowledge base" (the "common sense" knowledge). This technology relied on acquiring knowledge from domain experts in order to create "clones" of such experts (machines that would perform as well as the human experts). The limitation of expert systems was that they were "intelligent" only in one specific domain.

Gyorgy Polya may have been an indirect influence on the idea of using "common sense" or "heuristics" to match the "intelligence" of human experts. This Hungarian mathematician, born in what was then the Austro-Hungarian empire, was hired by Stanford in 1942 and became famous for his lessons on using intuition to solve mathematical problems, demonstrated in the book "How to Solve it" (1945). He basically founded a new discipline, the study of the nature and source of heuristics, which remained popular at Stanford.

Their "knowledge" was expressed in a formal language that can be subjected to logical inference: the language of first-order predicate logic, invented by mathematicians to express the relationships between objects. The beauty of expert systems is that, being based on first-order predicate logic, they can explain their conclusions: it is always possible to "backtrack" and retrace the chain of logical steps that led to the conclusions. After Dendral, most expert systems were written in LISP, the language based on Alonzo Church's lambda calculus invented by John McCarthy in 1958 when he was still at MIT, a vast improvement over Newell's IPL.

On the other hand, mathematical intuition seemed to constitute a higher level of intelligence than knowledge-based inference. Herbert Simon and his student Kenneth Kotovsky at the Carnegie Institute of Technology studied sequence extrapolation: how do you come up with the next number in a sequence? ("Human Acquisition of Concepts for Sequential Patterns", 1963). Walter Reitman, a colleague of Newell and Simon at the Carnegie Institute of Technology, built an early system of analogical reasoning, Argus ("An Information-processing Model of Thinking", 1964). Reitman was unhappy with the rigid, mechanical reasoning of the GPS and wanted to model the more chaotic approach to creativity as in the human mind. The GPS could not account for human creativity in fields such as art and music where the problem to be solved is not well-defined like in mathematical logic. It is in fact very ill-defined: which problems are we trying to solve when we compose a piece of music? Alas, Argus was written in the same, hopelessly complicated, programming language IPL. Analogical reasoning in a microdomain of geometry was the subject of a program called Analogy (1968), developed by Thomas Evans at the research laboratories of the Air Force near Boston ("A Heuristic Program to Solve Geometric-analogy Problems", 1968).

The Connectionists (Neural Networks)

Meanwhile, the other branch of Artificial Intelligence was pursuing a rather different approach: simulating what the brain does at the physical level of neurons and synapses. The symbolic school of John McCarthy and Marvin Minsky believed in using mathematical logic (i.e., symbols) to simulate how the human mind works; the school of "neural networks" (or "connectionism") believed in using mathematical calculus (i.e., numbers) to simulate how the brain works.

Since, in the 1950s, neuroscience was just in its infancy (medical machines to study living brains would not become available until the 1970s), computer scientists only knew that the brain consists of a huge number of interconnected neurons, and neuroscientists were becoming ever more convinced that "intelligence" was due to the connections, not to the individual neurons. A brain was viewed as a network of interconnected nodes, and our mental life as due to the way signals travel through those connections from the neurons of the sensory system up to the neurons that process those sensory data and eventually down to the neurons that generate action.

The neural connections can vary in strength from zero to infinite, and this is known as the "weight" of the connection. Change the weight of some neural connections and you change the outcome of the network's computation. In other words, the weights of the connections can be tweaked to cause different outputs for the same inputs. The problem for those designing "neural networks" consists in fine-tuning the connections so that the network as a whole comes up with the correct interpretation of the input; e.g. with the word "APPLE" when the image of an apple is presented. This is called "training the network". For example, showing many apples to the system and forcing the answer "APPLE" should result in the network adjusting those connections to recognize apples in general. This is called "supervised learning". The normal operation of the neural network is quite simple. The signals coming from different neurons into a neuron are weighed based on the weights of each input connection and then fed to an "activation function" (also known as the "nonlinearity") that decides what has to be the output produced by this neuron. The simplest activation function is a function that has a threshold, the "step" function: if the total input passes the threshold value, the neuron emits a one, otherwise a zero. This process goes on throughout the network. The network is usually organized in layers of neurons. The weights of the connections determine what the network computes. The weights change during "training" (i.e. in response to experience). Neural networks "learn" those weights during training. A simple approach to learning weights is to compare the output of the neural network to the correct answer and then modify the weights in the network so as to produce the correct answer. Each "correct answer" is a training example. The neural network needs to be trained with numerous such examples. Today's computers are powerful enough to deal with literally millions of examples. If the network has been designed well, the weights will eventually converge to a stable configuration: at that point the network should provide the correct answer even for instances that were not in the training data (e.g., recognize an apple that it has never seen before). The designer of the neural network has to decide the structure of the neural network (e.g. the number of layers, the size of each layer, and which neurons connect to which other neurons), the initial values of the weights, the activation function (the "nonlinearity"), and the training strategy. Both the initialization and the training may require the use of random numbers, and there are many different ways to generate random numbers. The term "hyperparameters" refers to all the parameters that the network designer needs to pinpoint. For example, the hyperparameters for modern convolutional neural networks include: the number of layers in the neural network, the activation function, the loss function, the kernel size and the batch size. There probably exists a best set of hyperparameters that optimizes the training of a neural network, but it varies case by case, and there is no easy deterministic way to discover it, hence the hyperparamets are usually set by trial and error. It may take months to come up with a neural network that can be trained.

Since the key is to adjust the strength of the connections, the alternative term for this branch of A.I. is "connectionism".

One of the most influential books in the early years of neuroscience was "Organization of Behavior" (1949), written by the psychologist Donald Hebb at McGill University in Montreal (Canada). Hebb described how the brain learns by changing the strength in the connections between its neurons. In 1951 two Princeton University students, Marvin Minsky and Dean Edmonds, simulated Hebbian learning in a network of 40 neurons realized with 3,000 vacuum tubes, and called this machine SNARC (Stochastic Neural Analog Reinforcement Computer). I wouldn't count it as the first neural network because SNARC was not implemented on a computer. In 1954 Wesley Clark and Belmont Farley at MIT simulated Hebbian learning on a computer, i.e. created the first artificial neural network (a two-layer network). In 1956 Hebb collaborated with IBM's research laboratory in Poughkeepsie to produce another computer model, programmed by Nathaniel Rochester's team (that included a young John Holland).

If there was something similar to the Macy conferences in Britain, it was the Ratio Club, organized in 1949 by the neurologist John Bates, a dining club of young scientists who met periodically at London's National Hospital to discuss cybernetics. McCulloch, who was traveling in Britain, became their very first invited speaker. Among its members was the neurologist William Grey-Walter, who in 1948 built two tortoise-shaped robots (better known as Elmer and Elsie) that some consider the first autonomous mobile robots. Turing was a member and tested the "Turing test" at one of their meetings. John Young was a member: in 1964 he would discover the "selectionist" theory of the brain. And, finally, another member was the psychiatrist Ross Ashby, who in 1948 actually built a machine to simulate the brain, the homeostat ("the closest thing to a synthetic brain so far designed by man", as Time magazine reported). The title of that paper became the title of his influential book "Design for a Brain" (1952). No surprise then that mathematical models of the brain proliferated in Britain, peaking just about in the year of the first conference on Artificial Intelligence: Jack Allanson at Birmingham University reported on "Some Properties of Randomly Connected Neural Nets" (1956), Wilfred Taylor built an associative memory at University College London ("Electrical Simulation of Some Nervous System Functional Activities", 1956), Raymond Beurle at Imperial College London studied "Properties of a Mass of Cells Capable of Regenerating Pulses" (1956), and Albert "Pete" Uttley at the Radar Research Establishment, a mathematician who had designed Britain's first parallel processor, wrote about "Conditional Probability Machines and Conditioned Reflexes" (1956). It is debatable whether, as argued by Christof Teuscher in his book "Turing's Connectionism" (2001), Turing truly predated neural networks (as well as genetic algorithms) in an unpublished 1948 paper, now known as "Intelligent Machinery", that was about "unorganized machines", i.e. random Boolean networks.

A purely cybernetic approach was well represented by Gordon Pask who, in Cambridge, was building special-purpose electro-mechanical automata such as Eucrates (1955), which simulated the interaction between tutor and pupil (i.e. a machine that can teach another machine), and in 1956 patented a machine called SAKI (which stood for "Self-Adaptive Keyboard Instructor") to train people to type on a keypunch. The eclectic Pask was also an artist who produced pioneering works of interactive art such as "MusiColour" (1953), a sound-activated light-show, and "Colloquy of Mobiles" (1968), an installation that allowed the audience to interact with five machines communicating among themselves via sound and light. In 1968 he went on to build even an electrochemical ear.

Frank Rosenblatt's Perceptron (1957) at Cornell University and Oliver Selfridge's Pandemonium (1958) at MIT popularized the new view of artificial neural networks: not knowledge representation and logical inference, but pattern propagation and automatic learning. The Perceptron, first implemented in software in 1958 on the Weather Bureau's IBM 704 and then custom-built in hardware at Cornell Aeronautical Laboratory, was the first trainable neural network (called "single-layer" even though it had two layers of neurons). The activation function was the same binary function (the "step function") used by the McCulloch-Pitts neuron but it had a learning rule (an algorithm for changing the weights, which was based on the errors that it made). The Perceptron's application was to separate data into two groups. The network of McCullouch-Pitts was a new way to build finite automata, an extension of logic. Rosenblatt's network, instead, was a new way to build learning algorithms, an extension of statistics. Its simple learning rule worked wonders, but the limitations of perceptrons were obvious to everybody and in the following years several studies found solutions. The British National Physical Laboratory (November 1958) organized a symposium titled "The Mechanisation of Thought Processes"; three conferences on "Self-Organization" were held in 1959, 1960 and 1962 which led to Rosenblatt publishing his report "Principles of Neurodynamics" (1962). However, nobody could figure out how to build a multilayer perceptron.

Pitts also worked with the MIT neurologist Jerome Lettvin (Pitts' best friend since 1938) and the Chilean biologist Humberto Maturana on their seminal study about the visual system of the frog ("What the Frog's Eye Tells the Frog's Brain", 1959). They discovered that the retina is more than a simple transmitter of impulses to the brain for the brain to analyze them: the retina includes neurons that already respond to specific features such as edges, lighting and movement. Some of these "feature detectors" of the frog's eye were nicknamed "bug detectors" because they specifically reacted to small, dark, moving objects. Something in that study shook Pitts' belief in Boolean logic as a model for the functioning of the brain (Pitts burned his unfinished doctoral dissertation and pretty much ended his career).

In 1960 Bernard Widrow and his student Ted Hoff at Stanford University built a single-layer network based on an extension of the McCulloch-Pitts neuron called Adaline (Adaptive Linear Neuron) and using a generalization of the Perceptron's learning rule, the "delta rule" or "least mean square" (LMS) algorithm (a way to minimize the difference between the desired and the actual signal), the first practical application of a "stochastic gradient descent" method to machine learning. This method had been introduced in 1951 for mathematical optimization by Herbert Robbins of the Univ of North Carolina ("A Stochastic Approximation Method ", 1951).

Widrow understood that both the activation and the learning had to be more complex, and that the latter depended on the former. The threshold activation function used in both the McCulloch-Pitts network and the Perceptron did not lend itself to significant learning. Widrow used their way of summing the inputs to a neuron (each input multiplied by its weight) but then picked a linear (not threshold) activation function so that the output of a neuron was not just "on" or "off". The advantage was that this function, while still very simple, lent itself to a more sophisticated learning algorithm. Widrow used the same principle of adjusting the weights/strengths of the connections to reduce the error (i.e., reduce the difference between desired and actual output), but he could now use gradient-descent learning to adjust the weights. (Technically speaking, each weight is changed according to the negative of the derivative of the error with respect to that weight, and a derivative cannot exist in the case of the threshold function, which is a discontinuous function, but it exists in the case of the linear function, which is continuous).

The Adaline was physically a small analog machine built by Hoff, which was even sold commercially by their startup. Trivia: Ted Hoff later joined a tiny Silicon Valley startup called Intel and helped design the world's first microprocessor.

The "gradient descent" method, discovered in 1847 by the French mathematician Augustin Cauchy, was first applied to control theory in 1960 by Henry Kelley at Grumman Aircraft Engineering Corporation in New York ("Gradient Theory of Optimal Flight Paths", 1960) and by Arthur Bryson at Harvard University ("A Gradient Method for Optimizing Multi-stage Allocation Processes", 1961). In 1955 the Hungarian physicist Dennis Gabor, the inventor of holography, had already devised how to employ the gradient-descent method for training the analog computer that his students were building at Imperial College London, but that research remained unpublished. The mathematical idea behind gradient-descent methods is simple: first one measures the global error in the performance of the network (desired output minus actual output), then one computes the derivative of such error with respect to the weight/strength of each connection, and finally one adjusts each weight/strength in the direction that decreases the error. Bottom line: neural networks can learn their weights (the weights of the connections between neurons) using the gradient-descent algorithm. That was a primitive form of "backpropagation".

At the IRE (Institute of Radio Engineers) convention of 1960 Marvin Minsky presented a lengthy paper titled "Steps Towards Artificial Intelligence" that was skeptic about reinforcement learning (quote: "I am not convinced that these statistical training schemes should play a central role in our models"). He would quickly become neural networks' nemesis.

Another important discovery that went unnoticed at the time was the first learning algorithms for multilayer networks, i.e. for "deep learning". They were published in 1965 by the Ukrainian mathematician Alexey Ivakhnenko in his book "Cybernetic Predicting Devices".

There were several projects in Europe, notably the Lernmatrix, created in 1961 by Karl Steinbuch (the engineer who, working at Standard Elektrik Lorenz, had designed Germany's first fully-transistorized commercial computer, the ER 56) and the (very similar) associative memory built by the chemist Christopher Longuet-Higgins' group at University of Edinburgh ("Non-holographic Associative Memory", 1969).

In retrospect, the program of neural networks (just like the program of cybernetics before it) was the program of how to bridge the gap between physiologists and engineers. One can see how a mathematical problem turned into a chemical problem and, eventually, into a neurological problem. Alan Turing's "The Chemical Basis of Morphogenesis" (1952), his only paper about chemistry, proposed a mechanism for the emergence of patterns in biological systems (like the stripes of the zebra), and that paper introduced nonlinear dynamics to study self-organizing processes. Raymond Beurle applied this approach to pattern formation in the brain and obtained the first neural field equations which described a wave of information propagating through the neural network ("Properties of a Mass of Cells Capable of Regenerating Pulses", 1956). The equations were improved by John Griffith at Cambridge University ("A Field Theory of Neural Nets", 1963) and then by Jack Cowan and Hugh Wilson at MIT, who obtained a model similar to the Hopfield networks of a decade later ("A Mathematical Theory of the Functional Dynamics of Cortical and Thalamic Nervous Tissue", 1973). Meanwhile, expanding Turing's theory, the Belgian chemist Ilya Prigogine established nonlinear non-equilibrium thermodynamics and, in a talk titled "Structure, Dissipation and Life" (delivered in June 1967 at the International Conferences On Theoretical Physics and Biology) described biological systems as "dissipative systems" which self-organize far from equilibrium. So indirectly Turing's legacy for connectionism was that he originated the thinking that led to employ nonlinear differential equations for neural networks.

Compared with expert systems, neural networks are dynamic systems, and predisposed to learning by themselves. "Unsupervised" networks, in particular, can discover categories by themselves; e.g., they can discover that several images refer to the same kind of object, a cat.

There are two ways to solve a crime. One way is to hire the smartest detective in the world, who will use experience and logic to find out who did it. On the other hand, if we had enough surveillance cameras placed around the area, we would scan their tapes and look for suspicious actions. Both ways may lead to the same conclusion, but one uses a logic-driven approach (symbolic processing) and the other one uses a data-driven approach (ultimately, the visual system, which is a connectionist system).

Expert systems were the descendants of the "logical" school that looked for the exact solution to a problem. Neural nets were initially viewed as equivalent logical systems, but actually represented the other kind of thinking, probabilistic thinking, in which we content ourselves with plausible solutions, not necessarily exact ones. That is the case of speech and vision, and of pattern recognition in general.

In 1965 Marvin Minsky and Samuel Papert of MIT began a vicious campaign against research in neural networks by circulating a technical manuscript that was eventually published as a book titled "Perceptrons" (1969), even though it was mostly about Adaline. It contained a devastating critique of neural networks that virtually killed the discipline. This came a decade after Noam Chomsky's review, of a book by Burrhus Skinner, had turned the tide in psychology, ending the domination of behaviorism and resurrecting cognitivism. Chomsky's campaign against behaviorism culminated in an article in the New York Review of Books of December 1971. Most A.I. scientists favored the "cognitive" approach simply for computational reasons, but those computer scientists felt somewhat reassured by the events in psychology that their choice was indeed wise.

Minsky's and Papert's proof came, by sheer coincidence, at the right time to avoid criticism: both Pitts and McCulloch died in 1969 (May and September), and Rosenblatt died in a boating accident in 1971.

Dave Block (a physicist who had worked with Rosenblatt at Cornell University) told Minsky that the limitations of the (single-layer) Perceptron could be easily overcome with multilayer neural nets, and, to be fair, Minsky and Papert accepted Block's criticism; but, unfortunately, Rosenblatt's learning algorithm did not work for multilayer nets.

In 1969 Stanford held the first International Joint Conference on Artificial Intelligence (IJCAI). Nils Nilsson from SRI presented Shakey. Carl Hewitt from MIT's Project MAC presented Planner, a language for planning action and manipulating models in robots. Cordell Green from SRI and Richard Waldinger from Carnegie-Mellon University presented systems for the automatic synthesis of programs (automatic program writing). Roger Schank from Stanford and Daniel Bobrow from Bolt Beranek and Newman (BBN) presented studies on how to analyze the structure of sentences. Connectionists were under-represented. The Artificial Intelligence magazine, founded in 1970, did not publish any paper on neural networks until 1989, when it published a survey by Geoffrey Hinton.

Nonetheless, the gradient method was perfected, to optimize multi-stage dynamic systems, by Bryson and his Chinese-born student Yu-Chi Ho in the book "Applied Optimal Control" (1969). At that point the mathematical theory necessary for backpropagation in multi-layer neural networks was basically ready. In 1970 the Finnish mathematician Seppo Linnainmaa invented "reverse mode of automatic differentiation", which has backpropagation as a special case. In 1974 Paul Werbos' dissertation at Harvard University applied Bryson's backpropagation algorithm to the realm of neural networks ("Beyond Regression", 1974). Werbos had realized that the "backpropagation" algorithm was a more efficient way to train a neural network than any of the existing methods. His discovery languished for several years because his background wasn't quite orthodox: his thesis advisor was the social scientist and cybernetic pioneer Karl Deutsch, and Werbos' algorithm of backpropagation was meant as a mathematical expression of the concept of "cathexis" that Sigmund Freud had introduced in his book "The Project for a Scientific Psychology" (1895). And the whole point of Werbos' research was to provide a better alternative to statistical analysis for long-range forecasting, in particular the forecasting of international affairs at the US Department of Defense.

Practitioners of neural networks also took detours into cognitive science. For example: Kaoru Nakano at University of Tokyo ("Learning Process in a Model of Associative Memory", 1971); James Anderson at Rockefeller University ("A Simple Neural Network Generating an Interactive Memory", 1972) in the laboratory of psychologist William Estes who had founded a whole new field with his manifesto "Toward a Statistical Theory of Learning" (1950); and Teuveo Kohonen in Finland ("Correlation Matrix Memories", 1972) used neural networks to model associative memories based on Donald Hebb's law. The neuroscientist Christoph von der Malsburg at the Max Planck Institute in Germany built a model for the visual cortex of higher vertebrates (" Self-organization of Orientation Sensitive Cells in Striate Cortex", 1973). The holy grail of neural networks was unsupervised learning: have the machine learn concepts from the data without human intervention. Several variations on Karl Pearson's decades-old method of "principal components analysis" were proposed, and significant contributions came from the science of signal processing. For example, Pete Uttley designed the Informon to separate frequently occurring patterns ("A Network for Adaptive Pattern Recognition", 1970). In 1975 the first multi-layered network appeared, designed by Kunihiko Fukushima in Japan, the Cognitron ("Cognitron - A Self-organizing Multilayered Neural Network", 1975). Stephen Grossberg at Boston University unveiled another unsupervised model, "adaptive resonance theory" ("Adaptive Pattern Classification and Universal Recoding", 1976), and some of his ideas anticipated Hopfield's continuous networks ("Contour Enhancement, Short Term Memory, and Constancies in Reverberating Neural Networks", 1973). And Shunichi Amari at University of Tokyo delivered the classic neural field equations that completed the work begun by Raymond Beurle in the 1950s ("Mathematical Theory on Formation of Category Detecting Nerve Cells," 1978; later expanded in "Field Theory of Self-organizing Neural Nets",

1983). The Italian-born Tomaso Poggio and the British-born David Marr (who was now at MIT) developed a nonlinear system for a specific case, that of binocular vision ("Cooperative Computation of Stereo Disparity", 1976). Therefore, by the mid-1970s significant progress had occurred (if not widely publicized) in neural networks.

Later, Walter Freeman at UC Berkeley (the same neurologist who in 1936 had performed the first "lobotomy" for psychiatric treatment), working in collaboration with the philosopher Christine Skarda, applied chaos theory to the study of brain processes ("How Brains Make Chaos in Order to Make Sense of the World", 1987). For the record, Skarda soon became a Buddhist philosopher convinced that our view of the brain is fundamentally wrong and has since spent more than 20 years in meditation retreat.

A much more stinging criticism of the logical school of A.I. could have come from neuroscience, a discipline that was beginning to use computer simulations. In 1947 Kacy Cole at the Marine Biological Lab near Boston pioneered the "voltage clamp" technique to measure the electrical current flowing through the membranes of neurons. Using that technique, in 1952 the British physiologists Alan Hodgkin and Andrew Huxley at Cambridge University built the first mathematical model of a spiking neuron, which also counts as the first simulation of computational neuroscience (for the record, they simulated the axon of the squid's brain). The Hodgkin–Huxley model is a set of nonlinear differential equations that approximates the electrical characteristics of neurons. The next major breakthroughs in the simulation of brain computation came respectively in 1962, when Wilfrid Rall at the National Institutes of Health simulated a dendritic arbor, and, then, in 1966, when Fred Dodge and James Cooley at IBM simulated a propagating impulse in an axon. Meanwhile, Donald Perkel at the RAND Corporation in Los Angeles had written computer programs to simulate the working of the neuron using one of the earliest computers, the Johnniac ("Continuous-time Simulation of Ganglion Nerve Cells in Aplysia", 1963). These simulations (by people who actually knew what a neuron looks like) bore little resemblance to the naïve digital neurons of the artificial neural networks.

Crucially, neuroscientists kept emphasizing the role of synapses: intelligence is not about the neuron but about the connections (the synapses) that create a network of neurons. Jean-Pierre Changeux in "Neuronal Man" (1985): "The impact of the discovery of the synapse and its functions is comparable to that of the atom or DNA". Joseph Ledoux in "Synaptic Self" (2002): "You are your synapses - they are who you are".

Cultural Background: The Knowledge Society

The Austrian-born economist Peter Drucker coined the term "knowledge worker" in his 1959 book "Landmarks of Tomorrow". Note that this was written at a time when there were very few electronic computers in the world. Drucker emphasized the "pattern" over the linear cause-effect paradigm (and, last but not least, predicted the decline of government and the collapse of Eastern Asian culture).

Another Austrian-born economist, Fritz Machlup of Princeton University, referred to universities as "knowledge industries" in "The Production and Distribution of Knowledge in the United States" (1962), which is also the book that popularized the expression "information society".

The expression "post-industrial society" was adopted by the Harvard University sociologist Daniel Bell, starting with the article "Notes on the Post-Industrial Society" (1967), to describe a society in which knowledge becomes the most valuable resource. Bell called it "knowledge society" in his 1973 book "The Coming of Post-Industrial Society" (page 212), a society that "is organized around knowledge".

In 1971 Japan adopted a national plan to become an "information society" by the year 2000. Coincidence or not, Japan never succeeded in software and services (the new knowledge economy) the way it did in cars and consumer electronics (the old physical industrial economy). Maybe there is a difference between information and knowledge, after all.

The Boom of Expert Systems

At the same time, expert systems were beginning to make inroads at least in academia, notably Bruce Buchanan's Mycin (1972) at Stanford for medical diagnosis and John McDermott's Xcon (1978) at Carnegie Mellon University (CMU) for product configuration, both written in LISP, and, by the 1980s, also in the industrial and financial worlds at large, thanks especially to many innovations in knowledge representation (Ross Quillian's semantic networks at CMU, Minsky's frames at MIT, Roger Schank's scripts at Yale University, Barbara Hayes-Roth's blackboards at Stanford University, etc).

A frame, which is a variant on Otto Selz's schema, is a structure that helps identify a situation; and, once we recognize the situation, the frame also tells us what we can do in and with that situation. A script, which is a social variant of Minsky's frame, represents stereotypical knowledge of situations as a sequence of actions and a set of roles. Once the situation is recognized, the script prescribes the actions that are sensible and the roles that are likely to be played. The script helps understand the situation and predicts what will happen in that situation. A script performs "anticipatory reasoning". Reasoning in ordinary daily life is not formal logical reasoning but instead "case-based" reasoning, a form of analogical reasoning in which each new situation is matched with known ones to predict what will happen next. Reasoning is "expectation-driven". There is a fundamental unity of cognitive phenomena such as perception, recognition, reasoning, understanding and memory: they occur at the same time, you can't have one without the other. Minsky and Schank were influenced by the "New Look" movement of Jerome Bruner and others: expectations determine what we perceive.

In 1975 the first A.I. startups had appeared: Leon Cooper of Brown University (the physicist who in 1957 co-developed the first theory of superconductivity and in 1981 would co-develop the BCM learning rule for neurons) co-founded Nestor with Charles Elbaum to develop neural-network technology; and Larry Harris of Dartmouth College founded Artificial Intelligence Corp to develop natural-language interfaces. In 1980 Edward Feigenbaum at Stanford co-founded the first major start-ups for expert systems: Intellicorp and Teknowledge. Brian McCune, an alumnus of the Stanford Artificial Intelligence Laboratory, was one of the founders in 1980 of Advanced Information and Decision Systems (AIDS) in Mountain View, later renamed Advanced Decision Systems. He and Richard Tong, a Cambridge University graduate, designed a concept-based text-retrieval system, Rubric, a progenitor of search engines. This A.I. boom did not extend much further than the USA: in 1973 a report submitted by James Lighthill, a Cambridge University mathematician, to the British government, titled "Artificial Intelligence - A General Survey", had pretty much killed A.I. research in Britain.

In the USA, instead, there were several projects of expert systems, especially in the medical field. In 1972 Saul Amarel's student Casimir Kulikowski and Sholom Weiss developed Casnet (CausalASsociational NETwork) at Rutgers University. In 1974 Harry Pople at University of Pittsburgh built Dialog (later renamed Internist) and in 1976 Stephen Pauker debuted PIP (Present Illness Program) at Tufts University. All the excitement about expert systems for medicine culminated in the first Artificial Intelligence in Medicine workshop at Rutgers University, convened in 1975 by Kulikowski.

Notable expert systems of the 1980s included: Prospector (1981), developed at SRI by the Shakey team (Nils Nilsson, Richard Duda, Peter Hart, etc) to help geologists in mineral exploration, written in a dialect of LISP (Interlisp) and running on a PDP-10; and DELTA (Diesel Electric Locomotive Troubleshooting Aid), deployed in 1983 by Piero Bonissone at General Electric for maintenance of locomotives, written in Forth and running on a PDP-11.

There was progress in knowledge-based architectures to overcome the slow speed of computers. In 1980 Judea Pearl introduced the Scout algorithm, the first algorithm to outperform alpha-beta, and in 1983 Alexander Reinefeld further improved the search algorithm with his NegaScout algorithm.

One factor that certainly helped the symbolic-processing approach and condemned the connectionist approach was that the latter uses much more complex algorithms, i.e. it requires computational power that at the time was rare and expensive.

(Personal biography: i entered the field in 1985 and went on to lead the Silicon Valley-based Artificial Intelligence Center of the largest European computer manufacturer, Olivetti, and i later worked at Intellicorp for a few years).

Rock Intermezzo: Welcome to the Machine

Rock music was perhaps the first art to envision the future of artificial intelligence. Klaus Schulze's "Cyborg" (1973), Pink Floyd's "Welcome to the Machine" (1975) and Kraftwerk's "Man Machine" (1978) evoked a world in which humanity is merging with machine intelligence. Sci-fi cinema was still mostly about space travel, time travel, alien invasions, dystopias, nuclear monsters, but the first intelligent computers popped up in JeanLuc Godard's "Alphaville" (1965), Stanley Kubrick's "2001 - A Space Odyssey" (1968) and Joseph Sargent's "Colossus - The Forbin Project" (1970), Michael Crichton's "Westworld" (1973) and Donald Cammell's "Demon Seed" (1977), and soon intelligent machines featured as supporting characters in several films including George Lucas' "Star Wars" (1977) and Ridley Scott's "Alien" (1979).

The Importance (or Redundance?) of Definitions

The founders of Artificial Intelligence never really defined it. In 1956 computers were considered "giant brains" because it was already obvious that, potentially, they were able to do things that humans couldn't do (for example, complex calculations). But it was also obvious that the computers of 1956 were not designed to do things that humans naturally do, even when they are not particularly intelligent: speaking, listening, seeing, recognizing objects and people, walking, grasping objects, and learning. The Turing Test was soon taken more seriously than Turing had asked for. In order to pass the Turing Test, a computer needs to be able to speak, listen, see, recognize objects, etc. So that became the indirect definition of Artificial Intelligence. Nobody came up with a proper definition, but it was understood that all those fields were difficult for traditional programs and something new was needed in the design of the computer or in the programming of the computer.

What is "intelligence"? Ask one thousand psychologists and you'll probably get one thousand definitions (don't ask philosophers because you'll get a 10-week class on the myriad theories of mind from "dualism" to "dual aspect monism"). What is "artificial"? Even that is not really obvious: is a spiderweb artificial or natural? If it's natural, then why would a computer be artificial? What makes something artificial instead of natural? One day John McCarthy defined Artificial Intelligence as "the science and engineering of making intelligent machines." Marvin Minsky defined it once as the "things that would require intelligence if done by men" (1968) and another time as "the ability to solve hard problems" ("Communication with Alien Intelligence", 1985). Socrates, who thought that conversations were pointless without proper definitions, must have turned in his grave. In 1982 Greg Chaitin added one last line to a to-do list in his essay "Goedel's Theorem and Information" and it reads: "Develop formal definitions of intelligence". A.I. practitioners became a little more articulate in the 2000s. Ray Kurzweil in his book "The Age of Spiritual Machines" (2000): "Intelligence is the ability to use optimally limited resources - including time - to achieve goals". Nils Nilsson wrote in his book "Quest for Artificial Intelligence" (2009): "Artificial intelligence is that activity devoted to making machines intelligent, and intelligence is that quality that enables an entity to function appropriately and with foresight in its environment." One of the last things that John McCarthy wrote before dying was: "We cannot yet characterize in general what kinds of computational procedures we want to call intelligent" (2007).

Where is the limit? There are many humans who don't speak, can't hear, cannot walk, etc, but we still consider them human beings. Philosophers love Artificial Intelligence because you can argue forever what constitutes "intelligence" and what does not. Today the term is so popular that it is applied to everything, even to programs that were already possible before 1956. I really don't know what the difference is between Computer Science and Artificial Intelligence: when does an algorithm become an A.I. algorithm? If A.I. is not popular, then no algorithm is called AI. If AI is popular (like today), then every algorithm is called AI! For example, ten years ago nobody called AI the statistical methods used in thousands of data-analysis programs. Now they are called "machine learning".

David Chapman, who has been working since 2010 on the hypertext book "Meaningness", is one of the people who fears that the discipline of Artificial intelligence is "doomed to recapitulate its own history, rediscovering the same dead ends over and over." However, the difference this time is that we are changing the definition of what Artificial Intelligence is. If the cuckoo clock were invented today, it would be tagged as "Artificial Intelligence".

But definitions do matter, because you may call the wrong expert. If you call "plumbing" the electrical wires of your house, you will call a plumber to fix an electrical "short". If all your firm needs is some statistical analysis and you call it "A.I.", you will hire an expert in so-called "deep learning" who will probably not solve your problem but charge you a lot of money.

On the other hand, we who grew up with A.I. may be too "puritanical". In September 2017 i had a meeting with Chinese venture capitalists who were using the "A.I." buzzword repeatedly. For the first 20 or 30 minutes i kept objecting "that's not A.I.!" but then i gave up: it was pretty obvious that they had a simple definition of A.I., i.e. A.I. is everything that is done by a machine. There is human intelligence, and there is machine intelligence (and maybe animal intelligence). The West has been fixated upon a (non) definition of intelligence that should apply equally to humans and machines, but these Chinese investors, who had missed 60 years of A.I., had come up with a much simpler and cleaner definition of A.I.: whatever chore is done by machines. They even mentioned the options/preferences tab of many applications as A.I.: the machine "learns" your preferences. I objected, of course, that the machine simply stores the preferences that i select; but they were hardly interested in splitting hairs. The distinction between "learning" and "storing" is philosophical. The net result is the same: some information is now available inside the machine that was not available before, and it is in a format that can be used for some practical application. That qualifies as "learning". This attitude is rapidly spreading among the new generations: they too missed the first 60 years of A.I.

I think that the motivation to start a whole new discipline came from the realization that the computer was ill-equipped for two of the most common of human activities: reasoning and recognizing. Computers were told what to do (were "programmmed"), and they were very good at obeying the instructions. But they were not good at doing those two things that all humans do all the time. Therefore those became the two pillars of A.I. research: how to build computers that can "reason" about a situation (typically, plan an action) and that can "recognize" things (typically, images and sounds). If you can't recognize patterns, you can't speak. If you can't reason, you can't answer a serious question. De facto, A.I. became the science of automated reasoners (e.g., expert systems, natural language processing and robotic arms) and recognizers (e.g., speech recognition and vision). I didn't say that reasoning and recognizing is all that humans do, but that may be all that is required to pass the Turing Test.

Just for argument's sake, let me tell you my rough definition of A.I. I honestly don't have a definition of "artificial intelligence", and that's because i don't have a definition of "intelligence" to start with, and, furthermore, because i never liked the term itself (if it's "artificial intelligence" than by definition it is not "intelligence" just like an "artificial leg" is not a "leg" although, hopefully, it works just as well as a leg); but i will try. A.I. is not automation and it is not Hollywood movies, it is something in between. A repetitive task does not require A.I. If it is repetitive, a simple algorithm can implement it. A philosopher can easily analyze that statement of mine and argue that every human behavior is repetitive if you look carefully enough, so even this concept can become a lengthy debate, but i show philosophers two simple cases so that they can come up with a better concept.. Who is the president of Russia? Vladimir Putin. Answering this question is only about information: a database contains the answer. Who will be the next president of Russia? This question is different: it requires in-depth knowledge of Russia, the ability to reason about it, and the answer won't be 100% certain. Or think of the questions "Where is Rome?" and "Where is Atlantis?" One is about information that can be found in a database of cities, whereas the other one requires knowledge of archeology and geology, and the answer won't be 100% certain. You don't need to be an expert to find out that Rome is in Italy, but only experts can tell you why the island of Thera (aka Santorini) is the likely location of Atlantis.

Philosophers make a big deal of the distinction between information and knowledge, but let's take the intuitive meaning of these two words: information is something that i can look up in an encyclopedia; knowledge is something that (currently) is only in the mind of experts and is really difficult to express in the formal language of computers. A.I. is not about processing information, but knowledge, so that a machine can answer questions such as "Who will be the next president of Russia?" and "Where is Atlantis?"

We can turn information into knowledge in two basic ways: we can elicit the knowledge from human experts who have acquired it over the years, or we can simulate the way a human expert creates knowledge: by trial and error. The knowledge-driven approach simulates the way the human mind thinks and the data-driven approach simulates the way the brain is structured. They sound like the same thing, but these two approaches lead to the development of wildly different technologies. One deals with symbols (just like human language is made of words), the other deals with numbers (just like the human brain is made of electrochemical signals).

Consciously or subconsciously, the discipline of Artificial Intelligence came to encompass all the technologies needed for a machine to pass the Turing Test that back then seemed impossible to program in the Von Neumann way (instruction by instruction): speech recognition, computer vision, natural language processing, reasoning, learning, common sense. There was an entire range of applications that didn't seem feasible with the Von Neumann architecture. This included tasks for which the algorithm does not exist, i.e. "expert" tasks (a medical encyclopedia is not equivalent to an experienced physician); tasks for which the algorithm exists but it is useless because "heuristics" prevails (you don't need to calculate the effects of hot temperature on your skin in order to conclude that you shouldn't touch boiling water"); tasks for which a deterministic algorithm is not possible because they involve uncertain quantities (e.g. "Brazil will win the next world cup"); tasks for which the algorithm would be too complicated (e.g., designing a cruise ship). In order to tackle these categories of problems, the new discipline had to invent new techniques to program computers in a different way. With hindsight, these categories of problem can be divided in two groups: one can be solved using logical (exact) reasoning and the other is usually approached with probabilistic reasoning. For example, we can calculate logically the best move of chess, whereas we can only be reasonably certain (not absolutely certain) that the object in the room is an apple (nobody would be shocked if it turned out to be a painted baseball or a giant strawberry). For example, we can determine what action is the best when we have all the facts, whereas we can only guess what the speaker intended to say when the speaker has a strong foreign accent or if we couldn't hear clearly all the words that were uttered. In most cases the latter prevails: we settle for a plausible explanation of what happened, aware that it may not be what we initially thought it was. Sometimes our thinking is exact (logical), but in most cases it is probabilistic. Speech, vision, language and common sense fall into the latter category. Time would tell that the aspects related to exact reasoning were not the problem: the real problem were all the aspects of human behavior that belong to inexact reasoning. Most of what we do daily is based on inexact reasoning: language is ambiguous (otherwise philosophers and lawyers would not have a job), vision is tentative (especially at a distance and in conditions of poor visibility), judgement is almost always probabilistic, and even "infallible experts" make mistakes. The human world is not an exact world. It is, in fact, a wildly inexact world in which "plausible" reasoning prevails over mathematical reasoning.

Today A.I. (as defined above) encompasses four main classes of technologies: computer vision (object detection, face recognition, scene analysis), natural language processing (speech recognition, automatic translation, discourse analysis, sentiment analysis), reasoning (exact and inexact inference, common sense, planning), and learning (which ranges from mere classification to theory formation).

There is progress in all of them, as long as we don't expect miracles, just like there is progress in every field of technology: lithium batteries, dishwashers, bicycle tyres, water heaters, nail clippers...

In retrospect, the funny thing about A.I. is that it was not obvious at all to establish what was difficult and what was easy. Originally many A.I. scientists thought that the ultimate sign of "intelligence" was the ability to solve mathematical problems (especially since most of them were mathematicians). But anything that can be automated will be automated. So, given that logic had been largely automated during the 20th century, it was actually not difficult for a computer to solve logical problems, such as mathematical theorems or chess. However, "reasoning" in general is not trivial, and is still an unsolved problem, because most of our "reasoning" is not logical at all. Most of our reasoning is better called "guessing". Our daily lives are not about proving mathematical theorems but about guessing what to wear, what to do about a contract, whom to call to get help, whether it is best to visit mother this weekend or next month, whether it is cheaper to buy a car this year or next year, and so on. Reasoning in the real world involves so many uncontrollable and unpredictable factors that it is inherently imprecise and uncertain. We mostly engage in "plausible" reasoning, not mathematical reasoning.

And, amazingly, out of this messy inexact reasoning we also learn. We are capable of a virtually infinite number of tasks. The steam engine cannot be programmed: it can do only one thing. The digital electronic computer can be programmed and it can therefore do many things, but needs a program for each task. In order to match human intelligence, A.I. also needs to build machines that can program themselves.

To start with, we don't have a viable definition of "intelligence". The concept of a general intelligence was formalized by British psychologist Charles Spearman in his book "The Abilities of Man, their Nature and Measurement" (1927), but instead Louis Thurstone of University of Chicago in his book "Primary Mental Abilities" (1938) criticized the notion of a general intelligence. This theory of multiple intelligences was embraced decades later by the psychologist Howard Gardner, then at Boston's Veterans Administration Hospital, in his book "Frames of Mind" (1983). Anecdote: when i asked a psychologist "did any of these definitions succeed?" he asked me to define "succeed..." The Turing Test has been used to define "intelligence" as "what humans do", but that is hardly a definition because it depends on which human you use as a model of intelligence (and which human you use as the judge). Nor do we have a viable definition of intelligence that is applicable to all kinds of systems: humans, animals, machines (and maybe other natural systems).

But the lack of a definition often benefits the field, because success and failure are measured against the definition: the least precise the definition, the easiest to claim success. My guess is that one of these days the singularists will simply change the definition of "singularity" and declare that it is already here, even ahead of time (except that obnoxious skeptics like me will then show that, based on the new definition, it has always been here).

My Chinese friends couldn't care less about my definition of A.I. They only want to know if the program (call it A.I. or not) is useful. As Xiaoping Deng famously said when he launched the capitalist reforms in China, it doesn't matter whether the cat is white or black as long as it catches mice.

A Brief History of Cognitive Science/ Part 1 (Note: There is no Part 2)

A conceptual revolution took place in psychology thanks to the German school of "gestalt" psychology, which was influenced by the Austrian philosopher Christian von Ehrenfels ("On the Qualities of Form", 1890) and the Austrian physicist Ernst Mach ("The Analysis of Sensations", 1886). The likes of Max Wertheimer, ("Experimental Studies on the Perception of Movement", 1912), Kurt Koffka ("The Growth of the Mind", 1921), Wolfgang Koehler "The Mentality of Apes" (1925) and Kurt Goldstein "The Organism" (1939) favored a holistic approach to studying the mind at a time when psychology in the USA was dominated by behaviorism. Hence the adage: "The whole is greater than the sum of its parts". During this time, the German psychologist Otto Selz proposed that solving a problem entails recognizing the situation via a known "schema" and filling the gaps in the schema ("On the Laws of the Orderly Thought Process", 1913). A schema organizes past experience and helps us deal with new experience.

Cognitive Psychology underwent a rapid mutation at about the same time that Computer Science was born.

Just like the history of computers begins in England, with Alan Turing and the Colossus, so does the history of Cognitive Science, and it is as intertwined with World War II: in 1944 Frederic Bartlett, author of "Remembering" (1932), and Kenneth Craik, author of "The Nature of

Explanation" (1943), two of most influential books in the field, were asked to set up the Applied Psychology Unit at Cambridge University.

Just like the US counterpart to England in computers was mainly Boston, so Boston was the other primordial pole of Cognitive Science: in 1952 Jerome Bruner set up the Cognition Project at Harvard University, and later published "A Study of Thinking" (1956). Bruner was one of the founders of the "New Look" movement whose tenet was that cognition and perception are tightly integrated: you see what you expect to see based on the concepts that are in your mind, and those concepts have in turn been molded by what you have seen ("On Perceptual Readiness" 1957). In 1961 another leader of the movement, Leo Postman, established the Institute of Human Learning at UC Berkeley.

Carnegie Mellon University (called Carnegie Institute of Technology until 1967) is where the two disciplines, A.I. and Cognitive Science, intersected from the beginning: Simon's and Newell's Logic Theorist (1956) pioneered both. Boston's MIT produced Noam Chomsky's "Syntactic Structures" (1957) while Cambridge produced Donald Broadbent's "Perception and Communication" (1958), Broadbent having succeeded Bartlett at the Applied Psychology Unit. In 1960 George Miller, author of "The Magical Number Seven, Plus or Minus Two" (1956), and Jerome Bruner set up the Center for Cognitive Studies at Harvard, coincidentally located in the old house of the great psychologist William James and one of the first laboratories where psychologies were given a computer (a PDP-4); and George Miller coauthored "Plans and the Structure of Behavior" with Eugene Galanter and Karl Pribram (all three were alumni of Stanley Smith-Stevens' Psycho-Acoustic Laboratory at Harvard).

In 1965 two major new schools appeared: Donald Michie set up the Experimental Programming Unit at Edinburgh and George Mandler (born in Austria) set up the Center for Human Information Processing (CHIP) at UC San Diego. The Harvard graduate Ulric Neisser (born in Germany) wrote the book that named the new discipline, "Cognitive Psychology" (1967), while at University of Pennsylvania.

David Marr at Cambridge University published three influential papers proposing computational theories for the cerebellum (1969), the neocortex (1970), inspired by David Hubel and Torsten Wiesel's "feature detectors", and the hippocampus (1971).

In 1972 San Diego and Carnegie Mellon produced influential books, namely Don Norman's and Peter Lindsay's "Human Information Processing" and Newell's and Simon's "Human Problem Solving", and Endel Tulving (born in the Soviet Union) compiled "The Organization of Memory", that contained his article on "semantic memory" and that established University of Toronto as a major center of research. The San Diego school, dominated by the LNR research group (Lindsay, Norman and David Rumelhart), worked on the "active structural network," inspired by the work of Ross Quillian and Charles Fillmore. Norman and Rumelhart published the paper "Active Semantic Networks as a Model of Human Memory" (1973) and then the book "Explorations in Cognition" (1975). (Trivia: Don Norman would become a Silicon Valley celebrity as a designer of human-computer interfaces after joining Apple in 1993).

A potentially serious flaw in the program of cognitive science is that it studied mostly humans, and not animals.

"The major difference between rats and people is that rats learn from experience" (Burrhus Skinner).

A Brief History of Logical Reasoning

We now think of mathematical logic as having a system, based on some axioms and definitions, that grows via theorems and proofs.

Most philosophers begin the history of Logic with Aristotle, but the computer age should perhaps recognize Ramon Llull in Spain as the man who started a new way of thinking about logic. Llull invented the volvelle, a paper computer that was incorporated in manuscripts so that readers could perform computations by simply turning circles and dials. Imagine a book that, in each chapter, contains a smartphone loaded with the apps that can solve the problems discussed in that chapter. His book on Logic was the "Ars Generalis Ultima" or "Ars Magna" (1305). Its declared intent was to provide logical arguments to convert Muslims to Christianity.

Three centuries later, a number of thinkers toyed with the power of a hypothetical universal language of logic that could represent and solve any problem in formal ways. They wrote books such as George Dalgarno's "Art of Signs" (1661) and John Wilkins "An Essay towards a Real Character and a Philosophical Language" (1668), both in England. The most famous was a book written by a German philosopher, Gottfried Leibniz, who also invented calculus: "De Arte Combinatoria" (1666) that discussed an "algebra of thought" based on a formal calculus of reasoning and on a universal language.

Two centuries later, the British mathematician George Boole began the program that would fulfill Leibniz's dream: mathematical logic was born in 1847 with Boole's "The Mathematical Analysis of Logic". Boole's stated goal was to found a science of reasoning, and his strategy consisted in expressing logic in the language of symbolic algebra. Hence, Boole's approach became known as an algebra of logic. Boole invented binary logic, the logic of true and false propositions that can be combined to obtain other true and false propositions. Realizing that all calculations can be represented as binary operations, and that his calculus of symbols and operations seemed capable of solving all logical problems, he titled his 1854 book "The Laws of Thought". Propositional logic was improved by another British mathematician, Augustus De Morgan, who expanded Boole's system to relations in "On the Syllogism Number IV, and on the Logic of Relations" (1860). The US philosopher Charles Peirce further expanded this notion in "Description of a Notation for the Logic of Relatives" (1870).

In 1874 the German mathematician Georg Cantor realized that most objects used by mathematicians are sets and thus founded set theory that works with operations such as union and intersection ("On a Property of the Collection of All Real Algebraic Numbers", 1874). For example, he reduced the mathematical function to a set consisting of ordered pairs. Set theory is now the foundation for all branches of mathematics. Cantor was the first mathematician to study the infinite. His set theory dealt with infinite sets (or "trans-finite" sets) and even showed that there can be many different kinds of infinities (another scientific finding that has a profound psychological impact on science fiction). Cantor realized that the set of natural numbers (1, 2, 3, 4,...) is the smallest kind of infinite set, and it has as many members as half of itself. That provided him with a definition of infinity: a set is infinite if it has a many members as a subset of itself. The set of irrational numbers (the numbers that are not fractions of natural numbers, such as pi and the square root of 2), instead, is "more infinite" than the set of natural numbers (it has a higher "cardinality"). In general, the set of all subsets of any infinite set is always bigger than the set itself.

Gottlob Frege's "Begriffsschrift/ Concept-script" (1879), subtitled "a formal language for pure thought modeled on that of arithmetic", picked up Leibniz's ambition to design a calculus of reasoning and a universal language, a scientific language to describe mathematical proofs, which, until then, had been mostly based on intuitive notions. He grounded mathematics in a new language of logic that went beyond propositional logic (the logic of "If all Greeks are humans and if all Socrates is a Greek, then Socrates is a human"). Frege formalized functions (which he expressed as "predicates"), variables and quantifiers ("every", "some", etc). His predicate logic used variables (such as "x" and "y") to express properties (such as Greek (x) is true when x = Socrates but false when x = Piero). Frege reduced meaning to truth: the meaning of something is whether the logical expression $p(x,y,...)$ is true or not. For example, capital (France, Paris) is true which means that Paris is indeed the capital of France. Frege introduced universal and existential quantifiers to express whether properties applied to all or only one member of a set. The leadership in logic started shifting towards Germany.

For the record, Peirce's paper "On the Algebra of Logic" (1885) independently introduced the quantifiers (such as "some" and "all"). The Italian mathematician Giuseppe Peano refined Frege's universal language in his "Arithmetices Principia" (1889) that streamlined a notation consisting of logical operators, relation symbols and quantifiers. In 1900 Peano's formal language was adopted and extended by the British philosopher Bertrand Russell, but in 1901 the latter discovered what became known as Russell's Paradox (first published in 1903 in his "Principles of Mathematics"): does the barber who shaves all and only men who do not shave themselves shave himself? Russell showed Frege that there was a contradiction in his logical system (originally Russell wrote it on a postcard). Russell divided logic into three parts: the calculus of propositions, the calculus of classes, and the calculus of relations.

Meanwhile, in 1900 a new German star, David Hilbert, gave a lecture titled "Mathematical Problems" at the International Congress of Mathematicians in Paris in which he emphasized the axiomatic method: solving a mathematical problem amounted to little more than to rewrite it in a formal manner. Until then the mathematical proof of a theorem had been thought of as a matter of intuition, almost a divine enlightenment. Hilbert transformed it into a completely different, and much more formal, concept: the axioms (the premises) are a finite sequence of symbols that can be turned into other finite sequences of symbols by applying the rules allowed by logic in a finite number of steps until a sequence of symbols is obtained that represents the proof of the theorem. Hilbert turned mathematical proofs into mechanical procedures, and procedures that, strictly speaking, did not even require mathematical ingenuity. There is no need to know what you are doing, as long as you apply the admissible rules and keep producing valid sequences of symbols. If at some point you obtain the proof, you succeed... even if you have no idea of what the theorem was about and what the proof shows. There is no need for intuition and no need for understanding. Three of the problems that he posed to the world's mathematicians defined the three main fields of mathematics of the 20th century: set theory, mathematical logic, and computability theory. And, thanks to Hilbert, Goettingen became the top university of mathematical logic in the world.

But Russell's paradox was a stumbling block to any further discussion on Logic. In 1908 two ways of avoiding (if not solving) the paradox were proposed: Russell invented "type theory" ("Mathematical Logic as Based on the Theory of Types") and Ernst Zermelo, another Goettingen alumnus, axiomatized "set theory" ("Investigations in the Foundations of Set

Theory"). In 1910 Russell coauthored "Principia Mathematica" with Alfred Whitehead, the culmination of the program started by Frege and Peano. In 1917 Hilbert delivered a lecture in Zurich titled "Axiomatic Thinking" in which he urged the application of the axiomatic method to all branches of science. In the winter of that year he taught a course in Goettingen, "Principles of Mathematics and Logic," that distilled his mature axiomatic method for mathematical logic. In particular, Hilbert worked out a subsystem of his functional calculus that he called "restricted functional calculus" and that today is called "first-order predicate calculus". Determined to found mathematics on perfect logical foundations, in 1920 Hilbert proposed a research project that became known as "Hilbert's program".

First-order logic was important because it was a formal way to talk about the real world, which is made of objects, and objects that relate to other objects. A formula in first-order logic says something about some relationship among some objects. So far first-order logic had been treated as a subset of logic, but in 1922 the Norwegian mathematician Thoralf Skolem rewrote set theory based on first-order logic ("Some Remarks on Axiomatized Set Theory"), which proposed that first-order logic was "all" of logic. Feeling that he was closing in on a final theory of mathematics, in 1928 Hilbert published "Principles of Mathematical Logic", co-written with Wilhelm Ackermann. Hilbert posed the "Entscheidungsproblem/ Decision Problem" (1928) at yet another international conference: does an algorithm exist that can always prove whether a logical statement is true or false based on the axioms?

In 1907 the Dutch mathematician Bertus Brouwer had started his program of "intuitionism", believing that any mathematical proof had to be constructive. In particular, he had the powerful intuition (sorry for the pun) that a proposition is not necessarily either true or false: it could also just be unprovable.

The Austrian mathematician Kurt Goedel graduated in 1930 with a dissertation that proved the completeness of first-order predicate calculus (roughly, that "anything true is provable"), but then his two incompleteness theorems ("On Formally Undecidable Propositions of Principia Mathematica and Related Systems", 1931) proved that the perfection demanded by Hilbert for axiomatic systems (that contain at least arithmetic) was impossible. Goedel showed that a formula exists that cannot be proven true or false (a formula that states its own unprovability); and that was enough to demolish the castles of logical systems built by Russell and Hilbert. Roughly, his completeness theorem proved that anything true is provable, but his incompleteness theorem proved that some statements cannot be proven true or false. No, the two theorems don't contradict each other. (John Von Neumann had made the same discovery, but never published it because Goedel beat him by a few weeks, and Von Neumann was instrumental in publicizing the importance of Goedel's theorems). Goedel used the natural numbers to encode mathematical statements and proofs, including proofs about natural numbers themselves. A natural number can encode a statement about itself or even an encoded proof about itself. It was then easy for Goedel to construct self-referential formulas. And that's where Hilbert's program collapsed. But Goedel also left us the gift of self-referential formulas. (Trivia: Goedel also discovered a solution to Einstein's equations of General Relativity which permits time travel, one of the most popular themes of science fiction).

Zermelo's axiomatic set theory prevailed over Russell's type theory because Zermelo's followers developed meta-logical tools such as model theory (Alfred Tarski) and proof theory (e.g., Gerhard Gentzen's theorem of 1936).

In 1933 the Polish mathematician Alfred Tarski reached conclusions similar to Goedel's: he showed that the concept of "truth" leads to logical contradictions. Tarski published "The Concept of Truth in Formalized Languages", an influential theory of truth: a sentence such as "snow is white" is true if and only if snow is white. It sounds trivial, but it was the first formal definition of truth that separated language and metalanguage. For example, "I am lying" does not separate the language and the metalanguage and appears to be a contradiction... until you realize that there are two levels at which that sentence operates. Truth is relative to the model in which it is interpreted.

Hilbert's ideas were proliferating. "On the Consistency of Arithmetic", published in July 1931 by the French mathematician Jacques Herbrand (who alas died at the age of 23 shortly after completing this paper) introduced the concept of a (general) recursive function made famous by Goedel's lecture "On Undecidable Propositions of Formal Mathematical Systems" of 1934. In 1932 Alonzo Church at Princeton created a method for defining functions called the "lambda-calculus" ("A Set of Postulates for the Foundation of Logic", 1932), and in April 1935 (one year before Turing's analogous paper) delivered a lecture about "An Unsolvable Problem of Elementary Number Theory" in which he proved that the Entscheidungsproblem is undecidable, and also proved that his class of lambda-definable functions and Goedel's class of recursive functions are identical. Alonzo Church's "foundation of logic" was the first logic programming language. After Stephen Kleene and Barkley Rosser proved that it led to a paradox, the logic was abandoned, but the lambda-calculus was rescued by Church's theory of types, a formulation of type theory that was simpler and more general than Russell's ("A Formulation of the Simple Theory of Types", 1940).

Finally, in 1936 the British mathematician Alan Turing, using an argument similar to Goedel's incompleteness theorem, invented the Turing machines that can carry out calculations by manipulating symbols on a tape ("On Computable Numbers, with an Application to the Entscheidungsproblem", 1936) and also proved independently what Church had just proven. Turing also showed that lambda-calculus and his own system of "machines" are equivalent. Bottom line: Herbrand, Goedel, Church and Turing had discovered the same thing. This came to be known as the Church-Turing thesis after Church's student Stephen Kleene published the book "Mathematical Logic" (1967): a function is lambda-computable if and only if it is Turing computable if and only if it is (general) recursive.

The time was fertile for thoughts of such kind: René Magritte's painting "The Treachery of Images" (1929) consists of a pipe and the caption "This is not a pipe", and James Joyce's novel "Finnegan's Wake" (1939) is an infinite loop (it ends with the same words that it begins). In Jorge-Luis Borges's tale "The Circular Ruins" (1940) each person is a dream in another person's mind, and in Maurits Escher's drawing "Drawing Hands" (1948) two hands are drawing each other on the same piece of paper. Miguel de Unamuno is a character in his own novel "Mist" (1914), André Gide's novel "The Counterfeiters" (1925) contains a novelist Édouard writing a novel titled "The Counterfeiters", and Aldous Huxley's novel "Point Counter Point" (1928) contains a novelist, Phillip Quarle, who wants to write a still untitled novel that looks just like "Point Counter Point". Trivia: other authors who will appear as characters in their own novels will include Gabriel García Márquez in his "One Hundred Years of Solitude" (1967) and Kurt Vonnegut in his "Breakfast of Champions" (1973). And readers will appear in Julio Cortazar's "Continuity of the Parks" (1963), that features the reader as the victim of the murder described in the story; in John Barth's "Frame-tale" (1968), that features the reader as invited to cut the pages of the book in order to create a Moebius strip, an infinite loop of the sentence "Once upon a time there was a story that began"; and in Italo Calvino's novel "If on a Winter's Night a Traveler" (1979), whose protagonist is the reader reading "If on a Winter's Night a Traveler".

Bela Bartok employed palindromes in his "Music for String Instruments, Percussion and Celeste" (1936), especially in the third movement, as well as in his 5th String Quartet (1934), as did Alban Berg in his opera "Lulu" (1935), particularly the second movement, the "Ostinato". A palindrome is a sentence whose reverse is also a sentence, such as "Was it a car or a cat I saw?" and "Never odd or even", but in music it's a phrase that is the same when played forward or backward. The protagonist of Buster Keaton's film "Sherlock Jr" (1924) is a projectionist who falls asleep and becomes a character in the movie that he is projecting. The protagonist of Orson Welles' film "Citizen Kane" (1941) walks down a hall whose walls are covered with mirrors, his image endlessly repeated, and the viewer realizes only later that the camera has been following one of the reflections, not the real man.

The "mise en abime", i.e. an image that contains a smaller copy of itself which contains a smaller copy of itself etc, became popular in advertising. For example, the magazine Cosmopolitan of April 1948 had a cover depicting a woman holding a copy of that magazine's very issue. Trivia: one of the most famous album covers in the history of rock music is the cover of Pink Floyd's "Ummagumma" (1969), designed by Storm Thorgerson, a cover containing itself containing itself etc.

The epic of 20th century logic was perhaps a reflection of a bigger shift in the way humans think about reality.

Hilbert died in 1943, the year when neural networks were born (Warren McCulloch's and Walter Pitts' binary neuron) and unaware that two years earlier his fellow-countryman Konrad Zuse had built the first Turing-complete machine, the Z3 programmable electromechanical computer. Turing died in 1954, two years before the first conference on Artificial Intelligence. Russell died in 1970, having witnessed an A.I. program, the Logic Theorist, prove 38 theorems of the "Principia Mathematica", but in his last years he was much more concerned with human stupidity than artificial intelligence (he wrote "Has Man a Future?" in 1961 despairing that the human race was about to annihilate itself with nuclear weapons). Goedel spent the last years of his life working on "rotating universes" in which people can time-travel to the past and on an ontological proof of God's existence (possibly the only mathematical mistake he ever made in his life). He died in 1978, just one year before Douglas Hofstadter published the book "Goedel, Escher, Bach" in which he envisioned Goedel's mathematical recursion and self-reference as the clues not only to understanding intelligence but even consciousness itself. Goedel seemed to have remained largely indifferent to Artificial Intelligence even after the British philosopher John Lucas published "Minds, Machines and Goedel" (1959), a "proof" based on Goedel's own incompleteness theorem that A.I. is impossible (an argument retold by the British physicist Roger Penrose in his 1989 book "The Emperor's New Mind").

A Brief History of Computers

The computer was born in the middle of World War II.

A German civil engineer named Konrad Zuse built the first programmable computer at the end in 1941, the "Z3", the first hardware implementation of the Turing machine. It was programmable and digital, but it still employed electro-mechanical relays. At the end of 1943 the Colossus debuted, a machine designed by telephone engineer Tommy Flowers in London. This was a digital electronic computer for the British army to decipher the secret code of the Germans.

A joint project between IBM and Harvard University yielded what IBM called ASCC and Harvard called Harvard Mark I, designed by Howard Aiken at Harvard and built by Claire Lake at IBM in February 1944. It was the first computer programmed by punched paper tape, but it still employed electro-mechanical relays, just like Zuse's Z3.

In 1946 the ENIAC, or "Electronic Numerical Integrator and Computer", was unveiled, built by John Mauchly and Presper Eckert at University of Pennsylvania. It didn't have the stored program, so it had to be reconfigured for each task, but it had a memory.

The idea of a stored program (making the computer a general-purpose device) was popularized by John Von Neumann's 101-page document titled "First Draft of a Report on the EDVAC" of June 1945, the result of discussions with the ENIAC team. Von Neumann (an Austrian-Hungarian Jew, real name Janos Neumann) had become famous in Europe by providing mathematical foundations to Quantum Mechanics in 1932, and in 1943 had joined the Los Alamos team working on the atomic bomb in New Mexico.

The first computer with a stored program was the "Manchester Baby", officially named Manchester Small-Scale Experimental Machine (SSEM), that ran the first program in June 1948 at University of Manchester, followed by the Manchester Mark 1 of April 1949. In February 1951 British defense contractor Ferranti unveiled its commercial version, the Ferranti Mark I, the first programmable electronic computer that could actually be purchased and used by a customer.

The second stored-program electronic computer was the Electronic Delay Storage Automatic Calculator (EDSAC), that debuted in May 1949, built by Maurice Wilkes at Cambridge University. Alan Turing's Automatic Computing Engine (ACE) was never completed, but an incomplete prototype was deployed in May 1950.

The third stored-program electronic computer, and the first to be actually deployed in the USA, was the SEAC, a scale-down version of the EDVAC and also the first to use semiconductors instead of vacuum tubes (May 1950).

The first commercial computers in the USA were the Univac (1951), built by the ENIAC team, and the IBM 701 (1952).

Von Neumann was personally involved in the design of a computer at the Institute for Advanced Study (IAS), built by Julian Bigelow (one of the co-authors with Norbert Wiener of the historical paper "Behavior, Purpose and Teleology" that launched Cybernetics in 1943). This IAS computer, operational in 1951, was used at the Los Alamos laboratories to simulate the hydrogen bomb before it was tested on an atoll of the Marshall Islands. Several "clones" of the IAS machine were built throughout the nation for research purposes: the Illiac at University of Illinois (1952), the Maniac, also at Los Alamos (1952), and the Johnniac at Rand Corporation in Los Angeles (1953).

The transistor had been invented in 1947 but up until then the electronic computers mostly used the cumbersome and unreliable vacuum tubes. The first fully transistorized computers were Bell Labs' Tradic (1954) and MIT's TX-0 (1955), just before the first Artificial Intelligence conference. Summarizing:

Zuse 3 (1941): Turing-complete, electromechanical, no stored program, no memory

Colossus (1943): electronic, not Turing-complete, no stored program, no memory

Harvard Mark 1 and IBM ASCC (1944): not Turing-complete, electromechanical, no stored program, decimal, no memory

ENIAC (1946): Turing-complete, electronic, no stored program, decimal (memory: delay lines)

Manchester Baby (1948): Turing-complete, electronic, stored-program (memory: Williams tubes)

Manchester Mark 1 (1949): Turing-complete, electronic, stored-program (memory: Williams tubes and drum memory)

Cambridge EDSAC (1949): ditto (memory: delay lines)

Pilot ACE (1950): ditto (memory: delay lines)

SEAC (1950): ditto (memory: delay lines) with semiconductor logic

Ferranti (1951): first commercial Turing-complete electronic stored-program computer (memory: Williams tubes and drum memory)

Univac (1951): ditto (memory: delay lines)

IAS (1951): ditto but not commercial (Williams tubes)

IBM 701 (1952): ditto, commercial (memory: Williams tubes)

Bell Labs Tradic (1954): fully transistorized

MIT TX-0 (1955): fully transistorized.

The first A.I. program, Newell's and Simon's Logic Theorist, was coded by Cliff Shaw at Rand Corporation and ran in August 1956 on a Johnniac.

"From so simple a beginning endless forms most beautiful and most wonderful have been, and are being evolved" (Charles Darwin in "On the Origin of Species")

Machine Learning before Artificial Intelligence

The mathematics of pattern recognition and classification existed before the invention of the digital computer but clearly the digital computer made it practical.

A system of pattern recognition operates on a dataset. If the dataset has been manually labeled by humans, the system's learning is called "supervised". If the dataset consists of unlabeled data, the system's learning is "unsupervised".

In supervised learning the system has to learn a model (basically, a generalization) so it will be able to correctly categorize future instances of each category. In this case "learning" means: to act based on patterns in the data. For example, recognize apples or forecast the effects of chess moves. In unsupervised learning the system has to discover patterns in the data, i.e. categories. For example, after watching millions of videos of cats and dogs, it may divide them into two groups (without being told what a cat and a dog are).

Supervised learning is about recognition, classification, prediction. Unsupervised learning is about clustering. What they have in common is that they are both methods of generalization.

An instance is described, mathematically speaking, by a vector of features. For example, the vector of an image contains features such as edges, shape, color, and texture.

The two fields that studied machine learning before it was called "machine learning" are statistics and optimization.

Statistical methods for pattern recognition are sometimes a century old. British statistician Karl Pearson invented "principal components analysis" in 1901 (unsupervised), popularized in the USA by Harold Hotelling ("Analysis of a Complex of Statistical Variables into Principal Components", 1933), and then "linear regression" in 1903 (supervised). He was a fascinating character who at the same time predated Albert Einstein's Relativity by more than a decade in "The Grammar of Science" (1892) and, alas, predated Adolf Hitler's ideas on the genetic inferiority of Jews. Contributions to pattern recognition came from different quarters: Ronald Fisher (one of the founders of population genetics) in Britain invented "linear discriminant analysis" in 1936; Joseph Berkson invented the most popular "logistic regression" method in 1944 working in a Minnesota clinic; the "k-nearest-neighbors" (KNN) classifier (aka "minimum distance classifier", aka "proximity algorithm") was invented in 1951 by Evelyn Fix and Joseph Hodges at the US Air Force School of Aviation Medicine in Texas; the Bell Labs physicist Stuart Lloyd invented "k-means clustering" in 1957 for signal processing; etc.

Linear classifiers were particularly popular, such as the "naive Bayes" algorithm, first employed in 1961 for text classification by Melvin Maron at the RAND Corporation and (the same year) by Marvin Minsky for computer vision (in "Steps Toward Artificial Intelligence", 1960); and such as the Rocchio algorithm invented by Joseph Rocchio at Harvard University in 1965.

Naive Bayes uses an approximation of Bayes' theorem that may sound wildly arbitrary (that the effects of a cause do not influence each other, which is like saying that the history of the world is a simple hierarchy of causes and effects); but it turns out that Naive Bayes works incredibly well in most cases, even defying its own limitations, as proven by the Portuguese-born Pedro Domingos at UC Irvine ("Beyond Independence", 1996). Nearest-neighbor methods are the simplest of these algorithms, and became very popular after a theorem by by Thomas Cover of Stanford University and his former student Peter Hart showed that they were also more reliable than it was apparent ("Nearest Neighbor Pattern Classification", 1967), although the world had to wait almost 40 years for a theorem by Kamalika Chaudhuri and Sanjoy Dasgupta of UC San Diego in order to fully grasp the mathematical properties of nearest-neighbor methods that makes them so efficient ("Rates of Convergence for Nearest Neighbor Classification", 2014).

A linear classifier uses a linear function to map the possible features of the data to a set of possible labels. Frank Rosenblatt's Perceptron of 1957 was also a linear classifier, whereas multilayer neural networks are nonlinear classifiers. Another kind of nonlinear classifier is "decision trees analysis", notably Iterative Dichotomiser 3 (ID3) invented by Ross Quinlan at University of Sydney in Australia ("Induction of Decision Trees", 1985) and in 1993 expanded to C4.5, which became the benchmark for supervised learning.

These statistical methods were widely employed in computer science. Popular textbooks such as "Introduction to Statistical Pattern Recognition" (1972) by Keinosuke Fukunaga of Purdue University and "Pattern Classification and Scene Analysis" (1973) by Richard Duda and Peter Hart of SRI International provided comprehensive summaries. Linear classifiers remained popular for text classification and, during the dot.com boom, became the method of choice for writing "recommender systems". Paul Resnick built an early one, called GroupLens, at MIT in 1994 based on a KNN algorithm (it was used to recommend articles on the Usenet).

None of this was marketed as Artificial Intelligence. Its roots were in statistics.

In the age of "big data" there is a tendency to focus on the data. However, it is important to remember that we frequently learn something from just one example. Show a banana to a child and the child will probably be able to recognize any future banana. Learning is not about "data". Learning is about the ability to form and use concepts. It just so happens that computers are good at processing data, so we invented a mathematics that allows them to overcome the limitations of not having concepts; i.e. the mathematics that allows them to do with information what we do with knowledge.

"Where is the Life we have lost in living? Where is the wisdom we have lost in knowledge? Where is the knowledge we have lost in the information?" ("The Rock", Thomas Stearns Eliot).

Probabilities, Markov Chains, Monte Carlo Methods

In 1494 an Italian friar, Luca Pacioli published the first book ("Summa de Arithmetica, Geometria, Proportioni et Proportionalita") that mentions the "problem of points", considered by many the inspiration for the development of a mathematical theory of probability. (Pacioli, who later in Milan became a friend and housemate of Leonardo da Vinci, also published "De Viribus Quantitatis", the first book of recreational mathematics, including tricks of prestidigitation). In 1564 the Italian mathematician, inventor and physician Geronimo Cardano, son of another friend of Leonardo da Vinci in Milan, wrote a book ("Liber de Ludo Aleae") on games of chance, basically a gambling manual, that would be published only a century later. (Cardano published more than 200 books and invented dozens of mechanical devices, besides popularizing the magic trick of the magic coloring book, but, ironically, his son became an addicted gambler). The first person to give us a rigorous treatment of probability was the Dutch astronomer Christian Huygens, who in 1657 published "De Ratiociniis in Ludo Aleae". (Huygens discovered the rings of Saturn, invented the pendulum clock and built countless mechanical automata for the royal court). In 1689 the Swiss mathematician Jacob Bernoulli finished his treatise "Ars Conjectandi" (published posthumously), a collection of problems of probability that stands as the first textbook on the subject. (This Bernoulli, not to be confused with the Johann Bernoulli of infinitesimal calculus, with whom he founded the calculus of variations, and not to be confused with the Daniel Bernoulli of the Bernoulli principle in fluid dynamics, belonged to an incredible family of mathematicians). The French mathematician Abraham de Moivre, exiled in Britain, wrote an even more popular textbook, "The Doctrine of Chances" (1718), the first one not to be written in Latin. The British priest, theologian and amateur mathematician Thomas Bayes died in 1761 before he could publish his main achievement, a formula on how to calculate the probability of an event, a formula that today we know as Bayes' Theorem. (His "Essay towards Solving a Problem in the Doctrine of Chances" was read to the Royal Society 1763). The French mathematician Pierre Laplace had become a living legend applying Newtonian gravitation to the solar system (Isaac Newton himself, unable to prove mathematically many details of the motion of the Sun, the planets and the moons, had conceded that God's intervention was periodically required to keep the planets from crashing). In 1796 Laplace had even discovered "black holes" (more than a hundred years before Einstein's Relativity). In 1812 Laplace published his textbook "Analytical Theory of Probabilities" with the version of Bayes' theorem that we use today; and also called it what it is, i.e. "probabilities". (Two years later, in his "Philosophical Essay on Probabilities", Laplace articulated the principle of causal determinism: if one "at a certain moment would know all forces that set nature in motion, and all positions of all items of which nature is composed", that person would be able to predict the future and wouldn't need probabilities).

The input to Bayes' theorem is the "prior" probability, which has to be decided either by statistical analysis or by subjective belief. Bayes' theorem then helps us calculate the "posterior" probability while evidence accumulates.

Markov chains extended the theory of probability in a new direction, to sequences of linked events.

Markov chains were first described in 1906 by the Russian mathematician Andrey Markov. The goal of statistical classifiers and neural networks is to identify something (classify it). The goal of Markov chains is different: they use probability theory for guessing what the next element in a sequence could be. Markov analyzed Alexander Pushkin's lengthy poem "Eugene Onegin" using this method, trying to guess the next letter from the previous one. Little did he know that his method would become one of the most popular "guessing" methods of the century, applied to everything from physics to economics and from genetics to sociology and even to gambling (one can interpret the state of the Markov chain as the fortune of a gambler). Markov chains can only be used when the process is a Markov process: the current state of the system is always dependent on the immediate previous state only. In 1936 Andrey Kolmogorov expanded the mathematics in such a way that Markov chains could be used for any continuous process, not just for sequences of discrete events. In 1948 Claude Shannon, the father of information theory, used Markov chains to produce a sentence, simply based on the distribution of alphabetical letters in English words, and came up with the sentence "in no ist lat whey cratict foure birs grocid". The similarity with real English convinced him that communication systems are Markov processes. In 1953 Shannon also built a machine based on Markov chains that could play a game against humans, the "mind-reading machine".

The popularity of Markov chains grew rapidly. The very first piece of computer music (music generated by software), the "Illiac Suite" premiered in August 1956 on the Illiac computer at University of Illinois, was generated by Markov chains programmed by Lejaren Hiller and Leonard Isaacson. The Greek composer Iannis Xenakis, who at the time was still an engineer in the studio of the architect Le Corbusier in Paris, used Markov chains to compose the electroacoustic piece "Analogique A-B (1958-59). In 1964 Hiroshi Kawano, a Japanese pioneer of computer art, started using Markov chains in his paintings.

Possibly the most influential application of the Markov chain is the Google search engine (1998).

A Markov chain prescribes only one action for each state, and there is no reward for the result of that action. When a Markov chain is equipped with multiple actions and rewards, it is called "Markov decision process". Finding a solution for a Markov decision process is not easy. In 1957 Richard Bellman of RAND Corporation introduced the "value iteration" method (in his book "Dynamic Programming"), and in 1960 Ronald Howard of MIT published the "policy iteration" method (in his book "Dynamic Programming and Markov Processes").

Markov chains are useful in an ideal world in which you can know the states of the system. Markov chains are "observed Markov models". In the real world it is rarely possible to know the state of a given system. Instead, we get indirect information. For example, the same word can be pronounced in many different ways by different speakers. In this case there is a whole distribution of sounds that corresponds to that state. In the case of two homophones like "write" and "right" there is one sound that correspondes to two states. In such cases we have a Markov process with unobserved (or "hidden") states, states that can only be observed indirectly. Instead of a regular Markov chain we need to use a "hidden Markov model". We can directly compute the probability of the next state in a Markov chain because we can observe the current state, but, in the case of "hidden" states, the procedure is more complicated and consists in steps that get repeated to achieve a better and better estimate. A particular hidden Markov model is defined by "transition probabilities" and "emission probabilities". The model generates two sequences: a path of states determined by transition probabilities, and the observed sequence due to the emission probability of each state in the path. The state path is a hidden Markov chain. An observed sequence (e.g. a sequence of vocal sounds) can be due to many different state paths (e.g. to many different words). The inference consists in determining ("decoding") the underlying cause of the observed sequence, i.e. the most likely sequence of hidden states that account for those observations; The decoding is usually done with an algorithm proposed by Andrew Viterbi 1967; i.e. the Viterbi algorithm finds the most probable path that generates the observed sequence. Another possible use of the hidden Markov model is to predict the next observation in the sequence.

The hidden Markov model is a Bayesian network that has the sense of time and can model a sequence of events. It was invented in 1966 by Leonard Baum, a cryptographer working at the Institute for Defense Analyses in Princeton.

The obscure algorithms called Markov Chain Monte Carlo (MCMC) are among the most important mathematical discoveries of the 20th century because they have had literally thousands of practical applications, especially in physics, solving problems that were considered impossible to solve in a reasonable time. In fact, a special issue of Computing in Science & Engineering (January 2000), edited by Francis Sullivan, included one of them, the "Metropolis algorithm", among the ten most important algorithms of the 20th century. (Sullivan wrote in his introduction that "great algorithms are the poetry of computation").

Starting in 1934 the Italian nuclear physicist Enrico Fermi developed techniques of "statistical sampling" to model the motion of neutrons. (In 1942, after moving to the USA, Fermi built the world's first nuclear reactor). Given the practical impossibility of calculating what happens to every neutron, Fermi found a way to generalize to the whole population the results obtained from a sample. Fermi didn't give a name to the techniques that he was using. In 1946 the Polish-born mathematician Stanislaw Ulam came up with a method to simulate and estimate (not calculate exactly) a nuclear explosion, and he nicknamed it "Monte Carlo" method because it involves a form of (mathematical) gambling. This was an example of intractable mathematical problems, "intractable" because the time required to find the exact solution grows exponentially. In these cases an approximate solution is better than no solution. John Von Neumann, who was involved in the design of the ENIAC, understood that this method was perfectly suited for the electronic computer. In 1948 his wife Klara Von Neumann and Nicholas Metropolis of the Los Alamos national laboratory programmed the ENIAC to run the Monte Carlo method, first on fusion and then on fission problems. Ulam and Metropolis published the first paper on the Monte Carlo method in 1949 (titled "The

Monte Carlo Method"). Over the years a variety of Monte Carlo algorithms have been developed.

At the time the Los Alamos National Laboratory was working on the atomic bomb, its numerous physicists, including Von Neumann and Metropolis, were developing mathematical methods to deal with nuclear experiments, and using the first electronic computers to perform the calculations. Nicholas Metropolis was trying to compute the equilibrium state of a collection of atoms at a certain temperature, a problem that requires to calculate very complicated integrals. These integrals cannot be approximated even with the Monte Carlo method. Metropolis led the development of a new computer, nicknamed MANIAC (Mathematical Analyzer, Numerical Integrator and Computer), his team member Marshall Rosenbluth invented a new algorithm and his other team member Arianna Rosenbluth implemented it on the Maniac ("Equation of State Calculations by Fast Computing Machines", 1953). This algorithm, now erroneously known as the "Metropolis algorithm", was the first Markov Chain Monte Carlo (MCMC) method, a method combining the Monte Carlo method and a Markov chain: an MCMC algorithm builds a convergent Markov chain whose limit is the desired probability distribution. (For the record: Edward Teller posed the mathematical problem, Marshall Rosenbluth solved it, Arianna Rosenbluth implemented it, Metropolis was their boss and the fifth co-author did not do anything).

Meanwhile, mathematicians and philosophers had been arguing over the meaning of probabilities. You can view a probability as an objective number (the number of times that you get heads and tails when you flip a coin) or as a subjective number (my belief that Brazil will win the next world cup). The British economist John Maynard Keynes at Cambridge University published "A Treatise on Probability" (1921) in which he argued against the subjective approach. A little later two independent minds came up with the opposite view. Frank Ramsey, a mathematical prodigy at Cambridge University defended subjective probability in his essay "Truth and Probability" (1926) just before dying at the very young age of 26. The Italian statistician Bruno DeFinetti studied subjective probability aiming to use it for predictive inference, and explained his approach in a talk given in 1928 in Bologna at the International Congress of Mathematics. His famous motto was: "Probability does not exist", meaning that there is no objective probability, just subjective ones. John von Neumann and Oskar Morgenstern mentioned in passing the need for a theory of subjective probability in their book "Theory of Games and Economic Behavior" (1944). The British mathematician Jimmie Savage (born Leonard Savage), who during World War II had worked with von Neumann at the Institute for Advanced Study in Princeton, carried out that task at University of Chicago and published the seminal book "Foundations of Statistics" (1954).

Others were working on both the mathematical and philosophical foundations of probability. Andrey Kolmogorov in his book "Foundations of Probability Theory" (1933) did for probability what Frege and Peano had done for arithmetic: he gave it logical foundations, based on just three axioms. Rudolf Carnap, a student of Gottlob Frege in his native Germany and a teacher of Walter Pitts at University of Chicago, studied the philosophical foundations of probability and induction in "Logical Foundations of Probability" (1950), and was probably the first to explore the relationship between probability and first-order predicate logic.

Bayes' theorem (actually due to Pierre-Simon Laplace in 1812 but based on Thomas Bayes' equation of 1761) basically calculates the probability of belonging to a class given some facts. For example, what is the probability that you are a lawyer if you live in the USA? Or, to take another problem, if you test positive for a disease, what are the odds that you actually have that disease? Bayes' theorem says that the "conditional" probability of A occurring given B equals the probability that B occurs given A (a simple statistical fact) multiplied by the probability of A and divided by the probability of B (also simple statistical facts). The theorem is also used to calculate "posterior" probability from a "prior" probability: the conditional probability that event A occurs given that even B has occurred.

Bayesian models can represent complex problems by joining together a series of these theorems. Bayesian models have "latent" or "random" variables: variables that cannot be observed, such as a column of empty data. Layers of latent variables create a hierarchy of concepts.

This kind of hierarchical Bayesian reasoning was defended, at a time when Bayes' theorem was rather unpopular, in a paper titled "Rational Decisions" (1952) by the British mathematician Jack Good (born John Irving Good), who during World War II had worked at Bletchley Park with Alan Turing and who would later start singularity thinking with the essay "Speculations Concerning the First Ultraintelligent Machine" (1964).

"All models are wrong", famously said the British statistician George Box, in 1976, but some "cunningly chosen parsimonious models often do provide remarkably useful approximations".

First Steps Towards an Artificial General Intelligence

There were many side tracks that didn't become as popular as expert systems and neural networks.

At the famous conference on A.I. of 1956 there was a third proposal for A.I. research. The Boston-based mathematician Ray Solomonoff presented

"An Inductive Inference Machine" for machine learning, which sketched a universal general-purpose learner. Induction is the kind of learning that allows us to apply what we learned in one case to other cases. His method used Bayesian reasoning, i.e. it introduced probabilities in machine learning.

The problem of induction is to discover the theory (or, if you prefer, the cause) that explains the facts. If a number of people are sick and show the same symptoms, we want to find out what is the disease. We look for explanations to what happens in our world. In other words, induction answers the primordial question: "Why?" If we find the cause, we should be able to predict the future.

There are usually many possible causes, each of which explains the facts: those are hypotheses. In fact, there are an infinite number of them, although some are extremely unlikely (e.g., i could be writing this sentence because an evil force has hijacked my brain and is dictating to me what should be written).

The "Epicurean principle" is to consider all hypotheses that explain the facts. Mathematically speaking, facts are data and each hypothesis is an algorithm that consumes data and produces data: if it produces the data that we observe, then the algorithm represents a valid hypothesis, i.e. it could be the explanation that we are looking for.

Algorithms can be represented by Turing machines. Therefore the problem of induction is to find the best Turing machine that is compatible with our data. If there are many, we need a method to pick one. Solomonoff chose to use "Occam's razor", a medieval principle that basically picks the simplest explanation, i.e. the simplest algorithm, i.e. the simplest Turing machine, i.e. the least complex one.

This requires a definition of "complexity" in Turing machines, which are sequences of zeroes and ones. Solomonoff ("A Formal Theory of Inductive Inference", 1964) and the Russian mathematician Andrei Kolmogorov ("Three Approaches to the Quantitative Definition of Information", 1965) proposed a measure of an algorithm's complexity: the shortest possible description of it; or, equivalently, the least number of bits necessary to write its Turing machine. In today's jargon: the shortest program that computes it. For example, the complexity of "pi" is the ratio between a circumference and its diameter. Occam's razor can now be formulated mathematically, for example: a hypothesis that is one bit shorter is twice as likely to be the true hypothesis.

That is how Solomonoff ranked hypotheses, using Turing Machines to represent hypotheses and "algorithmic information theory" to quantify their complexity. Each hypothesis/ algorithm/ program can be assigned a "prior probability" proportional to its simplicity. Simpler hypotheses/ programs are assigned greater prior probability.

Bayes' theorem is then used to promote programs whose outputs match the facts/data. This way the system learns to correctly predict anything using the minimum amount of data.

Solomonoff's inductive inference program is a universal predictor, and it's the best that one can build because all possible solutions are contained in Solomonoff's giant factory of programs, including anything that you can think of.

Unfortunately, Solomonoff induction cannot be computed because it requires infinite calculations. Then it became a matter of finding approximations to Solomonoff's perfect algorithm. First came the Ukrainian mathematician Leonid Levin, who had studied under Kolmogorov, with a tractable but still impractical approximation ("The Complexity of Finite Objects and the Development of Concepts of Information and Randomness by Means of the Theory of Algorithms", 1970). Incidentally, Levin later proved the existence of "NP-complete" problems, one of the great theorems of 20th century mathematics ("Universal Sequential Search Problems", 1972). Solomonoff himself worked on a "scientist's assistant" capable of solving two general classes of problems ("The Application of Algorithmic Probability to Problems in Artificial Intelligence," 1986).

Androids Dream of Electric Sheep (The Birth of Robots)

The word "robot" first appeared in Karel Capek's science-fiction theatrical play "R.U.R" (which stands for Rossum's Universal Robots) of 1920, but it referred to artificial humans built in a factory (like the replicants in "Blade Runner"), and "robota" means "serf labor" because these "robots" are used as slaves in the plot. So these robots were really the descendants of Mary Shelley's creature in "Frankenstein" (1818), or of Carlo Collodi's wooden boy in "The Adventures of Pinocchio" (1883) and Frank Baum's tin-man in "The Wonderful Wizard of Oz" (1900). Today's robots are rather the descendants of the mechanical automata that were built over the centuries in Europe, the Middle East and China, the most famous being Jacques de Vaucanson's "Flute Player" (1737) and Pierre Jaquet-Droz's "The Writer" (1768). The protagonist of Jacques Offenbach's "The Tales of Hoffmann" (1881) falls in love with Olympia, a mechanical doll. The first fictional robot could well be Thea von Harbou's Futura in "Metropolis" (1925), the novel on which Fritz Lang's film was based. It was, however, preceded in cinema by Burton King's and Harry

Grossman's Automaton in the serial "The Master Mystery" (1920), coincidentally shown in the same year as Capek wrote his play, and by Andre' Deed's movie "L'Uomo Meccanico/ The Mechanical Man" (1921).

Engineers were slow at catching up with novelists and filmmakers. The first actual machine called "robot" and displayed to a public was Eric Robot, built by William Richards for London's Exhibition of the Society of Model Engineers (1928), with the characters "RUR" emblazoned on its chest, but it was operated by two men. Westinghouse introduced some of the early machines that were marketed as a mechanical man or woman (rarely as a "robot"). Roy Wensley at the Pittsburgh laboratories built Herbert Televox (1928), capable of lifting the telephone receiver, reacting to three sounds and activating appliances accordingly, soon followed by the similar Katrina Van Televox (1930). After Wensley quit, Joseph Barnett at the Ohio laboratories built Willie Vocalite (1932), capable of responding to vocal signals as well as to light signals through photoelectric cells and of turning on and off any appliance to which it was connected. Westinghouse's publicity claimed that it could answer the phone, vacuum the floor and make coffee, all wild exaggerations. These artifacts (largely useless) were not sold but were simply toured around the USA to promote Westinghouse's appliances. Joseph Barnett's Elektro the Moto-Man was a sensation at the New York World's Fair of 1939: it "walked" across the stage and "replied" when asked to tell the story of its creation. In reality it had rollers on its feet, it responded to light impulses (no matter which words were uttered while flashing the light), and it only spoke the speeches recorded on eight 78-rpm records, each played by a different turntable; but it could smoke cigarettes and blow up balloons, enough to entertain the children. Also famous was Roll-oh, a robot shown by General Motors in a short film at New York World's Fair of 1940, but in reality just a stunt to publicize Chevrolet's "high-tech" cars..

In 1954 George Devol designed the first industrial robotic arm, Unimate, which, manufactured by Joseph Engelberger, was first delivered to a General Motors factory in New Jersey in 1959. Joseph Engelberger, a fan of Isaac Asimov, marketed George Devol's invention as a "robot", although the dull repetitive Unimate arm was hardly the intelligent and mobile robot that Asimov, Capek and Hollywood had in mind. This would create endless confusion in Artificial Intelligence about what constitutes a robot.

Unimate soon had a competitor: in 1962 the American Machine and Foundry (AMF) introduced a programmable robotic arm called Versatile Transfer Machine, or VersaTran, designed by Harry Johnson and Veljko Milenkovic (which also appears in Douglas Trumbull's 1972 movie "Silent Running").

In 1961 Claude Shannon's student Heinrich Ernst at MIT built a computer-controlled hand (capable of finding object by touch, grasping them and putting them into a box), the MH-1, controlled via the MIT's TX-0 computer, one of the first fully transistorized computers. And computer vision was inaugurated in 1963 at MIT by the dissertation of Lawrence Roberts ("Machine Perception of Three-Dimensional Solids").

The first major improvement over William Grey-Walter's Elmer and Elsie of 1949 was the wheeled robots developed at Johns Hopkins University by Leonard Scheer and John Chubbuck: first Mod I in 1961 (nicknamed "Ferdinand") and then Mod II three years later (nicknamed "Beast"). They were programmed to survive in their environment by finding electrical outlets: Mod I by touch only and Mod II also via a video system. They were equipped with sonars to avoid obstacles. Bernard Roth, who had studied kinematics (the mathematical theory of rigid body motions) at Columbia University, was one of John McCarthy's early coworkers at the Stanford Artificial Intelligence Lab in 1963 and decided to acquire the robotic arm with seven degrees of freedom developed at Rancho Los Amigos Hospital in southern California to help disabled patients (and that, contrary to many reports, was not controlled by a computer). Roth's students would include Donald Pieper (whose seminal thesis was "The Kinematics of Manipulators under Computer Control", 1968), Victor Scheinman, Brian Carlisle, Bruce Shimano, Kenneth Salisbury, and Lou Paul (who would write the seminal book "Robot Manipulators", 1981). Trivia: in 1971 the same Bernie Roth created the "Designer in Society" class at Stanford that launched "design thinking".

The robot Shakey, a project started in 1969 at the Stanford Research Institute (SRI) by Charles Rosen's team, represented the vanguard of autonomous vehicles. The robot's reasoning was done in Cordell Green's logic system QA3, written in LISP. By 1971 Shakey made the leap to a more powerful machine (a PDP-10) and the team had made some valuable contributions to the field: the "STRIPS planner", developed by Richard Fikes and Nils Nilsson, the "Hough transform" for computer vision, developed by Richard Duda and Peter Hart, and the "A* heuristic search" algorithm (that would remain the most used algorithm in its class for half a century).

Computation as Deduction

The idea of turning computation into deduction (into a form of theorem proving) harkens back to Kurt Goedel's "completeness theorem" in Austria (his PhD thesis "On the Completeness of the Calculus of Logic", 1929) and Jacques Herbrand in France (his PhD thesis "Research on the Theory of Demonstration", 1929).

Coming on the heels of these pioneering conferences on machine intelligence, in 1957 Alfred Tarski (who was then at UC Berkeley) organized a five-week long retreat at Cornell University that brought together 85 logical mathematicians and computer scientists, the first time that a large number of computer scientists came together with logicians. Although the logicians generally did not appreciate the fact that such computers as the IBM 704 and the scientific programming language FORTRAN were changing the world, that conference laid the foundations for automated reasoning. For example, a young Richard Friedberg, a summer intern at IBM, explained how to design a learning machine, a forerunner of genetic programming (published as "A Learning Machine", 1958).

The fundamental algorithm for automated deduction (for proving automatically theorems of first-order predicate logic) is the "resolution principle", a uniform proof procedure developed in 1964 by Alan Robinson while at the Argonne National Laboratory ("A Machine-Oriented Logic Based on the Resolution Principle", 1965). Cordell Green's QA3 at Stanford (1969) was one of the first resolution-based systems for reasoning. A few years later Robert Kowalski University of Edinburgh showed how to use this algorithm to perform computation ("Predicate Logic as Programming Language", 1974), therefore completing the process started by Goedel and Herbrand. Kowalski also helped Alain Colmerauer's team in France to develop the programming language ProLog (1972). Meanwhile, two schools of logic programming were developing, represented on the West Coast by Cordell Green at Stanford ("Application of Theorem Proving to Problem Solving", 1969) and on the East Coast by Carl Hewitt at MIT with his Planner system (also 1969). A subset of Planner was later used by Terry Winograd to implement his SHRDLU program. In 1973 Hewitt invented the actor model for the design of computer networks ("A Universal Modular Actor Formalism for Artificial Intelligence", 1973).

Richard Waldinger at CMU developed PROW (PROgran Writing), a theorem prover able to generate LISP programs from input-output pairs, one of the first applications of Robinson's resolution method ("A Step Toward Automatic Program Writing", 1969).

Inductive logic programming was pioneered by Gordon Plotkin at University of Edinburgh (his PhD thesis "Automatic Methods of Inductive Inference", 1972), building on the work of Mark Gold at the RAND Corporation ("Language Identification in the Limit", 1967, a famous paper on language learnability). Inductive logic programming attempts to infer logic programs from positive and negative examples. The name "inductive logic programming" was coined in 1991 by Donald Michie's former student Stephen Muggleton while at the Turing Institute in Glasgow.

Cordell Green was experimenting with automatic programming, software that can write software the same way a software engineer does, namely the QA3 system that could write simple programs. This had a practical application. As computers were getting more complex, they were also becoming less reliable, and testing them for quality assurance was becoming more important. One way to test them was to generate thousands of test programs, to simulate the expected results and then compare the simulation with the actual performance. The roots of combinatorial testing came from the field of statistics, from venerable textbooks such as Ronald Fisher's "The Design of Experiments" (1935) and William Cochran's and Gertrude Cox's "Experimental Designs" (1957), but the automatic generation of combinatorial test programs for quality assurance was an attempt to have computers write programs for other computers.

For the record, neural networks are actually a form of program synthesis: a neural network is an algorithm capable of changing itself in response to some data; a computer that programs itself. In fact, a common definition of program synthesis is: given a dataset of input-output pairs, produce a program whose behavior generates those pairs. That's precisely what a neural network does to itself.

"Begin at the beginning," the King said, very gravely, "and go on till you come to the end: then stop." (Lewis Carroll, "Alice in Wonderland").

The first Failure: Machine Translation

The discipline of Machine Translation actually predates Artificial Intelligence. Warren Weaver had been the chair of the Applied Mathematics Panel of the US Office of Scientific Research and Development during World War II, and was now director of the Natural Sciences Division of the Rockefeller Foundation, and still exerted some influence on government agencies. He had been impressed at the success of cryptography in deciphering enemy code using machines to analyze the frequency of letter patterns. In 1946 he discussed with British computer pioneer Andrew Booth the possibility of using the same technique to translate languages. In March 1947 he mentioned this idea in a letter to Norbert Wiener and finally, while he was in New Mexico, in July 1949 sent a "memorandum" simply titled "Translation" to about 30 important friends about using the digital electronic computer (that had just been invented) for language translation. As he wrote: ""It is very tempting to say that a book written in Chinese is simply a book written in English which was coded into the Chinese code". That started research in machine translation as an evolution of cryptography.

Meanwhile, Harry Huskey at UCLA had just built one of the earliest computers, the SWAC, and decided to use it for machine translation. In May 1949 the New York Times ran the first article ever on the new discipline when it described a demo of the SWAC (quote): "a new type of electric brain calculating machine capable not only of performing complex mathematical problems but even of translating foreign languages".

Weaver's memorandum set in motion several labs. For example, Abraham Kaplan at the RAND Corporation published the first paper on resolving ambiguity ("An Experimental Study of Ambiguity and Context", 1950). More importantly, MIT appointed the Israeli philosopher Yehoshua Bar-Hillel to lead research on machine translation. Bar-Hillel toured all the labs in 1951, and in 1952 organized the first International Conference on Machine Translation at MIT.

In 1954 Leon Dostert's team at Georgetown University and Cuthbert Hurd's team at IBM demonstrated a machine-translation system running on a 701 computer, using a vocabulary of 250 words and six rules of syntax. It was one of the first non-numerical applications of the digital computer. For example, the IBM 701 translated "Myezhdunarodnoye ponyimanyiye yavlyayetsya vazhnim faktorom v ryeshyenyiyi polyityichyeskix voprosov" into "International understanding constitutes an important factor in decision of political questions." In 2016 Google translates it as "International understanding is an important factor in the decision of political issues".

William Locke and Donald Booth collected all the historical papers in "Machine Translation of Languages", published in 1955, one year before the first Artificial Intelligence conference.

Noam Chomsky's "Syntactic Structures" (1957) galvanized the field because it showed an elegant way to formalized language as a "code".

Less appreciated was a fact emphasized by Chomsky's teacher, the linguist Zellig Harris at University of Pennsylvania: words that occur in similar contexts tend to have similar meanings. His article "Distributional Structure" (1954), in turn based on ideas already published by the structural linguists Edward Sapir at Yale University and Leonard Bloomfield at University of Chicago, anticipated ideas of deep learning and support vector machines. The British linguist John-Rupert Firth quipped: "You shall know a word by the company it keeps" (1957). This was the prehistory of "vector semantics": that the meaning of a word is determined by the distribution of words that surround it. (Trivia: Harris' wife was Princeton University's physicist Bruria Kaufman, who was Albert Einstein's main assistant).

Alas, in 1958 Bar-Hillel published a "proof" that machine translation is impossible without common-sense knowledge.

Nonetheless, in 1959 the philosopher Silvio Ceccato started a project in Italy funded by the US military, and in 1961 published his theory in the book "Linguistic Analysis and Programming for Mechanical Translation". Unfortunately, Ceccato's machine was destroyed in 1965 by communist demonstrators.

Peter Toma started working on machine translation at Caltech in 1956 and in 1958 moved to Georgetown University. He demonstrated his Russian-to-English machine-translation software, SYSTRAN in 1964.

David Hays conducted research on machine translation at the RAND Corporation starting in 1955. He popularized the "dependency grammar" developed in the 1930s by the French linguist Lucien Tesniere, and in 1967 published the first textbook of computational linguistics, "Introduction to Computational Linguistics".

Philip Stone at Harvard developed the General Inquirer (running on an IBM 7090 in 1961), the archetype of what would be called "sentiment analysis" in text understanding. But there was little opinionated text available before the explosions of user-generated content on the World Wide Web, so sentiment analysis didn't stage much of a progress until the 2000s. Sentiment analysis had been pioneered by (of all people) novelist Kurt Vonnegut: his 1946 master dissertation in anthropology, that was rejected by University of Chicago, spoke of the "emotional arc of a story".

In 1961 Melvin Maron, a philosopher working at the RAND Corporation, suggested a statistical approach to analyze language (technically speaking, a "naive Bayes classifier"), an approach that was initially ignored by the linguistic community.

In 1962 IBM demonstrated (at the World's Fair in Seattle) the first speech-recognition device, Shoebox, developed by William Dersch at IBM's San Jose laboratories.

Mortimer Taube, inventor of the most popular indexing and retrieval method for libraries, wrote in his book "Computers and Common Sense" (1961) that something can be automated only after it is formalized. First you turn a process into mathematics, then you can build a machine that performs that process. He argued, however, that formalizing human language makes little sense because a formalized language is a code, not the language that we speak.

The first practical implementations of natural language processing were conversational agents such as Daniel Bobrow's Student (1964), Joe Weizenbaum's Eliza (1966) and Terry Winograd's Shrdlu (1972), all from MIT, as well as LUNAR (1973), a system created by William Woods at nearby Bolt Beranek and Newman to answer questions about moon rocks.

Stanford psychiatrist Kenneth Colby developed Parry and, during the International Conference on Computer Communications of October 1972 in Washington, Vint Cerf (who two years later would publish the TCP protocol) staged the first chatbot to chatbot conversation ever: Stanford and MIT ran, respectively, Parry and Eliza over the Arpanet (soon to be renamed Internet).

Alas, in 1966 an advisory committee, the Automatic Language Processing Advisory Committee (ALPAC), featuring linguists from Harvard University, Cornell University, University of Chicago, the Carnegie Institute of Technology, as well as David Hays from the RAND Corp, and led by John Pierce of Bell Labs, issued a report titled "Computers in Translation and Linguistics" (the "ALPAC Report") that caused a dramatic reduction in funding for machine translation programs. (For the record, Pierce was an engineer who in 1946 worked with Claude Shannon and Bernard Oliver on the "pulse-code modulation", PCM, without which we wouldn't have digital audio in computers, and who in 1947 supervised the trio who invented the transistor, and in fact he was the one who gave it the name).

"Science progresses one funeral at a time" (Max Planck).

The Origins of Digital Life

In 1936 Alan Turing had shown how to build a computing machine. John Von Neumann wanted to build a self-replicating machine, a machine capable of creating a copy of itself.

In 1944 Erwin Schroedinger, one of the founders of Quantum Mechanics, published a book titled "What is Life?" in which he mused that chromosomes must contain the instructions to build the future organism as well as the machinery to execute them. Von Neumann differed from Schroedinger in that he separated the code and the machinery. Von Neumann came up with the design for what came to be known as a "cellular automaton" after discussing his idea with Stanislaw Ulam at Los Alamos National Laboratory, and in 1948 delivered a lecture titled "The General and Logical Theory of Automata" at a symposium in Pasadena (published in 1951).

Von Neumann's "universal constructor" consists of three parts: a description of itself, a decoder that constructs a machine based on that description, a copier that inserts a copy of the description inside the new machine. He took from Kurt Goedel's incompleteness theorem of 1931 the concept of storing a description of the organism within the organism, i.e. of storing the instructions for constructing an organism inside the organism itself. Von Neumann proved that a cellular automaton doesn't need to be a

Turing machine; and that a cellular automaton could implement a Turing machine. Note that Von Neumann came up with the vision of this self-replicating machine several years before James Watson and Francis Crick discovered the self-replicating process of DNA (1953).

Von Neumann's cellular automaton was never built in his lifetime. The first self-reproducing cellular automaton would be implemented much later, in 1994, by Renato Nobili of University of Padova in Italy and Umberto Pesavento of Princeton University.

Refining an idea pioneered by the German engineer Ingo Rechenberg at the Technical University of Berlin in his dissertation "Evolution Strategies" (1971), John Holland at University of Michigan introduced a different way to construct programs by using "genetic algorithms" (1975), the software equivalent of the rules used by biological evolution: instead of writing a program to solve a problem, let a population of programs evolve (according to some algorithms) to become more and more "fit" (better and better at finding solutions to that problem). His thesis advisor was Arthur Burks, who in 1946 had worked with John VonNeumann at the Institute for Advanced Study in Princeton on the theory of automata. In 1976 Richard Laing at the same university introduced the paradigm of self-replication by self-inspection ("Automaton Models of Reproduction by Self-inspection") that 27 years later would be employed by Jackrit Suthakorn and Gregory Chirikjian at Johns Hopkins University to build a rudimentary self-replicating robot ("An Autonomous Self-Replicating Robotic System", 2003). It took a decade for these ideas to be appreciated. The first international conference on genetic algorithms was held in 1985 at CMU.

"Each writer creates his precursors" (Jorge-Luis Borges).

Another Failure: Machine Learning

Machine learning lay mostly dormant until the 1970s. Earl Hunt, then a psychologist at UCLA, had developed a Concept Learning System, first described in his book "Concept Learning" (1962), for inductive learning, i.e. learning of concepts. In 1975 that was extended by Ross Quinlan, into Iterative Dichotomiser 3 (ID3) at University of Sydney in Australia. Patrick Winston's dissertation at MIT with Marvin Minsky was "Learning Structural Descriptions From Examples" (1970) that introduced "difference networks".

Polish-born Ryszard Michalski at University of Illinois built the first practical system that learned rules from examples, AQ11 (1978).

Meanwhile, John Anderson at CMU, had been developing since 1973 his own cognitive architecture, called ACT*.

Another school was started at Stanford by Bruce Buchanan, who had worked on Dendral, following his paper "Model-directed learning of production rules" (1978). His student Tom Mitchell graduated with a thesis on "Version Spaces" (1978), a model-based method for concept learning (as opposed to Michalski's data-based method).

After the pioneering work of Herbert Simon and Walter Reitman at Carnegie Institute of Technology, reasoning by analogy was studied at MIT by Patrick Winston ("Learning and Reasoning by Analogy", 1980), and at CMU by Jaime Carbonell, who had graduated from Roger Schank's case-based reasoning systems ("Learning and Problem Solving by Analogy", 1980) and who in 1981 organized the first conference on machine learning (held at CMU); and also by Ken Forbus at Northwestern University, who developed the Structure Mapping Engine (SME), based on the theories of psychologist Dedre Gentner ("Structure-mapping", 1983). Decades later, Douglas Hofstadter and the French psychologist Emmanuel Sander would argue that analogy is the foundation of all thinking in their book "Surfaces and Essences" (2013), but by then the A.I. community would be absorbed in deep learning.

In 1981 Allen Newell and Paul Rosenbloom at CMU formulated the "chunking theory of learning" to model the so-called "power law of practice", and in 1983 John Laird and Paul Rosenbloom started building a system called Soar that implemented chunking.

Then came the "explanation-based learning systems" such as Lex2 (1986), developed by Tom Mitchell at CMU, and Kidnap (1986), developed at University of Illinois by Gerald DeJong, whose dissertation at Yale University had been the Frump system of natural language processing based on Roger Schank's scripts; and the "learning apprentice systems" such as Leap (1985) by Tom Mitchell at CMU and Disciple (1986) by Yves Kodratoff in France.

An influential theory of learning was the "probably approximately correct" (PAC) model of learning, introduced in 1984. Leslie Valiant at Harvard University ("A Theory of the Learnable", 1984) considered inductive learning as the process of deducing a program for performing a task: the learner must select the best generalization function (the "hypothesis") for the data at hand out of a class of possible functions (the "hypothesis space"), a task traditionally done by hand but here automated using computational complexity theory. Some PAC ideas were predated by Vladimir Vapnik in Russia ("Estimation of Dependences Based on Empirical Data", originally published in Russian in 1979).

None of these attempts proved successful at building programs that learn. There seemed to be something about learning that still eluded both computer scientists and cognitive psychologists.

"Education is paradoxical in that it is largely composed of things that cannot be learned" (Roberto Calasso), or, if you prefer, "Education is what survives when what has been learned has been forgotten" (Burrhus Skinner).

Another Failure: Common Sense

Common sense, besides learning, was another missing ingredient. Humans employ naturally several forms of inference that are not deduction, and therefore are not exact. In general, we specialize in "plausible reasoning", not the "exact reasoning" of mathematicians. Finding exact solutions to problems is often pointless: it would take too long. If a tiger attacks you, you don't start calculating the most efficient trajectory: you would be dead by the time you finished your calculations. This became a popular subject of research after the publication of "Plausible Reasoning" (1976) by the German-born philosopher Nicholas Rescher at University of Pittsburgh and of "Logic and Conversation" (1975) by the British-born philosopher Paul Grice at UC Berkeley.

Most of our statements are actually uncertain. "The sky is blue" is obviously just an approximation; so is "blood is red". My height is actually not 171 cm: it is probably something like 171.46234782673...cm. Hence, in 1965 the Azerbaijani-born mathematician Lofti Zadeh invented "fuzzy logic" at UC Berkeley. In classical logic an entity either belongs or doesn't belong to a set. In fuzzy logic an entity has a degree of membership in a set. I belong to both the set of tall people (to some degree) and to the set of short people (to some degree). Following Zadeh's classic paper "Outline of a new approach to the analysis of complex systems and decision processes" (1973), a number of mathematicians invented "fuzzy inference systems": Ebrahim "Abe" Mamdani at University of London in 1975, Yahachiro Tsukamoto at the Tokyo Institute of Technology in 1979, Michio Sugeno at the Tokyo Institute of Technology in 1985.

Almost everything we say has a margin of uncertainty and approximation. Even if we don't know Bayes' theorem, we use probabilities all the time (and in most cases it is not the "probability" that mathematicians use). We unconsciously side with the French physicist Pierre Duhem: the certainty that a proposition is true decreases with any increase of its precision. I am certain of being 171 cms high until you ask me to be more precise: then i am less certain whether i am 171.1 cms tall or 171.2 or 171.3 or...

We are also very good at changing our conclusions: if you made plans to have dinner at a restaurant and it turns out that the restaurant has gone out of business, you effortlessly change your plans. Hence in 1979 Drew

McDermott at Yale University worked out "Nonmonotonic Logic" and John McCarthy at Stanford published "Circumscription".

We normally deal with objects, not with elementary particles or waves. The world that we encounter in our daily lives is a world of objects, and we intuitively know how to operate with objects. For example, water can certainly have all sorts of temperatures, but the important thing is that at a certain temperature it freezes and at a certain temperature it boils. We deal with "qualities" (such as "hot" and "cold") rather than with quantities (such as 32.6 degrees Celsius and -4 degrees Celsius). And these "qualities" are "fuzzy": my height is both short and tall, depending on the people around me. To some extent i am short and to some extent i am tall. There are other simple laws of causality connecting our actions and our objects that don't require any knowledge of theoretical physics. Hence, Pat Hayes in Britain published the "Naive Physics Manifesto" (1978), and "qualitative reasoning" was pioneered by two theses published at MIT, first Johan DeKleer ("Causal and Teleological Reasoning in Circuit Recognition", 1979), who had worked on the Sophie project with Brown and Burton at BBN, and then Kenneth Forbus ("Qualitative Reasoning about Physical Processes", 1981). In 1984 Doug Lenat started the project Cyc to catalog commonsense knowledge. (I have written a lengthy survey of these commonsense theories in my other book "Thinking about Thought").

There was a sense that "intelligence" without common sense is not intelligence, or, worse, it is plain dangerous. I would add that a sign of common sense is the other sense, the sense of humor. Machines can't laugh. "A black cat crossing your path signifies that the animal is going somewhere" (Groucho Marx).

Can an A.I. System win a Nobel Prize?

In 2014 there were more than 28,000 peer-reviewed journals. In 2009 University of Ottawa calculated that 50 million papers had been published since 1665, and 2.5 million new scientific papers are published every year. In 2014 John Ioannidis at Stanford University estimated that between 1996 and 2011 the number of scientists who published a paper was about 15 million, of which about 1% (or 150,608) publish one paper per year. In 2016 more than 1.2 million papers were published in life science journals alone, on top of the 25 million already in print. On the other hand, a survey of scientists conducted by Carol Tenopir and Donald King at University of Tennessee found that on average a scientist reads about 264 papers per year. That means that a biomedical scientist, over a normal career of about 50 years, will have read 13,200 papers out of 50-100 million, a tiny fraction. Meanwhile, a new article is being published every 30 seconds. For many biomedical topics a researcher can find tens of thousands of citations. For example, more than 70,000 papers have been published on the tumor suppressor p53. The need for tools that can scan all those papers, analyze data, carry out experiments and formulate hypotheses (the job of a scientist) is greater than ever.

In a 2003 talk at the American Association for Artificial Intelligence, Lawrence Hunter of University of Colorado proposed a revised Turing Test that would use the publication of a paper in a peer-reviewed scientific journal as a better test for human-level intelligence.

In October 2016 Hiroaki Kitano of Sony, president of the Systems Biology Institute (and founder of the Robocup competition), delivered a talk titled "Artificial Intelligence to Win the Nobel Prize and Beyond" in which he proposed a grand, new challenge for AI: to develop an A.I. system capable of scientific research in the biomedical sciences that can discover something worthy of the Nobel Prize. He asked: "What is the single most significant capability that Artificial Intelligence can deliver?" He argued that the future of humankind depends on scientific discovery just like in the past. Therefore that's the field in which new technology can provide the biggest benefits.

A kinder challenge for Artificial Intelligence would be to build a system that can discover something worthy not of the Nobel Prize but simply of a patent: build a program that will discover or invent something for which the patent office will accept the patent application. In 2015 almost three million people worldwide filed a patent for an invention. It can't be that difficult.

Nature has a unique way to hide herself from the human brain, but the human brain occasionally can find her out: the human brain collects data about her behavior, inspects the correlations among the data, and then occasionally devises a theory that explains the data. The job of the scientist is not trivial. As John Banville tells it in his novel "Kepler" about the German astronomer: "He was a blind man who must reconstruct a smooth and infinitely complex design out of a few scattered prominences that gave themselves up, with deceptive innocence, under his fingertips". What Johannes Kepler did was to renounce a postulate that had stood for thousands of years: that the planets move at a constant speed and make circles. Nicolaus Copernicus failed to explain the motion of the planets, and so did Galileo Galilei and so did Tycho Brahe. They all stood by the postulate that the planets must move at a constant speed and in circles. Kepler used data, lots of data, his own tables plus Tycho Brahe's tables, the work of decades of observations and calculations; but those data were not simply used to recognize patterns: Kepler realized that the dogma of uniform circular motion was false, that planets don't move in circles and that their speed changes as they move along their ellipse. Newton and Einstein did something similar: they threw out postulates that had been accepted by everybody for centuries, for example that the natural state of things must be rest and that space must be flat.

A popular joke in Silicon Valley is: what the contract says and what the engineer does are never the same. But that turns out to be a good thing. If humans simply did what the contract requires, and nothing more, there would be no progress. Luckily the engineer does something that was not in the contract, and ends up inventing the microprocessor or the World-wide Web. Indian classical music (such as the raga genre) is not written down because it is not about composing but about understanding, assimilating, and reinterpreting (what Westerners normally summarize as "improvising"). Two performances of the same raga are never the same, even if performed by the same musicians. Technological innovation is a similar process. The whole secret of evolution is to make variations, not exact copies.

"The worst labyrinth is not that intricate form that can entrap us forever, but a single and precise straight line" (Jorge-Luis Borges)

Attempts at creating programs that can do what a scientist does, such as a program capable of scientific discovery, date back from the early days of A.I., to at least 1962, when Saul Amarel at RCA Laboratories in New Jersey published the paper "An Approach to Automatic Theory Formation" (1962). In the same year Thomas Kuhn, a philosopher at UC Berkeley, published "The Structure of Scientific Revolutions" (1962) in which he argued that the history of science is a history of "paradigm shifts", of sudden realizations that old theories were wrong and a whole new way of thinking is required, a conceptual restructuring. Human intelligence came to rule the world because it is capable of these paradigm shifts, whereas other animals keep repeating the same way of thinking. A debate was soon raging on this topic, started by the Austrian philosopher Karl Popper with his book "The Logic of Scientific Discovery" (1965), which basically claimed that there is no logic of scientific discovery, to which Herbert Simon at Carnegie Mellon University (CMU) responded with "Models of Discovery and Other Topics in the Methods of Science" (1977). In that year his student Pat Langley unveiled Bacon, a system capable of discovering scientific laws.

Simon was intrigued by the cognitive process that leads to novel, creative ideas. This process involves the discovery of natural laws from experimental data, i.e. the formation of theories that explain those data, and then the design of experiments to confirm those theories. His former student Edward Feigenbaum started work on Dendral (DENDRitic

ALgorithm) in 1965 at Stanford to help chemists identify organic molecules (in collaboration with Nobel laureate Joshua Lederberg, founder of Stanford's department of genetics). This was a system that reasoned about knowledge provided by human experts. Then in 1970 Feigenbaum and Bruce Buchanan conceived Meta-Dendral, a system to learn the knowledge that Dendral needed to do its job. This was the first system for hypothesis formation. Meta-Dendral's machine-learning algorithms were generalized by Tom Mitchell in his 1978 dissertation and became his "version spaces". In 1978 Feigenbaum and Buchanan also launched Molgen, a system to plan experiments in molecular genetics, developed by their students Peter Friedland and Mark Stefik.

Again, not much came out of these attempts, probably because, surprisingly, we don't really know how we come up with bright ideas.

Just think of the key role of serendipity. A recent example is the discovery of the CRISPR gene-editing method: it was an accidental by-product of a yogurt company trying to improve its product. Scientists discovered that the tiny bacteria used to ferment the lactose in milk to make yogurt have an "immune system" for fighting virus attacks that consists in editing DNA. Basically, scientists discovered that nature was capable of editing DNA and then copied the method for completely different purposes (even to engineer babies). How do you teach machines to understand the importance of what you accidentally discover when you were looking for something else?

"The real voyage of discovery consists not in seeking new landscapes, but in having new eyes" (Marcel Proust)

A Premise to the History of Artificial Intelligence

Surprisingly few people ask "why?" Why did the whole program of A.I. get started in the first place? What is the goal? Why try and build a machine that behaves (and feels?) like a human being?

There were and there are several motivations. I believe the very first spark was pure scientific curiosity. A century ago an influential German mathematician, David Hilbert, outlined a program to axiomatize mathematics as a sort of challenge for the world's mathematicians. In a sense, he asked if we can discover a procedure that will allow anybody to solve any mathematical problem: run that procedure and it will prove any theorem. In 1931 Kurt Goedel proved his Theorem of Incompleteness, which was a response to Hilbert's challenge. It concluded: "No, that's not possible, because there will always be at least one proposition that we cannot prove true or false"; but in 1936 Alan Turing offered his solution, now known as the Universal Turing Machine, which is as close as we can get to Hilbert's dream procedure. Today's computers, including your laptop, your notepad and your smartphone, are Universal Turing Machines. And then the next step was to wonder if that machine can be said to be "intelligent", i.e. can behave like a human being (Turing's Test), can have conscious states, and can be even smarter than its creator (the Singularity).

The second motivation was purely business. Automation has been a source of productivity increase and wealth creation since ancient times. The rate of automation accelerated during the industrial revolution and it still is an important factor in economic development. There isn't a day when a human being isn't replaced by a machine. Machines work 24 hours a day and 7 days a week, don't go on strike, don't have to stop for lunch, don't have to sleep, don't get sick, don't get angry or sad. Either they function or they don't. If they don't, we simply replace them with other machines. Automation was pervasive in the textile industry way before computers were invented. Domestic appliances like dishwashers automated household chores. Assembly lines automated manufacturing. Agricultural machines automated grueling rural chores. That trend continues. As i type, machines (sensing cameras hanging from traffic lights remotely connected to the traffic division of a city) are replacing traffic police in many cities of the world to direct traffic (and to catch drivers who don't stop at red lights).

A third motivation was idealistic. An "expert system" could provide the service that the best expert in the world provides. The difference is that the human expert cannot be replicated all over the world, the expert system can. Imagine if we had an expert system that clones the greatest doctors in the world such that we could make that expert system available for free to the world's population (rich or poor), 24 hours a day, 7 days a week.

Paraphrasing something that Douglas Hofstadter wrote in his book "Fluid Concepts and Creative Analogies" (1995), there are two types of research on "intelligent" machines. One approach (the engineering approach) is interested in practical results, and will use whatever technology works to get the same result as the result obtained by humans, or an even better one. The other approach has little to do with practical results and instead studies the nature of intelligence and creativity. The engineer simply builds a machine that solves the problem, like the printing press and the steam engine. These are inventions that change the lives of millions of people. They do not pretend to simulate the human mind. The student of human intelligence, instead, wants to discover how we think. Our brains are slower; and they need to eat and sleep; and they get sick and get distracted; and they need to pay bills and taxes; but they still find their own way to solve a problem. For example, Hofstadter emphasized that "computer chess programs have taught us something about how human chessplayers play - namely, how they do not play". Nonetheless, these programs can beat any chess champion, hence they achieve the desired result; but they don't tell us much about how a chess champion can play chess so well without having the colossal database and the processing speed of a computer.

In particular, raw speed is not the point: you can build machines that do the same things that we do and do them better but they don't do it the way we do it. Airplanes fly faster than birds, but airplanes don't flap their wings like birds do. What is the goal? To build a machine that does something better than humans? That's why every machine was invented. If you want to call it "machine intelligence", then you have to also call clocks and dishwashers "machine-intelligent". Or is the goal to build a machine that is "intelligent" the way humans are, i.e. that exhibits human intelligence? In that case clocks and dishwashers don't qualify as "intelligent", and it is not clear if anything other than a human being would.

"The best material model for a cat is another cat, or, preferably, the same cat" (Arturo Rosenblueth , cofounder of cybernetics).

Steps toward the A.I. Winter

Unfortunately, the discipline of Artificial Intelligence soon became a tragicomedy of exaggerations and misunderstandings, i.e. of natural stupidity.

1958: A New York Times article (8 July 1958) reports a press conference in which Rosenblatt states "The Perceptron is the embryo of an electronic computer that (the US Navy) expects will be able to walk, talk, see, write, reproduce itself and be conscious of its existence"

1958: Bar-Hillel publishes a "proof" that machine translation is impossible

1965: Herbert Simon predicts that "Machines will be capable, within 20 years, of doing any work a man can do"

1966: The ALPAC Report causes reduction in funding for machine translation research

1969: Minsky and Papert's "Perceptrons" kills research in neural networks in the USA

1970: Marvin Minsky to Life Magazine: "In from three to eight years we will have a machine with the general intelligence of an average human being"

1972: Richard Karp shows there are many problems that can probably only be solved in exponential time

1973: The Lighthill Report kills A.I. in Britain

1980s: Wildly exaggerated Fifth Generation scare. The Fifth Generation Computer Systems was an initiative by Japan's Ministry of International Trade and Industry, launched in 1982, to create a supercomputer particularly suited to symbolic A.I., and A.I. scientists spoke of an impending apocalypse for the USA if it didn't embrace the concept and counterbalance it the way it did with the Sputnik.

As Saint Bonaventure said in the 13th century: "The higher a monkey climbs, the more you see of its behind".

Brain Simulation and Intelligence

Behind the approach of neural networks is the hidden assumption that intelligence, and perhaps consciousness itself, arises out of complexity. This is a notion that dates back at least to the British neurophysiologist William Grey-Walter who in 1949, before the age of digital computers, was already designing early robots named Machina Speculatrix using analogue electronic circuits to simulate brain processes. More recently, David Deamer at UC Santa Cruz has calculated the "brain complexity" of several animals ("Consciousness and intelligence in mammals: Complexity thresholds", 2012).

In 1990 Carver Mead at Caltech described a "neuromorphic" processor, a processor that emulates the human brain.

All the "intelligent" brains that we know are made of neurons. Could the brain be made of ping-pong balls and still be as intelligent? If we connect a trillion ping-pong balls do we get a conscious being? What if the ping-pong balls are made of a material that conducts electricity? What if i connect them exactly like the neurons are connected in my brain: do i get a duplicate of my consciousness or at least a being that is as "intelligent" as me? The hidden assumption behind neural networks is that the material doesn't matter, that it doesn't have to be neurons (flesh), at least insofar as intelligence is concerned; hence, a purist of connectionism would argue that a system made of a trillion ping-pong balls would be as intelligent as me, as long as it duplicates exactly what happens in my brain.

The Body

A lot of what books on machine intelligence say is based on a brain-centered view of the human being. I may agree that my brain is the most important organ of my body (i'm ok with transplanting just about any organ of my body except my brain). However, this is probably not what evolution had in mind. The brain is one of the many organs designed to keep the body alive so that the body can find a mate and make children. The brain is not the goal but one of the tools to achieve that goal.

Focusing only on mental activities when comparing humans and machines is a categorical mistake. Humans do have a brain but don't belong to the category of brains: they belong to the category of animals, which are mainly recognizable by their bodies. Therefore, one should compare machines and humans based on bodily actions and not just on the basis of printouts, screenshots and files. Playing a match of chess with the world champion of chess is actually easy. It is much harder for a machine to do any of the things that we routinely do in our home (that our bodies do). Playing chess is actually much easier than playing soccer with a group of children.

Furthermore, there's the meaning of action. The children who play soccer actually enjoy it. They scream, they are competitive, they cry if they lose, they can be mean, they can be violent. There is passion in what we do. Will an android that plays decent soccer in 3450 (that's a realistic date in my opinion) also have all of that? Let's take something simpler, that might happen in 50 or 100 years: at some point we'll have machines capable of reading a novel; but will they understand what they are reading? Is it the same "reading" that i do? This is not only a question about the self-awareness of the machine but about what the machine will do with the text it reads. I can find analogies with other texts, be inspired to write something myself, send the text to a friend, file it in a category that interests me. There is a follow-up to it. Machines that read a text and simply produce an abstract representation of its content (and we are very far from the day when they will be able to do so) are useful only for the human who will use it.

The same considerations apply to all the corporeal activities that are more than simple movements of limbs.

The body is the reason why i think the Turing Test is not very meaningful. The Turing Test locks a computer and a human being in two rooms, and, by doing so, it removes the body from the test. My test (let's immodestly call it the Scaruffi Test) would be different: we give a soccer ball to both the robot and the human and see who dribbles better. I am not terribly impressed that a computer beat the world champion of chess (i am more impressed with the human, that it took so long for a machine with virtually infinite memory and processing power to beat a human). I will be more impressed the day a robot dribbles better than Lionel Messi.

In fact, already in 1994 Minoru Asada's Lab at Osaka University and Manuela Veloso's student Peter Stone at CMU had been working on soccer-playing robots, and in 1997 Japanese scientists (mainly Hiroaki Kitano) launched RoboCup, the "Robot Soccer World Cup". Twenty years later you can watch very funny videos of the most recent robocup games. If you remove the body from the Turing test, you are removing pretty much everything that defines a human being as a human being. A brain kept in a jar is not a human being: it is a gruesome tool for classrooms of anatomy.

(I imagine my friends at the nearest A.I. lab already drawing sketches of robots capable of intercepting balls and then kicking them with absolute precision towards the goal and with such force that no goalkeeper could catch them; but that's precisely what we don't normally call "intelligence", that is precisely what clocks and photocopiers do, i.e. they can do things that humans cannot do such as keeping accurate time and making precise copies of documents, and that is not yet what Lionel Messi does when he dribbles defenders).

"Trying to understand perception by understanding neurons is like trying to understand a bird's flight by studying only feathers" (David Marr, 1982)

Intermezzo: We may Overestimate Brains

The record for brain size compared with body mass does not belong to Homo Sapiens: it belongs to the squirrel monkey (5% of the body weight, versus 2% for humans). The sparrow is a close second.

The longest living beings on the planet (bacteria and trees) have no brain.

Childhood

A machine is born adult. It's the equivalent of you being born at 25; and never aging until the day that an organ stops working. One of the fundamental facts of human intelligence is that it comes via childhood. First we are children, then we get the person that is writing (or reading) these lines. The physical development of the body goes hand in hand with the cognitive development of the mind. The developmental psychologist Alison Gopnik has written in "The Philosophical Baby" (2009) that the child's brain is wildly different from the adult brain (in particular the prefrontal cortex). She even says that they represent two different types of Homo Sapiens, the child and the adult. They physically perform different functions. Whether it is possible to create "intelligence" equivalent to human intelligence without a formation period is a big unknown.

Alison Gopnik emphasized the way children learn about both the physical world and the social world via a process of "counterfactuals" (what ifs): they understand how the (physical and psychological) worlds function, then they create hypothetical ones (imaginary worlds and imaginary friends), then they are ready to create real ones (act in the world to change it and act on people to change their minds). When we are children, we learn to act "intelligently" on both the world and on other people. Just like everything else with us, this function is not perfect. For example, one thing we learn to do is to lie: we lie in order to change the minds around us. A colleague once proudly told me: "Machines don't lie." That is one reason why i think that science is still so far away from creating intelligent machines. To lie is something you learn to do as a child, among many other things, among all the things that are our definition of "intelligence".

The Influence of Computer Vision

The history of computer vision began in 1963 with the PhD dissertation of Claude Shannon's student Lawrence Roberts at MIT ("Machine Perception of Three Dimensional Solids," 1963). This was the same Roberts who in 1966 would launch the Arpanet project, later renamed Internet.

Face recognition was pioneered by Woody Bledsoe at his Palo Alto consulting firm Panoramic Research, that mainly worked on military projects ("The Model Method in Facial Recognition", 1964). In 1966 he moved to SRI Intl where he trained Peter Hart, who later worked on Shakey the Robot (and who in 1973 coauthored the textbook "Pattern Classification and Scene Analysis" with Richard Duda).

In 1968 Adolfo Guzman at MIT built programs to detect the constituent objects of a scene ("Computer Recognition of Three-Dimensional Objects in a Visual Scene", 1968). Max Clowes ("On Seeing Things", 1971) at University of Sussex and David Huffman at UC Santa Cruz ("Impossible Objects as Nonsense Sentences", 1971) independently discovered methods to interpret pictures of polyhedra (solids such as cubes and pyramids), and Alan Mackworth at University of Sussex developed a program to interpret line drawings as polyhedral scenes ("Interpreting Pictures of Polyhedral Scenes", 1973). Computer vision was mostly about recognizing objects within a picture, and initially the prevailing method was to compare the regions of the picture with templates of typical objects. Martin Fischler and Robert Elschlager at Lockheed's Palo Alto Research Laboratory expanded this method with "stretchable templates" ("The Representation and Matching of Pictorial Structures", 1973). Takeo Kanade graduated from Kyoto University in 1973 with the world's first automated face recognition system ("Picture Processing System by Computer Complex and Recognition of Human Faces", 1973).

By coincidence or not, both Huffman and Kanade were pioneers of mathematical origami that consists in folding a two-dimensional piece of paper into a three-dimensional object. For example, Takeo Kanade, who was now at CMU, published the article "A Theory of Origami World" (1980).

In 1977 General Motors demonstrated Sight-I, a machine-vision system for inspecting integrated circuits, developed by Michael Baird, who at Georgia Institute of Technology had written one of the early theses on machine vision ("A Paradigm for Semantic Picture Recognition," 1973), working in Lothar Rossol's laboratory. In September 1978 Rossol organized at General Motors a symposium on computer vision, possibly the first of its kind in the world). In 1979 his team demonstrated Consight, "a vision-controlled robot system for transferring parts from belt conveyors", based on a simple scheme of light sensors, soon installed at a plant near Niagara Falls. In 1979 SRI's chief A.I. scientist Charles Rosen founded Machine Intelligence Company in Atherton to commercialize SRI's machine-vision technology that could recognize industrial parts as they moved along a conveyor belt. In 1981 Automatix, the new venture of Victor Scheinman (of Stanford Arm and PUMA fame), co-founded with Philippe Villers of CAD pioneer Computervision, introduced Robovision for arc welding, the first commercial robot equipped with a vision system, a technology borrowed from SRI's Shakey group. In 1982 the American Robot Corporation was founded in Pittsburgh and later renamed American Cimflex before merging with A.I. pioneer Teknowledge of Palo Alto. In 1982 General Motors and Japan's Fanuc established the joint venture GMF Robotics. In 1984 Adept, founded by Brian Carlisle and Bruce Shimano, i.e. the Vicarm engineers who had worked with Scheinman on PUMA, introduced a rival robotic arm also equipped with machine-vision, the first one to incorporate the direct-drive technology invented by Takeo Kanade at CMU in 1981.

Then David Marr's epochal posthumous book "Vision" (1982) was published. It summarized his research at MIT that the mind understands a scene through a three stage process: a primal two-dimensional sketch that contains the basic components of the scene; a 2.5 dimension sketch of the scene that also contains depth, an idea that the British-born Marr originally developed with the Italian-born Tomaso Poggio ("A Theory of Human Stereo Vision", 1977) and was partially preceded by Parvati Dev's neural model at University of Massachusetts ("Perception of Depth Surfaces in Random-dot Stereograms", 1975); and then, the final three-dimensional model. Computer vision became a popular subject. One limitation was that all the classic algorithms only dealt with straight lines.

Jitendra Malik at Stanford ("Interpreting Line Drawings of Curved Objects", 1985) was one of the young scientists who studied how to deal with curved objects. In 1987 Lawrence Sirovich and Michael Kirby, mathematicians at Brown University, used principle component analysis (i.e., linear algebra) to transform images of faces into mathematical vectors called "eigenfaces" ("Low-dimensional Procedure for the Characterization of Human Faces", 1987). This method constituted the basis of the "sliding window" approach of Matthew Turk and Alex Pentland at MIT ("Eigenfaces for Face Detection/Recognition", 1991). In 1987 another cognitive scientist, Irving Biederman at State University of New York in Buffalo, published an influential article to explain how we recognize objects, arguing that objects can be broken down into basic geometric solids called "geons" ("Recognition-by-components, a theory of human image understanding", 1987).

After graduation Malik moved to UC Berkeley, where he founded an important school of computer vision, and one of his students was Pietro Perona, who, in turn, moved to the California Institute of Technology (CalTech) and built another important group in computer vision that over the years refined Perona's "constellation models" for object detection. They included; Thomas Leung ("Finding Faces in Cluttered Scenes Using Labeled Random Graph Matching", 1995), Michael Burl ("Localization via Shape Statistics", 1995), Markus Weber ("Unsupervised Learning of Models for Recognition", 2000), Fei-fei Li ("A Bayesian Approach to Unsupervised One-Shot learning of Object Categories", 2003), Rob Fergus ("Object Class Recognition by Unsupervised Scale-Invariant Learning", 2003), etc. It was here that in 2003 Fei-fei Li built the dataset of images Caltech 101 that later evolved into ImageNet.

Back Propagation - A Brief History of Artificial Intelligence/ Part 2

Knowledge-based systems did not expand as expected: the human experts were not terribly excited at the idea of helping construct digital clones of themselves, and, in any case, the clones were not terribly reliable.

Expert systems also failed because of the World-wide Web: you don't need an expert system when thousands of human experts post the answer to all possible questions. All you need is a good search engine. That search engine plus those millions of items of information posted (free of charge) by thousands of people around the world do the job that the "expert system" was supposed to do. The expert system was a highly intellectual exercise in representing knowledge and in reasoning heuristically. The Web is a much bigger knowledge base than any expert-system designer ever dreamed of. The search engine has no pretense of sophisticated logic but, thanks to the speed of today's computers and networks, it "will" find the answer on the Web. Within the world of computer programs, the search engine is a brute that can do the job once reserved to artists.

Note that the apparent "intelligence" of the Web (its ability to provide all sorts of questions) arises from the "non-intelligent" contributions of thousands of people in a way very similar to how the intelligence of an ant colony emerges from the non-intelligent contributions of thousands of ants.

In retrospect a lot of sophisticated logic-based software had to do with slow and expensive machines. As machines get cheaper and faster and smaller, we don't need sophisticated logic anymore: we can just use fairly dumb techniques to achieve the same goals. As an analogy, imagine if cars, drivers and gasoline were very cheap and goods were provided for free by millions of people: it would be pointless to try and figure out the best way to deliver a good to a destination because one could simply ship many of those goods via many drivers with an excellent chance that at least one good would be delivered on time at the right address. The route planning and the skilled knowledgeable driver would become useless, which is precisely what has happened in many fields of expertise in the consumer society: when is the last time you used a cobbler or a watch repairman?

The motivation to come up with creative ideas for A.I. scientists was due to slow, big and expensive machines. Now that machines are fast, small and cheap, the motivation to come up with creative ideas is much reduced. Now the real motivation for A.I. scientists is to have access to thousands of parallel processors and let them run for months. Creativity has shifted to coordinating those processors so that they will search through billions of items of information. The machine intelligence required in the world of cheap computers has become less of a logical intelligence and more of a "logistical" intelligence.

The 1980s also witnessed a progressive rehabilitation of neural networks, a process that would turn exponential in the 2000s.

One important center of research was located in southern California. In 1976 the cognitive psychologists Don Norman and David Rumelhart (two members of the LNR research group) founded the Institute for Cognitive Science at UCSD and hired a fresh British graduate, Geoffrey Hinton, who had studied with Christopher Longuet-Higgins and who also happened to be the great-great-grandson of the founder of binary logic, George Boole. Soon, UC San Diego became a hotbed of research in neural networks. In June 1979 Hinton and James Anderson organized a symposium at UC San Diego on associative memory, which later became the book "Parallel Models of Associative Memory" (1981), attended by Norman, Rumelhart, McClelland, Sejnowski, Jerry Feldman from Rochester University (who in

1982, in collaboration with fellow Rochester scientist Dana Ballard, would publish the report "Connectionist models and their properties" that would popularize the term "connectionism"), Scott Fahlman from CMU (a specialist in semantic networks but today more famous for inventing in 1982 the "smiley" emoticon), Stuart Geman from Brown University (who would go on to develop "Gibbs sampling"), and others. In early 1982, inspired by Raj Reddy's Hearsay project at CMU, Rumelhart, Hinton, Jay McClelland, Paul Smolensky, and the biologists David Zipser and Francis Crick (of DNA fame) formed the PDP (Parallel Distributed Processing) research group of psychologists and computer scientists. After six months Hinton, the original organizer, moved to CMU (where he organized a summer workshop that introduced him to the ideas of Terrence Sejnowski on the Boltzmann machine) while McClelland moved to MIT. Soon the two were reunited at Carnegie Mellon where a second PDP group was spawned. The San Diego group would go on to include David Zipser's student Ronald Williams, Michael Jordan and Jeffrey Elman.

Neural networks were rescued in 1982 by the CalTech physicist John Hopfield, who described a new generation of neural networks ("Neural Networks and Physical Systems with Emergent Collective Computational Abilities", 1982). Hopfield designed a network in which all connections are symmetric, i.e. all neurons are both input and output neurons. It is a "recurrent" network because the effect of a neuron's computation ends up flowing back to that neuron. Until then the most popular architecture had been the "feedforward" kind: the output of a layer of neurons does not affect the same layer but only the layers that are downstream. Feedback networks, instead, can have all sorts of upstream repercussions. Recurrent networks can be very difficult to analyze, but Hopfield's networks have symmetric connections and the neurons are binary neurons, and in that case the network dynamics can be described with what physicists call an "energy function": one can measure the "energy" of each state of the network, and training the network is equivalent to lowering the energy. The network has learned something when it reaches an energy minimum. The memory of something is an energy minimum of this neural net.

These neural networks were immune to Minsky's critique. Hopfield's key intuition was to note the similarity with statistical mechanics. Statistical mechanics translates the laws of Thermodynamics into statistical properties of large sets of particles. The fundamental tool of statistical mechanics (and soon of this new generation of neural networks) is the Boltzmann distribution (actually discovered by Josiah-Willard Gibbs in 1901), a method to calculate the probability that a physical system is in a specified state.

Hopfield's original network used binary neurons a` la McCulloch-Pitts, but two years later Hopfield showed that his results held also in the case of continuous neurons ("Neurons with Graded Response have Collective Computational Properties like those of Two-state Neurons," 1984), reaching conclusions similar to those reached by Mike Cohen and Stephen Grossberg at Boston University for symmetric neural networks, but reached following a mathematical conjecture ("Absolute Stability of Global Pattern Formation and Parallel Memory Storage by Competitive Neural Networks", 1983).

In the same year (1982) Teuvo Kohonen popularized the "self-organising map" (SOM), soon to become the most popular algorithm for unsupervised learning ("Self-organized Formation of Topologically Correct Feature Maps," 1982), borrowing the architecture used by Christoph von der Malsburg in Germany to simulate the visual cortex ("Self-organization of Orientation Sensitive Cells in the Striate Cortex", 1973).

At the time there were important hubs of interdisciplinary research. CalTech combined the talents of Carver Mead, John Hopfield (who had arrived in 1980 from Princeton University) and Richard Feynman in a course titled "The Physics of Computation" in the Fall of 1981, and Minsky was one of the guest lecturers. The most important club exploring the intersection of biology and physics was the Helmholtz Club, started in September 1982 at UC Irvine (between Los Angeles and San Diego) by Francis Crick (co-discoverer of the DNA's double helix and then at the Salk Institute in San Diego), Vilayanur Ramachandran (back then a postdoc at UC Irvine), Gordon Shaw (a former Stanford physicist who had turned neurobiologist after moving to UC Irvine), David van Essen (a neuroscientist at CalTech), Joaquin Fuster (a neurophysiologist from UCLA), and John Allman (a neurobiologist at CalTech). The club lasted for twenty years, including new members such as Carver Mead (the Caltech computer scientist), philosopher Patricia Churchland (after she moved to UC San Diego in 1984), Terry Sejnowski (after he moved to the Salk Institute in 1988), plus neuroscientists from UC Los Angeles, the Salk Institute and University of Southern California, psychologists from UC Santa Barbara and UC San Diego, etc.

Footnote: Neural Networks and the Math of Glass

Hopfield based his model on the physics of spin glasses. He wasn't the first one to make the connection between neurons and spins. The mathematician Neville Temperley at the Atomic Weapons Research Establishment in Britain was perhaps the first to realize that a network of neurons behaves like a network of spins ("Memory - The Analogy with Ferromagnetic Hysteresis", 1955). Almost twenty years later the idea was resurrected by Bill Little at Stanford University (the physicist who in 1964 had shown that superconductivity can happen at room temperature), who published a study about the analogy between spin systems and both short-term and long-term memory, between the firing or not firing of a neuron and the up and down spin states of an atom ("The Existence of Persistent States in the Brain", 1974).

In 1975 the physicists Sam Edwards and Philip Anderson at Cambridge University had solved the problem of the disordered states of the so-called "spin glass", the ultimate disordered system ("Theory of Spin Glasses", 1975) and the Italian physicist Giorgio Parisi had found a more general solution ("Mean Field Theory for Spin Glasses", 1980). Hopfield extended their theory of spin glasses to neural networks, a fact even better clarified by the Israeli physicists Daniel Amit, Hanoch Gutfreund and Haim Sompolinsky at the Hebrew University in Israel ("Storing Infinite Numbers of Patterns in a Spin-Glass Model of Neural Networks", 1985)

While known since ancient times, glass is one of the most mysterious materials. In fact, physicists can't even decide whether it is a liquid or a solid. Even when it is shaped in smooth and elegant structures, glass is an example of disordered material. Glass is stuck, or "quenched," in a low-temperature disordered state. Physics knows well how to study disordered states caused by high temperatures, when atoms move frantically; but glass is disordered at low temperatures. The atoms of a glass are distributed at random locations but standing still. When water becomes ice, it undergoes a phase transition. What is even more puzzling about glass is that a liquid becomes glass without undergoing a phase transition: glass is, in a sense, a supercooled liquid. A glass is a liquid that gets more and more viscous while it is being cooled until it eventually stops flowing.

Spin glasses are materials that exhibit the same kind of "quenched" disorder, the opposite of "annealed" disorder. (The "glass" in the name is a misnomer: they are called that way only because of the analogy with the disorder of glass). They are systems far from equilibrium whose study has provided much of the mathematics used to study complex systems, for example the methods to solve combinatorial optimization problems.

Any real-world system is disordered, or, better, has a component of disorder. If we "quench" a system, the state is instantaneously changed to a permanent state far from equilibrium with the environment (imagine dipping an incandescent metal into a bucket of ice). If we "anneal" the system, the state is changed gradually so that the system is almost in equilibrium with its environment at all times (e.g., if we cool the hot metal slowly). Quenched disorder is frozen in time, annealed disorder evolves in time (e.g., it can reverse itself). Quenched disorder is much harder to model mathematically than annealed disorder. Nonetheless, the applications are intriguing, from protein folding to neural networks. For example, Arkadiusz Jedrzejewski and Katarzyna Sznajd-Weron at Wroclaw University in Poland described how the two main psychological theories of person's behavior are theories of disorder: respectively, the theory that the situation is more important to determine a person's behavior is a theory of annealed disorder whereas the theory the personality traits are more important is a theory of quenched disorder ("Person-Situation Debate Revisited", 2017). The "person-situation debate" in social psychology pitted the trait theorists, such as Hans Eysenck at King's College London, who wrote "Dimensions of Personality" (1947), and Raymond Cattell at University of Illinois, who wrote "Personality" (1950), and one of the first psychologists to use a computer, the Illiac (operational in 1952), against the situationists, such as Walter Mischel at Stanford, whose book "Personality and Assessment" (1968) started the whole riot, and Richard Nisbett at University of Michigan ("The Trait Construct in Lay and Professional Psychology", 1980).

Needless to say, a spin glass is not a brain: there is virtually nothing in our brain that is similar to a spin glass. Alas, there is little that is symmetric and mathematically elegant inside our brain.

But i digress...

Bayes Reborn - A Brief History of Artificial Intelligence/ Part 3

The Hopfield network proved Minsky and Papert wrong but it has a problem: it tends to get trapped into what mathematicians call "local minima". Two improvements of Hopfield networks were proposed in a few months: the Boltzmann machine and backpropagation.

The Boltzmann machine was inspired by the physical process of annealing. At the same time that Hopfield introduced his recurrent neural networks, Scott Kirkpatrick at IBM introduced a stochastic method for mathematical optimization called "simulated annealing" ("Optimization by Simulated Annealing", 1983), which uses a degree of randomness to overcome local minima. This method was literally inspired by the physical process of cooling a liquid until it achieves a solid state. As temperature decreases, the probability tends to concentrate on low energy states according to the Boltzmann distribution, a staple of thermodynamics.

In 1983 the cognitive psychologist Geoffrey Hinton, formerly a member of the PDP group at UC San Diego and now at CMU, and the physicist Terry Sejnowski, a student of Hopfield at Princeton University but now at

Johns Hopkins University, invented a neural network called the Boltzmann Machine that used a stochastic technique to avoid local minima, basically a Monte Carlo version of the Hopfield network ("Optimal Perceptual Inference", 1983). They replaced Hopfield's deterministic neurons with probabilistic neurons, and then used simulated annealing to find low energy states with high probability. The Boltzmann Machine (with no layers) avoided local minima and converged towards a global minimum. The learning rule of a Boltzmann machine is simple, and, yet, that learning rule can discover interesting features about the training data. In reality, the Boltzmann machine is but a case of "undirected graphical models" that have long been used in statistical physics: the nodes can only have binary values (zero or one) and they are connected by symmetric connections. They are "probabilistic" because they behave according to a probability distribution, rather than a deterministic formula. But there was still a major problem: the learning procedure of a Boltzmann machine is painfully slow. And it was still haunted by local minima in the case of many layers.

In 1982 David Parker at Stanford Linear Accelerator rediscovered backpropagation and called it "learning logic", and later published an influential report at MIT ("Learning Logic", 1985). In 1985 the young Yann LeCun too, at the Electrotechnic and Electronic School of Paris, rediscovered backpropagation ("A Learning Scheme for Asymmetric Threshold Network", 1985). Werbos had worked out the algorithm as an extension of statistical regression but they realized that it could be applied to training multi-layer neural networks, and therefore overcome Minsky's criticism. Paul Werbos himself was now explicitly proposing backpropagation as a method for neural networks ("Applications of Advances in Nonlinear Sensitivity Analysis", 1982).

Helped by the same young Hinton (now at Carnegie Mellon) and by Ronald Williams, both former buddies of the PDP group, the mathematical psychologist David Rumelhart optimized, in 1986, backpropagation for training multilayer (or "deep") neural networks using a "local gradient descent" algorithm that would rule for two decades, de facto a generalized delta rule ("Learning Representations by Back-propagating Errors", 1986; retitled "Learning Internal Representations by Error Propagation" as a book chapter). Rumelhart popularized the paradigm that would rule in multi-layer networks, the paradigm already used by Rosenblatt, Widrow, Ivakhnenko, Werbos, Fukushima and Hopfield, i.e. that the functioning of a neural network consists of two phases. During the feedforward phase the network uses some algorithm (that depends on an activation function, typically a sigmoid) to compute the output based on the input. This involves applying the algorithm to every neuron in the network, moving forward layer by layer. Then the error is computed as the difference between the desired output and the actual output. During the feedbackward phase, another algorithm (typically a gradient-descent algorithm) adjusts the weights of the network according to a function of the error, going backwards layer by layer, aiming to reduce the error. The difference between Widrow's method and Werbos-Rumelhart's method is that Widrow's method is linear and that limits the learning process in the case of many layers, whereas Werbos-Rumelhart's method benefits from a nonlinear activation function. The generalized delta rule used an S-shaped sigmoid function. Unlike Widrow's function, the sigmoid function introduced non-linearity in the neural network. Other than that, Werbos-Rumelhart's method was similar to Widrow's method: weights are changed proportionally to the negative gradient of the error; but the nonlinear approach changes each weight a tiny step in the direction to reduce network error, and so the network learns in a more gradual and stable way. To be fair, the term "backpropagation" was already in Rosenblatt's book "Principles of Neurodynamics" (1962), and even the general method (but not an actual algorithm).

The sigmoid was not a detail. The neurons of our brain are not as deterministic as Pitts thought: a neuron may actually spike even if the input is not strong enough, and may not spike when the input is strong enough. There is a probabilistic, not deterministic, correlation between input and output. And then there is a point at which the probability of spiking flattens out, a sort of saturation point. Mathematically, this is expressed by the sigmoid function, a curve that first goes up exponentially and then goes flat also exponentially. Engineers love the sigmoid because it shows up in multiple areas: from the phase transitions of water (the way it turns into ice or steam) to electromagnetism (notably, the transition from zero to one of computer bits). Many processes happen very fast but then start happening very slowly. The sigmoid even represents phenomena in economics, in sociology and in psychology. For example, population tends to grow exponentially, but then growth starts declining also exponentially. The straight lines of statistics and the step functions of early neural networks are approximations of the sigmoid: stretch the sigmoid, and it becomes a straight line; squeeze the sigmoid, and it becomes a step.

Error backpropagation is a very slow process and requires huge amounts of data; but backpropagation provided A.I. scientists with an efficient method to compute and adjust the "gradient" with respect to the strengths of the neural connections in a multilayer network. (Technically speaking, backpropagation is gradient descent of the mean-squared error as a function of the weights).

The world finally had a way (actually, two ways) to build multilayer neural networks to which Minsky's old critique did not apply.

Note that the idea for backpropagation came from both engineering (old cybernetic thinking about feedback) and from psychology. The first success story of backpropagation was NETtalk, a three-layer neural network designed in 1986 by Sejnowski at and programmed by George Miller's student Charlie Rosenberg that successfully learned to pronounce English text.

At the same time, another physicist, Paul Smolensky of University of Colorado, introduced a further optimization, the "harmonium", better known as Restricted Boltzmann Machine ("Information Processing in Dynamical Systems", 1986) because it restricts the kind of connections that are allowed between layers. The learning algorithm devised by Hinton and Sejnowski is very slow in multilayer Boltzmann machines but very fast in restricted Boltzmann machines. Multi-layered neural networks had finally become a reality. The architecture of Boltzmann machines makes it unnecessary to propagate errors, hence Boltzmann machines and all their variants do not rely on backpropagation.

These events marked a renaissance of neural networks. In 1986 Hinton and Sejnowski organized the first "Connectionist Summer School" at CMU (attended by McClelland, David Willshaw, who was a neurologist at University of Edinburgh, Dave Touretzky, and future stars such as Dana Ballard, Yann LeCun, the future guru of convolutional networks, and Gerald Tesauro, the future creator of TD-Gammon), a few weeks later Rumelhart and McClelland published the two-volume "Parallel Distributed Processing" (1986) and the International Conference on Neural Networks was held in San Diego in 1987. San Diego was an appropriate location since in 1982 Francis Crick, the British biologist who co-discovered the structure of DNA in 1953 and who now lived in southern California, had started the Helmholtz club with UC Irvine physicist Gordon Shaw (one of the earliest researchers on the neuroscience of music), Caltech neurophysiologist Vilayanur Ramachandran (later at UC San Diego), Caltech neurosurgeon Joseph Bogen (one of Roger Sperry's pupils in split-brain surgery), Caltech neurobiologists John Allman, Richard Andersen, and David Van Essen (who mapped out the visual system of the macaque monkey), Carver Mead, Terry Sejnowski and David Rumelhart. (Sad note: Rumelhart's career ended a few years later due to a neurodegenerative disease).

There were other pioneering ideas.

Unsupervised learning is closely related to the problem of source separation in electrical engineering, a problem that consists in discovering the original sources of an electrical signal. Jeanny Herault and Christian Jutten of the Grenoble Institute of Technology in France developed a method called "independent component analysis", a higher-order generalization of principal components analysis ("Space or Time Adaptive Signal Processing by Neural Network Models", 1986), later refined by Jean-Francois Cardoso at the CNRS in France ("Sources Separation Using Higher Order Moments", 1989) and by Pierre Comon at Thompson in France ("Independent Component Analysis", 1991).

In 1986 Ralph Linsker at IBM research labs in Yorktown Heights published three unsupervised models that can reproduce known properties of the neurons in the visual cortex ("From Basic Network Principles to Neural Architecture", 1986). Linsker later developed the infomax method for unsupervised learning that simplified independent component analysis ("Self-Organization in a perceptual network", 1988), which recycled one of Uttley's ideas ("Information Transmission in the Nervous System" (1979). The infomax principle is basically to maximize the mutual information that the output of a neural network processor contains about its input and viceversa.

David Zipser at UC San Diego came up with the "autoencoder", apparently from an unpublished idea by Hinton of the previous year ("Programming Networks to Compute Spatial Functions", 1986), although the term was first used in print by Suzanna Becker in 1990. An autoencoder is an unsupervised multi-layer neural network that is trained by backpropagation to output the input, or a very close approximation of it. In other words, an autoencoder tries to predict the input from the input. In other words, it is trying to learn the identity function. This sounds trivial, but in some cases the middle (hidden) layer ends up learning interesting facts about the data. The drawback of backpropagation is that it requires "supervision": someone has to tell the network what is the desired network (has to "train" the network). An autoencoder basically converts the supervised learning into an unsupervised learning. When one forces the network to produce the input as output, the network learns how to map all training inputs to themselves, which results in learning a representation of the training sample. If the middle layer of the network contains fewer units than the input layer, the network has to produce a compact representation, and ends up learning interesting facts about the data. Autoencoders are powerful models for capturing characteristics of data. The technique was employed by David Zipser to compress images ("Learning Internal Representations from Gray-Scale Images", 1987) and to compress speech waves ("Discovering the Hidden Structure of Speech", 1987). One can also view the action of an autoencoder as a way to store an input so that it can be subsequently retrieved as accurately as possible, i.e. the function of an autoencoder is to create a representation of the input (to "encode" the input) that allows the network to later retrieve it in a very accurate form.

One powerful effect of doing this layer after layer would be understood only two decades later and originate the boom of "deep learning".

Dana Ballard at University of Rochester predated the so-called "deep belief networks" as well as "stacked autoencoders" by 20 years when he used unsupervised learning to build representations layer by layer ("Modular Learning in Neural Networks", 1987).

Linsker's infomax was an early application of Shannon's information theory to unsupervised neural networks. A similar approach was tried by Mark Plumbley at Cambridge University ("An Information-Theoretic Approach to Unsupervised Connectionist Models", 1988).

An accelerated version of gradient descent was developed by the Russian mathematician Yurii Nesterov at the Central Economic Mathematical Institute of Moscow ("A Method of Solving a Convex Programming Problem", 1983), and it would become the most popular of the gradient-based optimization algorithms (known as "Nesterov momentum"). Later, Ning Qian at Columbia University showed similarities between Nesterov's theory and the theory of coupled and damped harmonic oscillators in physics ("On the Momentum Term in Gradient Descent Learning Algorithms", 1999).

Soon, new optimizations led to new gradient-descent methods, notably the "real-time recurrent learning" algorithm, developed simultaneously by Tony Robinson and Frank Fallside at Cambridge University ("The Utility Driven Dynamic Error Propagation Network", 1987) and Gary Kuhn at the Institute for Defense Analysis in Princeton ("A First Look at Phonetic Discrimination Using a Connectionist Network with Recurrent Links", 1987), but popularized by Ronald Williams and David Zipser at UC San Diego ("A Learning Algorithm for Continually Running Fully Recurrent Neural Networks", 1989). Paul Werbos, now at the US Department of Energy in Washington, expanded backpropagation into "backpropagation through time", i.e. backpropagation applied to sequences such as a time series ("Generalization of Backpropagation with Application to a Recurrent Gas Market Model", 1988); and variations on backpropagation through time include: the "block update" method pioneered by Ronald Williams at Northwestern University ("Complexity of Exact Gradient Computation Algorithms For Recurrent Neural Networks", 1989), the "fast-forward propagation" method by Jacob Barhen, Nikzad Toomarian and Sandeep Gulati at CalTech ("Adjoint Operator Algorithms for Faster Learning in Dynamical Neural Networks", 1991), and the "green function" method by Guo-Zheng Sun, Hsing-Hen Chen and Yee-Chun Lee at University of Maryland ("Green's Function Method for Fast On-Line Learning Algorithm of Recurrent Neural Networks", 1992). All these algorithms were elegantly unified by Amir Atiya at CalTech and

Alexander Parlos at Texas A&M University ("New Results on Recurrent Network Training", 2000).

These were carefully calibrated mathematical algorithms to build neural networks to be both feasible (given the dramatic processing requirements of neural network computation) and plausible (that solved the problem correctly).

Nonetheless, philosophers were still debating whether the "connectionist" approach (neural networks) made sense. Two of the most influential philosophers, Jerry Fodor and Zenon Pylyshyn, wrote that the cognitive architecture cannot possibly be connectionist ("Connectionism and Cognitive Architecture", 1988) whereas the philosopher Andy Clark at University of Sussex argued precisely the opposite in his book "Microcognition" (1989). Paul Smolensky at University of Colorado ("The Constituent Structure of Connectionist Mental States", 1988), Jordan Pollack ("Recursive Auto-associative Memory", 1988) and Jeffrey Elman ("Structured Representations and Connectionist Models", 1990) proved how neural networks could do precisely what Fodor thought they could never do, and another philosopher, David Chalmers at Indiana University, closed the discussion for good ("Why Fodor and Pylyshyn Were Wrong", 1990).

This school of thought merged with another one that was coming from a background of statistics and neuroscience. Credit goes to Judea Pearl of UC Los Angeles for introducing Bayesian thinking into Artificial Intelligence to deal with probabilistic knowledge ("Reverend Bayes on Inference Engines", 1982). Ray Solomonoff's universal Bayesian methods for inductive inference were finally vindicated. Pearl's Bayesian network is a formal way to represent how events depend on each other, i.e. a way to turn into graphical models and mathematical formulas a story that has a lot of mysteries.

A kind of Bayesian network, the Hidden Markov Model, was already being used by A.I., particularly for speech recognition.

Technically speaking, a Bayesian network is a "directed" graphical model: the interaction between its events has a directionality, there are causes and there are effects. A "Markov network" (also called "Markov random field") is an "undirected" graphical model, and can can represent some dependencies that a Bayesian network cannot. Furthermore, Bayesian networks are acyclic, whereas Markov networks may be cyclic. Markov networks had already been studied in the 1970s by mathematicians such as the Austrian-born Frank Spitzer at Cornell University and Andrey Kolmogorov's student Roland Dobrushin at Moscow State University.

Neural networks and probabilities have something in common: neither is a form of perfect reasoning. Classical logic, based on deduction, aims to prove the truth. Neural networks and probabilities aim to approximate the truth. Neural networks are "universal approximators", as proven first by George Cybenko in 1989 at University of Illinois ("Approximation by Superpositions of a Sigmoidal Function", 1989) and by Kurt Hornik at the Technical University in Austria, in collaboration with the economists Maxwell Stinchcombe and Halbert White of UC San Diego ("Multilayer feedforward networks are universal approximators", 1989). Cybenko and Hornik proved that neural networks can approximate any continuous function of the kind that, de facto, occurs in ordinary problems. Basically, neural networks approximate complex mathematical functions with simpler ones, which is, after all, precisely what our brain does: it simplifies the incredible complexity of the environment that surrounds us although it can only do it by approximation. Complexity is expressed mathematically by nonlinear functions. Neural networks are approximators of non-linear functions. The fact that a nonlinear function can be more efficiently represented by multilayer architectures with fewer parameters became a motivation to study multilayer architectures.

Footnote: A Brief History of the Gradient

Training a neural network means minimizing its error. The neural network needs two algorithms for this: one algorithm to minimize the "gradient" in a series of steps and one algorithm to compute the "gradient" at each step. Backpropagation is the algorithm to compute the gradient. The algorithm used in conjunction with backpropagation is an optimization method because minimizing the error is another way of saying that we want to find the minimum of a complex (non-linear) function. There are many such optimization methods. The need originally arose after Newton published his equations of gravitation. It was mainly astronomers who needed to find the minima of functions. The British mathematician John Wallis published the so-called "Newton's method" in his "Treatise of Algebra both Historical and Practical" (1685), the book that introduced the calculus of limits, but it was the French mathematician Augustin Cauchy in 1847 who invented the first efficient method of this kind. His "gradient-descent" method (or "steepest-descent method") enabled astronomers to calculate the orbit of an astronomical object without having to solve Newton's differential equations of motion, but instead using (simpler) algebraic equations.

The idea is quite intuitive if one thinks of a function as a curve in space similar to the skyline of a mountain range, with peaks and valleys. The goal is to descend from the top of a mountain to the parking lot. The fastest way at every step (if you are not scared of heights) is to always take the steepest way down: follow the direction of the slope downhill. The problem is that there are many cases in which this takes you to a valley, e.g. a glacier nestled between several mountains, where the strategy fails: you can't go any further down. Or, mathematically speaking, the gradient at any local minima is zero, so it doesn't tell you how to continue optimizing the function. That is the problem of "local minima".

A very popular gradient method for neural networks, employed in conjunction with backpropagation, is the "stochastic gradient descent" method, pioneered by Herbert Robbins in 1951, mainly because it is much more efficient, a fact of great importance when you are training a neural network with a large dataset; but also to reduce the chances of getting stuck in local minima.

Backpropagation is what is known in calculus as a "chain rule" and it computes the gradient at every step, after which the optimization method decides which step to take next. The iterative application of gradient computation and optimization results in the progressive modification of the "weights" that connect neurons.

There are also gradient-free methods of optimization, notably simulated annealing and genetic algorithm, which, in fact, have been studied by A.I. since the early days.

Much of A.I. reasoning consists in "finding" a solution in space of solutions. This task is closely related to the mathematical problem of optimization in which all possible states of the system have an objective function and the goal is to find the state with the maximum objective value. For example, hill-climbing search consists in a loop that continuously moves towards increasing value. Gradient descent is a close relative of hill climbing. This strategy is obviously naïve because hill-climbing will stop when it can't increase the value anymore, but that could just be a sub-peak of a mountain that is much higher (it could be a "local minimum"). The hill-climbing algorithm is fooled by any local minima and, additionally, it doesn't know which way to go if it enters a plateau (in which each direction yields the same value). Many methods have been proposed to improve simple search algorithms like hill climbing: Scott Kirkpatrick's simulated annealing in 1983, Tabu search, published in 1986 by Fred Glover at University of Colorado ("Future Paths for Integer Programming and Links to Artificial Intelligence", 1986), and GRASP (greedy randomized adaptive search procedure), introduced by Thomas Feo of University of Texas and Mauricio Resende of UC Berkeley ("A Probabilistic Heuristic for a Computationally Difficult Set Covering Problem", 1989).

Searching a "tree" of possible actions/solutions remained one of the fundamental chores of any "intelligent" system, and the A* algorithm remained the foundation of this kind of search. However, A* and all its variants are methods of "offline" search: even before taking the first step, they already plan the complete sequence of steps to achieve the goal, i.e. all the way from the start state to the goal state. Real-time search methods, instead, plan only a few first steps. This approach is more practical in real-world situations. Several real-time algorithms have been developed that are also variations of A*, notably the Learning Real-Time A* (LRTA*), introduced by Richard Korf at UCLA ("Real-Time Heuristic Search", 1987) and Real-time A* algorithms. The former improves its performance over time but it is less efficient than the latter.

"If you do not change direction, you may end up where you are heading" (Lao Tze)

Reinforcement Learning - A Brief History of Artificial Intelligence/ Part 4

In 1946 Adriaan de Groot, a Dutch psychologist who was also a chess master, had showed in his analysis "Thought and Choice in Chess" that expert chess players do not spend more time looking for alternative moves than novice players: the experts are good at "guessing" the best next move, not necessarily at calculating it. Nevertheless, programs designed to play games kept using "deep" searches. Goro Hasegawa's board game Othello (a sensation in Japan since its introduction in 1973) was a particular favorite of the A.I. community. In 1980 Peter Frey at Northwestern University organized the first "human versus machine" Othello tournament that featured the then world-champion Hiroshi Inouie. Paul Rosenbloom's Iago (1982) at CMU, which later (1986) evolved at the hands of Sanjoy Mahajan and Kai-fu Lee into Bill, a program that beat the top US master, as well as Michael Buro's Logistello (1997) at NEC Research Institute, that defeated the then world-champion Takeshi Murakami, played Othello like masters by employing "deep" searches into the space of all possible moves. Trivia: Kai-fu Lee's 1988 dissertation would be Sphinx, the first speaker-independent speech-recognition system, and Kai-fu Lee would later become a famous Chinese venture capitalist.

There was another way for algorithms to learn how to play games.

In 2016 reinforcement learning became popular thanks to Google DeepMind and their AlphaGo go-playing program, developed by Aja Huang's team (Huang was a former student of Coulom). Reinforcement learning is a very old idea. We've always known that it works. Unfortunately it works for learning only one thing (in AlphaGo's case playing weiqi). Once the machine has learned to do that one thing it is terribly difficult to train it for doing something else. It is also a bit silly because the time required for the machine to learn is so colossal that reinforcement learning has rarely been applied to mission-critical applications. Playing a game is the kind of application in which reinforcement learning makes sense: let the machine play against itself and after a few months it will become a champion. While you can use it wherever you like, reinforcement learning really makes sense only in cases where the rules never change and you can wait a few months. Learning how to cross a street, for example, may not be a suitable application because the conditions change all the time and the learning machine is playing against cars. It would have to die thousands of times before it learns to cross the street. It is possible but one wonders if it's not easier to simply write a procedure on how to cross a street: the learning time is a few milliseconds and the machine doesn't have to keep dying all the time.

Neural networks can be supervised or unsupervised. Supervised networks learn from a dataset of positive instances, and the typical application has been object recognition in computer vision. Unsupervised networks cluster objects together by similarity, simulating the formation of concepts. Reinforcement learning is neither supervised nor unsupervised: it improves performance by direct interaction with the environment. Reinforcement learning tasks can be described mathematically as Markov decision processes. Reinforcement learning is about maximizing a value function and/or a policy.

Reinforcement learning works with a combination of positive and negative rewards, i.e. reward and punishment, just like you teach children ("trial-and-error"). The "value function" assigns rewards and punishments to various states of the system. The "policy" function determines which "next move" is more likely to get the maximum reward from the value function. The "Q-function" is the combination of the value and policy functions. For mathematicians, "reinforcement" denotes an objective function that has to be maximized. Some algorithms modify the policy directly and are called "actor-only", whereas other algorithms work on the value function and are called "critic-only". The role of the neural network, in general, is to approximate the policy function when it is too complex.

Reinforcement learning was invented even before the field was called "Artificial Intelligence". Its roots harken back to the beginning of neuroscience and to the era of behaviorism in psychology. It was already described in 1911 by Edward Thorndike of Columbia University: "The greater the satisfaction or discomfort, the greater the strengthening or weakening of the bond." Conditioning was described in 1926 by Ivan Pavlov in Russia: "the magnitude and timing of the conditioned response changes as a result of the contingency between the conditioned stimulus and the unconditioned stimulus". In 1938 Burrhus Skinner, while at University of Minnesota, coined the expression "operant conditioning" for the learning process in both humans and animals, both acting in order to obtain rewards and avoid punishments. Donal Hebb in his book "The Organization of Behavior" (1949), written while he was at the Yerkes National Primate Research Center in Atlanta (1942-46), formulated a simple rule (now known as "Hebbian learning") for what happens between neurons in the brain: the bond between neurons that fire together gets stronger, the bond between neurons that don't fire together gets weaker; i.e. simultaneous firing of afferent and efferent elements. (Hebb, influenced by Pavlovian conditioning, had already proposed a similar learning rule in his master thesis "Conditioned and Unconditioned Reflexes and Inhibition", submitted to McGill University in Canada in 1932). In 1949 Claude Shannon suggested using an evaluation function to help a computer learn how to play chess ("Programming a Computer for Playing Chess", 1949). Reinforcement learning was the topic of Marvin Minsky's PhD dissertation in 1954 at Princeton University ("Neural Nets and the Brain Model Problem"), where he was a student of mathematician Albert Tucker of the "prisoner's dilemma" fame. Reinforcement learning was first used in 1959 by Samuel's checkers-playing program. The program learned quite well from excellent players, but playing against bad players caused its performance to decline, so eventually Samuel allowed it to play only against champions. In 1961 British wartime code-breaker, Alan Turing's cohort and molecular biologist Donald Michie at University of Edinburgh built a device (made of matchboxes!) to play Tic-Tac-Toe called MENACE (Matchbox Educable Noughts and Crosses Engine) that learned how to improve its performance. In 1976 John Holland at University of Michigan introduced classifier systems, which are reinforcement-learning systems with a credit-assignment algorithm inspired by Samuel's checkers player ("Adaptation", 1976).

Bernard Widrow modified his (supervised) Adaline algorithm to produce a reinforcement learning rule ("Punish/Reward - Learning with a Critic in Adaptive Threshold Systems", 1973), while Ian Witten at University of Essex in Britain was working on a reinforcement-learning system capable of long-term strategizing ("Human Operators and Automatic Adaptive Controllers", 1973). In 1977 Witten implemented the first actor-critic method ("An Adaptive Optimal Controller for Discrete-time Markov Environments", 1977), although the term was introduced only later by Andrew Barto ("Neuronlike Elements That Can Solve Difficult Learning Control Problems", 1983). This was an important step in reinforcement learning because the same model emerged from experiments on animal learning and from the study of the brain (specifically, the basal ganglia).

Reinforcement learning was resurrected in 1978 by Andrew Barto's student Richard Sutton at University of Massachusetts ("Single Channel Theory", 1978). They applied ideas published by the mathematician Harry Klopf at the Air Force research laboratories in Boston in his 40-page report "Brain Function and Adaptive Systems" (1972): the neuron is a goal-directed agent and an hedonist one; neurons actively seek "excitatory" signals and avoid "inhibitory" signals. Sutton expanded Klopf's idea into the modern "temporal-difference method" ("Toward a Modern Theory of Adaptive Networks", 1981). Temporal-difference methods had already been used in Arthur Samuel's checkers player of 1959, Paul Werbos' "heuristic dynamic programming" ("Advanced Forecasting Methods for Global Crisis Warning and Models of Intelligence", 1977), and in Ian Witten's actor-critic architecture of 1977, and would be reenacted in John Holland's "bucket brigade" algorithm for his new classifier systems ("Properties of the Bucket Brigade", 1985). These algorithms assign credit based on the difference between temporally successive predictions.

Neuroscience was reaching similar conclusions. Eric Kandel's group at Columbia University, which included Robert Hawkins, was studying the cellular mechanisms for habituation in higher invertebrates ("Is There a Cell-Biological Alphabet for Simple Forms of Learning?", 1984). Trivia: an Austrian influenced by Sigmund Freud's psychoanalysis, Kandel spent his life researching how memory is implemented in the cortex of the brain, and, in particular, he was the one who proved experimentally Hebbian learning when studying sea slugs ("Behavioral Biology of Aplysia", 1979). He also published one of the earliest textbooks of neuroscience, "Principles of Neural Science" (1981).

However, both mathematicians and neuroscientists were unhappy with Hebb's original rule, which was both biologically implausible (it doesn't explain how neurons and computationally unstable. In 1981 Elie Bienenstock, Leon Cooper and Paul Munro published a new learning rule for neurons, now known as "the BCM rule", a modified Hebbian learning rule. Three decades later, Nicolas Brunel at University of Chicago would find physiological evidence supporting this rule ("Inferring Synaptic Plasticity Rules from Spike Counts", 2015). In 1973 Maurice (Martin) Taylor at the Defence Research Medical Laboratories in Toronto came up with another learning rule ("The Problem of Stimulus Structure in the Behavioural Theory of Perception", 1973) that was rediscovered as STPD (Spike-Timing-Dependent Plasticity) in the 1990s by Henry Markram at the Max Planck Institute.

Sutton already employed differential Hebbian learning rules in 1981, but these rules were best outlined by Bart Kosko at Verac in San Diego ("Differential Hebbian Learning", 1986) and by Klopf himself ("A Drive-reinforcement Model of Single Neuron Function", 1986).

Ronald Williams was Rumelhart's and Hinton's coauthor in the seminal paper on backpropagation and at the time was already working on reinforcement learning ("Reinforcement Learning in Connectionist Networks", 1986). After moving to Northeastern University, he developed the REINFORCE algorithms (which, believe it or not, stands for "REward Increment = Non-negative Factor times Offset Reinforcement times Characteristic Eligibility"), a class of reinforcement learning algorithms for neural networks with stochastic neurons whose main virtue is that they are simple to implement ("Simple Statistical Gradient-following Algorithms for Connectionist Reinforcement Learning", 1992).

All the studies on reinforcement learning since Michie's MENACE converged together in the Q-learning algorithm invented in 1989 at Cambridge University by Christopher Watkins. The Q-learning algorithm was a temporal-difference method that simultaneously optimized value function and policy ("Learning from Delayed Rewards", 1989). Watkins basically discovered the similarities between reinforcement learning and the theory of optimal control that had been popular in the 1950s thanks to the work of Lev Pontryagin in Russia (the "maximum principle" of 1956) and Richard Bellman at RAND Corporation (the "Bellman equation" of 1957). Trivia: Bellman is the one who coined the expression "the curse of dimensionality" that came to haunt the field of neural networks.

Unfortunately, the two main temporal-difference methods, i.e. Sutton's adaptive heuristic critic algorithm and Watkins' Q-learning algorithms, are both very slow. Then, a few years later, Long-ji Lin at CMU proposed a way to speed up both by letting the learning agent remember and rehash its past experiences, specifically by replaying to the learning algorithm the sequence of past experiences backwards, a method that came to be known as "experience replay" ("Self-improving Reactive Agents Based on Reinforcement Learning, Planning and Teaching", 1992).

REINFORCE's "likelihood-ratio method" started the vogue for "policy gradient methods" that greatly improved reinforcement learning: policy gradient methods optimize policies by gradient descent methods. They were employed, for example, in skill learning by Barto's student Vijaykumar Gullapalli at University of Massachusetts ("Acquiring Robot Skills via Reinforcement Learning", 1994). Later Sham Kakade at University College London ("A Natural Policy Gradient", 2002) applied Shunichi Amari's natural policy gradient to reinforcement learning, an idea improved for high-dimensional movement by Jan Peters and Stefan Schaal at University of Southern California ("Reinforcement Learning for Humanoid Robotics", 2003). Trivia: Amari was one of the pioneers of the stochastic gradient method ("Theory of Adaptive Pattern Classifiers", 1967) and founder of the field of information geometry ("Differential-geometrical Methods in Statistics", 1985).

The first success stories of Reinforcement Learning were in robotics and this was because robots need to discover how to behave in a given environment through trial-and-error interactions. For example: the Obelix robot built by Jonathan Connell and Sridhar Mahadevan at IBM in 1991; the Sarcos humanoid built in 1996 by Stefan Schaal when he was at ATR in Japan; and the autonomous helicopter built by Andrew Bagnell and Jeff Schneider at CMU in 2001.

The first moderate success of reinforcement learning outside robotics came in 1992 when Gerald Tesauro at IBM unveiled a neural network, TD-Gammon, that learned to play better and better at the board game Backgammon ("Programming Backgammon Using Self-teaching Neural Nets", 1992). Gerald Tesauro had already taught a neural network to play backgammon in 1987 when he was at University of Illinois, in collaboration with Terry Sejnowski of Johns Hopkins University, but that used plain supervised learning (against a dataset of 3202 ranked board positions). In 1994 Sejnowski used TD-learning to train a network to play weiqi/go and in 1995 Sebastian Thrun at University of Bonn in Germany used it to train a network to play chess. In 1994 the program Chinook, designed by Jonathan Schaeffer at University of Alberta in Canada, won the world championship of checkers (13 years later Schaeffer would prove mathematically that Chinook is impossible to beat), but it used old-fashioned heuristics (i.e., logic thinking). Trivia: the first program to beat a world champion of backgammon was Hans Berliner's BKG 9.8 in 1979, running on a PDP-10 at CMU and connected by satellite to the robot Gammonoid in Monte Carlo.

For the record, a simpler version of Q-learning is State-Action-Reward-State-Action (SARSA), the subject of Gavin Rummery's dissertation at Cambridge University ("Online Q-learning Using Connectionist Systems", 1994).

Meanwhile in neuroscience "reinforcement" came to denote the effect of the neurotransmitter dopamine in the basal ganglia of the brain, notably in the work of Wolfram Schultz at University of Fribourg in Switzerland ("Reward-related Signals Carried by Dopamine Neurons", 1995), of James Houk at Northwestern University ("A model of how the Basal Ganglia Generates and Uses Neural Signals that Predict Reinforcement", 1995), and of Kenji Doya at the Okinawa Institute of Science and Technology in

Japan ("Temporal Difference Learning in Continuous Time and Space", 1996).

In 1998 Saso Dzeroski in Slovenia and Luc DeRaedt in Belgium wed Q-learning with inductive logic programming and obtained relational reinforcement learning, which recasts the Q-function of Watkins' Q-learning as a first-order regression tree ("Relational Reinforcement Learning", 1998).

At around this time history records the first major successes of evolutionary algorithms. For example, David Fogel and Kumar Chellapilla in San Diego developed Blondie24, an evolutionary algorithm to optimize a neural network: a population of programs that played each other in checkers and evolved safeguarding the best performing ones.

Despite all the progress, reinforcement learning suffers from serious computational problems. For example, the "credit assignment problem": a robot trying to achieve a goal has to make many moves whose reward is zero, and therefore it would not have the motivation to continue.

Don't you wish that reinforcement learning was used more often in society? Think of traffic fines. If a police officer catches you driving over the speed limit, you get fined; but maybe you never violated that law before. The system does not reward you for all the times that you observed the speed limit: it only punishes you for the one time when you were driving faster. It doesn't even matter if you were driving below the speed limit for a long time: the system doesn't count that time, it only counts the few minutes when you were driving faster. There is no reward for good behavior, only punishment for bad behavior.

Humans too learn via reinforcement learning but it is important to remember that humans can also reinvent the game. In 2014 the media publicized the "Pay It Forward Movement" that arose out of nowhere: strangers started paying for the stranger next in line behind them. Famously, at a Starbucks drive-thru one customer after the other kept paying for the next customer for eleven hours. While not clear who really started it, i like this version. A woman was ordering her coffee and the man in the car behind her started shouting and honking because he felt that she was too slow. She responded by telling the cashier that she was going to pay for the man's order. Then she drove away. The impatient man (we would normally say "the jerk") drove up to the window, ordered his coffee and was surprised to hear that he didn't have to pay because the very woman whom he had insulted had chosen to pay for him. Needless to say, this completely changed his psychological state: instead of being furious, he probably felt ashamed; instead of being in a wild egocentric state, only concerned about his own schedule, he felt like caring for the stranger behind him, and decided to pay too for a stranger. That unknown woman had used the exact opposite of reinforcement learning. She had used a clever strategy to alter the game. She had de facto invented a new game: instead of trying to win the old game, she invented a new game at which she won right away (assuming that her goal was to calm down the stranger).

Convolutional Neural Networks - A Brief History of Artificial Intelligence/ Part 5

Another thread in "deep learning" originated with "convolutional neural networks" (CNNs), a kind of hierarchical multilayer network invented in 1979 by Kunihiko Fukushima in Japan. Fukushima's Neocognitron was directly based on the 1958 studies of the cat's visual system by two Harvard neurobiologists, David Hubel, originally from Canada, and Torsten Wiesel, originally from Sweden ("Receptive Fields of Single Neurones in the Cat's Striate Cortex", 1959). They proved that visual perception is the result of successive transformations, or, if you prefer, of propagating activation patterns: the first layer of neurons connected to the retinas detects simple features like edges, while higher layers combine these features to detect more and more complex shapes such as round objects, shapes of faces, etc. They discovered two types of neurons: simple cells, which respond to only one type of visual stimulus and behave like convolutions, and complex cells. Fukushima's system was a multi-stage architecture that mimicked those different kinds of neurons.

The same study by Hubel and Wiesel inspired "scale-invariant feature transform" (SIFT), developed by David Lowe at University of British Columbia, that would remain the most popular algorithm in computer vision for two decades ("Object Recognition from Local Scale-Invariant Features", 1999). Lowe's original paper states: "These features share similar properties with neurons in inferior temporal cortex that are used for object recognition in primate vision."

In 1989 Alex Waibel at CMU pioneered a new kind of neural network, the "time-delay" neural network ("Phoneme Recognition Using Time-Delay Neural Networks", 1989). He was working on speech recognition, i.e. on classifying phonemes, and speech signals tend to be continuous, i.e. it is not clear where a phoneme begins and ends. Time-delay neural networks introduced delays in the activation function and organized the layers around clusters, each cluster focused only on small regions of the input. His team developed one of the first multi-lingual speech-to-speech translation systems, named Janus, in collaboration with Japan's ATR and

Germany's Siemens ("A Speech-to-speech Translation System using Connectionist and Symbolic Processing Strategies", 1991).

Despite all the progress, multilayer neural nets still could not compete with traditional learning approaches such as SVMs. The first major success of neural networks came in 1989 when Yann LeCun, a former Hinton assistant at Toronto University but now at Bell Labs, applied backpropagation to convolutional networks to solve the problem of recognizing handwritten numbers ("Handwritten Digit Recognition with a Back-Propagation Network", 1989) and obtained his first convolutional neural network, later nicknamed LeNet-1, which evolved into LeNet-4 (with the fellow Frenchmen Leon Bottou and Yoshua Bengio) when in 1993 the National Institute of Standards and Technology released its dataset of 60,000 handwritten digits. Within a few weeks the French Canadian Patrice Simard (also at Bell Labs) engineered a booster version. Convolutional networks, influenced by time-delay networks, were the first success story of deep learning. In the next few years, LeCun's team applied them to face detection ("Original Approach for the Localisation of Objects in Images, ", 1994) and then to reading cheques ("Gradient-based Learning Applied to Document Recognition", 1998), the latter being nicknamed LeNet-5.

By this time the architecture had stabilized in a sequence of convolutions and "pooling layers" (a more general kind of activation function), inspired by Fukushima's Neocognitron.

Actually, the first success story in processing bank cheques were the "graph transformer networks" developed by Leon Bottou working with Bengio and LeCun ("Document Analysis with Transducers", 1996).

The seven-layer LeNet-5 represented a major improvement in computational efficiency (at a time when GPUs didn't help yet) and its architecture would remain a reference model for a decade. This architecture consisted of three components: a convolution to extract features from the image, a pooling stage to reduce the size of the representation, and a non-linearity stage in the form of either a sigmoidal activation function or a hyperbolic-tangent (or "tanh") activation function (instead of the Perceptron's step function). Mathematically speaking, the convolution is a linear operation, and some nonlinear function must be introduced to make the neural network work. Note that the feature detectors detected the presence of a feature but ignored its location inside the image. Therefore the location of a feature did not affect the classification. In other words, two eyes, a nose and a mouth would be recognized as a face even if the mouth was placed between the eyes and the nose.

The problem with LeCun's network was that Werbos-style backpropagation took almost three days to train the network for such a simple application. Clearly, this approach would not work for more complex recognition tasks.

Other neural networks were used to detect faces in 1994 by Kah-Kay Sung and Tomaso Poggio at MIT ("Example-based Learning for View-based Human Face Detection", 1994) and in 1996 by Takeo Kanade, now at CMU ("Neural Network-Based Face Detection", 1996). Detecting faces is a significantly more difficult task than recognizing faces: a face can be hidden in the middle of a very messy scene. Once you know it is a face, then it is relatively easier to find which person has the most similar face. Until 1991, deep convolutional networks were used for recognizing isolated two-dimensional hand-written digits. Progress in recognizing three-dimensional objects had to wait until Juyang Weng's team at Michigan State University developed the Cresceptron, that used an improved technique called "max-pooling" ("Cresceptron, a Self-organizing Neural Network which Grows Adaptively", 1992).

Convolutional neural networks are not recurrent: they are feed-forward networks.

A convolutional neural network consists of several convolutional layers. Each convolution layer consists of a convolution or filtering stage (the "simple cell"), a detection stage, and a pooling stage (the "complex cell"), and the result of each convolutional layer is in the form of "feature maps", and that is the input to the next convolutional layer. The last layer is a classification module.

The detection stage of each convolutional layer is the middleman between simple cells and complex cells and provides the nonlinearity of the traditional multi-layer neural network. Traditionally, this nonlinearity was provided by a mathematical function called "sigmoidal", but in 2011 Yoshua Bengio ("Deep Sparse Rectifier Networks") introduced a more efficient function, the "rectified linear unit" (ReLu), also inspired by the brain, that have the further advantage of avoiding the "gradient vanishing" problem of sigmoidal units.

Every layer of a convolutional network detects a set of features, starting with large features and moving on to smaller and smaller features. Imagine a group of friends subjected by you to a simple game. You show a picture to one of them, and allow him to provide a short description of the picture to only another one and using only a very vague vocabulary; for example: an object with four limbs and two colors. This new person can then summarize that description in a more precise vocabulary to the next person; for example a four-legged animal with black and white stripes. Each person is allowed to use a more and more specific vocabulary to the next person. Eventually, the last person can only utter names of objects, and hopefully correctly identifies the picture because, by the time it reaches this last person, the description has become fairly clear (e.g. the mammal whose skin is black and white, i.e. the zebra).

(Convolution is a well-defined mathematical operation that, given two functions, generates a third one, according to a simple formula. This is useful when the new function is an approximation of the first one, but easier to analyze. You can find many websites that provide "simple" explanations of what a convolution is and why we need them: these "simple" explanations are a few pages long, and virtually nobody understands them, and each of them is completely different from the other one. Now you know where the term "convoluted" comes from!)

Footnote (or Part 5.2): Recurrent Neural Networks

Neural networks, up to this point, were good at recognizing patterns (e.g., that this particular object is an apple) but not for learning events in time. Neural networks, in principle, have no sense of time. Recurrent neural networks (and, in general, connectionist models) have a rudimentary sense of persistence, of "memory". The question soon arose of how to represent time in connectionist models, in particular related to natural language processing. Traditional approaches to representing time basically used a spatial metaphor for time: create a representation in space of the sequence of events. The difficulty in using neural networks for analyzing language is that the connectionist representation of a neural network has none of the compositionality of syntax. However, while the connectionist representation is not syntactically compositional, it can still be functionally compositional. Michael Jordan extended recurrent neural networks to analyses of sequences by feeding the output into a "context" layer which is then fed into the middle layer; i.e. the network is fed with the input of a step and with the output of the previous step ("Attractor Dynamics and Parallelism in a Connectionist Sequential Machine", 1986). Following in his footsteps, Jeffrey Elman at UC San Diego invented what he called a "simple recurrent neural network" or SRNN, which is actually another sophisticated model for processing sequences and not just patterns ("Finding Structure in Time", 1990). If recurrent neural networks can operate on sequences, it means that they can model relationships in time, something crucial for processing natural language.

Hava Siegelmann of Bar-Ilan University in Israel and Eduardo Sontag of Rutgers University proved an encouraging theorem: that recurrent neural networks can implement Turing machines if the weights of the connections are rational numbers; and that recurrent neural networks are even more powerful than a universal Turing machine if the weights are real numbers, i.e. if the network is analogic ("On the Computational Power of Neural Networks", written in 1992 but published only in 1995). Kenichi Funahashi and Yuichi Nakamura at Toyohashi University of Technology in Japan proved that recurrent neural networks are universal approximators, i.e. they can approximate any dynamical system ("Approximation of Dynamical Systems by Continuous Time Recurrent Neural Networks", 1993). Christian Omlin and Lee Giles of NEC Research Institute in Princeton proved that recurrent neural networks can approximate arbitrary finite state machines ("Constructing Deterministic Finite-State Automata In Sparse Recurrent Neural Networks", 1994).

Typical applications of RNNs are: image captioning, that turns an image into a sequence of words ("sequence output"); sentence classification, that turns a sequence of words into a category ("sequence input"); and sentence translation (sequence input and sequence output). The innovation in RNNs is a hidden layer that connects two points in time. In the traditional feed-forward structure, each layer of a neural network feeds into the next layer. In RNNs there is a hidden layer that feeds not only into the next layer but also into itself at the next time step. This recursion or cycle adds a model of time to traditional backpropagation, and is therefore known as "backpropagation through time".

Neural networks were being designed to solve one specific problem at a time, but in 1990 Robert Jacobs at University of Massachusetts introduced the "mixture-of-experts" architecture that trains different neural networks simultaneously and lets them compete in the task of learning, with the result that different networks end up learning different functions ("Task Decomposition Through Competition in a Modular Connectionist Architecture", 1990). Similarly, Tom Mitchell's student Rich Caruana at CMU showed that multiple neural networks learning tasks in parallel could benefit from sharing what they learned: it might be easier to learn several tasks at the same time than to learn them separately ("Multitask Learning, a Knowledge-based Source of Inductive Bias", 1993). This would also be shown by Ronan Collobert's and Jason Weston's unified architecture for natural language processing of 2008.

Another elusive goal of machine learning was to equip machines with "transfer learning", i.e. the ability to transfer what they learn on a task to another task, and therefore achieve "multitask learning", the ability to learn more than just one task. In 1991 Satinder Singh of University of Massachusetts published "Transfer of Learning by Composing Solutions of Elemental Sequential Tasks" and Lorien Pratt of the Colorado School of Mines published "Direct Transfer Of Learned Information Among Neural Networks". By then, Tom Mitchell's group at CMU had become the world's center of excellence in transfer learning, as documented by his German student Sebastian Thrun's "Learning One More Thing" (1994), which introduced the idea of knowledge transfer in order to learn multiple tasks. But not much has improved since Sebastian Thrun and Lorien Pratt curated the book "Learning to Learn" (1998).

Thrun's "lifelong learning" (learning multiple consecutive tasks) would remain a challenge for neural networks. A few years before Thrun's manifesto, Michael McCloskey and Neal Cohen at Johns Hopkins University had shown that the process of learning a new task abruptly and completely erases what a neural network has learned before, a phenomenon that came to be known as "catastrophic forgetting" ("Catastrophic Interference in Connectionist Networks", 1989).

Jonathan Baxter at the Australian National University extended Valiant's PAC model of learning to the case of multiple related learning tasks, a case in which the learner is provided with a family of hypothesis spaces instead of just one hypothesis space ("A Model of Inductive Bias Learning", 2000). This was an interesting mathematical theory but it had to wait one generation to yield practical results.

A general problem of neural networks with many layers ("deep" neural networks), and of RNNs in particular, is the "vanishing gradient", already described in 1991 by Josef "Sepp" Hochreiter at the Technical University of Munich ("Investigations on Dynamic Neural Networks", 1991) and more famously in 1994 by Yoshua Bengio ("Learning Long-Term Dependencies with Gradient Descent is Difficult"). The expression "vanishing gradient" refers to the fact that the computations for each new layer become less and less clear. It is a problem similar to calculating the probability of a chain of events: if you multiply a probability between 0 and 1 by another probability between 0 and 1 and so on several times in a row, the result is always zero, even in the case in which all those numbers expressed probabilities of 99%. A network with many layers is difficult to train because the "weights" of the last layer end up being too weak.

New unsupervised learning algorithms included imax by Geoffrey Hinton's student Suzanna Becker at University of Toronto ("A self-organising neural network that discovers surfaces in random dot stereograms", 1992) and Geoffrey Hinton's "wake-sleep" algorithm ("The Wake-sleep Algorithm for Unsupervised Neural Networks", 1995). Ralph Linsker's infomax was further simplified by Anthony Bell in Terrence Sejnowski's lab at the Salk Institute which improved infomax ("A Non-Linear Information Maximization Algorithm that Performs Blind Separation", 1995). Then Shunichi Amari at Riken in Japan realized that the infomax algorithm could be improved by using the so-called "natural gradient", a gradient that considers the old and the new states of a neural network as two distributions of probabilities, and then uses the so-called Kullback-Leibler divergence to measure the distance between these two distributions ("Neural Learning in Structured Parameter Spaces", 1996, later expanded in "Natural Gradient Works Efficiently in Learning", 1998). This was the rare application of differential geometry to statistics because it deals with Riemannian geometry instead of Euclidean geometry, just like Albert Einstein's theory of general relativity.

Meanwhile in 1996 David Field and Bruno Olshausen at Cornell University had invented "sparse coding", an unsupervised technique for neural networks to learn the patterns inherent in a dataset. Sparse coding helps neural networks represent data in an efficient way that can be used by other neural networks. Autoencoders have a propensity to diverge but sparse autoencoders fixed that problem. The idea came from the very way in which the primary visual cortex works. When a stimulus (e.g. a sound or an image) causes the activation of only a small number of neurons, this pattern of activation represents a "sparse coding" of the stimulus, which turns out to be an efficient way to represent stimuli: minimal computation and still the ability to reconstruct the input image. Autoencoders trained with a sparsity constraint exhibit similar benefits: they are simple and allow stacking multiple layers. Their vogue began with successful implementations by Andrew Ng's group at Stanford ("Self-taught Learning - Transfer Learning from Unlabeled Data", 2007) and by LeCun's group at New York University ("Fast Inference in Sparse Coding Algorithms with Applications to Object Recognition", 2008).

Some explored methods that were "semi-supervised", such as the "self-training" method devised by David Yarowsky at University of Pennsylvania ("Unsupervised Word Sense Disambiguation Rivaling Supervised Methods", 1995), which incorporated a model's own predictions into training (but, of course, the model is then unable to correct its own mistakes), Avrim Blum's and Tom Mitchell's "co-training" at CMU ("Combining labeled and unlabeled data with co-training", 1998), and later Zhi-Hua Zhou's "tri-training" at Nanjing University, which averages three independently trained models ("Exploiting Unlabled Data Using Three Classifiers", 2005).

Machine learning was doing just about fine without any need to model the time dimension: support vector machines, logistic regression, and traditional as well as convolutional neural networks can recognize patterns. There are, however, applications that require a sequential analysis, i.e. that need to model the time dimension. Examples are speech recognition, image captioning, language translation, and handwriting recognition. Hidden Markov models have been used to model time but they become computationally impractical for modeling long-range dependencies (when you have to remember not just a few earlier steps but many earlier steps).

In 1997 Sepp Hochreiter and his professor Juergen Schmidhuber (by now director of the Istituto Dalle Molle di Studi sull'Intelligenza Artificiale, or IDSIA, in Switzerland) came up with a solution: the Long Short Term Memory (LSTM) model. In this model, the unit of the neural network (the "neuron") is replaced by one or more memory cells. Each cell functions like a mini-Turing machine, performing simple operations of read, write, store and erase that are triggered by simple events. The big difference with Turing machines is that these are not binary decisions but "analog" decisions, represented by real numbers between 0 and 1, not just 0 and 1. For example, if the network is analyzing a text, a unit can store the information contained in a paragraph and apply this information to a subsequent paragraph. The reasoning behind the LSTM model is that a recurrent neural network contains two kinds of memory: there is a short-term memory about recent activity and there is a long-term memory which is the traditional "weights" of the connections that change based on this recent activity. The weights change very slowly as the network is being trained. The LSTM model tries to retain also information contained in the recent activity, information that traditional networks only use to fine-tune the weights and then discard.

LSTM recurrent neural nets represented a significant quantum leap in neural networks because they capture temporal dependencies, such as those necessary to understand video, text, speech and movement.

LSTMs were later improved for tasks such as handwriting recognition and speech recognition by Schmidhuber's student Alex Graves, who developed "connectionist temporal classification" or CTC ("Connectionist Temporal Classification", 2006).

At this point one is tempted to rewrite the history of neural networks as rediscovering the elements of programming languages. First there was the simple McCulloch-Pitts neuron, a computational unit that produces a sum of products, coupled with an activation function (such as the sigmoidal) that plays the role of the "IF" statement of programming languages. Hopfield's recurrent networks introduced the equivalent of the "loop". Finally, LSTMs added the equivalent of the variable of a programming language, a way to store a value.

LeCun's convolutional nets had solved the problem of how to train deep feedforward neural nets: Long Short Term Memory solved the problem of how to train deep recurrent neural nets.

At the same time that LSTM were introduced, in 1997 Mike Schuster and Kuldip Paliwal at the Advanced Telecommunications Research

Institute in Japan discovered that "bidirectional recurrent neural networks" can achieve a similar feat of persistence.

In the 2010s several other kinds of neural networks coupled with some long-term memories, similar to LSTMs, have been proposed: "Neural Turing Machines" (2014, by Alex Graves at Google's DeepMind in Britain) and "Memory Networks" (2014, by James Weston at Facebook's labs in New York). Just like LSTMs, these are neural networks that can perform complex inference. This seems to be a progressive shift towards a form of hybrid computing that may reconcile neural networks and the old knowledge-based school.

Lukasz Kaiser and Ilya Sutskever at Google introduced Neural GPUS, an improvement of Alex Graves' Neural Turing Machines, to synthesize an algorithm from its input-output examples ("Neural GPUs Learn Algorithms", 2015).

Footnote: Probability in the age of Neural Networks

The fundamental limitation of neural networks is that they need to be "trained" with thousands if not millions of examples before they can "learn". A child can usually learn a new concept from a single example, creating a generalization that she will be able to apply to similar objects or situations ("one-shot generalization").

Bayesian thinking interprets knowledge as a set of probabilistic (not certain) statements and interprets learning as a process to refine those probabilities. As we acquire more evidence, we refine our beliefs. In 1996 the developmental psychologist Jenny Saffran showed that babies use probability theory to learn about the world, and they do learn very quickly a lot of facts. So Bayes in the 18th century may have stumbled onto an important fact about the way the brain works, not just a cute mathematical theory.

Probabilistic induction was one of the very first proposals for artificial intelligence, notably Solomonoff's "Inductive Inference Machine" of 1956; and Judea Pearl's Bayesian reasoning ("Reverend Bayes on Inference Engines", 1982) provided the toolbox for probabilistic computation. Their motivation was the same as that which had originally motivated Fermi and Ulam: probabilistic reasoning is needed when the exact algorithm is too complicated and would result in unacceptable response time. In these cases it is preferrable to find an approximate (but quick) solution. In between, a generalization of subjective probability, the "theory of evidence", was developed by Arthur Dempster at Harvard University ("Upper and Lower Probabilities Induced by a Multivalued Mapping", 1967) and Glenn Shafer at Princeton University, who published the book "A Mathematical Theory of Evidence" (1976).

Morris DeGroot of CMU published "Optimal Statistical Decisions" in 1970, but the book that revived probabilistic reasoning was the English translation of DeFinetti's book "Theory of Probability" (1970), many years after his 1928 talk.

Meanwhile, in 1970 Keith Hastings at University of Toronto generalized the Metropolis algorithm ("Monte Carlo Sampling Methods using Markov Chains and their Applications", 1970).

Another popular MCMC algorithm, "Gibbs sampling", was developed in 1984 by the brothers Stuart and Donald Geman (respectively at Brown University and University of Massachusetts) who were working on computer vision ("Stochastic Relaxation, Gibbs Distributions, and the Bayesian Restoration of Images", 1984); Judea Pearl soon introduced Gibbs sampling in his Bayesian networks, also known as "belief networks" and "causal networks" ("Evidential Reasoning using Stochastic Simulation", 1987).

MCMC became especially popular after Adrian Smith at University of Nottingham, in collaboration with Alan Gelfand, showed how these algorithms can be used in a wide variety of cases ("Sampling-based Approaches to Calculating Marginal Densities", 1990). Just then the statistical software BUGS (Bayesian inference Using Gibbs Sampling) was becoming available for Bayesian inference using Markov chain Monte Carlo (MCMC): it had been developed since 1989 at University of Cambridge by Andrew Thomas, working under David Spiegelhalter. In 1993 the International Society for Bayesian Analysis met in San Francisco for its first conference. Finally, Radford Neal of University of Toronto published a report titled "Probabilistic Inference Using Markov Chain Monte Carlo Methods" (1993) that said it all, and in 1996 published the book "Bayesian Learning for Neural Networks".

Meanwhile, in 1983 Geoffrey Hinton at CMU and Terry Sejnowski at Johns Hopkins University had introduced the Boltzmann machine that was still evolving. Because the computational cost of the undirected models is so high, the Boltzmann machine used an approximation method that was de facto already Gibbs sampling. In 1992 Radford Neal at University of Toronto added "direction" to the connections (that were originally undirected, i.e. symmetric) of the Boltzmann machine in order to improve the training process. Pearl had described belief nets to represent expert knowledge (knowledge elicited from humans). Neal showed that belief nets can learn by themselves. In 1995 Hinton and Neal worked with the British neuroscientist Peter Dayan, who was keen on finding a similarity with Hermann von Helmholtz's theory of perception, and designed the

Helmholtz Machine ("The Helmholtz Machine", 1995) for which they invented the "wake-sleep" algorithm for unsupervised learning ("The Wake-sleep Algorithm for Unsupervised Neural Networks", 1995). The wake-sleep algorithm, conceived purely from neuroscience, trains top-down and bottom-up probabilistic models (i.e. generative and inference models) against each other in a multilayer network of stochastic neurons. It is a form of "variational Bayesian learning": it approximates Bayesian inference when it becomes intractable, as is typically the case in multilayer networks.

Meanwhile, the Swedish statistician Ulf Grenander (who in 1972 had established the Brown University Pattern Theory Group) fostered a conceptual revolution in the way a computer should describe knowledge of the world: not as concepts but as patterns. His "general pattern theory" provided mathematical tools for identifying the hidden variables of a data set. Grenander's pupil David Mumford studied the visual cortex and came up with a hierarchy of modules in which inference is Bayesian, and it is propagated both up and down ("On The Computational Architecture Of The Neocortex II", 1992). The assumption was that feedforward/feedback loops in the visual region integrate top-down expectations and bottom-up observations via probabilistic inference. Basically, Mumford applied hierarchical Bayesian inference to model how the brain works.

Hinton's Helmholtz machine of 1995 was de facto an implementation of those ideas: an unsupervised learning algorithm to discover the hidden structure of a set of data based on Mumford's and Grenander's ideas.

Another important idea originated in non-equilibrium thermodynamics (thermodynamics was that studies irreversible processes). Christopher Jarzynski at the Los Alamos National Laboratory used a Markov chain to gradually convert one distribution into another via a sequence of intermediate distributions ("Nonequilibrium Equality for Free Energy Differences", 1997), an idea that Radford Neal at University of Toronto turned into yet another annealing method ("Annealed Importance Sampling", 2001).

Footnote: What Didn't Use Neural Networks

Despite the theoretical progress in neural networks, the success stories of A.I. of those years did not use neural networks. In 1991 Thaddeus Beier of Silicon Graphics and Shawn Neely of Pacific Data Images invented a method to generate fake images (used in Michael Jackson's video "Black and White"). In 1994 Ernst Dickmanns' self-driving car drove more than 1,000 kms near the airport Charles-de-Gaulle in Paris. In May 1997 the IBM supercomputer "Deep Blue", programmed by Feng-hsiung Hsu (who had started building chess-playing programs in 1985 while at CMU) beat the world's chess champion, Garry Kasparov. In 1998 Hector Garcia-Molina's student Sergey Brin and Terry Winograd's student Larry Page at Stanford published the paper "The Anatomy of a Large-scale Hypertextual Web Search Engine" to document the search engine (and in particular their novel PageRank algorithm) developed for the digital-libraries project of the university, a search engine named "Google". And you can continue all the way to 2011, when IBM's Watson debuted on a TV show and Apple released Siri, without encountering neural networks in the headlines.

Intermezzo: The GPU and the Spring of A.I.

Neural networks had fallen out of favor again in the 1990s. In fact, in 2000 the organizers of the Neural Information Processing System (NIPS) conference discouraged the submission of papers on neural networks while encouraging instead submission of papers on SVMs.

A.I. was rescued by videogames. Videogames require very fast analysis of images. An image is made of millions of pixels. Analyzing one pixel at a time is not a good way to analyze an image quickly. Therefore the industry invented the GPU, the graphical processing unit, that analyzes all the pixels in parallel. It turns out that neural networks work in a similar way: the nodes of a layer behave like the pixels of an image, and must ideally be processed in parallel. Technically speaking, the layer of a neural network is a matrix and the training of a neural network consists in a series of matrix multiplications. In 2001 David McAllister's student Scott Larsen at University of North Carolina implemented matrix multiplication on a GPU. They turned the GPU, originally invented for rendering images, into a fast numeric calculator, i.e. a scientific instrument. In 2005 Patrice Simard's team at Microsoft pioneered the use of GPUs, in particular, for machine learning. In 2007 Nvidia, the most famous GPU manufacturer, released the high-level programming language CUDA (which originally meant Compute Unified Device Architecture), an extension to the C programming language, to help programmers implement matrix multiplications in their GPUs. It was perfect timing because Hinton and Bengio had just figured out the math to train "deep" neural networks. In 2009 Hinton's students Abdelrahman Mohamed and George Dahl at University of Toronto ("Deep Belief Networks Using Discriminative Features for Phone Recognition", 2011) seemed to show that the so-called "deep belief networks" were better than hidden Markov models at speech recognition but in reality it showed that amazing results could be achieved using GPUs (in their case an Nvidia Tesla S1070). In 2010 Dan Ciresan of

Schmidhuber's team at IDSIA built a nine-layer neural net on an Nvidia GTX 280 graphic processor.

The GPU made multi-layered neural networks possible. One could argue that A.I. in the 2010s succeeded in things such as speech/image recognition and game-playing that A.I. was already doing well in 30 years earlier, except that the performance was ridiculous; and therefore the key moment in the development of A.I. was the idea of using GPUs to accelerate the training of neural networks, and then exploit the full power of multi-layer ("deep") neural networks. Alas, the limitations of 30 years earlier remained: A.I. systems lacked common sense and learned in unnatural ways.

Deep Learning: A Brief History of Artificial Intelligence/ Part 6

The field of neural networks did not take off until 2006, when Yee-Whye Teh at the National University of Singapore and Geoffrey Hinton at University of Toronto developed deep belief networks (DBNs), a fast learning algorithm for Restricted Boltzmann Machines ("A Fast Learning Algorithm for Deep Belief Nets" and "Reducing the Dimensionality of Data with Neural Networks", both 2006). Until then scientists had failed to find an effective way to train "deep" architectures (multilayer neural networks). Deep belief networks can be pre-trained in an unsupervised way, one layer at a time. Hinton's algorithms worked wonders when used on thousands of parallel processors. That's when the media started publicizing all sorts of machine-learning feats.

Deep belief networks are layered hierarchical architectures that stack Restricted Boltzmann Machines (RBMs) one on top of the other, each one feeding its output as input to the one immediately higher, with the two top layers forming an associative memory. The features discovered by one RBM become the training data for the next one. De facto, Hinton's deep belief networks are probabilistic models that consist of multiple layers of probabilistic reasoning.

Technically speaking, Hinton's original "deep belief networks" were really "sigmoid belief networks" and only in 2009 did Hinton shift from sigmoid networks to deep Boltzmann machines

Hinton had discovered how to create neural networks with many layers. One layer learns something and passes it on to the next one, which uses that something to learn something else and passes it on to the next layer, etc.

Before 2006 deep networks performed worse than shallow networks.

DBNs are still limited in one respect: they are "static classifiers", i.e. they operate at a fixed dimensionality. However, speech or images don't come in a fixed dimensionality, but in a (wildly) variable one. They require "sequence recognition", i.e. dynamic classifiers, that DBNs cannot provide. One method to expand DBNs to sequential patterns is to combine deep learning with a "shallow learning architecture" like the Hidden Markov Model.

Hinton and Teh had solved the fundamental problem of multilayer networks: supervised training of deep networks is extremely difficult. Hinton and Teh figured out that it was efficient to start with unsupervised learning in which the network learns some general features. This can be repeated as many times (layers) as desired, using the output of an unsupervised learning as the input to the next one. Supervised learning takes place only on the last layer, without affecting the previous layers. Note that Hinton's work was inspired by the experiments on cats by the young Colin Blakemore at UC Berkeley. These experiments revealed that the visual cortex of the cat develops in stages, one layer at a time, as the cat is exposed to the world ("The Representation of Three-dimensional Visual Space in the Cat's Striate Cortex", 1970).

Hinton's and Teh's 2006 method for Deep Belief Nets came to be known as "greedy layer-wise unsupervised pre-training". Greedy algorithms are those that at every step choose the move that seems to be the most promising. Many greedy methods have been developed for classic optimization problems. Hinton and Teh had invented one to optimize the pre-training of multilayer neural networks.

A few months later Yoshua Bengio at University of Montreal introduced another method, the stacked autoencoder ("Greedy Layer-Wise Training of Deep Networks", 2007). Both techniques follow the same general procedure: they train each layer of the neural network via an unsupervised learning algorithm, taking the features produced at a level as input for the next level. The neural network learns a hierarchy of features one level at a time, and eventually it is fine-tuned in supervised fashion (backpropagation and gradient-based optimization), i.e. the last layer is a classifier.

Every neural network is, ultimately, a combination of "encoder" and "decoder": the first layers encode the input and the last layers decode it. For example, when my brain recognizes an object as an apple, it has first encoded the image into some kind of neural activity (representing shape, color, size, etc of the object), and has then decoded that neural activity as an apple.

Whereas Hinton pre-trained the network using restricted Boltzmann machines, Bengio showed that the learning algorithm of the autoencoder could also be employed as well for greedy layer-wise unsupervised pre-training. Instead of a sequence of restricted Boltzmann machines, Bengio used a sequence of autoencoders.

A stacked autoencoder is a neural network consisting of multiple layers of (sparse) autoencoders in which the outputs of each autoencoder is wired to the inputs of the successive autoencoder. Each autoencoder is designed to learn a representation of the input. Eventually, these representations are fed into a neural network in supervised learning. Therefore, just like the restricted Boltzmann machines in Hinton's Deep Belief Nets, a stacked autoencoder learns something about the distribution of data and pre-trains the neural network that has to operate on those data. The outputs of the pre-training networks (whether stacked autoencoders or restricted Boltzmann machines) are finally passed to the real classifier (such as SVM or logistic regression). Hinton and Teh used a stochastic approach. Bengio used a deterministic approach (a restricted Boltzmann machine can be viewed as a sort of probabilistic autoencoder). More importantly, autoencoders are easier to understand, implement and tweak. The difference with convolutional neural networks is that Hinton's and Bengio's methods work with global representations of the input, whereas LeCun's method focuses on spatially-close units (which is why convolutional neural networks are mostly used in image recognition). Autoencoders are much more general.

Alas, stacked autoencoders appeared to be not quite as accurate as deep belief networks.

Pascal Vincent of Bengio's group at Montreal later introduced the "denoising autoencoder" ("Extracting and Composing Robust Features with Denoising Autoencoders, 2008), a simple autoencoder with noise added to the training data, and he proved that such a denoising autoencoder works better at some tasks. Now stacked (de-noising) autoencoders were finally competitive with deep belief networks.

In other words: to make the reconstruction more challenging, one can use sparsity or noise. The denoising autoencoder applies noise to the input image before trying to reconstruct the clean input. The point of making the reconstruction more challenging is that the autoencoder tends to learn a more valuable representation of the data.

A third kind of step-wise pre-training of a neural network was developed by Jason Weston at NEC Labs in Princeton ("Deep Learning via Semi-supervised Embedding", 2008).

Meanwhile, Leon Bottou, now at NEC Labs in Princeton, worked on machine learning with large-scale datasets, and made "stochastic gradient descent" the optimization method of choice ("Large-Scale Machine Learning with Stochastic Gradient Descent", 2007). The stochastic gradient descent method is a drastic simplification of Rumelhart's gradient descent algorithm. The "gradient descent" method ruled until 2010 when James Martens at University of Toronto adapted a powerful optimization method, "Hessian-free optimization" (named after the 19th century German mathematician Ludwig Hesse who invented – Incomplete sentence), to neural networks ("Deep Learning via Hessian-free Optimization", 2010). Mathematicians realized that chances of success were greatly enhanced for training even very large and deep neural networks as long as they could use a lot of data. Meanwhile, in 2011 Michael Jordan's student John Duchi at UC Berkeley invented AdaGrad, a method to accelerate the learning process of the stochastic gradient descent algorithm, which would remain for a while the most popular of these "accelerators", replacing Nesterov's momentum. Its main competitors were Matt Zeiler's Adadelta ("An Adaptive Learning Rate Method", 2012), and later Adam by Diederik Kingma of University of Amsterdam and Jimmy Ba of University of Toronto ("A Method for Stochastic Optimization", 2015).

Later studies, for example by Andrew Ng's group at Stanford ("Rectifier Nonlinearities Improve Neural Network Acoustic Models", 2013) showed that rectified linear units were much important to the success of deep learning than the unsupervised layers of Hinton's and Bengio's original architectures. This generation showed that, after all, the backpropagation of the 1990s did work. It was just a matter of using large training datasets, powerful GPUs and the right kind of non-linearity.

Therefore, many scientists contributed to the "invention" of deep learning and to the resurrection of neural networks. But the fundamental contribution came from Moore's Law: between the 1980s and 2006 computers had become enormously faster, cheaper and smaller. A.I. scientists were able to implement neural networks that were hundreds of times more complex, and able to train them with millions of data. This was still unthinkable in the 1980s. Therefore what truly happened between 1986 (when Restricted Boltzmann machines were invented) and 2006 (when deep learning matured) that shifted the balance from the logical approach to the connectionist approach in A.I. was Moore's Law. Without massive improvements in the speed and cost of computers deep learning would not have happened. Deep learning owes a huge debt of gratitude to the supercharged GPUs (Graphical Processing Units) that have become affordable in the 2010s.

Credit for the rapid progress in convolutional networks goes mainly to mathematicians, who were working on techniques for matrix-matrix multiplication and made their systems available as open-source software. The software, such as UC Berkeley's Caffe, used by neural-network designers, reduces a convolution to a matrix-matrix multiplication. This is a problem of linear algebra for which seasoned mathematicians had provided solutions. Initial progress took place at the Jet Propulsion Laboratory (JPL), a research center in California operated by the California Institute for Technology (CalTech) for the space agency NASA. Charles Lawson was the head of Applied Math Group at JPL since 1965. Lawson and his employee Richard Hanson developed software for linear algebra, including software for matrix computation, which was to be applied to astronomical things like gravitational fields. In 1979, together with Fred Krogh, an expert in differential equations, they released a Fortran library called Basic Linear Algebra Subprograms (BLAS). By 1990 BLAS 3 incorporated a library for matrix-matrix operations called General Matrix to Matrix Multiplication (GEMM), largely developed at Britain's Numerical Algorithms Group (instituted in 1970 as a joint project between several British universities and the Atlas Computer Laboratory). The computational "cost" of a neural network is mainly due to two kinds of layers: the layers that are fully-connected to each other and the convolutions. Both kinds entail massive multiplications of matrices; literally millions of them in the case of image recognition. Without something like GEMM no array of GPUs could perform the task.

A landmark achievement of deep-learning neural networks was published in 2012 by Alex Krizhevsky and Ilya Sutskever from Hinton's group at University of Toronto: Krizhevsky wrote an extremely fast implementation of two-dimensional convolutions on a GPU (a network soon nicknamed AlexNet) which demonstrated that deep learning (using a convolutional neural network with five convolutional layers and Bengio's rectified linear unit) outperforms traditional techniques of computer vision after processing 200 billion images during training (1.2 million human-tagged images from ImageNet plus thousands of computer-generated variants of each). Deep convolutional neural networks became the de facto standard for computer-vision systems.

At the same time Google had decided to invest into neural networks and in 2011 had funded a research team led by Stanford professor Andrew Ng (real name Enda Wu) that came to be known as the "Google Brain" group.

In June 2012 Andrew Ng's group at Stanford (mainly Quoc Le and Marc'Aurelio Ranzato) demonstrated an unsupervised neural network (a nine-layer network in which three layers were autoencoders) that recognized cats in still frames of videos (not quite videos, but 10 million unlabeled images taken from videos): nobody had trained the network, i.e. nobody had told the network that there are cats in the world and how they look like ("Building High-level Features Using Large Scale Unsupervised Learning", 2012).

Why did it take until 2012? Before the annus mirabilis of 2012 multilayer neural network had to learn too many parameters from too few labeled examples. In other words, on one hand the computers were not powerful enough and on the other hand the training datasets were not large enough. In 2012 computers became powerful enough (and Google was willing to pay for them) just when datasets had become large enough.

In 2013 Dan Ciresan of Schmidhuber's team at IDSIA achieved near-human performance in recognizing Chinese handwriting, and in 2015 Jun Sun's team at Fujitsu's Chinese laboratories surpassed human performance.

In 2013 Google hired Hinton and Facebook hired LeCun; and in 2015 Yoshua Bengio joined Element AI, a Canadian startup founded by two veterans of the A.I. business, Nicolas Chapados and Francois Gagne, to provide customized A.I. services to businesses.

Krizhevsky's AlexNet, basically a deeper LeNet-style convolutional net, consisted of five convolutional layers followed by three fully-connected layers. It made use of Bengio's rectified linear units for the non-linear part, instead of a tanh or sigmoidal activation function. It was trained on Nvidia's GTX 580. In December 2013 Yann LeCun's laboratory at New York University (Pierre Sermanet, Rob Fergus and others) announced an AlexNet improvement called Overfeat.

So far deep learning had been used to classify images. The next step was to detect objects inside images. Until then the prevailing methods had been David Lowe's SIFT and Gabriela Csurkas' Bag-of-Words. In 2013 Jitendra Malik's student Ross Girshick at UC Berkeley introduced Region-based Convolutional Neural Networks (or R-CNN) whose "supervised pre-training domain-specific finetuning" paradigm quickly became the new standard for object detection ("Rich Feature Hierarchies for Accurate Object Detection and Semantic Segmentation", 2013).

Oxford's 16-layer VGG-16, developed in 2013 by Andrew Zisserman's student Karen Simonyan at Oxford University's Visual Geometry Group (VGG), winner in 2014 of the ImageNet challenge, showed that multiple 3×3 convolutions in sequence could efficiently emulate the larger convolutions of AlexNet ("Very Deep Convolutional Networks for Large-Scale Image Recognition", 2014). The advantage was not only accuracy but also smaller GPUs.

Meanwhile, Shuicheng Yan's team at the National University of Singapore was generalizing the linear convolution operation of the convolutional network with a general nonlinear function approximator (i.e., with a neural network itself, e.g. with a multilayer perceptron) while simplifying the architecture (and reducing the number of parameters) and thus obtaining the architecture called "Network in Network" or NiN ("Network In Network", 2014).

Later in 2014 Christian Szegedy at Google figured out a way to further optimize VGG-16 via a module inspired by the Network in Network model and called "inception module", thereby obtaining what was nicknamed the GoogLeNet or simply Inception ("Going Deeper with Convolutions", 2014). To save memory and computational costs, GoogLeNet replaced the fully-connected layer of LeCun's archetype with "global average pooling". Inception's first layer was a "depthwise separable convolution". This kind of convolutional layer, invented in 2013 by Laurent Sifre while interning at Google (and later the topic of his dissertation "Rigid-motion Scattering for Image Classification", 2014), can be trained more quickly and is less prone to the problem of "overfitting" than regular convolutions.

In 2016 Francois Chollet (of Keras fame) at Google replaced all Inception modules with depthwise separable convolutions and obtained Xception.

Sergey Ioffe and Christian Szegedy at Google introduced a major improvement over stochastic gradient descent called "batch normalization", a method, based on an old idea by LeCun ("Efficient Backprop", 1998), to increase the stability of a neural network and therefore accelerate the training of a deep network ("Batch Normalization", 2015).

Deep networks with many layers had an additional problem: the "degradation" problem. As one adds more layers to the network, one can incur increasingly hostile optimization issues with the result that a simpler network might function better than a deeper one. The problem was solved in 2015 first in the Highway Network built by Schmidhuber's team at IDSIA (probably the very first neural network with over 100 layers) and then in Microsoft's 152-layer Residual Net or ResNet, designed by Jian Sun's team (including Kaiming He), a network made of so-called "residual models" to avoid issues during training (and with the same "global average pooling" of GoogLeNet). Similar to Inception, and unlike AlexNet and VGG, ResNet is a collection of building blocks. It is another "network-in-network" architecture. ResNet proved that extremely deep networks can be trained using stochastic gradient descent. Kaiming He's 2016 revision of ResNet with "identity mappings" (i.e., functions of the "f(x) = x" kind) was also easier to train ("Identity Mappings in Deep Residual Networks", 2016). To prove his point, He built a 1001-layer ResNet.

R-CNN is slow because it consists of a sequence of convolutional layers, SVMs (the object detectors) and "bounding-box regressors". Training it is expensive in both space (storage) and time, and run-time object detection is slow too. The ResNet group at Microsoft (Kaiming He, Jian Sun, etc) came up with spatial pyramid pooling networks (SPPnet), a way to optimize the computation of the convolutional layers by sharing the features that they generate, thus accelerating the run-time response of R-CNN by ten or even 100 times ("Spatial Pyramid Pooling in Deep Convolutional Networks for Visual Recognition", 2014). They also innovated in that they accepted any image size. The other popular kinds of deep convolutional nets required a fixed input image size, i.e. cropped or warped input images. A few months later Ross Girshick at Microsoft came up with another alternative, Fast R-CNN, three times faster in training and ten times faster at run-time ("Fast R-CNN", 2015); but not quite "real-time". Another few months and, at the end of 2015, Shaoqing Ren, working with the same crowd (He, Girshick and Sun), unveiled Faster R-CNN, a two-module network conceived precisely to achieve real-time response: a Region Proposal Network that proposes regions for a Fast R-CNN detector (it suggests "where to look"), and also shares the convolutional features in order to optimize computation ("Towards Real-time Object Detection with Region Proposal Networks", 2015). Ali Farhadi's group at University of Washington found another way to optimize R-CNN, using regression methods: YOLO (You Only Look Once) ("Unified, Real-time Object Detection", 2016) is not as accurate as R-CNN, but it is simpler and faster. YOLO even learns representations of objects that can be generalized.

By combining different kinds of networks, one can do more than just recognize objects. For example, Ali Farhadi's student Roozbeh Mottaghi at the Allen Institute in Seattle built the Our Forces in Scenes dataset to train two parallel convolutional neural networks (two AlexNets), one to encode the layout of a scene and the other one to capture force information, and then a recurrent neural network that takes their output and predicts the motion of the object ("What Happens If...", 2016).

Robert Jacobs' "mixture-of-experts" architecture of 1990 was refined over the years as the need for parallel architectures increased. Ronan Collobert implemented it with support vector machines ("A Parallel Mixture of SVMs for Very Large Scale Problems", 2002), Volker Tresp at Siemens in Germany used Gaussian processes ("Mixtures of Gaussian Processes", 2001), and Babak Shahbaba at UC Irvine and Radford Neal at University of Toronto modeled the "expert" as a Dirichlet process, a concept introduced by Thomas Ferguson at UCLA in 1969, which is basically a distribution over distributions ("Nonlinear Models using Dirichlet Process Mixtures", 2009). Meanwhile, Yoshua Bengio had called for "conditional computation" to speed up the work of deep neural networks: ways to reduce the number of hidden units that need to be computed in a multi-layer neural network ("Deep Learning of Representations", 2013). Jeff Dean's group at Google used mixtures of experts to achieve that goal of conditional computation and therefore dramatically increase the capacity of deep networks ("Outrageously Large Neural Networks: The Sparsely-Gated Mixture-of-Experts Layer", 2017).

These networks kept evolving. In 2016 Kaiming He extended ResNet to 200 layers. Christian Szegedy produced Inception-v2 in 2015 and Inception-v3 and Inception-v4 in 2016. In the same year he also produced a hybrid Inception-ResNet. In both Szegedy was striving for a simpler implementation without losing accuracy.

In 2016 Antonio Criminisi's team at the Microsoft labs in Britain introduced conditional networks that wed the efficiency of decision trees with the accuracy of CNNs ("Decision Forests, Convolutional Networks and the Models in-Between", 2016).

The best performing neural network of the 2012 Large Scale Visual Recognition Challenge (ILSVRC), Toronto's AlexNet, had 8 layers. The best of 2014, Oxford's VGG-16, had 19 layers. The best of 2015, Microsoft's ResNet, had 152, i.e. it was eight times deeper than the VGG network of one year earlier (ResNet was the basis for the feature of real-time translation introduced in 2015 in Microsoft's Skype product, a largely disappointing product despite its very deep structure). The trend to go deeper was propelled by two factors: 1. the hardware was getting cheaper and more powerful (Nvidia's GPUs); and 2. the mathematical theory behind deeper networks was showing advantages. Ronen Eldan and Ohad Shamir at the Weizmann Institute of Science in Israel proved that (quote) "depth can be exponentially more valuable than width". Deeper networks seem to be generally more accurate and require fewer connections within each layer ("The Power of Depth for Feedforward Neural Networks", 2016).

However, in 2016, while building ResNetXt, Kaiming He and Ross Girshick (both now at Facebook) realized that repetition, not depth, was the key factor in the performance of networks like VGG-16 and ResNet that are built by simply stacking modules of the same topology on top of each other. They also noted that Inception's modules shared a "split, transform and aggregate" strategy. In fact, one can reimagine the operation of a single neuron in a network as a combination of three operations: splitting, transforming, and aggregating. Imagine that the single neuron is actually a network in itself and you get a function that aggregates all the transformations of the constituent neurons. The dimension of this function is the "cardinality". Their conclusion was that the depth of a neural network is only partially interesting for improving accuracy. There is a third factor that matters, besides width and depth: the "cardinality" ("Aggregated Residual Transformations for Deep Neural Networks",

2017). Their 101-layer achieved better accuracy than the 200-layer ResNet while reducing the complexity to half.

The need to deploy deep-learning architectures on computationally-limited devices, such as smartphones and self-driving cars, led to the study of more efficient architectures, such as Tsung-Wei Ke's "multigrid" at UC Berkeley, a "multiscale" architecture ("Multigrid Neural Architectures", 2017). Another multiscale architecture was developed by Kilian Weinberger's student Gao Huang at Cornell University, in collaboration with Facebook and Tsinghua University: MSDNet ("Multi-Scale Dense Networks for Resource Efficient Image Classification", 2018), which was an evolution of DenseNets ("Densely Connected Convolutional Networks", 2017), which employed dense connectivity to facilitate the re-use of features; an architecture similar to Scott Fahlman's "cascade-correlation learning architecture" of 1989.

Deep Belief Nets are probabilistic models that consist of multiple layers of probabilistic reasoning. Thomas Bayes' theorem of the 18th century is rapidly becoming one of the most influential scientific discoveries of all time (not bad for an unpublished manuscript discovered after Bayes' death). Bayes' theory of probability interprets knowledge as a set of probabilistic (not certain) statements and interprets learning as a process to refine those probabilities. As we acquire more evidence, we refine our beliefs. In 1996 the developmental psychologist Jenny Saffran showed that babies use probability theory to learn about the world, and they do learn a lot of facts very quickly. So Bayes had stumbled on an important fact about the way the brain works, not just a cute mathematical theory.

Since 2012 the applications of deep learning have multiplied. Deep learning has been applied to big data, biotech, finance, health care... Countless fields hope to automate the understanding and classification of data with deep learning.

One of the most challenging applications is to recognize what is going on in a picture: there's a story hidden in those pixels, a story in which the objects are in some kind of relation for some kind of reason. We can easily look at the picture and guess what people or animals are doing or which function the objects have. The first image-annotation system ("translating" images into sentences) was probably developed by Ryuichi Oka's team at Tsukuba Research Center in Japan ("Image-to-word Transformation Based on Dividing and Vector Quantizing Images with Words", 1999), followed by David Forsyth's team at University of Illinois ("Object Recognition as Machine Translation", 2002). Several refinements were proposed during the 2000s, notably by Larry Davis and his student Abhinav Gupta at University of Maryland ("Beyond Nouns", 2008).

In November 2014 simultaneously two groups (Fei-fei Li's team at Stanford and Oriol Vinyals' team at Google) announced that an A.I. system was capable of analyzing a scene and write a short description of it: "Stanford team creates computer vision algorithm that can describe photos" (Stanford's press release) and "A picture is worth a thousand (coherent) words: building a natural description of images" (posted on Google's research blog). Fei-fei Li and her Slovakian student Andrej Karpathy ("Deep Visual-Semantic Alignments for Generating Image Descriptions", 2014) had developed a mixed system of convolutional neural networks and recurrent neural networks, nicknamed Neural Talk (Karpathy was hired in 2017 by Tesla). Less publicized was the combination of recurrent neural networks and convolutional networks presented five months earlier at the International Conference on Machine Learning in Beijing by Pedro Pinheiro and Ronan Collobert from Switzerland's IDIAP ("Recurrent Convolutional Neural Networks for Scene Labeling", 2014).

By the end of the decade the typical convolutional neural network (CNN) looked like this: a feed-forward neural network with four kinds of layers, namely the convolution layers (that extract features), the ReLU layer, the pooling layer (that reduces the dimensionality) and a fully-connected layer (that classifies). The typical recurrent neural network (RNN) saved the output of a layer and fed it back to the input. In other words, an RNN was able to mix the current input with the previous inputs. A CNN took a fixed-size inputs (typically a two-dimensional matrix) and generated fixed-size outputs (for example, a number). An RNN was able to handle arbitrary input/output lengths. CNNs were typically used in classification problems (e.g. object recognition). RNNs were especially useful for processing sequential data (e.g. for speech recognition). The system generating the caption (description) of an image was a hybrid of CNN and RNN: the CNN classifies the image and the RNN generates the words.

Since 2012 all the main software companies have invested in A.I. startups: Apple (Siri in 2011; Boston's Perceptio and Britain-based VocalIq in 2015; UC San Diego's spinoff Emotient; Cambridge University's spinoff VocalIQ in 2015; CMU spinoff Turi in 2016, Stanford spinoff Lattice Data in 2017, Danish computer-vision startup Spektral in 2018), Google (Neven, 2006; Industrial Robotics, Meka, Holomni, Bot & Dolly, DNNresearch, Schaft, Bost, DeepMind, Redwood Robotics, 2013-14; API.ai and Moodstocks, 2016; Kaggle, 2017), Amazon (San Francisco's Kiva, 2012; New York's Angel.ai, 2016; San Diego's Harvest.ai, 2017), Yahoo (LookFlow, 2013), IBM (Colorado's AlchemyAPI, 2015; plus its own Watson project), Microsoft (Britain-based Switfkey, 2016; Silicon Valley's Genee and Canada's Maluuba, 2017; plus its own Project Adam), Facebook (Israel's Face.com, 2012; Silicon Valley's Wit.ai, 2015; Belarus-based Masquerade Technologies and Switzerland-based Zurich Eye, 2017), Twitter (Boston's Whetlab, 2015; Britain-based Magic Pony, 2016), Salesforce (MetaMind, founded in 2014 in Palo Alto by Richard Socher, and Stanford's PredictionIO in 2016), Intel (Nervana and Itseex in 2016), General Electric (Berkeley's Wise.io in 2017), etc.

Numenta's cofounder Dileep George founded Vicarious in 2010 to follow a similar route to Numenta's with a slightly different model, the "recursive cortical network" (RCN), which can generalize a bit more than a deep-learning network. In 2017 Vicarious's RCN, trained on 1,406 images outperformed a deep-learning network trained on 7.9 million images.

Clarifai, founded in 2013 in New York by Rob Fergus' student Matt Zeiler of New York University who had worked on multilayer "deconvolutional network" ("Adaptive Deconvolutional Networks for Mid and High Level Feature Learning", 2011), developed the neural network that won the ImageNet competition in 2013.

Transfer learning (a term coined by Edward Thorndike a century ago) is something that we humans do naturally: we can do many things, not just one. In most cases nobody "trained" us to do what we do: we learned it by "transferring" knowledge of how to do other tasks. We can easily apply knowledge about one task to completely different tasks. Neural networks, instead, are tailored and fine-tuned for just one specific application. In 2017 a Google team (Lukasz Kaiser, Aidan Gomez, Noam Shazeer, Ashish Vaswani, Niki Parmar, Llion Jones, Jakob Uszkoreit) announced MultiModel, a neural network that had learned how to perform eight tasks. It was basically a constellation of neural networks, each specialized in a task such as image recognition, speech recognition, translation and sentence analysis. This neural network was interesting mainly because its performance on one of the eight tasks improved not only when it was trained for that task but even when it was trained for one of the other seven tasks ("One Model To Learn Them All", 2017).

As deep learning became more popular, aspiring A.I. scientists started realizing the mindboggling complexity of designing a deep-learning algorithm for a real-world application. A few labs started using A.I. techniques to choose the best A.I. technique for a given application and a new discipline, automatic machine learning, was born: Kevin Leyton-Brown's Auto-Weka (2013) at University of British Columbia in Canada, based on Bayesian optimization; Frank Hutter's Auto-Sklearn (2015) at University of Freiburg in Germany, based on Bayesian optimization;

Randy Olson's Tree-based Pipeline Optimization Tool or TPOT (2015) at University of Pennsylvania based on genetic programming; Quoc Le's and Barret Zoph's AutoML (2017) at Google, based on reinforcement learning; to which Microsoft responded with Custom Vision Services (2017); and Auto-Keras, developed at Texas A&M University (which automatically searches for the best architecture and hyperparameters). And products from startups began to appear, for example OneClick.ai, sponsored in Seattle by Kai-Fu Lee (former president of Google China, founder of Microsoft Research Asia, and author of an atrocious bestselling book, "AI Superpowers"). Meanwhile, Sergey Levine's student Chelsea Finn at UC Berkeley worked out a general algorithm for meta-learning, i.e. for learning how to learn, "model-agnostic meta-learning" or MAML ("Model-Agnostic Meta-Learning for Fast Adaptation of Deep Networks", 2017).

Several platforms for deep learning have become available as open-source software: Torch (developed by Ronan Collobert at IDIAP in Switzerland in 2002); Caffe (developed by Yangqing Jia at UC Berkeley in 2013); TensorFlow (developed by Rajat Monga's team at Google in 2015); Chainer (developed by Seiya Tokui at Preferred Networks in Japan in 2015); Microsoft Cognitive Toolkit or CNTK (2016); Keras, a deep-learning library that runs on TensorFlow, CNTK and Theano (built in 2015 by Francois Chollet at Google); Pytorch (developed in 2016 by Soumith Chintala's team at Facebook); Gluon (introduced in 2017 by Microsoft and Amazon); Facebook's Detectron for object detection (2018); Google's Dopamine for reinforcement learning (2018); Nvidia's vid2vid for video-to-video synthesis (2018); the distributed deep-learning framework MXNet, born in 2015 from the fusion of three projects (CXXNET, developed by Bing Xu at University of Alberta, Minerva, developed by Zheng Zhang's group at New York University in Shanghai, and Purine, developed by Min Lin at National University of Singapore) and supported by Intel, Microsoft, CMU, MIT, University of Washington, and many others; the Open Neural Network Exchange (ONNX) for interoperability between this kind of machine-learning frameworks (introduced in 2017 by Facebook and Microsoft and supported by IBM, Intel, ARM, Qualcomm, and many others); etc. This open-source software multiplies the number of people who can experiment with deep learning.

The real protagonist of deep learning, however, is the GPU, the graphical processing unit, originally invented to boost the speed of videogames. This humble invention constitutes the physical engine of today's multilayer neural networks, which would not be feasible without it.

Trivia: ironically, the term "deep learning" was introduced in machine learning by Rina Dechter at the 1986 AAAI conference in the context of symbolic systems (which at the time were the "enemies" of neural networks).

Deep Reinforcement Leaning: A Brief History of Artificial Intelligence/ Part 7

The game of go/weiqi had been a favorite field of research since the birth of deep learning.

In 2006 Rémi Coulom in France applied Ulam's old Monte Carlo method to the problem of searching a tree data structure (one of the most common problems in computational math, but particularly difficult when the data are game moves), and thus obtained the Monte Carlo Tree Search algorithm, and then applied it to go/weiqi ("Bandit Based Monte-Carlo Planning", 2006). At the same time Levente Kocsis and Csaba Szepesvari developed the UCT algorithm (Upper Confidence Bounds for Trees algorithm), an application of the bandit algorithm to the Monte Carlo search, the bandit algorithm being a problem in probability theory first studied by Herbert Robbins in 1952 at the Institute for Advanced Study ("Some Aspects of the Sequential Design of Experiments", 1952).

These algorithms dramatically improved the chances by machines to beat go masters: in 2009 Fuego Go (developed at University of Alberta) beat Zhou Junxun, in 2010 MogoTW (developed by a Franco-Taiwanese team) beat Catalin Taranu, in 2012 Tencho no Igo/ Zen (developed by Yoji Ojima) beat Takemiya Masaki, in 2013 Crazy Stone (by Remi Coulom) beat Yoshio Ishida, and in 2016 AlphaGo (developed by Google's DeepMind) beat Lee Sedol. DeepMind's victory was widely advertised. DeepMind used a slightly modified Monte Carlo algorithm but, more importantly, AlphaGo taught itself by playing against itself. AlphaGo's neural network was trained with 150,000 games played by go/weiqi masters. AlphaGo had played 200 million games by January 2017 when it briefly debuted online disguised under the moniker Master; and in May 2017 it beat the world's master Ke Jie.

DeepMind had previously combined convolutional networks with reinforcement learning to train a neural network to play videogames: in 2013 Volodymyr Mnih and others had trained convolutional networks with a variant of Q-learning, the "asynchronous actor-critic" algorithm or A3C, in order to improve the policy function ("Playing Atari with Deep Reinforcement Learning", 2013). This Deep Q-Network (DQN) was the first in a progression of deep reinforcement learning methods developed by DeepMind, leading to AlphaGo and then AlphaZero.

Deep Q-learning suffers from some inherent limitations and requires a lot of computational power, which translates not only in cost but also in slow response time. DeepMind had rediscovered Lin's "experience replay" of 1992 as a remedy for the Atari-playing network. DeepMind then rediscovered the policy gradient methods for reinforcement learning (even though it was very different from the likelihood-ratio method of REINFORCE): for example, David Silver's "deterministic policy gradient" of 2014 and Nicolas Heess' "stochastic value gradient" of 2015. Volodymyr Mnih's team worked on asynchronous gradient-descent algorithms that run multiple agents in parallel instead of using "experience replay". In 2015 Arun Nair and others designed the General Reinforcement Learning Architecture (Gorila) that greatly improved the performance on those Atari games by performing asynchronous training of reinforcement learning agents (100 separate actor-learner processes).

DQN works well for videogames, which have a discrete action space, but it cannot be used for the continuous spaces of robot movement. Daan Wierstra's team at DeepMind (namely Timothy Lillicrap) developed an algorithm called "Deep Deterministic Policy Gradient" (DDPG) for continuous action spaces ("Continuous Control with Deep Reinforcement Learning", 2016) by extending David Silver's "deterministic policy gradient" or DPG (Deterministic policy gradient algorithms", 2014), which in turn was a variation on Roland Hafner's and Martin Riedmiller's "Neural Fitted Q Iteration with Continuous Actions" or NFQCA at University Freiburg ("Reinforcement learning in feedback control", 2011). However DDPG, just like NFQCA, was frequently unstable.

Ironically, few people noticed that in September 2015 Matthew Lai unveiled an open-source chess engine called Giraffe that uses deep reinforcement learning to teach itself how to play chess (at international master level) in 72 hours. It was designed by just one person and it ran on the humble computer of his department at Imperial College London. (Lai was hired by Google DeepMind in January 2016, two months before AlphaGo's exploit against the Go master).

In 2016 Toyota demonstrated a self-teaching car, another application of deep reinforcement learning like AlphaGo: a number of cars are left to randomly roam the territory with the only rule that they have to avoid accidents. After a while, the cars learn how to drive properly in the streets.

Poker was another game targeted by A.I. scientists, so much so that University of Alberta even set up a Computer Poker Research Group. Here in 2007 Michael Bowling developed an algorithm called Counterfactual Regret Minimization or CFR ("Regret Minimization in Games with Incomplete Information", 2007), based on the "regret matching" algorithm invented in 2000 by Sergiu Hart and Andreu Mas-Colell at the Einstein Institute of Mathematics in Israel ("A Simple Adaptive Procedure Leading to Correlated Equilibrium", 2000).

These are techniques of self-playing: given some rules describing a game, the algorithm plays against itself and develops its own strategy for playing the game better and better. It's yet another form of reinforcement learning, except that in this case reinforcement learning is used to devise the strategy from scratch, not to learn the strategy used by humans. The goal of CFR and its numerous variants is to approximate solutions for imperfect information games such as poker. CFR variants became the algorithms of choice for "poker bots" used in computer poker competitions.

In 2015 Bowling's team developed Cepheus and Tuomas Sandholm at CMU developed Claudico, that played the professionals at a Pittsburgh casino (Pennsylvania). Claudico lost but in 2017 Libratus, created by the same group, won. Libratus employed a new algorithm, called CFR+, introduced in 2014 by Finnish hacker Oskari Tammelin ("Solving Large Imperfect Information Games Using CFR+, 2014) that learns much faster compared with previous versions of CFR. However, the setting was absolutely unnatural, in particular to rule out card luck. It is safe to state that no human players had ever played poker in such a setting before. But it was telling that the machine started winning when the number of players was reduced and the duration of the tournament was extended: more players in a shorter time beat Claudico, but fewer players over a longer time lost to Libratus.

Deep reinforcement learning has three main problems: it requires a lot of data, which is unnatural (animals can learn something even if they saw it only once or twice), it fails all too easily in situations that differ just slightly from the situations for which the system has been trained, and its internal workings are largely inscrutable, which makes us uncomfortable to use them on mission-critical projects.

A different strand from DeepMind's A3C harkens back to the "relational Markov decision process", introduced by Carlos Guestrin ("Planning under Uncertainty in Complex Structured Environments", 2003); and to the Object-Oriented Markov Decision Process, introduced by Micheal Littman's student Carlos Diuk at Rutgers University ("An Object-Oriented Representation for Efficient Reinforcement Learning", 2008). The relational Markov decision process blends probabilistic and relational representations with the express goal of planning action in the real world. Marta Garnelo's "deep symbolic reinforcement" at Imperial College London ("Towards Deep Symbolic Reinforcement Learning",2016) and Dileep George's "schema networks" to transfer experience from one situation to other similar situations (Schema Networks", 2017) built upon these ideas, basically wedding first-order logic representation with deep reinforcement learning.

Yet another route led to Trust Region Policy Optimization (TRPO), introduced in 2015 by Pieter Abbeel's student John Schulman at UC Berkeley as an alternative to DeepMind's policy-gradient methods (at least for continuous control tasks). DeepMind itself contributed an alternative to policy-gradient methods: ACER developed by Ziyu Wang and others of Nando de Freitas' team ("Sample Efficient Actor-Critic with Experience Replay", 2016), a variant of the Retrace method developed by their colleague Remi Munos ("Safe and Efficient Off-policy Reinforcement", 2016). John Schulman himself improved TRPO with PPO, "proximal policy optimization", released by OpenAI in 2017 ("Proximal Policy Optimization Algorithms", 2017). DQN, A3C, TRPO, ACER and PPO were just the tip of the iceberg: algorithms for optimization of reinforcement learning multipled and evolved rapidly after the success of these techniques in learning to play games. Meanwhile DeepMind had improved the DQN algorithm into the Rainbow algorithm (first demonstrated in 2017). Rainbow and PPO became the reference standards.

"Policy search" methods, such as Marc Deisenroth's and Carl-Edward Rasmussen's PILCO (2011), were invented to make robots automatically learn new behavior through experience. These methods were used successfully to train both the vision and the motor components of a robot so that it can perform manipulation tasks. Ideally, the vision system should adapt to the goal of the task, but this cannot happen if the two are trained separately. Sergey Levine and his student Chelsea Finn at UC Berkeley developed a method for training the visual system and the control system of a robot at the same time instead of training them separately. In this way the vision system can adapt to the goals of the task, a much more efficient way to "see" things in the world, and, incidentally, the way all animals look at the world ("End-to-End Training of Deep Visuomotor Policies", 2016). Their "guided policy search" turned policy search into a kind of supervised learning whose training data are generated iteratively, which is an approach similar to "visual servoing", pioneered 25 years earlier by Bernard Espiau and Patrick Rives at INRIA in France ("A New Approach to Visual Servoing in Robotics", 1992), and of the "Bregman Alternating Direction Method of Multipliers" (2014) introduced by Huahua Wang and Arindam Banerjee at University of Minnesota. The difference is that Levine and Finn used a new convolutional architecture for both policies to automatically capture spatial information about the scene. Most successes in reinforcement learning had taken place in single-agent scenarios, with the advantage of an environment that remained largely unchanged. However, the real world is almost never a stationary environment because it involves multiple agents. Multi-agent learning is inherently more complicated. Traditionally, two extreme approaches were attempted. The decentralized-learning approach dates back to at least Independent Q-learning, developed by Ming Tan at GTE Labs in Boston ("Multi-agent Reinforcement Learning - Independent vs. Cooperative Agents", 1993), and was updated to DQN by Raul Vicente's students at University of Tartu in Estonia ("Multiagent Cooperation and Competition with Deep Reinforcement Learning", 2015). The centralized-learning approach dates back to at least Carlos Guestrin at Stanford ("Multiagent Planning with Factored MDPs", 2002), and later deep-learning implementations include Sainbayar Sukhbaatar's CommNet at New York University ("Learning Multiagent Communication with Backpropagation", 2016) and Alibaba's BicNet in China ("Multiagent Bidirectionally-Coordinated Nets", 2017). Neither approach was satisfactory. Eventually, a hybrid approach became popular. The paradigm of centralized training with decentralized execution was pioneered by Frans Oliehoek at University of Amsterdam ("Exploiting Locality of Interaction in Factored Decentralized POMDPs", 2008). Ryan Lowe of McGill University and Yi Wu of UC Berkeley called it "Decentralized Actor Centralized Critic" ("Multi-Agent Actor-Critic for Mixed Cooperative-Competitive Environments", 2017). Hybrid systems of this kind were developed by Jayesh Gupta at Stanford ("Cooperative Multi-agent Control Using Deep Reinforcement Learning", 2017), by DeepMind ("Value-Decomposition Networks For Cooperative Multi-Agent Learning", 2017) and by Shimon Whiteson's team at Oxford University, the QMIX method ("Monotonic Value Function Factorisation for Deep Multi-Agent Reinforcement Learning", 2018).

Progress in reinforcement learning for robots was helped by virtual environments in which one could test robot behavior. In 2016 OpenAI released the toolkit OpenAI Gym, and in 2017 OpenAI released Roboschool, a software environment to create real-world simulations for training robots. In 2017 Vinyals at DeepMind introduced the "StarCraft II Learning Environment" (SC2LE1), a reinforcement-learning environment based on the StarCraft II video game to test these systems.

A wide variety of reinforcement-learning algorithms existed by the end of the 2010s: Volodymyr Mnih's Advantage Actor Critic (A3C), John Schulman's PPO, DeepMind's Ape-X ("Distributed Prioritized Experience Replay", 2018), DeepMind's Impala ("Scalable Distributed Deep-RL with Importance Weighted Actor-learner Architectures", 2018), etc. More general frameworks for reinforcement learning included: Intel's Coach ("Reinforcement Learning Coach by Intel", 2017), Google's TensorFlow Agents ("Efficient Batched Reinforcement Learning in TensorFlow", 2017), OpenAI's Baselines (2017), Cambridge University's TensorForce ("A TensorFlow Library for Applied Reinforcement Learning", 2017), Eric Liang's RLlib at UC Berkeley (2018), etc.

If 2016 had been the year of reinforcement learning, by 2017 it was becoming an easy target for all sorts of attacks. First of all, several studies showed that other forms of machine learning could replicate the feats of DeepMind: Jeff Clune's team at Uber AI Labs ("Deep Neuroevolution: Genetic Algorithms Are a Competitive Alternative for Training Deep Neural Networks for Reinforcement Learning", 2017), Tim Salimans at OpenAI ("Evolution Strategies as a Scalable Alternative to Reinforcement Learning", 2017) as well as Ben Recht's students Horia Mania and Aurelia Guy at UC Berkeley ("Simple Random Search Provides a Competitive Approach to Reinforcement Learning", 2018) showed that much simpler genetic algorithms could do as well on most tasks. Secondly, Sham Kakade's student Aravind Rajeswaran at University of Washington showed that multi-layered neural networks are an unnecessary complication for robotic tasks of continuous control such as those of the OpenAI gym environment ("Towards Generalization and Simplicity in Continuous Control", 2017). For example, Nicolas Heess and others at Deep Mind used deep reinforcement learning to train a robot for walking, running, jumping and turning ("Emergence of Locomotion Behaviours in Rich Environments", 2017). The training required 64 CPUs running for over 100 hours. The video was impressive, but similar results had been obtained five years earlier by Emanuel Todorov's students Yuval Tassa and Tom Erez (both later hired by DeepMind) at University of Washington, running on much smaller and slower hardware using not reinforcement learning but a humbler method, online trajectory optimization, aka model predictive control, on their physics simulator MuJoCo ("Synthesis and Stabilization of Complex Behaviors through Online Trajectory Optimization", 2012).

The original reason to develop the complex algorithms of A.I. was not that it was impossible to play chess or weiqi/go with simpler algorithms but rather that it was plain silly to do so: it would have required colossal amounts of computation on that slow and expensive hardware. It seems like A.I. in the age of AlphaGo reached a point where its complex algorithms do require a colossal amount of computation; which seems to contradict the whole point of doing A.I. Nobody ever proved formally that simple algorithms cannot achieve superhuman performance at chess or go/weiqi or any other deterministic game. The availability of computational power reduces the raison d'être of A.I.'s sophisticated algorithms. What's the point of developing multilayer networks if simple architectures can do the same job? The only reason would be to speed up things, or run on cheaper hardware, but instead the opposite is happening: DeepMind and OpenAI have to throw more and more computational power at their algorithms. It was not surprising that, after the initial enthusiasm, A.I. scientists started revisiting the old simple algorithms that were dismissed 20 years earlier. Some of them might indeed prove that A.I. unnecessarily complicated things because we forgot that the stumbling block was not a faulty theory but only slow hardware.

One of the biggest problems for reinforcement learning is that it requires a reward function: the reward function really represents what we want the system to learn. The problem is that designing the correct reward function is extremely difficult in the real world. If the reward is not correct, reinforcement learning acts like an uncontrollable chain reaction.

Another unsolved problem in reinforcement learning is the problem of balancing exploration and exploitation. This is a general problem of mathematical optimization called "the multi-armed bandit problem". The classic example is the dinner: you can have dinner at your favorite restaurant, where you will almost certainly get a good meal, or try a new restaurant, with a chance that you'll discover a new favorite. This doesn't sound like a big deal. But apply the same reasoning to medical care and you can probably appreciate why it is a big deal. Exploitation consists in making the best decision that maximizes immediate return. Exploration consists in investing in gathering more knowledge about the environment, which in the long run might yield a higher return. Exploration is extremely expensive in reinforcement learning so it tends to be neglected. But in some cases there is a huge price to pay for it. That's why DQN could play Atari games so well but not Montezuma Revenge, a game that requires more exploration.

Historically speaking, we have learned that the classic A.I. problems can be divided in categories based on some factors, and it turns out that Go belong to the simplest category of A.I. problems. Problems that are deterministic, fully observable and known, discrete (a limited number of possible states) and static (the rules don't change) are actually "easy" to solve when we can afford to throw virtually infinite computational power at them. There is nothing in their formulation that makes it impossible to solve them: it is "difficult" to solve them simply because the number of possible solutions is very big. In the old days of slow, expensive hardware one would program the machine to look for the solution by simulating human intuition. In the age of cheap, fast hardware, one programs the machine to use brute force and simply try all possible combinations until one turns out to be the correct solution. The game of weiqi/go is one such problem: "difficult" because of the many many many possible moves, but "easy" for a machine that can try all of them.

The next step after weiqi/go was the multiplayer videogame Dota 2 of 2013, played daily by over one million people. In 2018 OpenAI announced a bot, OpenAI Five, that, like AlphaGo Zero, learned the game from scratch by self-play without any knowledge of how humans play it. This problem is neither fully observable/known nor discrete nor static. OpenAI Five trains by playing 180 years worth of games against itself every day, running on 256 GPUs and 128,000 CPU cores, using a separate LSTM for each hero. OpenAI Five uses proximal policy optimization. In 2018 OpenAi used the same algorithm to teach a robotic hand, Dactyl, to manipulate cubes. In October 2019 OpenAI demonstrated a robotic arm capable of solving the puzzle of Rubik's Cube. Other systems had solved Rubik's Cube before, and the robotic hand that they used was 15 years old, but OpenAI coupled its OpenAI Five's reinforcement learning algorithm with Automatic Domain Randomization (ADR), a new way to endlessly generate progressively more difficult environments. Traditionally, one would model the world and use that model for simulation. ADR, developed by Pieter Abbeel's student Josh Tobin at UC Berkeley in collaboration with OpenAI, allowed the system to be trained in increasingly more complicated simulations ("Domain Randomization for Transferring Deep Neural Networks from Simulation to the Real World", 2017). Rui Wang at Uber AI labs worked on a similar technique called Paired Open-Ended Trailblazer or POET ("Endlessly Generating Increasingly Complex and Diverse Learning Environments and Their Solutions", 2019). DeepMind responded with another evolution of AlphaGo called AlphaStar. In November 2019 DeepMind's AlphaStar, mainly the brainchild of David Silver and Orion Vinyals, won against 99.8% of its human opponents at the videogame StarCraft II. AlphaStar was trained from videos of human-played games and then improved by playing against itself. Training AlphaStar required 384 TPUs for 44 days, during which AlphaStar played the equivalent of 200 human years of Starcraft games. And, actually, DeepMind trained different AlphaStar versions for each of Starcraft's three alien races (each race requires different strategies). If the maps of the game are changed, AlphaStar needs to be re-trained from the beginning.

Shauharda Khadka of Kagan Tumer's group at Oregon State University combined reinforcement learning and genetic algorithms to teach a 3D humanoid how to walk ("Collaborative Evolutionary Reinforcement Learning", 2019): a population of neural nets are trained with reinforcement learning and then the top-performing neural nets are chosen in each generation to produce the next generation.

"That men do not learn very much from the lessons of history is the most important of all the lessons of history" (Aldous Huxley)

Intermezzo: What is Intelligence?

Philosophically, progress in trendy A.I. techniques such as reinforcement learning is quite shocking: they are actually expressed in quite simple mathematics. You can write the formulas in a few lines. Those formulas may look complicated to non-mathematicians but they are actually infinitely easier than, say, Einstein's equations of gravitation. Then you need to run these few lines millions of times over a large dataset of examples (e.g. of Atari games). And then these algorithms start behaving like masters of the game. Except that these algorithms don't even know the rules of the games. The Atari program was "learning" to play the games by looking at the pixels on the screen of the computer. That program has no idea what the rules of the game are, and no idea that it is a game. It is just a mathematical formula that gets repeated millions of times over thousands of examples. You can legitimately question whether this is "intelligence". And here the philosophers may divide in two schools: the ones who think that "intelligence" requires a full understanding of what you are doing, and those who think that, ultimately, all our "understanding" is simply a massive iteration of simple neural algorithms. The former group keeps hoping that one of these days we will find a game that cannot be solved by simply repeating an algorithm millions of times. So far we've been humbled by the machines: machines "learned" to play increasingly difficult games and became better than us even if the machines don't actually know the rules of the game (and don't even know that they are playing a game).

However, don't overrate the machine: the machine algorithm learns to play the game, and eventually beats the human masters, only if someone has designed the machine algorithm correctly. The machine is just the "learning agent" that will interact with its environment until it successfully solves the problem. The fundamental step in reinforcement learning is to capture the key features of the problem and shape (accordingly) the behavior of the "learning agent". This is done by a human expert, who these days frequently employs a "Markov decision process". Machines are becoming "learning agents", but not (yet) designers of learning agents.

And there is still a fundamental difference in the "learning agent" itself. We humans actually learn a lot more than just by "trial and error". Humans and machines are using two different approaches to learning a game. Humans employ a lot of common-sense knowledge and intuition (after all, the games have been designed by humans who share our knowledge of the world). The human approach is initially "instructional": someone told us how to play, or we can guess by ourselves in a few seconds how the game works. Reinforcement learning does not need to know "what" it is doing: it just needs to know what the goal is and what the possible actions are, and the machine's task is to "select" the best behavior to achieve the goal. The machine's approach is "selectional". A human player starts playing well in a few minutes. Reinforcement learning will eventually play very well but it may take hours, days or months to learn to play (depending on how fast the computer is). The way you learned to ride a bicycle is a mixture of these two approaches: your parents probably told you (instructed you on) how a bicycle works but then you had to keep trying until you got it right, every time adjusting your behavior to avoid falling and to improve stability (punishment and reward).

Reinforcement learning has always fascinated psychologists because it works only if the learning agent has a "holistic" understanding of the environment. An Atari videogame or the moves of weiqi constitute a very simple environment. Humans can apply reinforcement learning in much more complex environments.

There is another reason to be fascinated (not alarmed) by reinforcement learning algorithms. As Tambet Matiisen (University of Tartu in Estonia) wrote: "Watching them figure out a new game is like observing an animal in the wild".

Footnote: The Opaque Power of Neural Networks

The funny thing about multilayer networks is that nobody really knows why they work so well when they work well. Designing a multilayer network is still largely a process of "trial and error".

Like most nonlinear systems, multilayer neural networks are difficult to "understand". It is relatively easy to figure out what a linear algorithm does, although a computer may compute it millions of times faster than us, but a nonlinear algorithm is, to a large extent, inscrutable.

One could see this as a human limitation: we can see that the neural network works well, but we cannot understand how it is doing it. In reality, it is equally difficult for us to understand what animals are doing, and we cannot understand either what most humans are doing... unless they tell us what they are doing. So that is also a limitation by the machine: a neural network that cannot explain its behavior has a clear limitation compared with human beings who routinely explain what they are doing. Paraphrasing what the philosopher Daniel Dennett said in a 2017 interview, if the neural network "cannot do better than us at explaining what it is doing, then don't trust it." If the self-driving car returns from shopping one hour later than expected, it would be nice to ask: "Why are you so late? What happened?" Intelligence is not the performance of an action but the ability to explain all that happened during that performance other than the action itself. Imagine that a deep-learning system, proven to be more accurate than all the human experts combined, analyzes a CT scan of your body and tells you that you only have one month to live. The human experts may be less accurate but they can tell you why they think what they think. The neural network simply tells you that you have one month to live with no explanation. How does it feel? Have you become just a disposable record in a database of patient records? Explaining what is happening to you has become a waste of time? Perhaps more importantly: can you trust an opinion that comes with no explanation? Do you begin making arrangements for your funeral without knowing why you are going to die soon?

For a while, a popular technique was to exploit the "saliency map", as pioneered by Klaus-Robert Muller's group at the Technical University of Berlin ("How to Explain Individual Classification Decisions", 2010) and by Karen Simonyan and Andrea Vedaldi in Andrew Zisserman's lab at Oxford ("Deep Inside Convolutional Networks", 2014).

In 2013 Rob Fergus' student Matthew Zeiler at New York University introduced a visualization technique to get insight about the functioning of intermediate layers of deep convolutional networks ("Visualizing and Understanding Convolutional Networks", 2013). LeCun's student Anna Choromanska at New York University used the Physics of the spherical spin-glass model to explain why stochastic-gradient descent works so well in "deep" neural networks ("The Loss Surfaces of Multilayer Networks", 2015). In 2016 Wojciech Samek in Germany unveiled a method that he named "deep Taylor decomposition" (after the method invented by the British mathematician Brook Taylor in 1715 to approximate a function) to peek into the workings of deep neural networks (Deep Taylor Decomposition of Neural Networks", 2016). At about the same time Carlos Guestrin's team at University of Washington developed an algorithm called LIME (Local Interpretable Model-Agnostic Explanations) that can explain the predictions of any classifier ("Why Should I Trust You?", 2016). In 2017 David Gunning at DARPA (Defense Advanced Research Projects Agency) launched a four-year nation-wide project (involving ten research laboratories) to develop neural-network interfaces that can help ordinary users to understand how the neural network reached the conclusion it reached (the eXplainable A.I. or XAI program). For example, Mohamed Amer at SRI International was trying to visualize the inner workings of a neural network using the technique of generative adversarial networks. David Blau in the team of Joshua Tenenbaum and Antonio Torralba at MIT developed "GAN Dissection", a framework for visualizing and understanding the structure learned by a generative adversarial network ("GAN Dissection", 2018).

Been Kim at Google Brain developed the technique of "Testing with Concept Activation Vectors" (TCAV) for interpreting the behavior of a deep network, a way to ask the network how much a concept impacted its "reasoning" ("Interpretability Beyond Feature Attribution", 2018).

In 2015 Naftali Tishby of the Hebrew University in Israel offered an explanation based on Claude Shannon's theory of information. In 1999 Tishby, Fernando Pereira (then at Bell Labs and later hired by Google) and William Bialek (then at NEC Research Institute in New Jersey and later at Princeton University) had formulated the "information bottleneck method" for network optimization ("The Information Bottleneck Method", 2000). A network retains only what is essential because the process of "learning" is equivalent to squeezing information through a bottleneck: the network behaves like someone forced to retain only what is truly essential, or, better, relevant, and to discard the rest. The process works if the network chooses correctly what can be discarded. Tishby likes to say that "the most important part of learning is actually forgetting." They had also calculated the theoretical boundary of information bottleneck, i.e. the greatest degree of optimization that still retains the relevant information. In 2014 two physicists, David Schwab of Northwestern University and Pankaj Mehta of Boston University, discovered a striking similarity between Hinton's deep-learning algorithm and "block-spin renormalization", a routine mathematical method used in statistical physics to extract the relevant features of a system and determine which ones can be ignored ("An Exact Mapping Between the Variational Renormalization Group and Deep Learning", 2014). Invented in 1966 by Leo Kadanoff, it is used by physicists to describe a system in a statistical manner, without the need to know the exact state of all its particles, in particular at the so-called "critical point" of a physical system (like when water turns from liquid to vapor state). Renormalization is a mathematical way to describe what matters macroscopically of a system without bothering about the microscopic details that are not important for its macroscopic behavior (or, better, simply averaging over them). In 2015 Tishby and his student Noga Zaslavsky explained deep learning in terms of information bottleneck: a deep neural network works like an optimization algorithm that retains only the relevant information for data classification ("Deep Learning and the Information Bottleneck Principle", 2015). When a neural network is being trained with a dataset, at some point it enters a "compression" phase in which it starts shedding information, i.e. it starts "forgetting" what is not relevant in order to retain the capacity to learn more about what is relevant. If a neural network is trained with enough samples, it will converge to the Tishby-Pereira-Bialek bound. Alex Alemi at Google applied Tishby's information-bottleneck thinking to very deep neural networks ("Deep Variational Information Bottleneck", 2016).

Tishby's model of learning in a network is intriguing because it provides a link to theories of human memory. It has been obvious since at least British psychologist Donald Broadbent published "Perception and Communication" (1958) that the number of objects we see in a lifetime exceeds the number of neurons in the brain that would be needed to store them as images. Human memory must select what to remember and forget most of the stimuli that are perceived by the senses. Broadbent stated the principle of "limited capacity" of the brain (also known as the "filter theory") to explain how a limited-capacity system such as the brain can cope with the overwhelming amount of information available in the world. Tishby's theory can even provide a link to dreaming: Francis Crick, co-discoverer of the double-helix structure of DNA, once speculated that the function of dreams is to "clear the circuits" of the brain ("The Function of Dream Sleep", 1983). The brain, in the face of huge daily sensory stimulation, must understand what matters, understand what does not matter, remember what will still matter and forget what will never matter again.

Footnote: Non-deep Learning (Machine Learning in Statistics and Elsewhere)

Deep learning was not the only game in town. In fact, many nonlinear algorithms have been introduced by the field of machine learning, notably during the 1990s.

In fact, the most successful method was a method of statistical learning, the linear classifier called "support vector machine" (SVM). The original SVM algorithm was invented by the Soviet mathematician Vladimir Vapnik and Alexey Chervonenkis in 1963 (they originally called it "generalized portrait"), and improved by Tomaso Poggio at MIT (in 1975 he introduced the "polynomial kernel"), but lay dormant until in 1991 Isabelle Guyon at Bell Labs (where Vapnik had moved in 1990) adapted SVMs to pattern classification ("A Training Algorithm for Optimal Margin Classifiers", 1992) using the optimization algorithm called "minover" invented by the physicists Marc Mezard and Werner Krauth in France to improve Hopfield-style neural networks ("Learning Algorithms with Optimal Stability in Neural Networks", 1987). Guyon turned a linear algorithm into a nonlinear algorithm. Another European-born Vapnik collaborator at Bell Labs, Corinna Cortes, further improved an SVM into a "soft-margin classifier" ("Support-Vector Networks", 1995). Cortes used SVMs for handwritten digit recognition in 1995 at the same time that, in a nearby lab, Yann LeCun was working on LeNet-5, with results that were similar. Another "killer application" of SVMs (given the exponential increase of texts on the World-wide Web) was to classify texts into categories, as done by Thorsten Joachims ("Text Categorization with Support Vector Machines", 1998). A theorem proven by Thomas Cover at Stanford ("Geometrical and Statistical Properties of Systems of Linear Inequalities with Applications in Pattern Recognition", 1965) states that a complex pattern-classification problem (such as recognizing an image) is more likely to be linearly separable (i.e. to succeed) when cast in a high-dimensional space than in a low-dimensional space (e.g. in many dimensions than in two dimensions). A kernel is a function that transforms a two-dimensional data space into a very high dimensional space (the "feature space"). This function requires a lot of computation, but mathematicians discovered a "trick" to make it feasible (which is still called "kernel trick"). Most problems of categorization are nonlinear, a fact that makes them very hard to solve. The kernel trick allows one to transform a nonlinear problem into a linear problem, after which there are plenty of well-known linear classifiers that one can use on the feature space. This trick harkens back to a theorem proven in 1909 by the British mathematician James Mercer and adapted in 1964 by the Soviet mathematician Mark Aizerman. That's the trick that Guyon used to make Vapnik's original concept the most popular method of statistical learning.

It is not surprising that the performance of SVMs compared favorably with multi-layer neural networks because an SVM can be viewed as a three-layer perceptron with a hidden layer made of the support vectors.

Tin-kam Ho at Bell Labs improved decision tree analysis with the stochastic discrimination method developed in 1990 by Eugene Kleinberg (her advisor at State University of New York) and obtained "random decision forests" ("Random Decision Forests", 1995), perfected by Leo Breiman at UC Berkeley: a "forest" is a large set of decision trees.

Classification and Regression Trees or CART for short is a term introduced by Leo Breiman to refer to decision-tree algorithms that can be used for classification or regression predictive modeling problems.

SVM probably remains the most used method in machine learning even in the age of deep learning. The reason is computational: you can distribute easily an SVM algorithm over dozens of machines so it will learn quickly, whereas a deep learning algorithm (which in theory is more efficient) is difficult to distribute over dozens of machines. That's because SVMs are linear models whereas neural networks are nonlinear. As of 2017, Google is de facto the only organization that has been able to perform distributed computation of deep learning algorithms. This has more to do with the big-data infrastructure (such as MapReduce, that Google published in 2005) than with the "intelligence" of the system. If you don't distribute neural networks over multiple machines, the training may take weeks or months.

The 1990s also witnessed the first applications of Bayesian inference networks: the Quick Medical Reference - Decision Theoretic or QMR-DT project (1991), by Gregory Cooper's team at Stanford, and NASA's Vista project (1992), led by Matthew Barry and Eric Horvitz.

The training of a machine-learning algorithm works best when the numbers of positive and negative instances are roughly equal. Alas, this is not the case in the real world. For example, the dataset of medical images is almost always imbalanced: healthy people rarely get a radiological scan of their chest and therefore most chest CT scans are of patients with known diseases. If the training dataset consists of 10,000 medical images of people with a heart condition and of 10 healthy people, the classifier will tend to classify healthy people as at risk of a heart attack. And viceversa: if the training dataset consists of 10,000 medical images of healthy people and of 10 people with heart conditions, the classifier will tend to classify people at risk as perfectly healthy. This problem is extremely common in many fields. This is an old problem in machine learning systems (or simply classifiers). The common remedy is to employ an "ensemble method", i.e. to combine multiple (weak) learning algorithms into a (strong) learning algorithm. A number of techniques (or meta-algorithms) have been proposed since at least the 1990s: BAgging (Bootstrap Aggregating), proposed by Leo Breiman at UC Berkeley in 1994; SMOTE (Synthetic Minority Over-Sampling Technique), developed by Nitesh Chawla and others at University of South Florida in 2000; etc.

"Boosting" is a class of methods for improving the accuracy of both linear and nonlinear classifiers. Robert Schapire at Bell Labs proved their theoretical feasibility ("The Strength of Weak Learnability", 1990) and AdaBoost (Adaptive Boosting), developed in 1995 in collaboration with Yoav Freund, was the most popular incarnation. A booster is not actually a classifier itself, just a "booster": it turns a linear combination of weak learners into a strong learner. The weakest learner is random guessing; the next weakest learners are learners that are only slightly better than random guessing; and so forth; a strong learner is a very accurate learner. Then came: DataBoost by Hongyu Guo and Herna Viktor at University of Ottawa in 2004; BEV (Bagging Ensemble Variation) by Cen Li at Middle Tennessee State University in 2007; RUSBoost (Random Under-Sampling Boost) by Taghi Khoshgoftaar and others at Florida Atlantic University in 2008; etc.

Michael Jordan at MIT (a former student of David Rumelhart at UC San Diego) worked on graphical models (see his book "Learning in Graphical Models", 1998); Bernhard Schoelkopf in Germany specialized in "kernels" (see his book "Learning with Kernels", 2002); Carl-Edward Rasmussen in Britain used "Gaussian processes" (see his book "Gaussian Processes for Machine Learning", 2006).

Furthermore, despite the hype around deep learning, algorithms for real-time computer vision such as David Lowe's SIFT, OpenCV (Open Source Computer Vision), the library of computer-vision functions released in 1999 by Intel, HOG (Histograms of Oriented Gradients), published in 2005 by Navneet Dalal and Bill Triggs at INRIA in Paris (the National Institute for Research in Informatics and Automation), SURF (Speeded Up Robust Features), published in 2006 by Luc Van Gool's team at ETH Zurich (the Swiss Federal Institute of Technology), and ORB, released in 2011 by Gary Bradski's team at Silicon Valley startup Willow Garage, continued to prevail over convolutional neural networks. These algorithms had several advantages over neural networks: they are easier to implement, don't require as much processing power, and can be trained with a smaller set. Therefore they are generally preferred in real-time applications. They are, however, very different from how the human brain works.

In 2001 Paul Viola and Michael Jones at the Mitsubishi research laboratories in Boston developed a face detector that achieved fast processing via cascading classifiers ("Rapid Object Detection Using a Boosted Cascade of Simple Features", 2001).

The similarities between language parsing in natural language processing and scene analysis in machine vision had been known for decades. Gabriela Csurka at the Xerox Research Centre in France developed the equivalent of the "bag-of-words" technique for machine vision: the "bag-of-visual words" or "bag-of-features" technique, which improved Perona's "constellation of parts" method for object detection and would remain the most popular technique for image classification for at least a decade ("Visual Categorization with Bags of Keypoints", 2004).

For object detection, instead, the most popular method became "deformable part model" (DPM), developed by Pedro Felzenszwalb and David McAllester at University of Chicago, that won the PASCAL VOC competition in 2007 ("Discriminatively Trained, Multiscale, Deformable Part Model", 2008).

Trivia: Which is the Homeland of Deep Learning?

Fukushima is Japanese; LeCun, Bengio and Collobert are French; Hinton and Zisserman are British; Thrun, Schmidhuber, Hochreiter, Szegedyand Socher are German; Andrew Ng, Fei-fei Li, Yangqing Jia, Aja Huang, Kaiming He and Dong Yu are Chinese; Krizhevsky and Sutskever are Russian; Olshausen is Swiss; Goodfellow is Canadian. Then add Siegelmann from Israel, Daniela Rus from Romania, Quoc Le from Vietnam, Malik from India, Simonyan from Armenia, Karpathy from Slovakia, Abbeel from Belgium, Vinyals from Spain, and the DeepMind founders from Britain (Demis Hassabis, son of a Greek man and a Chinese woman, and Mustafa Suleyman, son of a Syrian man and a British woman) and from New Zealand (Shane Legg, a student of Schmidhuber in Switzerland).

What do they have in common? None of them was born in the USA. Many of them now work in the USA or for US companies, but they were not born there and many were not educated there. There was a period of time during which US universities (especially the heavy weights Stanford, MIT, Yale and Carnegie Mellon) shunned neural networks, and US-born students had little motivation to work on such an esoteric and useless subject. The flame of neural networks was kept alive in the countries where symbolic A.I. did not completely obscure neural networks, which is pretty much every country except the USA.

Japan has always had a strange attitude towards software and neural networks were probably more appealing in that country for being so simple to program (and so difficult to implement on the hardware). In 1982 Canada opened the Canadian Institute for Advanced Research (CIFAR), a center that was independent from the schools of thought in the USA, and whose founder, James Mustard, was certainly closer to brain studies than to software. European mathematicians were attracted by the mathematics of neural networks. Luckily, in 1988 the Italian philanthropist Angelo Dalle Molle established in Lugano (Switzerland) the Istituto Dalle Molle di Studi sull'Intelligenza Artificiale, another center insulated from academic pressures.

In 1991 the British physicist John Taylor (who had once delivered a lecture titled "A Model of Thinking Neural Networks", 1973, and who dabbled in paranormal phenomena) founded the International Conference on Artificial Neural Networks (ICANN), the first edition being held in Finland.

Whatever the reason, deep learning was the outcome of a truly distributed and international collaboration from which the USA was actually notably missing.

"Never underestimate the power of a small group of people to change the world. In fact, it is the only way it ever has" (Margaret Mead)

Footnote: Neural Networks as Vector Spaces

When a neural network is trained to "learn" some pattern, its neurons get organized in a rather geometric manner.

A neural network basically constructs a high-dimensional space in which the distance between two points mirrors the degree of relationship between two objects in the real world. For example, two words that tend to show up together in many sentences will be represented by two points very close to each other. These "points" in high-dimensional spaces are "vectors". This was the discovery made by Tomas Mikolov's team at Google in 2013, using the "skip-gram" method for constructing vector representations of words from analyzing large sets of text, a method now known as "word2vec" ("Distributed Representations of Words and Phrases and their Compositionality", 2013). The same approach can be used to analyze images or speech, and, again, turn a large set into a high-dimensional vector space. Now you can use an old mathematical tool called "vector arithmetic" and perform calculations on these vectors. Mikolov showed that, for example, one can perform this algebraic operation: "vec(king) - vec(man) + vec(woman)" and obtain "vec(queen)". Unfortunately (or luckily), this method ended up revealing embarrassing biases in the texts of our world. For example, in 2016 Tolga Bolukbasi at Boston University used a neural network to calculate "vec(father) - vec(doctor) + vec(mother)". You would expect the answer to be still "doctor" as there are many female doctors, but instead the answer (obvious once you see it) was "nurse": the line from male doctor to female doctor is not a straight one ("Man is to Computer Programmer as Woman is to Homemaker", 2016). As Einstein would put it, the vector space is warped!

Neural networks that are trained only on co-occurrence of words in the past texts of the world tend to amplify the biases of the past. Unfortunately, this bias is then passed on to the many "recommendation engines" that increasingly control people's private lives. The result is, for example, that males will be more likely to be presented an ad advertising a highly-paid Silicon Valley job than females will. Hence the need for "de-biasing" the algorithms used to train neural networks for word embeddings. James Zou's group at Stanford, for example, introduced a post-processing algorithm that removes some of the biases "Multiaccuracy", 2018).

The skip-gram model, just like the bag-of-words model, needs to be trained to minimize a "loss function". The problem is that probabilistic models of language like these are computationally prohibitive because they require the mathematical processing of an entire vocabulary, which consists of tens of thousands of words. Two main "short cuts" were in use: "hierarchical softmax", developed by Yoshua Bengio's student Frederic Morin at University of Montreal ("Hierarchical Probabilistic Neural

Network Language Model", 2005), and "noise-contrastive estimation", developed by Michael Gutmann and Aapo Hyvarinen at University of Helsinki ("Noise-contrastive Estimation", 2010). Mikolov used "negative sampling", a simplified variation of noise-contrastive estimation.

For the record, vectors were introduced in the 19th century book "The Barycentric Calculus" (1827), published by August Moebius, the same German mathematician who in 1858 discovered the Moebius strip, and were indirectly also popularized by James Maxwell's classic on electromagnetism, "Treatise on Electricity and Magnetism" (1873), and even more by Werner Heisenberg's version of Quantum Mechanics (1925) that used matrix and vector algebras instead of calculus (Erwin Schroedinger's version).

It turns out that they seem to be ideal mathematical tools not only for electromagnetic waves and quantum systems but also for brains.

"The task is...not so much to see what no one has yet seen; but to think what nobody has yet thought, about that which everybody sees" (Erwin Schroedinger).

From Recognizing to Creating: Generative Adversarial Networks (History/ Part 8)

Richard Feynman's last words written on his blackboard were: "What i cannot create, i do not understand." Machines may have become good (or, at least, better) at classifying objects in categories (i.e. in recognizing what an object is), but they still lag far behind in drawing an example of a category. There is a difference between recognizing a dog and drawing a dog. If you understood what a dog is, it should be easy for you to sketch what a dog looks like. You have a generative mind: you can classify an object in its category and you can draw a typical object of that category, an object that presumably is not like any specific object that you have seen (unless you are Giotto). In order to implement this "generative" behavior a new approach to machine learning was required.

A recurrent network was used to generate sequences by Hinton's student Ilya Sutskever ("Generating Text with Recurrent Neural Networks", 2011), but recurrent neural networks were clearly limited in their ability to look ahead. In 2014 Hinton's student Alex Graves at University of Toronto used an LSTM network, more efficient at storing and retrieving information than plain recurrent networks, to generate handwriting. You can enter a text at his webpage

"http://www.cs.toronto.edu/~graves/handwriting.cgi" (as of 2017) and the system will write it out in human-like handwriting ("Generating Sequences With Recurrent Neural Networks", 2014). This was an important first step.

"Turing Learning" was developed by Roderich Gross at University of Sheffield: it pits two algorithms against each other, one trying to classify the other while the other is trying to fool the former ("A Coevolutionary Approach to Learn Animal Behavior Through Controlled Interaction", 2013). In a similar fashion in 2014 Ian Goodfellow, one of Bengio's students at University of Montreal, invented "generative adversarial networks" (GANs), consisting of two neural networks that compete against each other, one trying to fool the other ("Adversarial Examples and Adversarial Training", 2014). Another member of the same lab, Mehdi Mirza, improved the idea with "conditional adversarial nets" ("Conditional Generative Adversarial Nets", 2014). In 2015 Alec Radford at Indico Data Solutions in Boston proved that an expanded version of GAN can generate perfectly valid images, except that they are not real ("Unsupervised Representation Learning with Deep Convolutional Generative Adversarial Networks", 2015).

GANs contain two independent neural networks that behave as adversaries: one (the "discriminator") tries to correctly classify the real images while the other (a.k.a. the "generator") produces fake images to fool the former (the "discriminator"); the generator needs to improve its ability to "fake" images while the discriminator needs to improve its ability to discriminate between fake ones and real ones. The images produced by the generator are only partially random because they have to resemble the real ones. The generator is trying to fool the discriminator while the discriminator is trying to not get fooled by the generator. As they evolve the respective skills, both tend towards the point where the counterfeits and the originals are indistinguishable. The process trains the discriminator to classify more and more accurately. It also, incidentally, trains the generator to produce highly realistic pictures of imaginary objects, which may represent an art in itself.

The wonders of GAN quickly lured legions of researchers. In 2016 a joint team of University of Michigan (Honglak Lee) and the Max Planck Institute in Germany (Bernt Schiele and Zeynep Akata) employed GANs to generate images from text descriptions ("Generative Adversarial Text to Image Synthesis", 2016). Antonio Torralba's student Carl Vondrick at MIT employed a GAN to predict the plausible evolution of a scene, i.e. to generate a video. This implies understanding what is going on in the scene and inferring what is reasonable to see happen next ("Generating Videos with Scene Dynamics", 2016). Alexei Efros' students at UC Berkeley (including Jun-Yan Zhu and Phillip Isola) created a neural network that can turn the picture of a horse into the picture of a zebra using GANs ("Unpaired Image-to-Image Translation using Cycle-Consistent Adversarial Networks", 2017). Then they used "conditional adversarial networks" to develop the "Pix2pix model" capable of generating images from sketches or abstract diagrams ("Image-to-Image Translation with Conditional Adversarial Networks", 2017). When they released the related Pix2pix software, it started a wave of experiments (many of them by professional artists) in creating images: sketch your desired handbag and the system displays what appears to be a real handbag and even colors it. The conditional adversarial network (cGAN) learns how to map an input image onto an output image, or, if you prefer, how to redraw an image with different attributes. Later in 2017 Ming-Yu Liu's team at Nvidia used a slightly different architecture for their image-to-image translation system UNIT (or UNsupervised Image-to-image Translation), i.e. variational autoencoders coupled with generative adversarial networks ("Unsupervised Image-to-Image Translation Networks", 2017). A few months later Jaakko Lehtinen's team at Aalto University in Finland published a paper showing how GANs can create photorealistic pictures of fake celebrities.

Within a few years, Alec Radford's original DCGAN (Deep Convolutional Generative Adversarial Network) begat legions of variants: Yann LeCun's Energy-based GAN or EGBAN (2016) at New York University; Léon Bottou's Wasserstein GAN or WGAN at Facebook (2017), Aaron Courville's WGAN-GP at Montreal Institute for Learning Algorithms (2017) that generated photorealistic images of bedrooms; Trevor Darrell's bidirectional GANs or BiGANs at UC Berkeley (2017); Luke Metz's BEGAN (Boundary Equilibrium GAN) at Google that generated photorealistic faces (2017); Alexei Efros' above-mentioned Cycle-Consistent Adversarial Networks or CycleGANs (2017) at UC Berkeley; Fernando de la Torre's HDCGAN (2018) at CMU for high resolution pictures; Aaron Courville's "adversarially learned inference model" (2017) that learns both a generation network and an inference network; etc. Han Zhang at Rutgers University, working with Ian Goodfellow of University of Montreal, employed Xiaolong's self-attention mechanism for long-range dependencies to build SA-GAN, that improved the quality of the generated images because, unlike earlier GANs, it generated details based on cues from all points of the image, not only the adjacent ones ("Self-Attention Generative Adversarial Networks", 2018).

In 2018 Andrew Brock (working with Karen Simonyan at DeepMind) demonstrated BigGAN, capable of generating even more photorealistic images (an "inception score" of 166.3 and a "Frechet inception distance" of 9.6 when trained on ImageNet, way better than all the predecessors).

Brock, however, had to use a lot of "brute force": each generated image was the result of between 24 and 48 hours of processing on 512 TPUs ("Large Scale GAN Training for High Fidelity Natural Image Synthesis", 2018). Aidan Clark in Karen Simonyan's team at DeepMind extended the architecture of BigGAN to video and obtained DVD-GAN ("dual video discriminator") that could produce videos up to 256x256 resolution and up to 48 frames ("Adversarial Video Generation on Complex Datasets", 2019).

An important insight came from the work of Matthias Bethge's students Leon Gatys and Alexander Ecker at University of Tubingen in Germany: they realized that a neural network trained to recognize an object tends to separate content and style, and the "style" side of it can be applied to other objects, therefore obtaining a version of those objects in the style that the network previously learned. They taught a neural network to capture the style of an artist and then generated fake paintings in the artist's style ("A Neural Algorithm of Artistic Style", 2015). The Finnish trio of Tero Karras, Samuli Laine, Timo Aila at Nvidia developed a new kind of GAN that generated even more photorealistic faces of people (who don't exist) ("A Style-Based Generator Architecture for Generative Adversarial Networks", 2018). Their "style-based generator" was inspired by Gatys' "style transfer" networks, refined and accelerated by Serge Belongie's student Xun Huang at Cornell University ("Arbitrary Style Transfer in Real-time with Adaptive Instance Normalization", 2017).

Dongdong Chen in Gang Hua's team at Microsoft Research Asia extended the concept of style transfer to real-time videos with their system COVST ("Coherent Online Video Style Transfer", 2017). Ting-Chun Wang in Bryan Catanzaro's team at Nvidia used generative adversarial networks to do the same thing ("Video-to-Video Synthesis", 2018). The Nvidia team then used that method to create a videogame that made silly headlines such as "Watch AI conjure up an entire city from scratch" when in reality they had only generated the graphics. The world of the videogame had been created with a popular game engine; and the quality of the AI-generated graphics was inferior to the rendering of traditional graphic tools.

The performance of style transfer was improved by Fei-fei Li's student Justin Johnson at Stanford Univ ("Perceptual Losses for Real-Time Style Transfer and Super-Resolution", 2016) and by Dmitry Ulyanov of Skoltech & Yandex in Russia and Andrea Vedaldi of Oxford University ("Instance Normalization", 2016). Style-transfer networks of CycloGAN's generation included DiscoGAN, developed by Taeksoo Kim at SK T-Brain in South Korea ("Learning to Discover Cross-Domain Relations with Generative Adversarial Networks", 2017), and DualGAN, developed by

Zili Yi at Memorial University of Newfoundland in Canada ("Unsupervised Dual Learning for Image-to-image Translation," 2017). The original style-transfer networks could only transfer one style per network. Within a few months, however, the first multi-style transfer networks emerged. A multi-style transfer network is a network that can learn the style of multiple painters. This was attempted, for example, by Vincent Dumoulin at Google Brain ("A Learned Representation for Artistic Style", 2017) and by Serge Belongie's student Xun Huang at Cornell University ("Arbitrary Style Transfer in Real-time with Adaptive Instance Normalization," 2017). Xinyuan Chen at Jiao Tong Univ employed a new kind of GAN called "adversarial gated network" (or "gated GAN") to achieve multi-style transfer in a single network, so she could render photographs in the style of famous painters such as Monet, Van Gogh and Cezanne ("Adversarial Gated Networks for Multi-Collection Style Transfer", 2019).

At the end of 2018 Nvidia then demonstrated GauGAN, a network capable of turning simple sketches into photo-realistic scenes.

GANs are probably more interesting as a model of human intelligence than the inventors realized. Competition is one of the key factors in evolution, and, in particular, in the evolution of the brain. Competition often ends up being collaboration: when two adversaries compete, they indirectly help each other improve. They induce a positive feedback loop on their skills. The fundamental case of competition is perhaps the relationship between the two sexes. Charles Darwin in "The Descent of Man and Selection in Relation to Sex" (1871) and Ronald Fisher in "The Genetical Theory of Natural Selection" (1930) had already pointed out that sexual selection could greatly accelerate evolution: the female chooses the male and therefore males are pressured to try and be chosen, and as more and more males qualify the female has to become choosier, pressuring the males to further improve, and so on in an endless positive feedback loop. Geoffrey Miller in "The Mating Mind" (2000) went beyond the tail of the peacock and the song of the thrushes. He speculated that language itself, and therefore mind, is created via a feedback loop of this kind: Miller views the human mind not as a problem solver, but as a "sexual ornament". The human brain's creative intelligence must exist for a purpose, and that purpose is not obvious: survival in the environment does not quite require the sophistication of Einstein's science or Michelangelo's paintings or Beethoven's symphonies. On the other hand, these are precisely the kind of things that the human brain does a lot better than other animal brains. The human brain is much more powerful than it needs to be. Miller explains the emergence of art, science and philosophy by thinking not in terms of survival benefits but in terms of reproductive benefits. Sexual selection shapes not only the animal world but also our own mind and our civilizations.

Alas, GANs are very difficult to train. As an alternative to GANs, Thomas Brox's student Alexey Dosovitskiy showed that one can train a convolutional net to generate images ("Learning to Generate Chairs, Tables and Cars with Convolutional Networks", 2016); and no adversarial training was used by Qifeng Chen at Stanford University to synthesize photorealistic images ("Photographic Image Synthesis with Cascaded Refinement Networks", 2017)

GANs are certainly impressive in the way they can generate realistic images of objects that don't exist. However, computer-based visual effects have been around in the entertainment industry since at least the 1980s, using methods invented by the likes of William Reeves at Lucasfilm (the "particle systems" method of 1983) and Alan Barr at CalTech (the "solid primitives" method of 1984). In 1981 a British firm launched the Quantel Paintbox workstation that quickly revolutionized television graphics. For example, a Quantel Paintbox was used for the visual effects of the video of Dire Straits' "Money For Nothing" (1985). Then, of course, there was Industrial Light & Magic, the visual effects studio founded in 1975 by film director George Lucas that produced the visual effects for the "Star Wars" series (created by Lucas in 1976), the "Indiana Jones" series (created by Steven Spielberg in 1981), the "Back to the Future" series (created by Robert Zemeckis in 1985), Robert Zemeckis' "Who Framed Roger Rabbit" (1988), etc. Another pioneer of computer animation, Pacific Data Images in Silicon Valley, created the morphing visual effects in the video for Michael Jackson's "Black or White" (1991) using a new algorithm developed by Thaddeus Beier of Silicon Graphics and Shawn Neely of Pacific Data Images itself.

Footnote: Simplifying the (Artificial) Brain

Convolutions are not fun systems to implement on machines. Convolutional neural networks demand large amounts of memory and computational power. Krizhevsky's AlexNet of 2012 used 61 million parameters. It performed 1.5 billion operations to classify an image. Deepface, one of the first systems to apply deep learning to face recognition, developed in 2014 by Yaniv Taigman of Facebook and Lior Wolf of Tel Aviv University (who both worked at Face.com before it was acquired by Facebook in 2012), classified human faces using 120 million parameters and was trained on the four million facial images of the Labeled Faces in the Wild (LFW) datase. Karpathy's Neural Talk system of 2014 used 130 million convolutional parameters and 100 million recurrent parameters to generate captions for images. The deep networks that came after these ones were orders of magnitude more complex and therefore even more demanding.

Several tricks were proposed to make it easier for these monsters to do their job. For example, Song Han at Stanford pioneered a technique of network compression that made it possible to fit these networks on existing chips ("Learning both Weights and Connections for Efficient Neural Networks", 2015), but it was neither easy nor cheap. In 2016 Han published "deep compression", a method to compress deep networks by an order of magnitude without losing accuracy, and designed a specific hardware architecture, Efficient Inference Engine (EIE). Han shrunk the memory requirement of AlexNet and VGG-16 by 35 times and by 49 times respectively while retaining the same accuracy. In 2016 Kurt Keutzer's group at UC Berkeley (including Forrest Iandola, founder of DeepScale.ai) in collaboration with Song Han published Squeezenet, that achieved the same accuracy of AlexNet with 50 times fewer parameters.

Another approach, "low-precision convolution networks", was pioneered by Yoshua Bengio's student Matthieu Courbariaux at Montreal University. His "BinaryConnect" (2015) was a "binarized neural network": it limited weight values to only a + or - value, thus improving dramatically the computational efficiency of the algorithm and reducing the amount of memory required. This project evolved into BinaryNet by Ran El-Yaniv's group at Technion in Israel ("Binarized Neural Networks", 2016). Another binarized neuronal net, called Xnor-net, was proposed in 2016 by Ali Farhadi's group at University of Washington, resulting in 58 times faster convolutional operations and 32 times memory savings. Meanwhile, also in 2016, Fengfu Li and Bo Zhang of the Institute of Applied Mathematics in Beijing published a study of ternary-weight networks (neural networks with weights that can have three values, +1, 0 and -1) showing that their performance was only slightly worse than in the high-precision counterparts ("Ternary Weight Networks", 2016). An even better result was achieved also in 2016 by Eriko Nurvitadhi of CMU, working in Debbie Marr's group at Intel Labs ("Accelerating Deep Convolutional Networks Using Low-precision and Sparsity", 2016).

Artificial and Natural Neural Networks: The Myth of Backpropagation

Despite the success of deep learning in so many fields, saying that artificial neural networks are similar to the neural networks of the brain is like saying that linear regression or multiplication are similar to the way the neural networks of the brain work. Artificial neural networks are just one approximation and optimization method that works pretty well in some cases.

They are mathematical procedures. They are not what happens in the brain. Neither backpropagation nor Boltzmann machines reflect what our brain does for unsupervised learning. Neuroscience has not discovered any biological mechanism for errors to be backpropagated any further than a single synapse. Backpropagation implements a precise, symmetric model of connectivity among neurons, which is not what we see in the brain. Our neurons are wildly interconnected, but the connections are far from symmetric. All the methods that evolved out of "gradient descent" share very little with the processes discovered by neuroscience in the brain.

One of the most criticized videos in the history of A.I. was probably Hinton's "Here's how the Brain Implements Backpropagation" (2015, now withdrawn). Oddly enough, Hinton himself had recognized in 1989 that backpropagation is biologically implausible because there is no evidence that synapses can propagate signals in the reverse direction or that neurons can propagate error derivatives backwards and at the same time propagate signals forward according to a nonlinear function ("Connectionist Learning Procedures" in Artificial Intelligence magazine). A Boltzmann machine is a fantastic mathematical technique and perhaps more closely resembles the working of the brain, but, again neuroscience has found no such formulas in the brain.

Yoshua Bengio's "Towards Biologically Plausible Deep Learning" (2016) begins with this sentence: "Neuroscientists have long criticised deep learning algorithms as incompatible with current knowledge of neurobiology".

More importantly, a lot of "machine learning" uses techniques that are not neural networks and still work very well. Usually, nobody claims that linear regression or SVM mirror the way the brain works.

The biological process that adjusts the strength of connections between neurons in the brain is called Spike-Timing-Dependent Plasticity (STDP), and might not be the only one. STDP is an extension of Hebbian learning first suggested in 1973 by the Canadian psychologist Martin Taylor ("The Problem of Stimulus Structure in the Behavioural Theory of Perception", 1973). Hebbian learning is the fact that a synapse gets stronger when a presynaptic spike occurs just before a postsynaptic spike often enough. Taylor envisioned a symmetric process that would weaken the synapse whenever the opposite happened often enough (anti-Hebbian learning). Basically, an input that is likely to be the cause of a post-synaptic output is made even more likely to contribute to that output in the future, whereas inputs that are certainly not the cause of the post-synaptic output are made less likely to contribute in the future.

Wulfram Gerstner at Federal Institute of Technology Lausanne (EPFL) in Switzerland (who in 1996 rediscovered the differential Hebbian rules) expressed this fact in terms of the competition between "good" and "bad" timings: good timing is when the presynaptic spike arrives before a postsynaptic spike, and viceversa for bad timing ("A Neuronal Learning Rule for Sub-millisecond Temporal Coding", 1996). Gerstner wrote a book that every A.I. scientist should read: "Neuronal Dynamics" (2014).

Footnote: The Problem with Pattern Recognition

Neural networks are very good at discovering a pattern and drawing conclusions from it. If all objects of a certain shape are called "apple", then a new object of approximately that shape is most likely also an "apple". If people take the umbrella when it rains, and it starts raining, then most likely you too will take an umbrella. The problem with pattern recognition is, quite simply, that it often leads to the wrong conclusions.

For example, during World War II the British analyzed the bombers returning from bombing campaigns over Germany and decided to reinforce them where they were taking most bullets. The Hungarian-born statistician Abraham Wald at Columbia University humbly disagreed: those aircrafts were the ones that had survived and returned to base. The aircrafts that had been downed by the enemy did not return to base and therefore were not analyzed to recognize the most important pattern: that bullets in the other parts of the aircraft caused the aircraft to crash. The original conclusion was based on the wrong assumption that German anti-aircraft fire hit only the parts of the aircraft that had bullet holes, when in fact it is logical to assume that anti-aircraft fire hits all parts of the aircraft randomly.

This is known as an "observer selection effect": when analyzing patterns, you are biased by what you see, and you may be ignoring the most important factors because you don't see them. You need to complement your analysis of the pattern with logical thinking; and logical thinking may show you that the pattern you see is actually misleading. (Trivia: ironically, Wald died in an air crash).

Footnote: The Truth about AlphaZero

AlphaGo of 2016 was trained (over several months) by supervised learning from human expert moves and by reinforcement learning from self-play. In 2017 DeepMind's new system, AlphaGo Zero, learned the rules of weiqi/go in three hours by simply playing against itself ("Mastering the Game of Go without Human Knowledge", 2017). After 40 days of training, it was able to beat any version of the data-trained

AlphaGo. Also impressive was the fact that AlphaGo Zero of 2017 consumed less power than AlphaGo of 2016: it ran on a single machine with 4 Tensor Processing Units (TPUs) whereas AlphaGo of 2016 was distributed over many processors for a total of 48 TPUs. Whereas previous editions of Alpha used two neural networks for value and policy, AlphaGo Zero combined policy and value into just one neural network, a network made of a stack of residual modules with ReLu non-linearity. This neural network knew nothing about Go. The real "intelligence" was embedded in the search algorithm that was the real self-playing actor (or, better, algorithm): it was yet another variant of the Monte-Carlo tree search using Ronald Howard's policy-iteration method of 1960. The neural network acted as this algorithm's memory of what it had learned. More importantly, it looked like AlphaGo Zero had in a few days not only learned the rules of weiqi/go but also independently rediscovered thousands of years of human knowledge about playing weiqi/go as well as discovered new knowledge (creative novel strategies that had never been used by human masters).

It worked because weiqi/go is a fully deterministic world, because each player has complete information about the state of the game (of the world), because the number of possible moves (of possible actions in the world) is finite, and because the effects on the world of any action can be predicted exactly. There is a huge dataset of games played by human masters, each game is relatively short (about 200 moves), and, last but not least, it is a case in which we can afford to lose thousands of times before the system starts doing the right thing (winning games). All of these conditions are quite rare in the real world. When it has to lift an object, a robot works in a world that is not deterministic, where the number of possible moves is infinite, where the full effect on the world of an action is unpredictable, where a simple torque may require thousands of adjustments, and where a mistake can cost millions of dollars. We can build a huge dataset of how humans grasp the object, but we have to do it for each object and probably each person uses a slightly different movement of the arm, the hand and the fingers.

The game of weiqi/go is a case in which the Markov assumption holds: the current state is all you need to know in order to determine the best next move. You don't care how you got to the current state. However, the real world is mostly a very different kind of game, in which the current state is influenced by all sorts of factors. A game like football is much harder than weiqi/go, and almost any transaction in society is much harder than football. AlphaGo is a toy, just a toy. It will become a serious challenger to human intelligence only if we turn our own world into a toy-kind of world, by removing all the complexity and leaving only Markov conditions. If, in our obsessed determination to structure our world, we manage to turn our daily lives into the equivalent of a weiqi/go game, then AlphaGo will surpass human intelligence (except that i am not sure i would still call it "human" intelligence).

For the record, AlphaGo of 2016 can only play go on a standard 19 x 19 board: change the size of the board, and AlphaGo doesn't know how to play go anymore. That is the problem of all algorithms that don't learn the rules but simply try to mimic human behavior: a parrot can only repeat a few words but cannot have a conversation, for the simple reason that it has no clue what a conversation is.

If AlphaGo took a normal I.Q. test, it would get a zero: it can only do one thing, and cannot even understand the questions (Even the very lenient I.Q. test carried out in 2017 by the Chinese Academy of Sciences, that didn't require the machine to understand the questions, found that AlphaGo had an I.Q. of 48).

No wonder that AlphaGo plays go better than any human being: it has nothing else to do. AlphaGo is spending its entire life simply playing that game against itself. No human being is so stupid to spend her or his life playing the same game against herself or himself and doing nothing else. We have better things to do. Even the most obsessed player is busy doing many things, from buying food to reading books like this one. The brain of that player is doing thousands of different things, and many of them at the same time (like when you drive in heavy traffic listening to the radio and cursing at the driver who just cut in front of you).

AlphaGo can only play weiqi, and it keeps playing it all the time, day and night, against itself. That's how it gets better and better. There is no name for a person who can only do one thing, and keeps doing it better and better because it does it all the time, nonstop, 24 hours a day, because no human being has ever exhibited that kind of neurological disease. We would certainly deliver such a person to a psychiatric hospital.

AlphaGo Zero is obviously not a case of artificial general intelligence but of very narrow intelligence: just like AlphaGo, it can only do one thing. However, it is an interesting case because it shows that in some cases a machine can learn to perform a task not only better than humans but in ways that are different from the way humans think. Whether this is different from what a clock or a TV set do is open to debate. A clock keeps time better than any human can: is it a narrow artificial intelligence? A TV set does something that no human can do: it broadcasts images. In fact, any commercially available program performs a task better than most if not all humans: what is truly different between AlphaGo Zero playing weiqi better than humans and your tax preparation software? The software engineer will reply: AlphaGo Zero learned by itself to perform its task instead of being programmed to perform that task. But that is playing with words: a software engineer architected AlphaGo, i.e. wrote software for AlphaGo Zero to do whatever it did to learn to play weiqi. A better answer is that AlphaGo Zero keeps improving itself at what it does, but that's what any reinforcement-learning system does, as well as other kinds of learning systems (e.g. evolutionary algorithms); and, honestly, there are millions of programs that improve themselves over time: any program that interacts with the environment can absorb information from the environment and use that information to improve its interaction with the environment (the recommendation algorithms of commercial websites like Amazon are a typical example). The "environment" in AlphaGo Zero is the rules of the game, which is actually a very simplified environment compared with the real world. The final answer is that AlphaGo Zero is an experiment in new ways to program machines to perform narrow tasks better than humans. For tax preparation a traditional program is much preferred. For weiqi, AlphaGo Zero is preferred.

This limitation to narrow intelligence did not escape AlphaGo Zero's creators who immediately set out to create a more general program. In less than two months (end of 2017) DeepMind readied a new program ("Mastering Chess and Shogi by Self-Play with a General Reinforcement Learning Algorithm", 2017) called AlphaZero that, using the same tactic as AlphaGo Zero, had learned to play other games, including chess. AlphaZero learned chess in four hours and DeepMind announced that it had crushed Stockfish. Stockfish is one of the most popular "chess engines", developed by Marco Costalba in 2008 in Italy as an evolution of Tord Romstad's Glaurung (Norway, 2004), maintained by an open-source community, and not using neural networks. The training of AlphaZero took nine hours on 5000 TPUs (Google's specialized processor). AlphaZero was not the first program to beat Stockfish: Komodo, first developed in 2010 by Don Dailey and Larry Kaufman, had done it a few months earlier; and Houdini, developed by Robert Houdart in Belgium in 2010, won that year's TCEC (Top Chess Engine Competition), the world championship of chess engines; not Stockfish (that ended third even behind Komodo). The difference is that Komodo and Houdini played fair and square, whereas AlphaZero played against a defanged version of Stockfish: the version of Stockfish used was not the most recent one, Stockfish was not allowed access to an opening book (it is optimized for that scenario), its hashtable memory was limited to 1 Gigabyte (it requires a lot more when running on 64 cores), and the games were played at a fixed time of 1 minute per move (Stockfish is designed to optimize time management). Furthermore, AlphaZero ran on 4 TPUs whereas Stockfish ran on 64 CPU cores. Google's TPU of 2017 boasted a performance of 180 TFLOPS, dozens of times more than the performance of 64 cores. It was certainly an impressive demonstration of machine learning, but all chess positions with seven pieces or less had been mathematically solved already in 2012 and using significantly less computing power (Convekta's Lomonosov Tablebases, the first complete seven-piece endgame database). And, of course, AlphaZero did not learn solely from self-play: someone had to tell AlphaZero the rules of the game.

In December 2018 DeepMind finally published the paper on AlphaZero (credited to David Silver, Thomas Hubert and Julian Schrittwieser). However, as it is often the case with AI, we are playing with words. The paper stated clearly: "We trained separate instances of AlphaZero for chess, shogi, and go"; i.e. each instance of AlphaZero learned to play either or shogi or go, and then, yes, each instance became the best at the one and only game it mastered. The paper, however, closed with the vision of "a general game-playing system that can learn to master any game", an ambiguous sentence that led readers (and the media) believe that AlphaZero could learn all these games at the same time. We dumb humans can learn an unlimited number of games without having to forget the ones we already learned: one human brain learns an unlimited number of things. AlphaZero, instead, can learn to play chess only if it forgets how to play go. Sure, you can have three AlphaZeros running in parallel, one playing chess, one playing go and one playing shogi. Put them into the same machine and you can claim that the machine has learned to play all three games, and it is a world champion in all three. But that's conceptually very different from saying that we have one artificial brain playing three different games: what we really have is three artificial brains, each capable of playing only one game. It's the difference between saying "my kitchen can cook, wash the dishes, and even expel the smell" instead of saying "my kitchen has an oven, a dishwasher and a fan". The kitchen as a whole is certainly better than me at each of those tasks, but its working is a far cry from what the human brain does.

The media were impressed that AlphaZero took only 4 hours to learn what humans had learned in 100 years, but the difference between 4 hours and 40 hours or 4,000 hours or 4 million hours is simply the speed of the processor. The question is whether it did learn by itself, or did not, all that knowledge. Whether it did it in one second or in one year is relevant for winning the game, but doesn't make a difference in the achievement. To wit, whether Einstein came up with Relativity in one day or one year makes no difference to the achievement. And there are plenty of machines that can do things a lot faster than me. I cannot run as fast as my car. It is nothing new that a machine can be faster than a human, and faster than previous generations of that machine.

It is important to remember that these are games. There is a finite board, there are only so many things that can happen, and a move has a limited number of effects. The real world is different from a board game. The board of the real world is infinitely bigger than a weiqi or chess board. The number of things that can happen to you in the real world is virtually infinite. And also virtually infinite can be the effects of your actions. A friend who studied law always reminds me the story of the man who throws a cigarette butt on the floor. What if the wind blows it on dry vegetation and the vegetation catches fire? And what if that fire causes a car to hit a post? And what if a wheel of that car rolls down the hill and hits and kills a child? And what if the mother of that child runs across the street without paying attention to the coming truck and gets killed too and in the process the truck swerves violently and crashes into a house killing six people? And so on and on. In the real world AlphaZero would have to learn an infinite number of rules, and learning them by "reinforcement" may just be impossible: reinforcement learning works when there is an immediate reward or punishment. Without any (real) intelligence, AlphaZero would have to try an infinite number of actions in order to find out how to do laundry: ring the bell, turn on the TV, read page 145 of this book, jump from the roof, and so on. Its algorithm would eventually find a reward for one specific action: put the clothes in the washing machine, close the lid, and press the start button. And even this action would have to be repeated successfully many times, with the proper sequence of pressing buttons, before AlphaZero accepts it as the correct operation to wash clothes. Even brute-force A.I. would require an immensely powerful computer for AlphaZero to carry out this simple operation that any human intelligence can perform in a few seconds.

Before you surrender to AlphaZero and its likely descendants, think how long it took you to learn how to play a game: three times? seven times? How long did it take you to become a good player? One hundred times? Now remember that these DeepMind algorithms had to play the game millions of times to learn what you learned in a few attempts. And in the meantime you probably learned also a lot of other things.

To assess AlphaZero's achievement, we have to decide (yet again), what we are looking for: are we looking for a tool that does something better than humans do or are we looking for human-like intelligence? Is AlphaZero better than humans at what it does? Sure, just like an airplane can fly and i cannot, and a car can move a lot faster than the fastest man, and a TV set can show the image of something happening far away. These are all examples of super-human achievements achieved by machines that use non-human methods (non-human "intelligence"?) So is a hammer, by the way, and so is a screwdriver. These are tools that can do things that we cannot do. Every machine exists because it can do something better than most humans can. If it cannot, then it's a toy for children.

Is Alphazero a case of (human) intelligence? No, of course not. We humans don't need to play millions of games to master a new game. We humans need to play a few times and we treasure advice from fellow humans. We are very different from AlphaZero. Can AlphaZero do something a lot better and faster than us? Sure, just like every computer program, just like every car, just like every machine in an assembly line.

Perhaps the most stunning fact about AlphaGo and its descendants is that they are based on old A.I. techniques. The ideas behind AlphaZero are so simple that, within one month, David Foster, co-founder of Applied Data Science, published instructions on "How to build your own AlphaZero A.I. using Python and Keras".

DeepMind's A.I. systems are certainly impressive research programs, but one has to put them in perspective. When you marvel at the power of AlphaGo or AlphaZero, ask yourself: "How many lives have DeepMind's A.I. systems saved so far?" Or, if you are the Wall Street kind of person: "What kind of economic revolution have DeepMind's A.I. systems caused so far?" To put things in perspective, exactly 100 years before DeepMind was founded, two German chemists, Fritz Haber and Carl Bosch, invented a process to produce the fertilizer ammonia: that humble invention caused a revolution in agriculture that fed billions of people and created a vast global economy.

As of 2018, AlphaGo has not been made available to the weiqi/go community. Ditto for AlphaGo Zero and AlphaZero.

Footnote: Dynamic Routing and Capsule Networks

In 2017 Geoffrey Hinton spoke of the shortcomings of convolutional neural networks in recognizing images (despite the media's euphoria for it) and came up with a new idea, the "capsule network" ("Dynamic Routing Between Capsules", 2017). He first proposed it eight years earlier when he wrote: "This paper argues that convolutional neural networks are misguided in what they are trying to achieve" ("Transforming Auto-encoders", 2011). One problem of "convnets" is that they are "translation invariant": they detect the co-existence of some features and ignore their relative position. If an image has two eyes, a nose and a mouth, it gets classified as a face, even when the eyes are placed below the mouth. But a face is not just an aggregate of some features but an ordered aggregate of such features: it does make a difference whether the mouth is above or below the nose! To overcome this limitation, Hinton designed layers that consisted not of individual networks, but rather of capsules, groups of functional networks. Each capsule is programmed to detect a particular attribute of the object being classified. (Of course, a traditional software engineer could just write a little bit of code to specify where the eyes are supposed to be in relation to the nose, but that would be truly old-fashioned).

The other problem of convolutional nets is that they are susceptible to "white-box adversarial attacks": one can easily embed a secret pattern into an image to make it look like something else to the neural network (but not to the human eye). Capsnets should be biologically more plausible. After a quarter of a century, Hinton rediscovered something that he had researched when he was still at UC San Diego surrounded by neuroscientists: a way to generate shape descriptions ("A Parallel Computation that Assigns Canonical Object-based Frames of Reference", 1981). That method was later improved by Bruno Olshausen, Charles Anderson, and David Van Essen at Caltech ("A Neurobiological Model of Visual Attention and Invariant Pattern Recognition based on Dynamic Routing of Information", 1993), and the result was a better way to represent an object in space. Anderson and Van Essen had been researching a computational model of visual attention ("Shifter Circuits", 1987), in particular the mechanism that regulates the flow of data within and between cortical areas. Olshausen worked on the hypothesis that the brain contains a population of control neurons whose only job is to route the data flow in the cortex. These neurons implement a process called "dynamic routing", the brain's equivalent of a computer's routing circuits. This provides a more plausible model of object recognition than the collections of loosely related "invariant features" proposed by Fukushima and LeCun with their convolutional networks. Furthermore, Olshausen's model seems to be a better approximation of the ventral stream in the visual cortex.

There is a fundamental difference between today's artificial neural networks and the brain: Today's artificial neural networks are sequences of flat layers, whereas the neocortex of the brain has both horizontal layers and vertical columns.

A neural network simply matches patterns. As such, it is prone to making very silly mistakes. Joshua Tenenbaum's team at MIT, in collaboration with IBM and DeepMind, combined deep learning and symbolic reasoning to create a neuro-symbolic concept learner (NS-CL) capable of learning about the world just as a child does: by looking around and talking ("The Neuro-Symbolic Concept Learner", 2019).

Lengthy and Boring Footnote: Variational Inference

Inference in probabilistic models is often intractable. The most common approximation algorithms were still based on Monte Carlo methods. An optimization-based alternative to the sampling-based Monte Carlo methods is variational inference, and the wake-sleep algorithm was a case of variational inference. In 1993 Geoffrey Hinton and Drew VanCamp at University of Toronto had already proposed a kind of variational inference (that they called "ensemble learning") for neural networks ("Keeping Neural Networks Simple by Minimizing the Description Length of the Weights", 1993). Variational inference and MCMC do the same job (approximate inference of posterior probability) but in wildly different ways, each with pros and cons.

Variational methods convert a complex problem into a simpler problem. A practical example in physics is the study of a many-body system, which is intractable until it is approximated as a one-body system. The mathematical tool is the calculus of variations that we owe to Swiss mathematicians: pioneered by the brothers Jacob and Johann Bernoulli in 1696, it was formalized as a distinct branch of calculus by Leonhard Euler in his book "Methodus Inveniendi Curvas Lineas" of 1744, and Euler also gave it a name in a 1756 lecture titled "Elementa Calculi Variationum". The calculus of variations (mainly developed by the German mathematician Karl Weierstrass in the 1860s) searches through a space of functions for the "best" function. Therefore the calculus of variations deals not with functions but with "functionals", functions of functions: the solution of the problem is not a number but a function. The calculus of variations, used for optimization purposes, is an alternative to Richard Bellman's dynamic programming. It turns out that this calculus also helps to approximate probabilistic inference, and that's what is referred to as variational inference.

After Hinton's attempts, variational inference for probabilistic models was pioneered by Michael Jordan at UC Berkeley ("Mean Field Theory for Sigmoid Belief Networks", 1996) and by Tommi Jaakkola at MIT ("Variational Methods for Inference and Estimation in Graphical Models", 1997), and was inspired by statistical physics, notably by the Italian physicist Giorgio Parisi ("Mean Field Theory for Spin Glasses", 1980).

Variational inference is about maximizing the "evidence lower bound" or ELBO. In other words, one can view problems of probabilistic inference (i.e. infer the value of something given the value of something else) as problems of optimization (find the values that minimize or maximize some function, in this case maximize the ELBO or, equivalently, minimize another quantity called the "Kullback-Leibler divergence"). Several methods have been proposed to maximize the ELBO.

Alex Graves (who had studied with Schmidhuber at IDSIA, but now at University of Toronto) applied the "minimum description length principle" to variational inference ("Practical Variational Inference for Neural Networks", 2011). He employed the "minimum description length principle" that had already been used to train an autoencoder by Hinton in 1993 ("Keeping Neural Networks Simple by Minimizing the Description Length of the Weights", 1993). This principle dates back to Jorma Rissanen in Finland ("Modeling by Shortest Data Description", 1978) and is basically a computational version of Occam's razor in which the best model for a dataset is the one that yields to the best compression of the data. In 2012 David Blei of Princeton University and Michael Jordan of UC Berkeley (and Jordan's student John Paisley) presented an alternative to the mean-field method, based on Robbins' 1951 stochastic optimization ("Variational Bayesian Inference", 2012). Then Matt Hoffman at Adobe applied this method to analyze large libraries of texts ("Stochastic Variational Inference", 2013). In 2013 David Blei and his student Rajesh Ranganath at Princeton University developed a "black box" variational inference algorithm, one that can be quickly applied to many models with little changes ("Black Box Variational Inference", 2014), which then evolved into the "hierarchical variational models" of 2016. In general, the expression "black box" is used when one analyzes a system without looking at its internal workings, only by studying its inputs and outputs.

Bayesian models classify data divide into "discriminative" and "generative". The former simply makes a prediction (for example, whether a sentence is German or English). The latter builds a model of the data (e.g. a model of the German and English languages) and therefore "understands" more of the data. A discriminative model learns the conditional probability ("the probability of y given x"), whereas a generative model learns the joint probability of the two events happening together. The former deals only with the existing set of data. The latter can generate missing data or compress the data. A generative classifier learns the "rule" that generates the data. Once it "understood" what the rule was, it could generate data that have not happened but could happen and belong to the same class as the ones that did happen. On the other hand, a discriminative classifier sticks to the observed data and yields a simpler rule. Discriminative models include: logistic regression, support vector machines (SVMs), nearest neighbor, conditional random fields, and traditional neural networks. Generative methods include naive Bayes, hidden Markov models, restricted Boltzmann machines, and generative adversarial networks. It is natural to assume that generative classifiers are better, but in 2001 the young Andrew Ng, when he was studying with Michael Jordan at UC Berkeley, compared linear regression and naive

Bayes and concluded that discriminative classifiers are usually a better choice ("On Discriminative vs Generative classifiers", 2001).

There are three kinds of generative models (or "density modeling") in the age of neural networks: generative adversarial networks (a game between two networks), variational autoencoders (methods that maximize the ELBO) and autoregressive models.

The probabilistic interpretation of autoencoders was pioneered by LeCun's student Marc'Aurelio Ranzato at New York University ("Efficient Learning of Sparse Representations with an Energy-based Model", 2007) and by Bengio's student Pascal Vincent at University of Montreal ("A Connection Between Score Matching and Denoising Autoencoders", 2011).

It was time to rediscover Hinton's "wake-sleep" approach. Variational autoencoders were introduced in 2013 by Max Welling and his student Diederik "Durk" Kingma at University of Amsterdam in the Netherlands ("Auto-encoding Variational Bayes", 2013). They then worked with Dutch data scientist Tim Salimans on bridging the gap between MCMC and variational inference ("Markov Chain Monte Carlo and Variational Inference", 2015). Salimans was soon hired by OpenAI to work on generative adversarial networks with Goodfellow and Radford, as well to improve the accuracy of variational autoencoders with Kingma, also hired by OpenAI ("Improving Variational Inference with Inverse Autoregressive Flow", 2016). A more general and efficient model of variational autoencoders was introduced in 2014 by Daan Wierstra's team at DeepMind, namely Danilo Rezende and Shakir Mohamed, by fusing elements of deep learning and of probabilistic inference, and they used it to generate realistic images ("Stochastic Backpropagation and Approximate Inference in Deep Generative Models", 2014). These two projects yielded a new class of powerful generative models, called "Deep latent Gaussian models" (DLGM). In 2015 Daan Wierstra's team at DeepMind unveiled the Deep Recurrent Attentive Writer or DRAW, which is a neural network built around a framework of variational autoencoders ("A Recurrent Neural Network for Image Generation", 2015). DRAW extended the variational autoencoder with two techniques ("progressive refinement" and "spatial attention") that greatly improved its efficiency, thereby enabling the generation of larger and more complex images. Systems like DRAW (that Wierstra named "sequential generative models") show that inference, generation and generalization are different aspects of the same process. In fact, the same Rezende-Mohamed team built a system capable of one-shot generalization, of generalizing a concept after just one encounter ("One-Shot Generalization in Deep Generative Models", 2016). Kihyuk Sohn, Honglak Lee, and Xinchen Yan of NEC Laboratories in Detroit built their hybrid "conditional variational autoencoder" VAEGAN (which stands for "Variational Autoencoder + Generative Adversarial Net") that can generate faces "conditioned" on parameters such as "young", "blonde", etc ("Learning Structured Output Representation using Deep Conditional Generative Models", 2015). The inventions of generative adversarial networks and of variational autoencoders were the events that sparked renewed interest in probabilistic thinking.

In 2015 Brendan Frey's student Alireza Makhzani at University of Toronto in collaboration with Goodfellow (then at OpenAI) developed "adversarial autoencoders", i.e. a combination of probabilistic autoencoder and generative adversarial networks to perform variational inference.

In 2018 Juergen Schmidhuber of IDSIA and David Ha (a former Wall Street managing director now working at Google) published a deep reinforcement learning algorithm ("World Models", 2018) that solved the "car racing" problem in which an agent has to drive a car along a racetrack as fast as possible. The solution consisted of three components: a variational autoencoder that creates a compact representation of the situation (the car relative to the environment, e.g. a bend approaching); an LSTM recurrent neural network with 256 hidden units that predicts the next situation based on the current actions (steering, accelerating and braking); and a densely connected single-layer neural network that chooses the next action, which is a combination of three actions (steering, accelerating and braking). Based on random interactions with the environment, the network builds a mental model of how the world works (i.e., its physical laws) and of how its own actions affect the state of the world. At this point the agent can learn the optimal driving strategy without actually driving. To be fair, PILCO (Probabilistic Inference for Learning Control), developed in 2011 by Marc Deisenroth at University of Washington and Carl-Edward Rasmussen at Cambridge University achieved similar results using a simpler method: a Gaussian process turns the data from the environment into a model of the system, and then uses this model to learn to perform complex control tasks like riding a unicycle.

Variational autoencoders led to an important generalization of neural networks. "Deep" neural networks with increasing number of layers, with smaller and smaller steps in between layers, clearly approximated a "continuum" of layers. David Duvenaud's student Ricky Chen at the University of Toronto studied continuous-depth neural networks that replaced the discrete sequence of layers with the continuous dynamics described by ordinary differential equations ("Neural Ordinary Differential Equations", 2018). The idea of continuous-time neural networks was not new. In fact it dated back 30 years to a theoretical speculation by LeCun,

Hinton and Sejnowski ("A Theoretical Framework for Back-propagation", 1988), and to a formulation by Barak Pearlmutter of Siemens labs at Princeton ("Gradient Calculations for Dynamic Recurrent Neural Networks", 1995); but Duvenaud was the first to actually implement it.

The third kind of generative models was mostly the work of Google's division DeepMind. Autoregression is a forecasting method in which future values of a time series are estimated solely based on a weighted sum (a linear combination) of past values of the series. In 2011 Iain Murray of University of Edinburgh and Hugo Larochelle of University of Toronto built NADE, which stands for "Neural Autoregressive Distribution Estimator", and used it to generate handwritten digits. In theory, the restricted Boltzmann machine is not suitable for estimating joint probabilities but they basically converted the restricted Boltzmann machine into a Bayesian network ("The Neural Autoregressive Distribution Estimator", 2011). In 2015 NADE evolved into MADE or "Masked Autoencoder for Distribution Estimation", that added a process called "masking" (similar to the widely used method of "dropout" training). Then, in 2014, Andriy Mnih and Karol Gregor of Wierstra's team at DeepMind designed a new kind of autoencoder, the "deep autoregressive network" or DARN, based again on Rissanen's "minimum description length principle" and they similarly used it to generate handwritten digits ("Deep Autoregressive Networks", 2014). The next autoregressive network to come out of DeepMind was PixelRNN, developed by a team led by Koray Kavukcuoglu (another former student of LeCun at New York University). Schmidhuber's student Alex Graves at the Technical University of Munich had introduced multidimensional Long Short-Term Memory networks in 2009 ("Offline Handwriting Recognition with Multidimensional Recurrent Neural Networks", 2009) and DeepMind designed an architecture of twelve two-dimensional LSTM layers with the novelty of convolutions applied both horizontally and diagonally ("Pixel Recurrent Neural Networks", 2016). The same team implemented PixelCNN ("Conditional Image Generation with PixelCNN Decoders", 2016) and then WaveNet (2016), based on PixelCNN, that can generate speech and music. A team at OpenAI (Tim Salimans, Diederik Kingma, Andrej Karpathy) then turned PixelCNN into an autoregressive model, PixelCNN++, that uses the previous nine generated pixels to calculate how to generate the next pixel.

Footnote of the Footnote: Hierarchical Bayesian Networks

Back to Bayesian inference. Another thread led away from neural networks. In 1999 Joshua Tenenbaum graduated at MIT with a dissertation on "A Bayesian Framework for Concept Learning". In 2001 he and his student Thomas Griffiths at Stanford University developed their "hierarchical Bayesian models" for inductive generalization at various levels of abstraction, i.e. learning higher-level concepts ("Structure Learning in Human Causal Induction", 2001). These are directed graphical models (like Pearl's Bayesian nets and unlike Boltzmann machines) implemented in conjunction with the Markov Chain Monte Carlo procedure.

The hierarchical Bayesian framework was later refined by Tai-Sing Lee of CMU ("Hierarchical Bayesian Inference In The Visual Cortex", 2003). These studies were also the basis for the widely-publicized "Hierarchical Temporal Memory" model of the startup Numenta, founded in 2005 in Silicon Valley by Jeff Hawkins, Dileep George and Donna Dubinsky; yet another path to get to the same paradigm: hierarchical Bayesian belief networks. The startup was charged with developing the theory presented by Silicon Valley entrepreneur Jeff Hawkins in his book "On Intelligence" (2004), ambitiously subtitled "How a new understanding of the brain will lead to the creation of truly intelligent machines".

In 2008 Alison Gopnik proposed the "Theory Theory" of how children learn new concepts, the idea being that children use the same approach that a scientist uses to develop a scientific theory. Tenenbaum's hierarchical Bayesian models were immediately identified as a plausible mathematical tool to mimic what happens in the child's mind. The neural network is also unable to do much with the concept that it learns. We generally use the concepts that we learn, we do it right away and we do it naturally. We can imagine a whole universe of relationships between the new concepts and other concepts, and we can explain those relationships. More importantly, we can take action based on the new concept. Joshua Tenenbaum, Charles Kemp (now at CMU) and Thomas Griffiths (now at UC Berkeley) wrote ("How to Grow a Mind", 2011) in which they argued that hierarchical Bayesian models underlie all of our cognitive life. In 2014 Tenenbaum, working with Brenden Lake of New York University and Ruslan Salakhutdinov of University of Toronto, used Bayesian reasoning instead of neural networks and devised a program that learns in a more human-like fashion, although only in the narrow domain of handwritten characters ("Human-level Concept Learning through Probabilistic Program Induction", 2014).

There is also a trend to think of transfer learning and multi-task learning (the two kinds of learning that are common among humans but so hard to realize in algorithms) as forms of generalization, after Tenenbaum and

Griffiths at Stanford University viewed Bayesian inference as the path to generalization ("Generalization, Similarity, and Bayesian Inference", 2001), following the ideas of Stanford psychologist Roger Shepard ("Toward a Universal Law of Generalization for Psychological Science", 1987).

The problem is that many probabilistic models are mathematically intractable. Surya Ganguli's student Jascha Sohl-Dickstein at Stanford rediscovered Jarzynski's 20-year-old idea of using a Markov process to gradually convert one distribution into another, in particular intractable distributions into tractable ones ("Deep Unsupervised Learning using Nonequilibrium Thermodynamics", 2015).

Alas, in 2016 Jim Crutchfield at UC Davis demonstrated that probabilistic induction is not suited for non-linear systems, no matter what ("Multivariate Dependence Beyond Shannon Information", 2016); and, in general, it doesn't take mathematicians to realize that humans don't think in probabilistic terms

Cultural Background: The Network Age

Is the brain a neural network? No, of course not. The brain is not just a network (and the neuron is not just a node) but we live in the age of the network. We think of everything in terms of networks. It is probably a consequence of liberal capitalism. When people lived under the dictatorship of kings and popes, the pyramid was the preferred topological model. Everything was supposed to be a hierarchy. For example, there was a hierarchy of nature with animals at the bottom, humans above them, spirits above humans, and a supreme god at the top. Now that we got rid of the hierarchy, we think of societies, economies and cities as networks. The network metaphor has become pervasive even in Physics and in Linguistics.

The millennial generation is often referred to as "the digital natives" but they are more properly the "network natives". Whether data, texts and images are digital or analogic would make no big different to their lives: does anyone know the difference between the new digital and the old analogic television? The difference is that their lives (they are told) are networked. They grow up thinking of networks: the encyclopedia is a network (Wikipedia), their social life is a network (Facebook), their government structures health care in "provider networks", public transportation is a network, etc.

The network is a modern invention. The Silk Road was actually a network of trade routes, but nobody called it "Silk Network". The so-called "Peutinger Map" at the Austrian National Library in Vienna (a medieval French copy of a Roman original dating probably from the age of Augustus) is a parchment scroll, 34-cm high and 675-cm long, representing the road network of the Roman Empire squeezed so as to look like a series of straight lines converging on Rome. When in 1820 Louis Becquey proposed to build a network of canals in France, his 75-page "Report to the King on the Interior Navigation" never used the word "network". Until recently, a machine was routinely showed as neatly divided into modules and the workings of the machine was routinely represented as multiple flows of material, i.e. lines; now any manual begins with a diagram that shows the network of processes.

The buzzword (and the ruling metaphor) of the 1980s was still modularity. The French sociologist Emile Durkheim's "The Division of Labor in Society" (1893) had hailed the ubiquity of modularity (not of networking) in all human societies. For Georg Simmel's "The Metropolis and Mental Life" (1903) city life led to the division of labor, whereas today almost all sociologists think of city life as creating networks. Modularity had been made popular in the 1920s by architects such as Le Corbusier, Walter Gropius and Buckminster Fuller, leading to the modular mass housing of the 1960s such as Moshe Safdie's "Habitat 67" in Toronto and Kisha Kurokawa's "Nagakin Capsule Tower" (1972) in Tokyo. Modularity was also made popular by computer science: the hardware (as per John Von Neumann's architecture) was represented as a set of modules, and programming languages such as Niklaus Wirth's Modula (1976) encouraged modularity in software. Finally, the metaphor of modularity infiltrated cognitive science with David Marr's "Vision" (1982) and Jerry Fodor's "The Modularity of Mind" (1983)

They didn't know it, but, indirectly and involuntarily, Emile Durkheim and Georg Simmel had founded "social network analysis", the study of patterns of social interactions, although it was only in 1954 that the term "social network" was coined (by the Australian anthropologist John Barnes in the article "Class and Committees in a Norwegian Island Parish"); and many consider "Who Shall Survive?" (1934), published by the Romanian psychiatrist Jacob Moreno, as the founding text of social network analysis.

As usual, it was Marshall McLuhan's book "Understanding Media" (1964), that popularized the adage "the medium is the message", that pioneered today's "network" metaphor: "It is a principal aspect of the electric age that it establishes a global network that has much of the character of our central nervous system." The emphasis on cities came from Edward Laumann at University of Chicago who wrote "Bonds of Pluralism - The Form and Substance of Urban Social Networks" (1973).

During the 1990s, with the invention of the World-wide Web, connectivity began to replace modularity as the ruling paradigm and now it's all about connectivity. The beneficial properties of network topologies are routinely hailed like the ideological dogmas at a meeting of a communist party. Publishers relish books such as Dutch sociologist Jan van Dijk's "The Network Society" (1991), Spanish sociologist Manuel Castels' "The Rise of the Network Society" (1996), Belgian sociologist Armand Mattelart's "Networking the World 1794-2000" (2001); or "Information Rules" (1998), subtitled "A strategic guide to the network economy", by UC Berkeley economists Carl Shapiro and Hal Varian, and "Platform Revolution" (2016), subtitled "How networked markets are transforming the economy", by Dartmouth College's Geoffrey Parker. In 2003 the Organisation for Economic Co-operation and Development (OECD) published a report titled "Networks of Innovation" that begins with the sentence: "OECD countries are increasingly characterised as network societies". In 2014 the "Oxford Handbook of Innovation Management" includes a chapter by Timothy Kastelle and John Steen on "Networks of Innovation" according to which "Networks are fundamental to understanding and managing innovation". There is a quasi-religious yearning to the prophecies of how the forces of connection will defeat the forces of division, a yearning expounded by books such as Parag Khanna's "Connectography - Mapping the Future of Global Civilization" (2016).

Social-network analysis rediscovered graph theory, invented by the Swiss mathematician Leonhard Euler in 1736 to solve the problem known as "The Seven Bridges of Koenigsberg", as well as the more recent "random graphs" introduced in 1959 by the Hungarian mathematicians Paul Erdos and Alfred Renyi.

In 1998 David Krackhardt and Kathleen Carley of CMU invented Dynamic Network Analysis (aka DNA, but not related the genetic thing) and preached that networks occur across multiple domains and at different levels (meta-networks or high-dimensional networks). In 1999 the Hungarian physicist Albert-Laszlo Barabasi at University of Notre Dame in Indiana focused on "scale-free" networks, i.e. networks whose distribution follows a power law: according to him these networks are ubiquitous in natural, technological and social systems. Barabasi founded the Center for Complex Network Research at Northeastern University (to study how networks emerge, what they look like, and how they evolve) and the Network Science Society; and wrote the book "Linked - The New Science of Networks" (2002). Quote: "Networks are present everywhere."

For the record, later studies, for example by Aaron Clauset and his student Anna Broido at University of Colorado, seemed to prove the opposite, that the power law is rare in real-world networks ("Scale-free Networks are Rare", 2018). But power laws had become a popular meme in their own after the physicist Kenneth Wilson at Cornell University had shown why they pop up in all sorts of phase transitions (regardless of the material involved) by linking them to a new mathematical object that the French mathematician Benoit Mandelbrot at IBM was studying at the same time for completely different reasons (see the legendary paper "How Long Is the Coast of Britain?" of 1967) and to which he would eventually give a name in his book "Fractals" (1975). Wilson's intuition marked the birth of "renormalisation group theory" which is now widely used in physics ("Renormalization Group and Critical Phenomena", 1971). Further boosting the popularity of power laws, one decade later the Danish physicist Per Bak observed power-law behavior in complex nonlinear systems and jumpstarted another whole new discipline (that took the name from another legendary paper, "Self-organized Criticality" of 1987).

The "network effect" is one of the most quoted "effects", although nobody really knows what it is. The first influential person to talk about the "network effect" was AT&T's president Theodore Vail in 1908: the value of a network is proportional to how many people use it. Similarly, David Sarnoff, the mogul of RCA from 1919 until 1971, stated that the value of a network in the broadcast industry is proportional to the number of viewers. In 1980 Robert Metcalfe, the inventor at Xerox PARC of the local-area networking technology Ethernet, stated that the value of a network is proportional to the square of the number of devices. In the age of the Internet and of social media, the network effect became a must in the business plan of Internet startups looking for venture capital, and also for honest economists trying to explain how trivial ideas like WhatsApp ended up being worth $18 billion.

No wonder then that, in our age, we should be thinking of the brain as a network. The idea originated in neuroscience around 1911 with Edward Thorndike. In a sense, the other school of A.I., the symbolic knowledge-based school, still belonged to a transitional era between the king and democracy, an era in which a code of laws (represented by mathematical logic) was driving all the processes. Maybe this school was doomed to fade away, regardless of its scientific merits, simply because it didn't fit well with the "network" paradigm.

However, the fact that the network has become the favorite topology of the 21st century doesn't mean that everything is indeed just a network. A century from now our descendants might laugh at our simplistic view of the brain the same way that today we laugh at René Descartes' view of the brain as a hierarchy.

Gerald Edelman and Giulio Tononi in their book "A Universe of Consciousness" (2000) explain that not only are there many different types of neurons in the human brain (more than 70 in the eye's retina alone), and that no two neurons are alike, and that neurons don't even fire at a constant rate, but also that there are different topologies in the brain: some regions are networks (notably the thalamo-cortical system, although this network is better viewed as a network of networks, a network of specialists), other regions are long loops (notably between the cortex, the cerebellum and the hippocampus) and other regions are fans (the nuclei responsible for categorization and action and that project into the whole brain). The various regions of the neocortex are organized into columns and layers. Communications between neurons can take place via more than 50 different kinds of neurotransmitters. In 2017 Ido Kanter, a physicist at Bar-Ilan University, published a study that contradicts the stereotype of how neurons communicate. Traditionally, we assumed that each neuron sums up all the signals from other neurons and, when this quantity reaches a threshold, the neuron fires its own signal to other neurons; but Kanter's team discovered that a neuron contains many independent excitable places, each acting as a threshold unit that sums up the incoming signals.

The neural networks of Artificial Intelligence resemble the neural networks of brains the same way that a car resembles a horse: the car can travel faster than a horse and offers a lot more comfort, but is not a horse.

A neural network needs to see thousands of images of an object before it can recognize that object. A human brain doesn't go through this kind of training. A human can learn to recognize and write a letter of a foreign alphabet after just seeing one example of it. I don't need to see a thousand different versions of the word "mouth" in Chinese: i have seen it once and memorized it, and now i can recognize it wherever i see it, in whichever size, color and shape, and even if the paint is faded and even if someone has scribbled over it. A neural network treats an image as a pattern of pixels and, after seeing many patterns that belong to the same thing, constructs the model that will help correctly classify future instances of that pattern. My brain simply remembers how to write the character: a square with two strokes at the bottom. Sometimes i simply remember a Chinese character by analogy with an object. The Chinese character for "enter" looks just like the letter lambda in Greek. If you straighten the top, it becomes the character for "person". And if you add a straight line where the two curves join, it becomes the character for "big". My brain can employ several different forms of reasoning to remember a character that a neural network can learn only after seeing thousands of them. More importantly, the human brain can learn quickly how to act in dangerous situations. Sure, every time we get better at avoiding danger, just like a neural network, but we normally survive the first one. A neural network trained to learn how to cross the street has to die thousands of times before it learns how to do it without being run over by cars.

Perhaps a better paradigm for the 21st century would be the one proposed by the German sociologist Niklas Luhmann in his milestone "Theory of Society" (1997), which is really a theory of communication: social systems are systems of communication. Luhmann found a connection between human society, cybernetics, autopoiesis (the process popularized by Chilean biologist Humberto Maturana) and the favorite math book of the counterculture, "Laws of Form" (1969), a quasi-mystical treatise written by the British mathematician George Spencer-Brown.

"Humans cannot communicate; not even their brains can communicate; not even their conscious minds can communicate. Only communication can communicate" (Niklas Luhmann).

The Robots are Coming – A Brief History of A.I./ Part 9

The story of robots is similar to the story of neural networks. Collapsing prices and increased speeds have enabled a generation of robots based on relatively old theory.

The first computer-controlled robotic arm was probably the "Case arm", developed at Case Institute of Technology in 1965, with five degrees of freedom. In 1968 Jerry Feldman's team at Stanford published the "Hand-eye System", a combination of television camera and "Rancho arm", both connected to a PDP-6 computer. The arm was capable of the sorting cubes by size and of stacking the cubes on top of each other. The computer-controlled "Stanford arm", with six degrees of freedom, was designed in 1969 by Stanford student Victor Scheinman, formerly a rocket scientist who had worked on NASA's Apollo missions and was now in charge of maintaining the "Rancho Arm". Attaching a computer (i.e. software) to a robotic arm meant that the arm was capable of performing more than just one repetitive action. Another computer-controlled industrial robotic arm was the Milacron T3 built in Cincinnati by Richard Hohn in 1973. Despite the primitive state of the field, the first International Symposium on Industrial Robots was already held in Chicago in April 1970. In September 1973 luminaries of robotics met in Udine (Italy) for a symposium titled "Theory and Practice of Robots and Manipulators" which became the most important conference on robotics, later abbreviated as RoManSy (Robot-Manipulated-Symposium). Between these two events a lot had happened. Several hardware "firsts" were achieved in the early 1970s: the marvel of the Hitachi Technology Fair of 1970 was Masakazu Ejiri's robot that assembled objects based on drawings; Freddy (1971), built by Donald Michie's group at University of Edinburgh, was the first robot using a videocamera to guide its behavior, followed by the SIRCH assembly robot developed by Alan Pugh and others at University of Nottingham (1972) whose goal was precise gripping driven by visual feedback; Kuka's Famulus (1973), from Germany, was the first robotic arm utilizing electric instead of hydraulic drives; David Silver's arm at MIT (1974) was the first robotic arm with touch sensors; and ASEA's IRB6 (1974), in Sweden, was the first robot controlled by a microprocessor (using Intel's 8008 microprocessor). Nachi had built Japan's first industrial robotic arms in 1969, and Toshio Kono had founded the robotics startup Dainichi Kiko in 1971, but the Japanese wave began in earnest in 1974 when Seiuemon Inaba's firm FANUC (Factory Automation Numerical Control), a Fujitsu spinoff at the foot of Mount Fuji, installed robotic arms for assembly in its factory. Hitachi introduced its first commercial robot, the arc-welding Mr Aros, in 1975, one of the first controlled by a microprocessor. Both Yaskawa and FANUC introduced their first robotic arms in 1977, respectively the Motoman L10 and the Model-1. In 1973 Scheinman founded his own company, Vicarm, and designed robotic arms for SRI, Jet Propulsion Laboratories (JPL) and MIT. Vicarm was acquired by Joseph Engelberger's Unimation and remained their California laboratory. He designed for them the Programmable Universal Machine for Assembly (PUMA) commissioned by General Motors: delivered to Lothar Rossol, the man who had founded the A.I. lab at General Motors, it was deployed in 1978. The success of PUMA encouraged many to start firms to make small electric robotic arms.

Assembly robots (robotic arms for assembly) spread in every industrialized country. In Italy, for example, Olivetti introduced its Sigma in 1975 and Digital Electronic Automation (DEA) its Pragma in 1979. Then in 1978 Hiroshi Makino at University of Yamanashi invented a new design for robots, called SCARA (Selective Compliance Assembly Robot Arm), that in 1981 a coalition of Japanese manufacturers backed as the standard for assembly robots that were to be simpler, smaller and faster. Sankyo Seiki (ironically a producer of music boxes) released the first commercial SCARA, Skilam (1981). IBM's first commercial robotic arm, the 7535 (1982), was simply the Sankyo Seiki robot. After FANUC demonstrated robots making parts for other robots (in 1981), in 1982 General Motors decided to set up a joint venture called GMFANUC (General Electric would follow suit in 1986).

Hobbyists were already building all sorts of anthropomorphic life-size domestic robots, and, if you believe the media, they were far ahead of academia and of the industry, from Claus Scholz's MM7 (1958), a Viennese robot that the press portrayed serving food and vacuuming, to Ben Skora's Arok (1975), a Chicago robot that, according to InfoWorld of 24 September 1984 (page 22), could take out the trash, walk the dog, serve drinks, and vacuum. Of course, the press exaggerated a bit back then, just like today.

The stimulus to develop mobile robots came from space exploration. In fact, a company called Space General Corporation had already built in 1961 a remotely-controlled vehicle for lunar exploration on behalf of the Jet Propulsion Laboratories (JPL). Jim Fletcher, the co-founder, would become the chief of NASA in 1971 and in 1972 in 1972 would begin the development of the Space Shuttle. NASA's Surveyor of 1966 and the Soviet Union's Lunokhod of 1970, which both explored the Moon surface (the Lunokhod drove about 39 kilometers in about five months), and NASA's Viking of 1976 that explored Mars showed the need for robotic explorers (controlled remotely from Earth). One simple problem: often it took days to simply move a rock. In 1973 the Hungarian-born Antal Bejczy at Jet Propulsion Laboratories (JPL) augmented the Stanford arm with several sensors (ranging sensors, tactile sensors and proximity sensors besides television cameras) and with an on-board minicomputer (General Automation SPC-16) connected with a remote PDP-1O over the Arpanet. Bejczy also published the equations for the kinematics of such an arm ("Robot Arm Dynamics and Control", 1974). Trivia: Bajczy in the 1990s would oversee NASA's project for robotic telesurgery, published in 1996, the predecessor of Intuitive Surgical's DaVinci, and then help design in 1997 NASA's Sojourner, the first rover to land on the Mars.

Shakey had set in motion the history of mobile robots and inspired researchers both at home and far away. Shakey was capable of moving only indoors, in a well-structured environment of straight edges. Eventually, Stanford students built a Shakey for the outdoors. The "Stanford cart" was originally built in 1961 by Jim Adams, a student coming from the Jet Propulsion Laboratory (JPL) interested in finding a way for scientists to maneuver a Moon rover equipped with a television camera; but this was not a robot, rather just a remote-controlled battery-propelled cart on four bicycle wheels. However, resurrected in 1966 by John McCarthy's student Rodney Schmidt, by 1971 it became an outdoors autonomous vehicle, albeit only able to follow a curving white line, and at a speed of about 10 meters per hour (0.01 km/h). In 1973 John McCarthy's new student Hans Moravec took over the project and eventually developed the first three-dimensional vision for the cart. After graduating from Stanford in 1980, Moravec moved to CMU where he worked on a robot that came to be known as the "CMU Rover", first presented in 1982. The Hilare robot, built in 1977 by Georges Giralt's team at the Laboratory of Analysis and Architecture of Systems, was the French Shakey. In 1981 Ruzena Bajcsy's student Russell Andersson at CMU built SCIMR (Self Contained Independent Mobile Robot). Incidentally, the Czech-born

Bajczy had studied with John McCarthy at Stanford. In 1982 Bart Everett completed Robart I at the Naval Postgraduate School in Monterey (his master thesis). Wheeled robots were obviously easier to build than legged robots.

The first robotic hand with three fingers (each with multiple joints) was designed in 1979 by Tokuji Okada at Niigata University in Japan. In 1981 the French surgeon Raoul Tubiana pulished a study ("The Architecture and Function of the Hand", 1981) showing that the human hand has 22 degrees of freedom that allow it to grasp objects. In 1982 Kenneth Salisbury at Stanford University built a robotic hand, in collaboration with the Jet Propulsion Laboratory (JPL), and proved the minimum number of degrees of freedom required for grasping an object: nine, that he implemented as three fingers with nine joints ("Kinematic and Force Analysis of Articulated Mechanical Hands", 1982). That started a race for more and more degrees of freedom. The first robotic arm that was actually capable of grasping an object was developed in 1984 at University of Utah by Stephen Jacobsen (a professor and entrepreneur whose firm was building mechanized dinosaurs for theme parks), a joint project with MIT.

The first robot-assisted surgery was performed in 1985 at the Memorial Medical Center in Long Beach (near Los Angeles) using a PUMA 560 programmed by Yik-san Kwoh. In 1988 Brian Davies at London's Imperial College designed the Probot and in 1992 Senthil Nathan used it to perform the first fully robotic surgery in history.

The robotic hand designed at the National Taiwan University (NTU) by Han-Pang Huang and Li-ren Lin in 1995 had 17 degrees of freedom. In 1997 the DIST hand designed at University of Genova by Andrea Caffaz had 16 degrees of freedom. The Robonaut hand for the International Space Station, designed in 1999 by Chris Lovchik of NASA in Texas under the supervision of Myron Diftler of Lockheed (who then became the leader of the project at NASA) had five fingers with 14 degrees of freedom. In 2011 the R2 would become the first human-like robot to become a permanent resident of the International Space Station.

If you polled ordinary people about what makes a robot a robot, most of them would probably reply that it has to walk like us. Walking is not trivial, so much so that it takes months for babies to learn it. Footstep planning for humanoid biped robots has become a separate discipline. For a long time the fundamental technique was the "semi-inverse method", or, better, the zero-moment-point (ZMP) method, invented by Miomir Vukobratovic and Davor Juricic in Serbia in 1969. In 1966 Ichiro Kato's laboratory at Waseda University had already started work on the robot series WL (the Waseda Leg) that in 1973, incorporating the ZMP method, provided the foundation for the Wabot (WAseda roBOT), the first real-size anthropomorphic walking robot. Wabot-2 of 1984 had 10 fingers and two feet, could read a normal musical score, and could play it on a keyboard. Renamed "Wasubot", in March 1985 it performed live with a symphony orchestra at the opening ceremony of the International Science and Technology Exposition in Tsukuba.

Kato's student Atsuo Takanishi at Waseda University was involved in the development of the biped WL-5 (1971), the foundation of the Wabot, and led the development of the WL-10R (1983) and of the WL-12RIII that was able to walk up a staircase (1990), and eventually, in 1996, of the family dubbed Wabian (WAseda BIpedal humANoid) with 35 degrees of freedom (the Wabian-RIV of 2007 would boast 43 degrees of freedom, of which six in each leg, seven in each arm, three in each hand, three in the waist and four in the neck). The WL-10R was the first successful implementation of dynamic walking. So far most bipeds had used static walking, which keeps the robot always balanced but it is more similar to the turning of a wheel than to the way animals walk.

Kato and Takanishi generated the gait by feedback. Hirofumi Miura and Isao Shimoyama at University of Tokyo, instead, generated the gait by feedforward for their series Biper that led to Biper-3 (1981) and Biper-4 (1983).

Two-legged robots were too difficult to realize. Scientists quickly realized what every zoologist and pediatrist knows: that it is easier to crawl on all fours than to walk on two legs. A four-legged robot, dubbed "Phony Pony", tethered to a remote minicomputer, was walking in 1966 at University of Southern California, built by Robert McGhee (before he moved to Ohio) and his student Andrew Frank. Ten years later, Yoji Umetani's student Shigeo Hirose at the Tokyo Institute of Technology built the four-legged robots Kumo-I (1976) and PV-II (1978) that began the Titan series of robots. Hirose had previously built the many-wheeled snake-like Active Cord Mechanism III or ACM III (1972), the first snake robot. It would be even easier to move around if one had six legs: four legs have "dynamic stability", six legs have "static stability" (there are always three on the ground that provide stability). Hence a few of the earliest legged robots had six legs, starting with the computer-controlled hexapod robot built in 1972 at University of Rome by Massimiliano Petternella and Serenella Salinari and with the Masha hexapod robot designed in 1976 at Moscow State University by a team including the Soviet neurophysiologist Victor Gurfinkel, a pupil of Nikolai Bernstein whose book "The Coordination and Regulation of Movement" (1940) had presented a scientific theory of locomotion. Hexapod robots were shown in 1977 by Robert McGhee at Ohio State University (the "Bionic Bug") and in 1983 by Marc Raibert at CMU (who in 1980 had founded their Leg Lab and who in 1992 would start Boston Dynamics). The first commercial hexapod was Odetics' Odex (1983).

The Japanese public (as well as the workforce) seemed more willing than any other country to accept the robots. While it had started by importing robots from the USA, by 1981 Japan had many more industrial robots than the USA (14,000 versus 4,200). Despite the widely publicized death of a factory worker named Kenji Urada, the second human killed by a robot (in July 1981), the gap kept widening. (The first human to be killed by a robot was Robert Williams in January 1979 at a Ford plant near Detroit). Japanese scientists were already experimenting with "domestic" robots such as Eiji Nakano's "nursebot" called Melkong at Tohoku University (first presented in 1981) and Susumu Tachi's robotic dog for blind people called Meldog at University of Tokyo (first presented in 1981). Dainichi Kiko was even manufacturing a waiter robot for restaurants (1983), deployed by restaurants both in Tokyo and in Pasadena. In July 1983 the Seibu chain of department stores became the first to sell robots to the public (Dainichi Kiko models) and introduced robots that could follow customers and carry their purchases. In 1982 the inventor Shunichi Mizuno unveiled a life-size New Monroe, the first of his sexy robots (or "cybots"). Japan is a country that had never seen a clock until the Spanish missionary Francis Xavier brought one as a gift in 1551 (four centuries later it would become the world's main exporter of watches), and a country that remained isolated from the West between 1639 (when the emperor kicked out all foreigners) and 1854 (when the USA forced Japan to sign the treaty of Kanagawa). In between those dates the only automata built in Japan were the karakuri mechanical dolls and puppets, especially the ones used in the Bunraku puppet theater, perhaps descendants of that Spanish clock. The Takayama festival of karakuri that still takes place today was started in 1652. Nonetheless, Japan adopted robots at lightning speed: first the industrial robots (which are mostly arms), then anthropomorphic robots, then robots for entertainment (culminating in Oriza Hirata's robot theater of the 2000s). Far from demonizing robots, Japan's culture adopted them. The USA had robots in cinema, whereas Japan had them in children's entertainment. Robots appeared in comic books (mangas) such as Gajo Sakamoto's "Tanku Tankuro" (1934), Osamu Tezuka's "Tetsuwan Atomu/ Astro Boy" (1951), Mitsuteru Yokoyama's "Tetsujin 28go/ Iron Man No 28" (1956) and "Jaianto Robo/ Giant Robot" (1967). There were even Hiroshi Fujimoto's and Motoo Abiko's cat robot "Doraemon" (1969), Go Nagai's drivable robot "Mazinger Z" (1972), and Akira Toriyama's girl robot "Dokuta Suranpu/ Dr Slump" (1980). There were mangas about cyborgs, such as Kazumasa Hirai's and Jiro Kuwata's "Eitoman/ 8 Man" (1963) and Shotaro Ishinomori's "Cyborg 009" (1964).

There were robots on television, both in cartoons (anime), such as Yoshiyuki Tomino's "Yusha Raidin/ Brave Raideen" (1975), "Voltes V" (1977) and "Gandamu" (1979), and in special-effects live-action shows (tokusatsu), such as Shotaro Ishinomori's "Kamen Rider" (1971), "Jinzo Ningen Kikaida/ Android Kikaider" (1972), "Robotto Keiji/ Robot Detective" (1973) and "Ganbare Robocon" (1974). Whereas inventors in the USA aimed for a (programmable) hybrid of personal computer and home appliance, such as Joseph Bosworth's RB5X (1982) and Mike Forino's Hubot (1984), Japanese companies launched the concept of robots as toys: Bandai, Takara, Namco, Tomy (the programmable Omnibot 2000 of 1984 that spoke, moved and carried objects), etc. Japanese attitudes towards robots may be influenced by their polytheistic, animistic religion Shinto (as Ichiro Kato said in interviews) or by Buddhist philosophy (as Masahiro Mori was in his book "The Buddha in the Robot").

The Italian cyberneticist Valentino Braitenberg, in his book "Vehicles" (1984), showed that no intelligence is required for producing "intelligent" behavior: all that is needed is a set of sensors and actuators. As the complexity of the "vehicle" increases, the vehicle seems to display an increasingly intelligent behavior. In 1985 Rodney Brooks, a young Australian-born MIT scientist who had cut his teeth on the "Stanford cart" project with Hans Moravec, presented at a robotics symposium in France a robot nicknamed Allen that used little or no representation of the world. He showed that one can know nothing, and have absolutely no common sense, but still be able to do interesting things if equipped with the appropriate set of sensors and actuators. This was proven in a more convincing manner in 1988 by the hexapod Genghis, driven by a network of 57 finite-state machines. In 1991 Brooks published an influential paper, "Intelligence without Representation", that argued against the knowledge-based approach of Shakey and STRIPS, and in 1993 he launched the project Cog, a robot with 21 degrees of freedom and a variety of visual, auditory, vestibular, kinesthetic, and tactile sensors that simulated how a child acquires skills by trial and error.

Nonetheless, in 1986 Jaime Carbonell (a former student of Roger Schank at Yale but now at CMU) began work on a planner called Prodigy, the first major improvement over the venerable STRIPS and a ten-year project that would involve students such as Steve Minton and Manuela Veloso. Just like STRIPS, Prodigy was applied to robot navigation. In 1994 Veloso even attempted to build soccer-playing robots.

Upon graduating in 1984, Moravec's student Chuck Thorpe was tasked with launching CMU's center for autonomous vehicles. He teamed up with Red Whittaker (the brain behind the mechanics) and in 1986 they produced the semi-autonomous NavLab 1 (a Chevrolet van fitted with several computers, a camera and a GPS unit), that drove around campus at a speed of 2 km/h. Whittaker also founded RedZone whose robots helped to clean up the nuclear disasters at Three Mile Island and Chernobyl. The breakthrough came when, in 1988, Dean Pomerleau at CMU pioneered the use of neural networks for road navigation, i.e. self-driving vehicles, and demonstrated ALVINN (which stands for "Autonomous Land Vehicle in a Neural Network").

In 1991 Shuuji Kajita and Kazuo Tani at the National Institute of Advanced Industrial Science and Technology (AIST) introduced the "linear inverted pendulum" method (LIPM) for planning the steps of a biped robot.

In 1986 the Japanese firm Honda had already decided to develop a walking robot. In 1993 Honda produced its first humanoid robot (the P1) and in 1996 demonstrated the P2 publicly. This was the biped robot that galvanized the community of roboticists.

It wasn't clear yet what was the best architecture for robots. The 1992 robot competition run by the American Association for Artificial Intelligence (AAAI) was won by a University of Michigan robot called CARMEL (which stands for Computer-Aided Robotics for Maintenance, Emergency, and Life support), developed by Terry Weymouth's students, followed by SRI's robot Flakey (developed by Kurt Konolige's team since 1985): the former's navigation employed a traditional hierarchical architecture whereas the latter adopted elements of Brooks' distributed architecture.

In 1993 Tom Mitchell's team at CMU (featuring Joseph O'Sullivan, Sebastian Thrun and Reid Simmons) started work on Xavier, that looked like a water heater but whose tiered architecture (integrated with Manuela Veloso's Prodigy planning system in 1995) represented a major advance. Thrun went on to design the museum tour-guide robot Minerva (1998) and Pearl (1998), a "nursebot" for elderly care facilities (a collaboration with University of Michigan and University of Pittsburgh), before becoming famous for his self-driving car.

The old methods of footstep planning were revolutionized by the incorporation of Monte Carlo techniques that randomly explore the space of possibilities while keeping track of progress. This led to the proliferation of "probabilistic roadmap methods" (or "sampling-based robot motion planning") the Randomized Path Planner designed by Jean-Claude Latombe in 1991 at Stanford later improved by students such as Lydia Kavraki (her Probabilistic Roadmap Planner of 1996) and David Hsu (his Expansive-Spaces Tree planner of 1997). Meanwhile, in 1998 Steven LaValle at Iowa State Univ came up with Random Tree Planner, improved in collaboration with Stanford student James Kuffner as the

Rapidly-exploring Random Trees (RRT) planner. Sampling-based planners ruled in the 2000s.

In 1997 NASA's Mars Pathfinder deployed on Mars the first roving robot, Sojourner. While not a biped robot at all, it injected confidence in the field of autonomous vehicles.

Others, building on Rodney Brooks' theories and Valentino Braitenberg's theories of embodied intelligence (the brain is embedded in a body, and the body is embedded in the environment), aimed at building "developmental robots", robots that grow up just like humans, acquiring knowledge as they interact with the world; notably: SAIL, built since 1998 by Juyang Weng at Michigan State University, and Darwin V, built in 1998 at UC San Diego around the neuroscientist Gerald Edelman's theory of neuronal group selection. Meanwhile, Dario Floreano at AREA Science Park in Italy, Francesco Mondada at IPFL in Switzerland and Stefano Nolfi at the National Research Council of Italy had established the field of evolutionary robotics. In 2000 Floreano and Nolfi published the book that gave the field its name.

It took almost 15 years of research at Honda, but in 2000 Toru Takenaka (a student of Masahiro Mori, the first scientist to predict that robots would some day become conscious) could at last unveil the humanoid robot ASIMO (Advanced Step in Innovative Mobility). At the same time, Yoshihiro Kuroki's team at Sony unveiled the third prototype of the Sony Dream Robot project, SDR-3X, which evolved into the SDR-4X of 2002 and into the SDR-4XII of 2003, later renamed QRIO (pronounced "curio"). And Hirochika Inoue's team at Tokyo University (that featured James Kuffner and Masayuki Inaba) developed the humanoid robots H6 (1999) and H7 (2000). In 2001 Japan launched the Humanoid Robotics Project (HRP) under the direction of the same Hirochika Inoue. For example, the remote-controlled HRP-1S was designed to operate with construction workers.

Cynthia Breazeal's emotional robot Kismet (1999) at MIT and Hiroshi Ishiguro's Actroid (2003) at Osaka University in Japan were little more than interesting psychological experiments. NEC's PaPeRo (2001) and Mark Tilden's biomorphic robot Robosapien (2004) were toys and probably discovered the largest market for robots: children. Aldebaran Robotics' cute Nao was in fact conceived for education.

Meanwhile, progress in footstep planning was coming from James Kuffner, first at University of Tokyo ("Footstep Planning Among Obstacles for Biped Robots", 2001) and then at CMU with Joel Chestnutt ("A Three-tiered Planner for Biped Navigation over Large Distances", 2004). James Kuffner moved to CMU in 2002 and remained the link with

Tokyo University, of which the Japanese Institute of Advanced Industrial Science and Technology (AIST) was an emanation.

In 2003 Klaus Loeffler demonstrated the robot Johnnie built at the Technical University of Munich in Germany. In 2005 Jun-ho Oh showed Hubo at the Korea Advanced Institute of Science and Technology (KAIST). Quadrupeds such as the Boston Dynamics' BigDog of 2005 were far less interesting because four legs makes them more similar to four-wheeled vehicles which we've had since the first wagon was built in prehistory.

Robots were dangerous machines. It was too common for workers to get injured by robots. Hence in 1996 Northwestern scientists Michael Peshkin and Ed Colgate designed "intelligent assist devices", later better known as collaborative robots or "cobots", for Prasad Akella's team at General Motors: these were robots that were safe for human workers. In 1997 Peshkin and Colgate founded Cobotics (later renamed CoMoCo). In 2004 KUKA released its first cobot, LBR 3. In 2005 Esben Ostergaard, Kasper Stoy and Kristian Kassow founded Universal Robots in Denmark that released the UR5 in 2008, and in 2008 Rodney Brooks and Ann Whittaker launched Heartland Robotics (later renamed Rethink Robotics) that released the most famous cobot yet, Baxter, in 2012.

In 2005 Toyota launched the Partner Robots project. The idea was to build robots for medical and elderly care. Unfortunately, they got a bad reputation because they debuted playing drums and trumpets at the 2005 World Expo, i.e. they were just toys; but in 2007 the next humanoid of the series, Robina (Robot as Intelligent Assistant), guided visitors through a museum and was being designed to become a medical nurse, a "nursebot". The first in-home trials took place in 2011, assisting people with limb disabilities. In 2012 Toyota renamed the project Human Support Robot (HSR) to produce robots that can assist people in their everyday activities.

This was the era of the first "domestic" robots, like Luna, conceived in 2011 by RoboDynamics in Santa Monica, and Jibo, designed again by Cynthia Breazeal.

Osamu Hasegawa's robot that learned functions it was not programmed to do (2011) and Rodney Brooks' hand-programmable robot "Baxter" (2012), i.e. the first "collaborative robots" or "cobots", look good on video but still look as primitive as Shakey in person (in 2018 Brooks' company Rethink Robotics stopped operations).. In 2005 the driver-less car Stanley developed by Sebastian Thrun at Stanford won DARPA's Grand Challenge, but that was in the middle of the Nevada desert.

Lola Canamero's Nao (2010) at University of Hertfordshire in Britain, a robot that can show its emotions, was followed by David Hanson's Sophia

(2015) in Hong Kong, a robot that can display more than 60 facial expressions and that in 2017 was granted Saudi Arabian citizenship.

A branch of robotics is preoccupied with the self-reconfigurable modular robot, a concept introduced by Toshio Fukuda in Japan with its CEBOT (short for "cellular robot") that was capable of reconfiguring itself ("Self Organizing Robots Based On Cell Structures", 1988). The leadership remained in Japan (for example, Satoshi Murata's modular robotic system M-TRAN of 1999) until Daniela Rus at MIT, inspired by the art of origami and a math theory by Erik Demaine, invented a robot that folds automatically ("Programmable Matter by Folding", 2010) which led to the self-configuring "M-blocks". Rus is also working on the Robot Compiler: someday we will be able to order a robot for a specific function and the Robot Compiler will 3D-print a custom robot for us.

Manufacturing plants have certainly progressed dramatically and can build, at a fraction of the cost, the tiny sensors and assorted devices that used to be unfeasible and that can make a huge difference in the movements of the robot; but there has been little conceptual breakthrough since STRIPS of 1969 (the "problem solver" used by Shakey). What is truly new are the techniques of advanced manufacturing and the speed of GPUs.

In fact, nothing puts the progress in A.I. (or lack thereof) better in perspective than the progress in robots. The first car was built in 1886. 47 years later (1933) there were 25 million cars in the USA, probably 40 million in the world, and those cars were much better than the first one. The first airplane took off in 1903. 47 years later (1950) 31 million people flew in airplanes, and those airplanes were much better than the first one. The first public radio broadcast took place in 1906. 47 years later, in 1953, there were more than 100 million radios in the world. The first television set was built in 1927. 47 years later (1974) 95% of households in the USA owned a TV set, and mostly a color TV set. The first commercial computer was delivered in 1951. 47 years later (1998) more than 40 million households in the USA had a computer, and those personal computers were more powerful than the first computer. The first (mobile) general-purpose robot was demonstrated in 1969 (Shakey). In 2016 (47 years later) how many people own a general-purpose robot? How many robots have you seen today in the streets or in your office?

In June 2016 MIT Technology Review had an article about robots that announced: "They're invading consumer spaces including retail stores, hotels, and sidewalks". Look around you: how many robots do you see in the grocery shop and how many robots do you see taking a stroll on the sidewalk? I'll take a wild guess: zero. That's the great robot invasion of

2016, which competes with Orson Welles' famous Martian invasion of 1938 (total number of Martians on the streets of the USA: zero).

Most of the robots that accounted for the $28 billion market of 2015 (Tractica's estimate) were industrial robots, robots for the assembly line, not intelligent at all). Then there are more than ten million iRoomba (the home robot introduced by Rodney Brooks' iRobot in 2002) but those only vacuum floors. Those robots will never march in the streets to conquer Washington or Beijing. They are as intelligent as your washing machine, and not much more mobile.

Willow Garage, founded in 2006 by early Google architect Scott Hassan, has probably been the most influential laboratory of the last decade. They popularized the Robot Operating System (ROS), developed at Stanford in 2007, and they built the PR2 robot in 2010. ROS and PR2 have created a vast open-source community of robot developers that has greatly increased the speed at which a new robot can be designed. Willow Garage shut down in 2014, and its scientists founded a plethora of startups in the San Francisco Bay Area committed to developing "personal" robots.

The field of "genetic algorithms", or, better, evolutionary computing, has witnessed progress that mirrors the progress in neural-network algorithms; notably, in 2001 Nikolaus Hansen introduced the evolution strategy called "Covariance Matrix Adaptation" (CMA) for numerical optimization of non-linear problems. This has been widely applied to robotic applications and certainly helped better calibrate the movements of robots.

There are more than 3,000 DaVinci robots in the hospitals of the world, and they have performed about two million surgeries since 2000, the year when Intuitive Surgical of Sunnyvale was allowed to start deploying it. But DaVinci is only an assistant: it is physically operated by a human surgeon. In 2016, however, Peter Kim of the Children's National Health System in Washington unveiled a robot surgeon, the Smart Tissue Autonomous Robot (STAR), capable of performing an operation largely by itself (although it took about ten times longer than a human surgeon). In 2015 Google and Johnson & Johnson formed Verb Surgical to build robot surgeons.

The most sophisticated robots are actually airplanes. People rarely think of an airplane as a robot, but that's what it is: it mostly flies itself, from take-off to landing. In 2014 the world's airplanes carried 838.4 million passengers on more than 8.5 million flights. In 2015 a survey of Boeing 777 pilots reported that, in a typical flight, they spent just seven minutes manually piloting the airplane; and pilots operating Airbus planes spent half that time.

Therefore robots as "co-pilots" (as augmentation, not replacement, of human intelligence) have been very successful.

The most popular robot of 2016 is, instead, Google's self-driving car (designed by Sebastian Thrun), but this technology is at least 30 years old: Ernst Dickmanns demonstrated the robot car "VaMoRs" in 1986 and in October 1994 his modified Mercedes drove the Autoroute 1 near Paris in heavy traffic at speeds up to 130 km/h. In 2012 Google's co-founder Sergey Brin estimated that Google would have autonomous cars available for the general public within five years, i.e. by 2017. This is what happens when you think you know the future while in reality you don't even know the past. (Incidentally, Google engineers still use the "miles" of the ancient imperial system instead of the kilometers of the metric system, a fact that hardly qualifies as "progress" to me).

Intermezzo: Will Robots Be Conscious?

Whenever we discuss A.I. and robots, someone has to ask the question "Will these machines become conscious?"

The food industry slaughters 60 billion farmed animals (mammals, birds and fish) every year: why in heaven are we concerned for the consciousness of robots when we are not concerned for the consciousness of mammals and birds, whose brain is so similar to ours?

Brute-force A.I.

Despite all the hoopla, to me machines are still way less "intelligent" than most animals. Recent experiments with neural networks were hailed as sensational triumphs because a computer finally managed to recognize cats in videos (at least a few times). How long does it take for a mouse to learn how a cat looks like? And that's despite the fact that computers use the fastest possible communication technology, whereas the neurons of a mouse's brain use hopelessly old-fashioned chemical signaling.

One of the very first applications of neural networks was to recognize numbers. Sixty years later the ATM (automatic teller machine) of my bank still cannot recognize the amounts on many of the cheques that i deposit, but any human being can. Ray Kurzweil is often (incorrectly) credited with inventing "optical character recognition" (OCR), a technology that dates back to the 1950s (the first commercial OCR system was introduced by David Shepard's Intelligent Machines Research Corporation and became the basis for the Farrington Automatic Address Reading Machine delivered to the Post Office in 1953, and the term "OCR" itself was coined by IBM for its IBM 1418 product). Buy the most expensive OCR software and feed it the easiest possible case: a well-typed page from a book or magazine. It will probably make some mistakes that humans don't make, but, more interestingly, now slightly bend a corner of the page and try again: any human can still read the text, but the most sophisticated OCR software on the market will go berserk.

For similar reasons we still don't have machines that can read cursive handwriting, despite the fact that devices with handwriting recognition features already appeared in the 1990s (GO's PenPoint, Apple's Newton). Most people don't even know that their tablet or smartphone has such a feature: it is so inaccurate that very few people ever use it. And, yet, humans (even not very intelligent ones) can usually read other people's handwriting with little or no effort.

What has significantly improved is image recognition and speech recognition. Fei-fei Li's 2014 algorithm generates natural-language descriptions of images such as: "A group of men playing frisbee in a park". This result is based on a large dataset of images and their sentence descriptions that she started in 2009, ImageNet. In the 1980s it would have been computationally impossible to train a neural network with such a large dataset. The result may initially sound astounding (the machine algorithm even recognized the frisbee) but, even with the "brute force" of today's computers, in reality this is still a far cry from human performance: we easily recognize that those are young men, and many other details. And Peter Norvig of Google showed at Stanford's L.A.S.T. festival of 2015 a funny collection of images that were wrongly tagged by the machine because the machine has no common sense.

We are flooded with news of robots performing all sorts of human tasks, except that most of those tasks are useless. On the other hand, commenting on the ongoing unmanned Mars mission, in April 2013 NASA planetary scientist Chris McKay told me that "what Curiosity has done in 200 days a human field researcher could do in an easy afternoon." And that is the most advanced robotic explorer ever built.

What today's "deep learning" A.I. does is very simple: lots of number crunching. It is a smart way to manipulate large datasets for the purpose of classification. It was not enabled by a groundbreaking paradigm shift but simply by increased computing power.

The "Google Brain" project started at Google in 2011 by Andrew Ng is the quintessential example of this approach. In June 2012 a combined Google/Stanford research team used an array of 16,000 processors to create a neural network with more than one billion connections and let it loose on the Internet to analyze millions of still frames of videos (it recognized that many of them had a similar feature, the shape of a cat). Given the cost, size and speed of computers back then, 30 years ago nobody would have tried to build such a system. The difference between then and now is that today A.I. scientists can use thousands of powerful computers to get what they want. It is, ultimately, brute force with little or no sophistication. Whether this is how the human mind does it is debatable. And, again, we should be impressed that 16,000 of the fastest computers in the world took a few months to recognize a cat, something that a kitten with a still underdeveloped brain can do in a split second. I would be happy if the 16,000 computers could just simulate the 302-neuron brain of the roundworm, no more than 5000 synapses that nonetheless can recognize with incredible accuracy a lot of very interesting things.

The real innovation in Ng's approach was the idea to use GPUs. That simple idea made it possible to train multi-layer neural networks. In fact, one could argue that the real turning point in the history of Artificial Intelligence came when Ng at Stanford ("Large-scale Deep Unsupervised Learning using Graphics Processors", 2009) and Schmidhuber at IDSIA ("Deep Big Simple Neural Nets Excel on Handwritten Digit Recognition", 2010) showed that fast processors (GPUs) and the dataset were more important than all the philosophical tweaking of architectures: "brute force" was more important than elegant math. They tackled well-known problems and showed that, without any theoretical improvement, these problems could be solved simply by throwing enough computational power and training data at the neural network.

The human brain consumes about 20 Watts per hour. I estimate that AlphaGo's 1920 processors and 280 GPUs consumed about 440,000 Watts per hour (and that's not including the energy spent during the training process). What else can AlphaGo do besides playing Go? Absolutely nothing. What else can you do besides playing games? An infinite number of things, from cooking a meal to washing the car. AlphaGo consumed 440,000 W to do just one thing. Your brain uses 20 W and does an infinite number of things. How would you call someone that has to use 20,000 times more resources than you to do just one thing? What AlphaGo did is usually called "stupidity" not "intelligence". Let both the human and AlphaGo run on 20 Watts and see who wins. If it takes 440,000 Watts to play Go, how many Watts will it take to do everything else that a go/weiqi master can do with his brain? Like driving a car, cooking a meal, jogging in the park, reading the news, chatting about literature with a friend, etc? A ridiculous number of machines will be needed to match the human capability, an amount of power perhaps exceeding the 15 terawatts that all nations combined consume. Perhaps it will take more machines that we can possibly build with all the materials available on the planet.

DeepMind's network that learned to play Atari videogames like a master was widely publicized. Less publicized was a study by Joshua Tenenbaum's student Pedro Tsividis at MIT, in collaboration with Harvard psychologists, which showed that humans can learn the same Atari videogames to the same level of Deep Mind's program in a few minutes, whereas DeepMind's program needs hundreds of hours of game-playing ("Human Learning in Atari", 2017).

Brute force is the paradigm that now dominates A.I. After all, by indexing millions of webpages, a search engine is capable of providing an answer to the vast majority of questions (even "how to" questions), something that no expert system came close to achieving.

One wonders if slow and cumbersome computers were a blessing for the scientific community of the 1960s because those archaic machines forced computer scientists to come up with creative models instead of just letting modern high-speed computers crunch numbers until a solution is found.

John McCarthy was right to complain that, once A.I. solves a problem, the world does not consider it "artificial intelligence" anymore. But he didn't realize why: because, so far, whenever A.I. solved a problem (e.g., playing chess better than a master), the world realized that the solution wasn't special at all, it was just a matter of implementing very demanding mathematics on very fast computers. If 1+1=2 is not A.I., then playing chess is not A.I. A.I. has become synonymous with "running computationally intensive programs on supercomputers". We are impressed by the results, but we (correctly) don't consider them A.I. for the simple reason that human intelligence is something else. The name of the discipline is misleading. Not our fault.

Brute-force A.I. doesn't Speak

Most disheartening has been the scant progress in Natural Language Processing (NLP) since 1970.

Following are examples of exchanges between a person and Terry Winograd's program SHRDLU at MIT in 1970:

Person: FIND A BLOCK WHICH IS TALLER THAN THE ONE YOU ARE HOLDING AND PUT IT INTO THE BOX.

Computer: BY "IT", i ASSUME YOU MEAN THE BLOCK WHICH IS TALLER THAN THE ONE i AM HOLDING.

Person: DOES THE SHORTEST THING THE TALLEST PYRAMID'S SUPPORT SUPPORTS SUPPORT ANYTHING GREEN?

Computer: YES, THE GREEN PYRAMID.

Person: WHAT COLOR IS IT?

Computer: BY "IT", i ASSUME YOU MEAN THE SHORTEST THING THE TALLEST PYRAMID'S SUPPORT SUPPORTS. RED

Person: DOES A STEEPLE

Computer interrupting: SORRY, i DON'T KNOW THE WORD "STEEPLE".

Person: A "STEEPLE" IS A STACK WHICH CONTAINS TWO GREEN CUBES AND A PYRAMID.

Computer: i UNDERSTAND.

Person: ARE THERE ANY STEEPLES NOW?
Computer: NO.

It beats some of today's "chatbots".

We virtually abandoned the idea of having machines understand and speak our language and resorted to the opposite: make humans speak like machines. That is what you do when you talk on the phone with a machine that asks you for numbers; and that is what you do when you talk to your smartphone's "assistant" according to the rules of that assistant. Nothing illustrates the gap between machine intelligence and human intelligence better than comparing how much of a language a toddler learns in two years and how little of any language all machines ever invented have learned in more than 60 years.

IBM's Watson, that debuted with much fanfare in 2011 on a quiz show competing against human experts, was actually not capable of understanding the spoken questions: the questions were delivered to Watson as text files, not as spoken questions (a trick which, of course, distorted the whole game).

The most popular search engines are still keyword-based. Progress in search engines has been mainly in indexing and ranking webpages, not in understanding what the user is looking for nor in understanding what the webpage says. Try for example "Hey i had a discussion with a friend about whether Qaddafi wanted to get rid of the US dollar and he was killed because of that" and see what you get (as i write these words, Google returns first of all my own website with the exact words of that sentence and then a series of pages that discuss the assassination of the US ambassador in Libya). Communicating with a search engine is a far (far) cry from communicating with human beings.

Products that were originally marketed as able to understand natural language, such as SIRI for Apple's iPhone, have bitterly disappointed their users. These products understand only the most elementary of sounds, and only sometimes, just like their ancestors did decades ago. Promising that a device will be able to translate speech on the fly (like Samsung did with its Galaxy S4 in 2013) is a good way to embarrass yourself and to lose credibility among your customers.

The status of natural language processing is well represented by antispam software that is totally incapable of understanding whether an email is spam or not based on its content while we can tell in a split second.

During the 1960s, following (and mostly reacting against) Noam Chomsky's "Syntactic Structures" (1957) that heralded a veritable linguistic revolution, a lot of work in A.I. was directed towards "understanding" natural-language sentences, notably Charles Fillmore's case grammar   at Ohio State University (1967), Roger Schank's conceptual dependency theory at Stanford (1969, later at Yale), William Woods' augmented transition networks (ATNs) at Harvard (1970), Yorick Wilks' preference semantics at Stanford (1973), and semantic grammars, an evolution of ATNs by Dick Burton at BBN for one of the first "intelligent tutoring system", Sophie (started in 1973 at UC Irvine by John Seely Brown and Burton). Unfortunately, the results were crude. Schank and Wilks were emblematic of the revolt against Chomsky's logical approach that did not work well in computational systems. Schank and Wilks turned to meaning-based approach to natural language processing.

Terry Winograd's SHRDLU and Woods' LUNAR (1973), both based on Woods' theories, were limited to very narrow domains and short sentences. Roger Schank moved to Yale in 1974 and attacked the Chomsky-ian model that language comprehension is all about grammar and logical thinking. Schank instead viewed language as intertwined with cognition, as Otto Selz and other cognitive psychologists had argued 50 years earlier. Minsky's "frame" and Schank's "script" (all variations on Selz's "schema") assumed a unity of perception, recognition, reasoning, understanding and memory: memory has the passive function of remembering and the active function of predicting; the comprehension of the world and its categorization proceed together; knowledge is stories. Schank's "conceptual dependency" theory, whose tenet is that two sentences whose meaning is equivalent must have the same representation, aim to replace Noam Chomsky's focus on syntax with a focus on concepts. We humans use all sorts of complicated sentences, some of them very long, some of them nested into each other. Little was done in discourse analysis before Eugene Charniak's thesis at MIT ("Towards a Model of Children's Story Comprehension", 1972), Indian-born Aravind Joshi's "Tree Adjunct Grammars" (1975) at University of Pennsylvania, and Jerry Hobbs' work at the SRI Intl ("Computational Approach to Discourse Analysis", 1976). Then a handful of important theses established the field. One originated from the SRI, Barbara Grosz's thesis at UC Berkeley ("The Representation and Use af Focus in a System for Understanding Dialogs", 1977). And two came from Bolt Beranek and Newman, where William Woods had pioneered natural-language processing: Bonnie Webber's thesis at Harvard: ("Inference in an Approach to Discourse Anaphora", 1978) and Candace Sidner's thesis at MIT ("Towards a Computational Theory of Definite Anaphora Comprehension in English Discourse", 1979).

In 1974 Marvin Minsky at MIT introduced the "frame" for representing a stereotypical situation ("A Framework for Representing Knowledge",

1974) and in 1975 for the same purpose Roger Schank, who had already designed MARGIE (1973, which, believe it or not, stands for "Memory, Analysis, Response Generation, and Inference on English"), in collaboration with Stanford student Chris Riesbeck, and psychologist and social scientist Robert Abelson at Yale introduced the script ("Scripts, Plans, and Knowledge", 1975). Schank's students built a number of systems that used scripts to understand stories: Richard Cullingford's Script Applier Mechanism (SAM) of 1975; Robert Wilensky's PAM (Plan Applier Mechanism) of 1976; Wendy Lehnert's question-answering system QUALM of 1977; Janet Kolodner's CYRUS (Computerized Yale Retrieval and Updating System) of 1978, that learned events in the life of two politicians; Michael Lebowitz's IPP (Integrated Partial Parser) of 1978, that in order to read newspaper stories about international terrorism introduced an extension of the script, the MOP (Memory Organization Packet); Jaime Carbonell's Politics of 1978, that simulated political beliefs; Gerald DeJong's FRUMP (Fast Reading Understanding and Memory Program) of 1979, an evolution of SAM for producing summaries of newspaper stories; BORIS (Better Organized Reading and Inference System) of 1980, developed by Lehnert and her student Michael Dyer, a story-understanding and question-answering system that combined the MOP and a new extension, the Thematic Affect Unit (TAU). Starting in 1978 these systems were grouped under the general heading of "case-based reasoning". Meanwhile, Steven Rosenberg at MIT built a model to understand stories based on Minsky's frames. In particular, Jaime Carbonell's PhD dissertation at Yale University ("Subjective Understanding", 1979) can be viewed as a precursor of the field that would be called "sentiment analysis".

It is important to realize that, despite the hype and the papers published in reputable (?) A.I. magazines, none of these systems ever worked. They "worked" only in a very narrow domain and they "understood" pretty much only what was hardwired into them by the software engineer. That's why they were never used twice. They were certainly steps forward in theoretical research, but very humble and very short steps. In 2017 Schank published on his blog an angry article titled "The fraudulent claims made by IBM about Watson and A.I." that started out with the sentence "They are not doing cognitive computing no matter how many times they say they are" but perhaps that's precisely what Schank was doing two generations earlier.

These computer scientists, as well as philosophers such as Hans Kamp in the Netherlands (founder of Discourse Representation Theory in 1981), attempted a more holistic approach to understanding "discourse", not just individual sentences; and this resulted in domain-independent systems such as the Core Language Engine, developed in 1988 by Hiyan Alshawi's team at SRI in Britain. Meanwhile, Melvin Maron's pioneering work on statistical analysis of text at UC Berkeley ("On Relevance, Probabilistic Indexing, and Information Retrieval", 1960) was being resurrected by Gerard Salton at Cornell University (the project leader of SMART, System for the Mechanical Analysis and Retrieval of Text, since 1965). This technique, true to the motto "You shall know a word by the company it keeps" (1957) by the British linguist John-Rupert Firth, represented a text as a "bag" of words, disregarding the order of the words and even the grammatical relationships. Surprisingly, this method was working better than the complex grammar-based approaches. It quickly came to be known as the "bag-of-words model" for language analysis. Technically speaking, it was text classification using naive Bayes classifiers. In 1998 Thorsten Joachims at University of Dortmund replaced the naive Bayes classifier with the method of statistical learning called "Support Vector Machines", invented by Vladimir Vapnik at Bell Labs in 1995, and other improvements followed. The bag-of-words model became the dominant paradigm for natural language processing but its statistical approach still failed to grasp the meaning of a sentence. Nor did it have any idea of why a sentence was where it was and what it did there. Barbara Grosz at SRI International built an influential framework to study the sequence of sentences, i.e. the whole discourse, the "Centering" system ("Providing a Unified Account of Definite Noun Phrases in Discourse", 1983), later refined when she moved to Harvard ("A Framework for Modelling the Local Coherence of Discourse", 1986, but unpublished until 1995).

Perhaps the first major progress in machine translation since Systran was demonstrated in 1973 by Yorick Wilks at Stanford. His system was based on something similar to conceptual dependency, "preference semantics" ("An Artificial Intelligence Approach to Machine Translation", 1973).

The method that did improve the quality of automatic translation is the statistical one, pioneered in the 1980s by Fred Jelinek's team at IBM and first implemented there by Peter Brown's team (the Candide system of 1992). When there are plenty of examples of (human-made) translations, the computer can perform a simple statistical analysis and pick the most likely translation. Note that the computer isn't even trying to understand the sentence: it has no clue whether the sentence is about cheese or parliamentary elections. It has "learned" that those few words in that combination are usually translated in such and such a way by humans. The statistical approach works wonders when there are thousands of (human-made) translations of a sentence, for example between Italian and English. It works awfully when there are fewer, like in the case of Chinese to English.

Neural Machine Translation

Yoshua Bengio at University of Montreal started working on neural networks for natural language processing in 2000 ("A Neural Probabilistic Language Model", 2001). Bengio's neural language models learn to convert a word symbol into a vector within a meaning space. The word vector is the semantic equivalent of an image vector: instead of extracting features from an image, it extracts the semantic features of the word to predict the next word in the sentence. Bengio realized something peculiar about word vectors learned from a text by his neural networks: these word vectors represent precisely the kind of linguistic regularities and patterns that define the use of a language, the kind of things that one finds in the grammar, the lexicon, the thesaurus, etc; except that they are not separate databases but just one organic body of expertise about the language. Firth again: "you shall know a word by the company it keeps".

In 2005 Bengio developed a method to solve the "curse of dimensionality" in natural language processing, the problem of training a network with the particular data that are vocabularies ("Hierarchical Probabilistic Neural Network Language Model", 2005). After Bengio's pioneering work, several others applied deep learning to natural language processing, notably Ronan Collobert and Jason Weston at NEC Labs in Princeton ("A Unified Architecture for Natural Language Processing", 2008), one of the earliest multitask deep networks, and capable of learning recursive structures. Bengio's mixed approach (neural networks and statistical analysis) was further expanded by Andrew Ng's and Christopher Manning's student Richard Socher at Stanford with applications to natural language processing ("Learning Continuous Phrase Representations and Syntactic Parsing with Recursive Neural Networks", 2010), which improved the parser developed by Manning with Dan Klein ("Accurate Unlexicalized Parsing", 2003). The result was a neural network that learns recursive structures, just like Collobert's and Weston's. Socher introduced a language-parsing algorithm based on recursive neural networks that Socher also reused for analyzing and annotating visual scenes ("Parsing Natural Scenes and Natural Language with Recursive Neural Networks", 2010).

However, Bengio's neural network was a feed-forward network, which means that it could only use a fixed number of preceding words when predicting the next one. Czech student Tomas Mikolov of the Brno University of Technology, working at Johns Hopkins University in Sanjeev Khudanpur's team, showed that, instead, a recurrent neural network is able to process sentences of any length ("Recurrent Neural

Network-based Language Model," 2010). An RNN transforms a sentence into a vector representation, or viceversa. This enables translation from one language to another: an RNN (the encoder) can transform the sentence of a language into a vector representation that another RNN (the decoder) can transform into the sentence of another language. (Mikolov was hired by Google in 2012 and by Facebook in 2014, and in between in 2013 he invented the "skip-gram" method for learning vector representations of words from large amounts of unstructured text data).

In 2013 Nal Kalchbrenner and Phil Blunsom of Oxford University attempted statistical machine translation based purely on neural networks ("Two Recurrent Continuous Translation Models"). introduced "sequence to sequence" (or "seq2seq") learning, a new paradigm in supervised learning, but the length of the output sequence of characters was limited to be the same length as the output.

Bengio's group (led by Kyunghyun Cho and Dzmitry Bahdanau on loan from Jacobs University Bremen) established the standard "encoder-decoder" model of machine translation: an encoder neural network reads and encodes a source sentence into a fixed-length vector, and then a decoder outputs a translation from the encoded vector ("Learning Phrase Representations using RNN Encoder-Decoder for Statistical Machine Translation", 2014). This project also introduced a simpler alternative to Long Short-Term Memory (LSTM) in recurrent architectures, later named "gated recurrent unit" (GRU).

The problem with Bengio's original encoder-decoder approach is that it represented all the information of the sentence into a fixed-length vector, a fact that obviously caused a decline in accuracy with longer sentences. Bahdanau then added "attention" to the encoder-decoder framework ("Neural Machine Translation by Jointly Learning to Align and Translate", 2014): attention is a mechanism that improves the ability of the network to inspect arbitrary elements of the sentence. This improved architecture really showed that neural networks could be applied to translating texts because the attention mechanism overcame the original limitation and enabled neural networks to process long sentences. The attention mechanism developed by Bahdanau came to be known as "RNNSearch" or simply "additive attention".

The desire to add "attention" skills to a neural network dates from the 1980s when neuroscience began to elucidate how the brain makes sense of visual scenes and so quickly. In 1986 the neuroscientists Christof Koch and Shimon Ullman proposed that the primate brain creates a visual "saliency map". A saliency map, basically, encodes the importance of each element in the visual space. This led in 1998 to the attention-based model of Laurent Itti, a student of Christof Koch at Caltech ("A Model of

Saliency-based Visual-attention for Rapid Scene Analysis", 1998). Attention was introduced in image recognition tasks by Volodymyr Mnih at DeepMind ("Recurrent Models of Visual Attention", June 2014) whose "recurrent attention model" (RAM) was applied few months later to object recognition by Jimmy Lei Ba at University of Toronto ("Multiple Object Recognition with Visual Attention", December 2014).

Looping back to the field of computer vision that had jumpstarted the field, this attention-based technique was used by Bengio's other student Kelvin Xu to automatically generate captions of images ("Show, Attend and Tell", 2015).

Bengio's student Kyunghyun Cho showed that the same architecture of gated recurrent neural networks, convolutional neural networks and attention mechanism dramatically improved performance in multiple tasks: machine translation image caption generation and speech recognition ("Describing Multimedia Content using Attention-based Encoder-Decoder Networks", 2015).

Some other attention mechanisms were introduced a few months later by Christopher Manning's student Minh-Thang Luong at Stanford ("Effective Approaches to Attention-based Neural Machine Translation", 2015), notably the "dot-product" (or multiplicative) mechanism. Luong's multiplicative attention proved to be much faster and more efficient than Bahdanau's additive attention.

"Attention" is, however, a misnomer: the purpose of human attention is to speed up the process, if at the cost of accuracy, whereas "attention" in neural networks is a complex algorithm that, at every step, looks back at the input (or, better, at the hidden state of the encoder, a layer that captures the significant dependencies). The advantage for a neural network is that it doesn't have to encode all information of the input into one fixed-length vector. "Attention" provides flexibility but not necessarily agility.

Phil Blunsom's group at Oxford University used an attention-augmented LSTM network (trained with almost 100,000 articles from the CNN and more than 200,000 articles from the Daily Mail websites) to read a text and then produce an answer to a question ("Teaching Machines to Read and Comprehend", 2015). This Attentive Reader was a generalization of Weston's memory networks for question answering. A couple of years later, Christopher Manning's student Danqi Chen, while interning at Facebook, developed DrQA ("Reading Wikipedia to Answer Open-Domain Questions", 2017), an evolution of the Attentive Reader but using a method called "distant supervision" invented by Dan Jurafsky's team at Stanford ("Distant Supervision for Relation Extraction Without Labeled Data", 2009).

Scene understanding (what is going on in a picture, which objects are represented and what are they doing) is easy for animals but hard for machines. "Vision as inverse graphics" is a way to understand a scene by attempting to generate it: what caused these objects to be there and in those positions? The program has to generate the lines and circles that constitute the scene. Once the program has discovered how to generate the scene, it can reason about it and find out what the scene is about. This approach reverse-engineers the physical process that produced the scene: computer vision is the "inverse" of computer graphics. Therefore the "vision as inverse graphics" method involves a generator of images and then a predictor of objects. The prediction is inference. This method harkens back to the Swedish statistician Ulf Grenander's work in the 1970s.

After DRAW, DeepMind (Ali Eslami, Nicolas Heess and others) turned to scene understanding. Their AIR ("Attend-Infer-Repeat", 2016) model, which was again a combination of variational inference and deep learning, inferred objects in images by treating inference as a repetitive process, implemented as a LSTM that processed (i.e., attended to) one object at a time.

Lukasz Romaszko at University of Edinburgh later improved this idea with his Probabilistic HoughNets ("Vision-as-Inverse-Graphics", 2017), similar to the "de-rendering" used by Jiajun Wu at MIT ("Neural Scene De-rendering", 2017).

Ali Eslami and Danilo Rezende at DeepMind developed an unsupervised model to derive 3D structures from 2D images of them via probabilistic inference ("Unsupervised Learning of 3D Structure from Images", 2016). Based on that work, in June 2018 they introduced a whole new paradigm: the Generative Query Network (GQN). The goal was to have a neural network learn the layout of a room after observing it from different perspectives, and then have it display the scene viewed from a novel perspective. The system was a combination of a representation network (that learns a description of the scene, counting, localizing and classifying objects) and a generation network (that produces a new description of the scene).

Understanding what is going on in a visual scene implies recognizing the various objects and their spatial relations (usually represented in a "scene graph"). Roei Herzig and Moshiko Raboh at Tel Aviv University used a new method, "architectural invariance", to create such scene graph ("Mapping Images to Scene Graphs with Permutation-Invariant Structured Prediction", 2018). The concept of "architectural invariance" was introduced by Ruslan Salakhutdinov's student Manzil Zaheer at Carnegie Mellon University in collaboration with Alexander Smola's team at

Amazon's labs in Palo Alto as they developed a deep network that could operate on generic sets, not just vectors of fixed dimensions ("Deep sets", 2017).

Some tried to bridge the world of deep learning with the world of variational inference. A multi-layer neural network is a kind of step-by-step procedure, although each step (each layer) is a complicated one. Each successive layer of the network learns some more and more complex or abstract feature. However, this process of discrete steps is not well-suited for processing continuous processes. How closely a neural network models the continuity of reality ends up depending on how many layers it has: adding more layers, in general, can increase the granularity of the model. But each layer adds a computational cost. Yiping Lu at Peking University showed that Kaiming He's ResNet, Xingcheng Zhang's and Zhizhong Li's PolyNet at the Chinese University of Hong Kong (2017), Gustav Larsson's FractalNet at University of Chicago (2017), a very deep architecture with no residuals, and Aidan Gomez's Reversible Residual Network (RevNet) of 2017, a variant of ResNet, can be interpreted as different kinds of numerical discretizations of differential equations ("Beyond Finite Layer Neural Networks - Bridging Deep Architectures and Numerical Differential Equations", 2017). At the same time, Lars Ruthotto of Emory University and Eldad Haber of University of British Columbia in Canada viewed deep learning as the problem of estimating the parameters of a system of nonlinear equations and expressed its computations in terms of ordinary differential equations ("Stable Architectures for Deep Neural Networks", 2017). In other words, deep learning can be reduced to the (discrete) method for approximating and solving (continuous) differential equations that the 18th century Swiss mathematician Leonhard Euler described in his book "Institutionum Calculi Integralis" (1770). Building upon these intuitions, David Duvenaud at University of Toronto proposed a new way to implement neurocomputation without using layers, a way to summarize its computation as differential equations with no concept of layer ("Neural Ordinary Differential Equations", 2018).

In 2014, at the same time that Cho and Bahdanau were refining the encoder-decoder framework, Ilya Sutskever, Oriol Vinyals and Quoc Le at Google solved the "sequence-to-sequence problem" of deep learning using a Long Short-Term Memory ("Sequence to Sequence Learning with Neural Networks"), so the length of the input sequence of characters doesn't have to be the same length of the output. Sutskever, Vinyals and Le trained a recurrent neural network that was then able to read a sentence in one language, produce a semantic representation of its meaning, and generate a translation in another language via another encoder-decoder architecture.

The crowning achievement of neural machine translation was Google's "dynamic coattention network" (DCN) of 2016, based on the Sutskever-Vinyals-Le model and on the attention technique pioneered by Dzmitry Bahdanau's BiRNN (bidirectional RNN) at Jacobs University Bremen in Germany to improve the speed of machine translation ("Neural Machine Translation by jointly Learning to Align and Translate", 2015). This Google neural translation machine consisted of a deep LSTM network with eight encoder layers and eight decoder layers.

Of course the question is whether these systems that translate one sentence into another sentence based on simple mathematical formulas are actually "understanding" what the sentence says. Kevin Knight's student Xing Shi at University of Southern California demonstrated that the vector representations of neural machine translation (their hidden layers) capture some morphological and syntactic properties of language ("Does String-Based Neural MT Learn Source Syntax?", 2016), and Yonatan Belinkov at MIT discovered even some semantical properties hidden in those vector representations ("Evaluating Layers of Representation in Neural Machine Translation on Part-of-Speech and Semantic Tagging Tasks", 2017).

In 2018 Xuedong Huang's team at Microsoft built a system that achieved human parity on the dataset newstest2017 and claimed that such a system was able to translate sentences of news articles from Chinese to English with the same quality and accuracy as a person. The team combined three techniques developed by Microsoft in China: dual learning (2016, in collaboration with Peking University), deliberation networks (2017, in collaboration with University of Science and Technology of China), and joint training (2018, again in collaboration with University of Science and Technology of China).

Recurrent neural networks had matured enough that in November 2016 Google switched its translation algorithm to a recurrent neural network, and the jump in translation quality was noticeable.

After the successful implementations by Sutskever and Bahdanau in 2014, the sequence-to-sequence modeling required by machine translation was implemented with recurrent neural networks: use a series of (bi-directional) recurrent neural networks to map an input sequence to a variable-length output sequence. Within two years, however, architectures for sequence-to-sequence modeling that were entirely convolutional were proposed by Nal Kalchbrenner, now at DeepMind, namely his ByteNet ("Neural Machine Translation in Linear Time", 2016), and by Jonas Gehring in the Facebook team of Yoshua Bengio's former student Yann Dauphin, namely ConvS2S ("Convolutional Sequence to Sequence Learning", 2017). As far as sequence-to-sequence modeling goes, there are at least two advantages of convolutional networks over recurrent ones.

One is that their computation can be parallelized, i.e. done faster. Secondly, multi-layer convolutional neural networks create hierarchical representations of the sequence, as opposed to to the chain structures created by recurrent networks, The lower layers of such hierarchies model local relationships (between nearby items of the sequence) and higher layers model non-local relationships (between distant items of the sequence). This architecture provides a faster path to relate elements that are in arbitrary positions of the sequence. The distance still matters, of course: the computational "cost" of relating two items increases exponentially with their distance in ByteNet and linearly in ConvS2S.

"Translation is not a matter of words only; it is a matter of making intelligible a whole culture" (Anthony Burgess)

Understanding this Book (or any Book)

Gerard Salton's SMART (1965) used a very simple algorithm to represent how words occur next to other words. Thirty years later, Curt Burgess and Kevin Lund at UC Riverside recast this simple idea in more abstract terms: lexical co-occurrence can be used to create high-dimensional spaces in which each word is a multi-dimensional "vector" whose coordinates are determined by its co-occurrence with each of the other words (Hyperspace Analogue to Language", 1995). The distances between word vectors expresses similarity in meanings. This was the beginning of a new approach to natural-language processing: vector-based models of word meaning.

Tom Landauer, a psychologist at University of Colorado, in collaboration with Susan Dumais of Bellcore in New Jersey, formulated "latent semantic analysis", according to which knowledge can be created from simply studying co-occurrence of data in a large corpus of texts ("A solution to Plato's Problem", 1997). They showed that latent semantic analysis, with no prior linguistic knowledge, acquired knowledge about the full vocabulary of the English language at about the same speed as schoolchildren. The meaning of a word was expressed as a vector calculated from the word's average effect on the meaning of phrases in which it occurs.

The premise of these vector-based models was "the distributional hypothesis": words frequently occurring in the same contexts must be related (semantically similar).

By using the distributional approach, one can easily discover that two words are synonyms, or that they belong to the same category.

The beauty of these models is, of course, their simplicity: they represent meaning simply by using "distributional" information, and, mathematically speaking, this representation is a vector that lends itself to simple calculations (the whole of linear algebra is about vector operations, or, equivalently, matrices). Usually, the dimensions of a vector space stand for words, but they can also stand for non-linguistic objects such as images.

The obvious problem of the distributional approach is polysemes, words that have several meanings (in the English language the most infamous example is "get").

Yoshua Bengio's neural probabilistic language model of 2003 popularized the distributional approach in the world of neural networks. A few years later, Ronan Collobert and Jason Weston at NEC Labs in Princeton demonstrated that, using convolutional neural networks, one could build a general architecture for natural-language processing, a field which until then had mainly focused on task-specific architectures ("A Unified Architecture for Natural Language Processing", 2008); a task-independent architecture for "sequence tagging" refined when Jason Weston was at Google ("Natural Language Processing almost from Scratch", 2011).

Beyond vector-based models of word meaning, one would like vector-based models of phrases. These are usually constructed by combining word vectors, following the pioneer work of Mirella Lapata and her student Jeff Mitchell at University of Edinburgh. ("Vector-based Models of Semantic Composition", 2008).

There are more sophisticated approaches to "semantic composition", notably the theory developed by Stephen Clark at Oxford University in collaboration with quantum physicist Bob Coecke and logical mathematician Mehrnoosh Sadrzadeh ("A Compositional Distributional Model of Meaning", 2008). They applied techniques from mathematical logic, category theory, and quantum physics to the distributional hypothesis.

Richard Socher, working with Andrew Ng and Christopher Manning at Stanford, used recursive neural networks to learn vector space representations for multi-word phrases and sentences ("Dynamic Pooling and Unfolding Recursive Autoencoders for Paraphrase Detection", 2011).

Tomas Mikolov, while at Google, invented Word2vec, an influential method to build vector representations of words from very large datasets of texts ("Efficient Estimation of Word Representations in Vector Space", 2013). Word2vec's skip-gram method predicts the neighbors of a given word in a random sentence; or, better, it provides the probability for every word in the vocabulary of being the neighbor of that given word. An extension of Word2vec called "FastText" was developed in 2015 by the team of Tomas Mikolov at Facebook and made available as open source ("Bag of Tricks for Efficient Text Classification", 2016). Meanwhile, a representation method similar to Word2vec called "GloVe" was introduced by Christopher Manning's students Jeffrey Pennington and Richard Socher ("GloVe: Global Vectors for Word Representation", 2014). These "word embeddings" (like Word2vec and GloVe) derive a map of how words relate to each other based on the configurations in which the words appear in large amounts of text.

The next step was to develop models that can not only detect the relative positions of words but also capture information about sentences. The "Skip-thoughts" method proposed by Ruslan Salakhutdinov's student Jamie/Ryan Kiros at University of Toronto is almost literally the equivalent for sentences of the skip-gram model for words: if the skip-gram model predicts the words surrounding a given word, the skip-thoughts method predicts the sentences surrounding a given sentence ("Skip-Thought Vectors", 2015). In 2018 Matt Gardner's group at the Allen Institute in collaboration with Luke Zettlemoyer's group at University of Washington introduced ELMo ("Embeddings from Language Models"), an alternative to GloVe and Word2vec ("Deep Contextualized Word Representations", 2018). ELMo, trained on a vast amount of texts, builds a representation of language use that can be transferred to a variety of natural-language processing tasks (question answering, summarizing, sentiment analysis, etc). The "Quick-Thoughts" method by Honglak Lee's student Lajanugen Logeswaran at University of Michigan is a variant of "Skip-Thoughts" that lends itself to faster training (An Efficient Framework for Learning Sentence Representations", 2018).

Most natural-language processing tasks require that one first computes a "baseline" by averaging a sentence's word vectors (the so-called "Bag-of-Words approach") or manipulating a linear weighted combination of the word vectors like the method proposed by Sanjeev Arora at Princeton University ("A Simple but Tough-to-Beat Baseline for Sentence Embeddings", 2016) or the one proposed by Andreas Rueckle at the Technical University of Darmstadt ("Concatenated p-mean Word Embeddings as Universal Cross-Lingual Sentence Representations", 2018).

A variation of Luong's "dot-product" attention mechanism, "self-attention", became important for all natural-language tasks. Self-attention is an attention mechanism that captures the relationships between two items of a sequence regardless of their distance within the sequence. Self-attention was invented by Mirella Lapata's student Jianpeng Cheng ("Long Short Term Memory Networks for Machine Reading", 2016), and then employed by Ankur Parikh at Google ("A Decomposable Attention Model", 2016), and by Socher's group at Salesforce ("A Deep Reinforced Model for Abstractive Summarization", 2017). Yoshua Bengio's student

Zhouhan Lin in Montreal ("A Structured Self-attentive Sentence Embedding", 2017) devised an elegant method to augment a bidirectional LSTM with a self-attention mechanism. Until then attention had been used as an add-on to a recurrent neural network. Google's "transformer" architecture, instead, disposed of the recurrent neural network altogether and relied solely on self-attention ("Attention Is All You Need", 2017). This approach pioneered a completely different way to analyze a sentence. A stack of encoders reads the input (of any length) and produces a representation of it using "self-attention", while a stack of decoders generates the output. The self-attention mechanism is really what makes the transformer different from an LSTM: self-attention models relationships between all the words of a sentence, regardless of their position, in order to produce the best possible representation. Google's transformer, a project started in 2017 by Jakob Uszkoreit at Google laboratories in Germany and mostly implemented by Ashish Vaswani and Noam Shazeer, boasted an improvement in translation quality, and it also seemed to learn grammatical facts on its own. Google's Transformer offered another advantage. During the previous year, a number of architectures had been proposed to reduce the computational cost of sequence-to-sequence modeling, notably: Nal Kalchbrenner's ByteNet and Facebook's ConvS2S. The computational cost of relating distant elements in a sequence increases exponentially in ByteNet and linearly in ConvS2S, but doesn't increase at all in Transformer: it remains constant.

Abhinav Gupta's student Xiaolong Wang at CMU, in collaboration with Ross Girshick and Kaiming He of Facebook, developed an algorithm superficially similar to he self-attention mechanism of Google's transformer but designed instead to capture long-range dependencies ("Non-local Neural Networks", 2018).

Meanwhile, those working on "visual question answering", i.e. answering questions about a visual scene, had been forced to adopt different strategies. Zichao Yang at CMU in collaboration with Microsoft ("Stacked Attention Networks for Image Question Answering", 2016) used neural attention, and Jiasen Lu at Virginia Tech introduced a new form of attention, "co-attention" ("Hierarchical Question-image Co-attention for Visual Question Answering", 2016), but Peng Wang and Qi Wu in Anton van den Hengel's team at University of Adelaide in Australia merged neural networks with the knowledge-based approach which had become almost anathema in the era of deep learning ("Explicit Knowledge-based Reasoning for Visual Question Answering", 2017). The catch here is to learning a "disentangled" representation that can be interpreted. One influential method to learn a "disentangled" representation was developed by Joshua Tenenbaum's team at MIT using a variational autoencoder ("Deep Convolutional Inverse Graphics Network", 2015).

Deep reinforcement learning models, such as AlphaGo, suffer from the inability to adapt to even minor changes in the task. DeepMind then turned to a 20-year-old idea, relational reinforcement learning, a form of inductive logic programming, and used self-attention to compute its relations ("Relational Deep Reinforcement Learning", 2018).

Word2vec and GloVe are both kinds of unsupervised learning: they skim through vast amounts of data and build a representation of language use based on the distribution of words with no help from humans. ELMo, Skip-Thoughts and Quick-Thoughts belong to the same category. Unsupervised learning of language use ruled until Alexis Conneau of Antoine Bordes' group at Facebook in France introduced InferSent (a bi-directional LSTM architecture) that outperformed Skip-Thought ("Supervised Learning of Universal Sentence Representations from Natural Language Inference Data", 2017).

That opened the gates to supervised architectures for language representation. Sandeep Subramanian in Yoshua Bengio's group at University of Montreal used a bidirectional "gated recurrent unit" (GRU) neural network ("Learning General Purpose Distributed Sentence Representations via Large Scale Multi-task Learning", 2018). Based on Jonathan Baxter's old theory of inductive bias learning, Subramanian assumed that training a network on multiple weakly-related tasks was useful to generate general representations of sentences, representations that encode the "inductive biases" of multiple models that can, therefore, be employed for novel tasks. Minh-Thang Luong at Google ("Multi-task Sequence to Sequence Learning", 2015) had pioneered the idea: he had trained a model on a set of weakly-related tasks, fusing sequence-to-sequence learning and multi-task learning. Another member of Google's team introduced two supervised models ("Universal Sentence Encoder", 2018), one based on Ashish Vaswani's "transformer" architecture at Google Brain ("Attention Is All You Need", 2017) and the other based on Mohit Iyyer's "deep averaging network" (DAN) architecture at University of Maryland ("Deep Unordered Composition Rivals Syntactic Methods for Text Classification", 2015).

Using a technique derived from Mikolov's skip-gram, Oriol Vinyals and Quoc Le at Google revolutionized the venerable branch of discourse analysis. They trained a recurrent network with a large set of chats between users and support technicians. This created the equivalent of a translation (or, better, of a sequence-to-sequence model): the question asked by a user has to be "translated" into the response of the support technician ("A Neural Conversational Model", 2015).

Then Oriol Vinyal used the same technique of machine translation to analyze images and create captions. The best architecture to represent images as vectors was the convolution neural network, so Vinyal used a convolution neural network as the image encoder ("Show and Tell - A Neural Image Caption Generator", 2015) and a decoder RNN turned that vector representation into sentences. It achieved the same feat of a neural network trained to describe a scene. The similarities between language parsing in natural language processing and scene analysis in machine vision had been known at least since Gabriela Csurka developed the "bag-of-visual-words" or "bag-of-features" technique. Ironically, the biggest success story of the "bag-of-words" model has been in image classification, not in text classification. In 2003 Gabriela Csurka at Xerox in France applied the same statistical method to images. The "bag-of-visual-words" model was born, that basically treats an image as a document. For the whole decade this was the dominant method for image recognition, especially when coupled with a support vector machine classifier. This approach led, for example, to the system for classification of natural scenes developed in 2005 at Caltech by Pietro Perona and his student Fei-fei Li.

Mikolov's method was also the basis for the R-Net developed by Microsoft's Chinese laboratories (Furu Wei and others) that in January 2018 won the Stanford reading-comprehension competition performing better (on some metric on some reading task) than humans.

Up to this point the technique for natural language processing included: the "bag-of-words" approach, in which sentence representations are independent of word order; the sequence models developed by Michael Jordan (1986) and Jeffrey Elman (1990) at UC San Diego; and models based on tree structures, in which a sentence's symbolic representation is derived from its constituents following a syntactic blueprint (the typical symbolic structure that results from this process resembles an inverted tree). The latter arose in the 1990s after a debate on representations in neural networks that started in 1984 when Geoffrey Hinton (then at CMU) circulated a report titled "Distributed Representations" about representations in which "each entity is represented by a pattern of activity distributed over many computing elements, and each computed element is involved in representing many different entities." The problem of representing tree structures in neural networks was solved by Jordan Pollack of Ohio State University who came up with the Recursive Auto-Associative Memory or RAAM ("Recursive Distributed Representations", 1990). A few years later Christoph Goller and Andreas Kuechler in Germany extended Pollack's RAAM so that it could be used for arbitrarily complex symbolic structures, e.g. any sort of tree structure ("Learning

Task-dependent Distributed Representations by Backpropagation Through Structure", 1995).

For question-answering systems James Weston (now at Facebook's labs in New York) developed "Memory Networks" (2014), neural networks coupled with long-term memories.

The trend towards more complex memory structures, required for analysis of lengthy text, led to a variant of memory networks that is trained end-to-end, namely Sainbayar Sukhbaatar's "end-to-end memory network", developed at New York University in collaboration with the Facebook group of Jason Weston and Rob Fergus ("End-to-End Memory Networks", 2015).

"Sequence tagging" (or "labeling") is the process of assigning each item in a sequence to a category, a process that is used in both natural language processing and bioinformatics. This process was traditionally implemented either with generative models such as the hidden Markov models employed in speech recognition or with the "conditional random fields" invented by John Lafferty (a former member of Fred Jelinek's group at IBM, now at CMU), working with Andrew McCallum and Fernando Pereira ("Conditional Random Fields", 2001). Collobert's technique constituted the first major innovation, and it was countered years later by the bi-directional LSTM with conditional random fields developed by Zhiheng Huang, Wei Xu and Kai Yu of Baidu ("Bidirectional LSTM-CRF Models for Sequence Tagging", 2015).

Collobert's neural-network architecture for NLP formed the basis for Soumith Chintala's "sentiment analysis" at New York University, that learned to categorize movie reviews as positive or negative ("Sentiment Analysis using Neural Architectures", 2015). Socher, who at Stanford had created "recursive neural tensor network" ("Recursive Deep Models for Semantic Compositionality Over a Sentiment Treebank", 2013), helped Christopher Manning's student helped Kai Sheng Tai to develop Tree-LSTM, a generalization of LSTMs to the tree structures used in natural language processing that further improved sentiment analysis taking advantage of the research started by Pollack 25 years earlier ("Improved Semantic Representations From Tree-Structured Long Short-Term Memory Networks", 2015). Sentiment analysis was also the objective of two projects in New England. In 2016 the Computational Story Laboratory of University of Vermont (led by Peter Dodds and Chris Danforth) used Teuvo Kohonen's Self Organising Map (SOM) to study what Kurt Vonnegut had termed the "emotional arcs" of written stories ("The Emotional Arcs of Stories are Dominated by Six Basic Shapes", 2016). In 2017 Eric Chu of MIT's Laboratory for Social Machines directed by Deb Roy (later hired by Twitter) used deep convolutional neural networks to infer the emotional content of videos and television shows by analyzing both the story, the facial expressions and the soundtrack, i.e. for both audio and visual sentiment analysis ("Audio-Visual Sentiment Analysis for Learning Emotional Arcs in Movies", 2017).

Alec Radford at OpenAI discovered a mysterious property of a particular kind of LSTM networks: trained (with 82 million Amazon reviews) to predict the next character in the text of Amazon reviews, the network develops a "sentiment neuron" that predicts the sentiment value of the review, i.e. it develops the ability to discover the sentiment of the text, and the sentiment neuron adjusts its value on a character-by-character basis ("Learning to Generate Reviews and Discovering Sentiment", 2017). On the other hand, the "deep forest" method developed by Zhi-Hua Zhou and Ji Feng at Nanjing University used the old-fashioned method of decision tree ensembles instead of neural networks and performed as well on sentiment analysis as the best neural networks ("Deep Forest", 2017).

A footnote on sentiment analysis. There are countless precursors, like Carbonell's dissertation of 1979 at Yale and Clark Elliott's PhD dissertation of 1992 at Northwestern University ("The Affective Reasoner", 1992), an implementation of Andrew Ortony's psychological theory (his "appraisal model" of 1988); but the discipline was truly born in 2002 with two studies: one by Peter Turney at the Institute for Information Technology of Canada ("Thumbs up or Thumbs down? Semantic Orientation Applied to Unsupervised Classification of Reviews", 2002) and one (a movie review classifier) and the other one by Bo Pang and Lillian Lee at Cornell University ("Thumbs up? Sentiment Classification using Machine Learning Techniques", 2002). Jeonghee Yi at IBM in San Jose (2003) was perhaps the first one to use "Sentiment Analysis" in the title of his paper.

Training a neural network requires a well-structured dataset. But a lot of real-world information comes in unstructured formats such as books, magazines, radio news, TV programs, etc. Hence the need for text-understanding, or reading-comprehension, technology. Understanding a text requires, first of all, determining what the real focus is. Hence a number of neural attention mechanisms were developed, mainly Jason Weston's memory networks at Facebook ("Memory Networks", 2014) and Richard Socher's dynamic memory networks at MetaMind in Palo Alto ("Ask Me Anything", 2015). Minjoon Seo in Ali Farhadi's group at the Allen Institute for Artificial Intelligence ("Bidirectional Attention Flow for Machine Comprehension", 2016) developed the Bidirectional Attention Flow (BiDAF) model, a new kind of "attention" technique, inspired by the "bi-attention" technique of by Dzmitry Bahdanau's BiRNN. Seo's architecture can model the context at different levels of granularity.

Just about at the same time, extensive datasets such as SQuAD, MARCO, CoQA and QuAC made it possible to train neural networks for reading comprehension tasks.

Weizhu Chen's team at Microsoft developed Reasonet in 2016, that combined memory networks with reinforcement learning, and FusionNet in 2017, that introduced a simpler attention mechanism called "History of Word". In 2017 the Reinforced Mnemonic Reader, developed in China jointly by Xipeng Qiu at Fudan University and the National University of Defense Technology, set a new record ("Reinforced Mnemonic Reader for Machine Reading Comprehension", 2017). Alas, recurrent neural networks are very slow both in training and inference a fact which prevents them from being deployed in real-time applications. In 2017 Quoc Le's team at Google collaborated with Wei Yu of CMU to develop QANET, a system for end-to-end natural-language processing that eschewed recurrent neural networks, and instead used convolutions to model local interactions and used self-attention to model global interactions. QANet showed impressive improvement in speed at question-answering tasks ("Combining Local Convolution with Global Self-Attention for Reading Comprehension", 2018).

Ali Farhadi's former student Mark Yatskar, now at the Allen Institute, and others updated BiDAF with self-attention and obtained BiDAF++ ("A Qualitative Comparison of CoQA, SQuAD 2.0 and QuAC", 2018). SDNet, developed by Xuedong Huang's group at Microsoft, was basically Google's BERT improved with the "context", i.e. with a history of the questions and answers that led the current question ("Contextualized Attention-based Deep Network for Conversational Question Answering", 2018). FlowQA, developed by Hsin-Yuan Huang and Wen-tau Yih at the Allen Institute, encoded the conversation history in a deeper manner ("Grasping Flow in History for Conversational Machine Comprehension", 2018).

Transfer learning for text classification was tried by many, and in particular by Andrew Dai and Quoc Le at Google ("Semisupervised Sequence Learning", 2015), who proposed fine-tuning a "language model". Their project was influential because before them LSTM networks had rarely been used for natural-language processing tasks despite their power to model sequential data. They showed that it was possible through a pretraining step. The problem is that their method was impractical: it required millions of documents in a given domain in order to achieve good performance in that domain.

It was the field of computer vision that first demonstrated the importance of transfer learning from large pre-trained models, for example fine-tuning models that were pre-trained on ImageNet (the classic approach pioneered by Fei-fei Li), a fact discussed in a famous paper by Hod Lipson's student Jason Yosinski at Cornell University ("How Transferable are Features in Deep Neural Networks?", 2014).

Another improvement came from the method of "hypercolumns", introduced by Jitendra Malik's student Bharath Hariharan at UC Berkeley ("Hypercolumns for Object Segmentation and Fine-grained Localization", 2015). This method too required some kind of pre-training, whether language modeling, paraphrasing, entailment or machine translation. Luke Zettlemoyer's team at University of Washington achieved some success using this method to build a language model ("Deep Contextualized Word Representations", 2018).

Jeremy Howard at University of San Francisco and Sebastian Ruder at the National University of Ireland combined techniques of transfer learning from computer vision, Hariharan's hypercolumns method and Dai's fine-tuning method to build a more general system to classify documents, called ULMFiT, that made transfer learning the new standard for natural-language processing ("Universal Language Model Fine-tuning for Text Classification", 2018). This was a three-layer LSTM architecture: first the language model is trained on a general corpus of texts to capture general features of the language; then such model is fine-tuned on a specific task using two new techniques ("discriminative fine-tuning" and "slanted triangular learning rates"); and finally the system is further fine-tuned using a new technique called "gradual unfreezing". These novel techniques remedied a well-known problem: neural networks are prone to catastrophic forgetting during fine-tuning.

Systems such as OpenAI's Generative Pre-trained Transformer or GPT ("Improving Language Understanding by Generative Pre-Training", 2018) and Google's Bidirectional Encoder Representations from Transformers or BERT ("Pre-training of Deep Bidirectional Transformers for Language Understanding", 2018) combined Google's "transformer" architecture with UMLFiT-style transfer learning.

Now researchers could choose between two strategies for applying pre-trained language models to specific tasks: feature-based (such as ELMo) and fine-tuning based (such as OpenAI's GPT and Google's BERT).

The importance of pre-trained representations (when using deep learning to study written text) is due to the fact that deep learning requires data for training, and there are few data available for written text. A text can be just about anything, it can have an infinite number of meanings and purposes. The available datasets that are task-specific have relatively few examples that can be used for training a deep network. Luckily, it turns out that one can pre-train a deep network on generic language data (typically, the texts found on the World-wide Web) and then specialize it for a narrow domain with impressive results, and it doesn't really matter that the fine-tuning for the special domain is done using "small" data. Pre-training can either be context-free (such as word2vec or GloVe) or contextual (such as ELMo and ULMFit). Contextual representations can be unidirectional or bidirectional. As of 2018, BERT was the most popular of the bidirectional contextual methods. Ironically, BERT used a technique called "cloze procedure", originally published in 1953 by the psychologist Wilson Taylor at University of Illinois for measuring reading skills and widely used in schools worldwide. It consists in deleting random words from a text and testing if you can still understand it. BERT is trained to predict the deleted words.

Within one year Jacob Devlin's BERT had been improved by multiple teams to achieve quasi-human performance on several tests, notably the GLUE (General Language Understanding Evaluation) test for reading-comprehension tasks introduced in 2018 by Sam Bowman of New York University: RoBERTa by Luke Zettlemoyer's team at Facebook and at the University of Washington ("A Robustly Optimized BERT Pretraining Approach", 2019), that pushed the GLUE benchmark to 88.5; FreeLB by Tom Goldstein's student Chen Zhu at the University of Maryland in collaboration with Jingjing Liu's team at Microsoft, that used so-called "adversarial training" to achieve 88.8 ("Enhanced Adversarial Training for Language Understanding", 2019); StructBERT by an Alibaba team in China, that incorporated language structures into pre-training and pushed the GLUE benchmark to 89.0 ("Incorporating Language Structures into Pre-training for Deep Language Understanding", 2019); and Albert (short for "A lite BERT") by Zhenzhong Lan at Google that achieved 89.4 ("A Lite BERT for Self-supervised Learning of Language", 2019). The media indulged in headlines such as "Machines beat humans at reading", but in reality within a few weeks some major papers that disproved BERT's skills at comprehending language, for example one by Timothy Niven and Hung-Yu Kao of Taiwan's National Cheng Kung University ("Probing Neural Network Comprehension of Natural Language Arguments", 2019) and one by Tal Linzen of Johns Hopkins University ("Right for the Wrong Reasons," 2019).

BERT's "pre-training" was based on autoencoding and was capable of modeling bidirectional contexts. This approach looked more promising than the main alternative: pretraining based on autoregressive language modeling, as in the case of OpenAI's GPT (GUID Partition Table). However, within one year the autoregressive approach was again in vogue thanks to the work of Ruslan Salakhutdinov's team at CMU that first developed an attention model named Transformer-XL ("Attentive Language Models Beyond a Fixed-length Context", 2019) and then unveiled an autoregressive pre-training method, XLNet, both collaborations with Google's scientist Quoc Le ("Generalized Autoregressive Pretraining for Language Understanding", 2019). XLNet was capable of modeling bidirectional contexts and outperformed BERT in several tasks.

The sequence-to-sequence (Seq2Seq) method introduced in 2013 by Nal Kalchbrenner and Phil Blunsom, and especially the attention-based Seq2Seq model introduced by Dzmitry Bahdanau, triggered a boom in systems for text summarization (abstractive summarization, not just reproducing a few significant sentences) because they can "write" and not only "read". Alexander Rush, Sumit Chopra and Jason Weston at Facebook developed Attention-Based Summarization ("A Neural Attention Model for Abstractive Sentence Summarization", 2015) and opened the floodgates, soon followed by Jiatao Gu's CopyNet at University of Hong Kong (Incorporating Copying Mechanism in Sequence-to-Sequence Learning", 2016), the Forced Attention Sentence Compression Model developed by Phil Blunsom's student Yishu Miao at Oxford University, that uses variational autoencoders ("Language as a Latent Variable", 2016), Read-Again Summarization developed by Raquel Urtasun's student Wenyuan Zeng at University of Toronto ("Efficient Summarization with Read-Again and Copy Mechanism", 2016), Ramesh Nallapati's SummaRuNNer in Bowen Zhou's group at IBM ("A Recurrent Neural Network Based Sequence Model for Extractive Summarization of Documents", 2017), etc. Most of these systems (Gu's, Miao's, Zeng's, Nallapati's) employed the "pointer network" conceived by Oriol Vinyals of Google and Meire Fortunato of UC Berkeley, an extension of the sequence-to-sequence model using Bahdanau's attention mechanism. A major step forward for longer-text summarization was the Pointer-generator Network or PGNet developed by Christopher Manning's student Abigail See at Stanford ("Get To The Point - Summarization with Pointer-Generator Networks", 2017) synthesizing all of these ideas.

Several startups began offering services of text analysis and summary: Narrative Science, founded in 2010 in Chicago by Northwestern University's professors Kristian Hammond and Larry Birnbaum (a student of Schank's at Yale University in 1986); Maluuba, founded in 2011 in Canada by two students of University of Waterloo, Sam Pasupalak and Kaheer Suleman, and acquired in 2017 by Microsoft; Semantic Machines, founded in 2014 in Berkeley by Dan Roth, UC Berkeley scientist Dan Klein and former Dragon executive Larry Gillick, and acquired by Microsoft in 2018; and MetaMind, founded in 2014 in Palo Alto by Richard Socher and acquired by Salesforce in 2016. But their narrative summaries only worked in very narrow domains under very friendly circumstances.

The results are still far from human performance. The most illiterate person on the planet can understand language better than the most powerful machine.

To be fair, progress in natural language understanding was hindered by the simple fact that humans prefer not to speak to another human in our time-consuming natural language. Sometimes we prefer to skip the "Good morning, how are you?" and get straight to the "Reset my Internet connection" in which case saying "One" to a machine is much more effective than having to wait for a human operator to pick up the phone and to understand your issue. Does anyone actually understand the garbled announcements in the New York subway? The advantage of machine communications is that they are clear. We make a long story about short stories. A human speaking to another human may say: "Can you please open the window as i'm having difficulties breathing. You know, i had asthma as a child". A human speaking to a machine only needs to say: "Alexa, open the window". Humans are weird: they like to understand the meaning of what is going on, not just be told what to do. Like it or not, humans can more effectively go about their business using the language of machines. For a long time, therefore, Natural Language Processing remained an underfunded research project with few visible applications. It is only recently that interest in "virtual personal assistants" has resurrected the field.

In order to realize how far we are from having machines that truly "understand" our language, think of a useful application that would greatly help civility in written conversations: the equivalent of a spelling checker for hostile moods. Imagine an app that, when you try to send an email, would warn you "The tone of this email is rude: do you really want to send it?" or "The tone of this email is sarcastic" or "The tone of this email is insulting". It is not difficult for a human to read "between the lines", to understand the hidden motivation of a message and, in particular, to understand when the writer is deliberately trying to hurt your feelings. Written hostilities can escalate quickly in the age of email and texting. The interesting fact is that we understand in a second that the tone of an email is not friendly, even when the email is a correct reply to our question or a positive comment to something we have done. When a friend was celebrating the killing of Osama bin Laden, i quipped "Yes, we are very good at assassinating people". You understand the sarcasm and indirect critique of US foreign policy, don't you? You may also understand that i was greatly annoyed by that operation, and, even more, by the fact that people were celebrating in the streets. We routinely get in trouble when we speak quickly because we say something that we "should not have said": this doesn't mean that what we said was false, but that we said it on purpose to cause harm and perhaps humiliate. Most of the time we regret doing it. Conversely, we are easily ticked off by the wrong tone in an email that was sent to us. We immediately understand the "tone" of an email, especially when it's meant to hurt or annoy us, and we are very good at forging a tone that will hurt or annoy somebody. We word our sentences according to our mood and to the mood we want to create in the other person.

When you chat with someone, you pay little attention to the grammatical structure of what you are saying (in fact, you make a lot of grammatical mistakes, interrupting your own sentences, restarting them, interjecting a lot of random noise such as "hmmm") but you pay a lot of attention at the dynamics of the conversation, which depends heavily on the tone of your and their voices (assertive, soothing, angry, etc). The importance of mood in understanding what is going on cannot be overstated.

The impact of mood on comprehension was already studied by Gordon Bower at Stanford in "Mood and Memory" (1981), with a tentative computational theory based on semantic networks, and Daniel Martins in France ("Influence of Affect on Comprehension of a Text", 1982). The collusion of emotion and cognition has been studied for a long time, from Richard Lazarus at UC Berkeley ("Thoughts on the Relations between Emotion and Cognition", 1982) to Joseph LeDoux at New York University (his book "The Emotional Brain", 1996) via multi-level theories of cognition-emotion interaction such as Philip Barnard's "interacting cognitive subsystems model" ("Interacting Cognitive Subsystems", 1985) at Cambridge University and Barnard's collaborator John Teasdale's model of nine cognitive subsystems at Oxford University ("Emotion and two kinds of meaning", 1993).

More studies have emerged in the 1990s about how emotional states influence what one understands, for example Joseph-Paul Forgas' "affect infusion model" ("Mood and Judgement", 1995) and Isabelle Tapiero's book "Situation Models and Levels of Coherence" (2007). But little progress has been made in computing moods. The one influential paper on the subject was written by a philosopher, Laura Sizer at Hampshire University ("Towards A Computational Theory of Mood", 2000).

The other thing that we humans can do effortlessly (and frequently abuse this skill) is to generate stories. Given something that happened or an article that we read or a television show that we watched, we can easily create a story to describe it. That's another thing that machines can't do in any reasonable fashion, despite decades of research. The pioneering systems of automatic story generation were the Automated Novel Writer, developed since 1971 at University of Wisconsin (a status report was published in 1973) by Sheldon Klein, who had already worked on automatic summaries at CMU ("Automatic Paraphrasing in Essay Format", 1965); James Meehan's story generator Tale-Spin ("The Metanovel", 1976), advised by Roger Schank at Yale, a program that generated stories about woodland creatures; and Michael Lebowitz's Universe at Columbia University ("Creating Characters in a Story-Telling Universe", 1984). Selmer Bringsjord of Rensselaer Polytechnic Institute in New York state and David Ferrucci of IBM started building Brutus in 1990 ("AI and Literary Creativity", 1999). Then came Scott Turner's Minstrel at UCLA ("Minstrel, a computer model of creativity and storytelling", 1992) and Rafael Perez y Perez's Mexica in Britain ("A Computer Model of Creativity in Writing", 1999). An illiterate four-year child is infinitely better than these systems at "narrating" a generic event.

In 2019 Alec Radford and Jeffrey Wu of OpenAI demonstrated an algorithm capable of creating convincing articles. In February 2019 the Guardian newspaper (the printed version) published an article written by GPT2 itself. (A few days later another Guardian article, penned by real human Hannah Jane Parkinson, screamed "AI can write just like me - Brace for the robot apocalypse", which, again, made us wonder who's more intelligent, human or machine). Called GPT2, OpenAI's system was based on Ashish Vaswani's "transformer" method (with a whopping 1.5 billion parameters) and it was trained on WebText (millions of web pages) to predict the next word in a text. The "distributional" approach had found its AlphaGo. The problem, of course, is that GPT2 created "fake" stories. Since it had no idea of what it was generating, it also had no idea whether these were real facts: it is merely a game in which GPT2 "predicts" what is the most likely word following the current one. It turns out that this game can produce a text that sounds authentic. GPT2 is not even great to create "fake news" because you can't really control the ideological orientation of the generated text. But GPT2 was indeed a major achievement because it achieved multitask learning without any explicit supervision. In other words, it was equally proficient at question answering, machine translation, reading comprehension, and summarization (Language Models are Unsupervised Multitask Learners", 2019).

Machine Translation too has disappointed. Despite recurring investments in the field by major companies, your favorite online translation system succeeds only with the simplest sentences, just like Systran in the 1970s. Here are some random Italian sentences from my old books translated into English by the most popular translation engine: "Graham Nash the content of which led nasal harmony", "On that album historian who gave the blues revival", "Started with a pompous hype on wave of hippie phenomenon".

In November 2016 the new Google Translate feature was widely publicized because it dramatically improved the machine-translation score called BLEU (bilingual evaluation understudy), introduced in 2002 by IBM. The new Google Translate was developed by Quoc Le (born in Vietnam), Mike Schuster (born in Germany), and Yonghui Wu (a Chinese-born veteran of Google's search engine). I tried it myself on simple sentences and the improvement was obvious. I tried it on one of my old music reviews written in Italian and the result is difficult to understand (maybe the original was too!) The biggest mistake: the Italian plural "geni" got translated as the plural of "gene" but in that context it is obviously the plural of "genius".

After successfully employing that recurrent neural network to improve Google's machine translation, Ilya Sutskever announced that: "all supervised vector-to-vector problems are now solved thanks to deep feed-forward neural networks" and "all supervised sequence-to-sequence problems are now solved thanks to deep LSTM networks" (at the 2014 Neural Information Processing Systems conference in Montreal). Unbridled optimism has always been A.I.'s main enemy.

Even if we ever get to the point that a machine can translate a complex sentence, here is the real test: "'Thou' is an Old English word". Translate that into Italian as "'Tu' e` un'antica parola Inglese" and you get an obviously false statement ("Tu" is not an English word). The trick is to understand what the original sentence means, not to just mechanically replace English words with Italian words. If you understand what it means, then you'll translate it as "'Thou' e` un'antica parola Inglese", i.e. you don't translate the "thou"; or, depending on the context, you might want to replace "thou" with an ancient Italian word like "'Ei' e` un'antica parola Italiana" (where "ei" actually means "he" but it plays a similar role to "thou" in the context of words that changed over the centuries). A machine will be able to get it right only when it fully understands the meaning and the purpose of the sentence, not just its structure.

(There is certainly at least one quality-assurance engineer who, informed of this passage in this book, will immediately enter a few lines of code in the machine translation program to correctly translate "'Thou' is an Old English word". That is precisely the dumb, brute-force, approach that i am talking about).

Or take Ronald Reagan's famous sarcastic statement, that the nine most terrifying words in the English language are "I'm from the government and i'm here to help". Translate this into Italian and you get "Le nove parole piu` terrificanti in Inglese sono `io lavoro per il governo e sono qui per aiutare'". Those are neither nine in the Italian translation (they are ten) and they are not "Inglese" (English) because they are now Italian. An appropriate translation would be "Le dieci parole piu` terrificanti in Italiano sono `io lavoro per il governo e sono qui per aiutare'". Otherwise the translation, while technically impeccable, makes no practical sense.

Or take Bertrand Russell's paradox: "the smallest positive integer number that cannot be described in fewer than fifteen words". This is a paradox because the sentence in quotes contains fourteen words. Therefore if such an integer number exists, it can be described by that sentence, which is fourteen words long. When you translate this paradox into Italian, you can't just translate fifteen with "quindici". You first need to count the number of words. The literal translation "il numero intero positivo piu` piccolo che non si possa descrivere in meno di quindici parole" does not state the same paradox because this Italian sentence contains sixteen words, not fourteen like the original English sentence. You need to understand the meaning of the sentence and then the nature of the paradox in order to produce an appropriate translation. I could continue with self-referential sentences (more and more convoluted ones) that can lead to trivial mistakes when translated "mechanically" without understanding what they are meant to do.

Translations of proverbs can be equally tricky. Take the Italian "Tra il dire e il fare c'e` di mezzo il mare", which is equivalent to the English "Easier said than done". In 2017 the most popular online translator renders it as "Between the saying and the sea there is the middle of the sea". Even the translation into Spanish fails (it is rendered as "Entre el dicho y el mar est el medio del mar") despite the fact that the equivalent Spanish proverb is very similar to the Italian ("Del dicho al hecho hay mucho trecho"). Our software engineer is now frantically entering a few lines of code in the online translator to make sure that this Italian proverb will be translated correctly in English and Spanish: alas, there are hundreds of languages and thousands of proverbs in each one, so the possible combinations are millions.

To paraphrase the physicist Max Tegmark, a good explanation is one that answers more than was asked. If i ask you "Do you know what time it is", a "Yes" is not a good answer. I expect you to at least tell me what time it is, even if it was not specifically asked. Better: if you know that i am in a hurry to catch a train, i expect you to calculate the odds of making it to the station in time and to tell me "It's too late, you won't make it" or "Run!" If i ask you "Where is the library?" and you know that the library is closed, i expect you to reply with not only the location but also the important information that it is currently closed (it is pointless to go there). If i ask you "How do i get to 330 Hayes St?" and you know that it used to be the location of a popular Indian restaurant that just shut down, i expect you to reply with a question "Are you looking for the Indian restaurant?" and not with a simple "It's that way". If i am in a foreign country and ask a simple question about buses or trains, i might get a lengthy lecture about how public transportation works, because the local people guess that i don't know how it works. Speaking a language is pointless if one doesn't understand what language is all about. A machine can easily be programmed to answer the question "Do you know what time it is" with the time (and not a simple "Yes"), and it can easily be programmed to answer similar questions with meaningful information; but we "consistently" do this for all questions, and not because someone told us to answer the former question with the time and other questions with meaningful information, but because that is what our intelligence does: we use our knowledge and common sense to formulate the answer.

Ludwig Wittgenstein in the "Philosophical Investigations" (published posthumously in 1953) wrote that "the meaning of a word is its use in the language". That statement launched a whole new discipline, now called "pragmatics", via John Austin's analysis of speech acts (starting with a lecture at Harvard University in 1955 that in 1962 became the book "How to Do Things with Words"), Paul Grice's "conversational maxims" ("Logic and conversation", 1975) and Dan Sperber's and Deirdre Wilson's "relevance theory" ("Relevance - Communication and Cognition", 1986). The term "pragmatics" was coined by Charles Morris, the founder of modern semiotics, in his book "Foundations of the Theory of Signs" (1938), which divided the study of language in three branches: syntax, semantics and pragmatics.

In the near future it will still be extremely difficult to build machines that can understand the simplest of sentences. At the current rate of progress, it may take centuries before we have a machine that can have a conversation like the ones I have with my friends on the Singularity. And that would still be a far cry from what humans do: consistently provide an explanation that answers more than it was asked.

A lot more is involved than simply understanding a language. If people around me speak Chinese, they are not speaking to me. But if one says "Sir?" in English, and i am the only English speaker around, i am probably supposed to pay attention.

The state of Natural Language Processing is well represented by the results returned by the most advanced search engines: the vast majority of results are precisely the kind of commercial pages that i don't want to see. Which human would normally answer "do you want to buy perfume Katmandu" when i inquire about Katmandu's monuments? It is virtually impossible to find out which cities are connected by air to a given airport because the search engines all return hundreds of pages that offer "cheap" tickets to that airport.

Take, for example, zeroapp.email, a young startup being incubated in San Francisco in 2016. They want to use deep learning to automatically catalog the emails that you receive. Because you are a human being, you imagine that their software will read your email, understand the content, and then file it appropriately. If you were an A.I. scientist, you would have guessed instinctively that this cannot be the case. What they do is to study your behavior and learn what to do the next time that you receive an email that is similar to past ones. If you have done X for 100 emails of this kind, most likely you want to do X also for all the future emails of this kind. This kind of "natural language processing" does not understand the text: it analyzes statistically the past behavior of the user and then predicts what the user will want to do in the future. The same principle is used by Gmail's Priority Inbox, first introduced in 2010 and vastly improved over the years: these systems learn, first and foremost, by watching you; but what they learn is not the language that you speak.

I often like to discuss with machine-intelligence fans a simple situation. Let's say you are accused of a murder you did not commit. How many years will it take before you are willing to accept a jury of 12 robots instead of 12 humans? Initially, this sounds like a question about "when will you trust robots to decide whether you are guilty or innocent?" but it actually isn't (i would probably trust a robot better than many of the jurors who are easily swayed by good looks, racial prejudices and many other unpredictable factors). The question is about understanding the infinite subtleties of legal debates, the language of lawyers and, of course, the language of the witnesses. The odds that those 12 robots fully understand what is going on at a trial will remain close to zero for a long time.

"I am for richness of meaning rather than clarity of meaning" (Robert Venturi, architect).

Boring Footnote: Semantic Analysis

Semantic parsing is different from the generative approach that Chomsky pioneered. Syntactic parsing wants to find out which one is the noun and which one the verb and so on: i.e. wants to build a tree that represents the grammatical structure of the sentence. Semantic parsing wants to turn a sentence into a logical representation, for example into a formula of first-order predicate logic. The advantage of this approach is that the logical representation lends itself to logical reasoning, i.e. automated processing by the computer. In 1970 Alfred Tarski's former student in philosophy, Richard Montague at UCLA, developed a formal method for mapping natural language into first-order predicate logic. Mark Steedman at University of Edinburgh introduced "combinatory categorial grammar" that treats verbs as functions ("Combinatory Grammars and Parasitic Gaps", 1987). Technically speaking, both employed a compositional semantics based on the lambda calculus invented in 1936 by Alonzo Church at Princeton University. Semantic parsing was applied to database queries by John Zelle and Raymond Mooney at University of Texas, who designed the system CHILL (Constructive Heuristics Induction for Language Learning), based on the learning methods of inductive logic programming ("Learning Semantic Grammars with Constructive Inductive Logic Programming", 1993). In 2005 Luke Zettlemoyer at MIT started developing a Steedman-style learning semantic parser ("Learning to Map Sentences to Logical Form", 2005). These approaches turn an utterance directly into a logical representation.

Probabilistic logic has been used to represent the meaning of natural language by Lise Getoor's student Matthias Broecheler at University of Maryland ("Probabilistic Similarity Logic", 2012), by Raymond Mooney's team at University of Texas, that merged Montague and Markov via Pedro Domingos' Markov logic networks ("Montague Meets Markov", 2013); and by Tom Mitchell for parsing of conversations, i.e. not just one sentence at a time but an entire discourse ("Parsing Natural Language Conversations using Contextual Cues", 2017).

The parser is supposed to learn how to map natural language sentences into logical representations of their meaning. The training data may consist of sentences coupled with lambda-calculus meaning representations, and the parser is expected to build a generalization that will help generate the logical representation of future sentences. Mooney's students built systems such as KRISP and WASP (2006) that use statistical machine learning to learn grammars. Mark Steedman's student Tom Kwiatkowski at University of Edinburgh introduced an intermediate representation to learn language-independent grammars ("Inducing Probabilistic CCG Grammars from Logical Form with Higher-order Unification", 2010).

The success stories of the 2010s were mostly in distributional semantics. "Frame semantics" was instead used by chatbots designed to answer simple questions (the Alexa kind of chatbot). In frame semantics the system is trained to identify the action, the object and the modifiers (space and time) of a sentence. Ask a chatbot to "find a flight to Beijing on friday" and the chatbot will break down the sentence into the action ("find"), the object ("flight") and the modifiers ("Beijing" and "friday"). The traditional, rule-based, "model-theoretic semantics" was largely abandoned because the rules, full of exceptions, are just too difficult to encode; hence we can't come up with good models of the language to drive the semantic analysis.

There were also the first attempts at "grounded semantics". Grounded semantics is the idea that a system should learn language (unsupervised) by simply "listening" to people talking and by engaging in conversations, i.e. via "indirect" supervision. The pioneering methods by Dan Roth's student James Clarke at University of Illinois ("Driving Semantic Parsing from the World's Response", 2010) and Michael Jordan's student Percy Liang at UC Berkeley ("Learning Dependency-Based Compositional Semantics", 2011) were limited to question-answering. Tom Mitchell's student Jayant Krishnamurthy at CMU ("Weakly Supervised Training of Semantic Parsers", 2012) and Mark Steedman's student Siva Reddy ("Large-scale Semantic Parsing Without Question-answer Pairs", 2014) used combinatory categorial grammar to learn a semantic parser. Unsupervised semantic analysis had been studied by Pedro Domingos' student Hoifung Poon at University of Washington ("Unsupervised SEMANTIC PARSING", 2009), who then moved to Microsoft, and a further step towards grounded semantics was taken by Ankur Parikh, working with Hoifung Poon at Microsoft, ("Grounded Semantic Parsing for Complex Knowledge Extraction", 2015).

None of these experiments has been particularly successful. Either our natural language is fundamentally not logical or we still haven't figured out its logic.

Footnote: A.I. Kills the Copyright

Since the invention of Wikipedia, i have been annoyed by how many Wikipedia articles plagiarize my research (in most cases without mentioning the source). Nothing we can do about it: it has always been the case that someone can simply rephrase someone else's research and make it his/her own. (Ironically, Wikipedia also has a page that disparages my research). If A.I. ever gets to the point that it can rephrase an article, this practice will skyrocket. Anybody can have her own A.I. bot that rephrases articles and can then present the rephrased articles as her own article. My bot could work all day and night, rephrasing millions of articles. My bot could scan all the news in the world and produce dozens of original articles every day that are simply plagiarizing the articles of my favorite news media: all the articles of the New York Times, of BBC News, of the New York Review of Books, of Art Magazine, and so on. It will be perfectly legal because each imitation will be different from the original, although clearly "inspired" by that original. Sitting comfortably in our chair, and hardly paying any attention, we will be able to compete with any news media in the world, with any magazine, and with any encyclopedia, dwarfing all of them with the multitude of imitations produced effortlessly by our summarizing and rephrasing bot.

Can you Hear me?

A brief summary of the field of speech recognition can serve to explain the infinite number of practical problems that must be solved in order to have a machine simply understand the words that i am saying (never mind the meaning of those words, just the words). A vast gulf separates popular books on the Singularity from the mundane daily research carried out at A.I. laboratories, where scientists work on narrow specialized technical details. Skip this chapter if you are bored by technical details, but trust me that the technical details are neither trivial nor few, and will keep several generations of engineers busy for a long time.

The history of speech recognition goes back at least to 1961, when IBM researchers developed the "Shoebox", a device that recognized spoken digits (0 to 9) and a handful of spoken words. In 1963 NEC of Japan developed a similar digit recognizer. Tom Martin at the RCA Laboratories was probably the first who applied neural networks to speech recognition ("Speech Recognition by Feature Abstraction Techniques", 1964). In 1970 Martin founded Threshold Technology in New Jersey which developed the first commercial speech-recognition product, the VIP-100.

Speech analysis became a viable technology thanks to conceptual innovations in Russia and Japan. In 1966 Fumitada Itakura at NTT in Tokyo invented Linear Predictive Coding ("One Consideration on Optimal Discrimination or Classification of Speech", 1966), a technique that 40 years later would be still used for voice compression in the GSM protocol for cellular phones; and Taras Vintsiuk at the Institute of Cybernetics in Kiev invented Dynamic Time Warping ("Speech Discrimination by Dynamic Programming", 1968), utilizing dynamic programming (a mathematical technique invented by Richard Bellman at RAND in 1953) to recognize words spoken at different speeds. Dynamic Time Warping was refined in 1970 by Hiroaki Sakoe and Seibi Chiba at NEC in Japan. Meanwhile, in 1969 Raj Reddy (who had been the first PhD graduate of the Stanford computer science department in 1966) founded the speech-recognition group at CMU and supervised three important projects: Harpy (Bruce Lowerre 1976), that used a finite-state network to reduce the computational complexity; Hearsay-II, that pioneered the "blackboard" in which knowledge acquired by parallel asynchronous processes gets integrated to produce higher levels of hypothesis, a blend of bottom-up and top-down processing (Rick Hayes-Roth, Lee Erman, Victor Lesser and

Richard Fennell, 1975); and Dragon, developed in 1975 by Jim Baker, who then moved to Massachusetts to start a pioneering company with the same name in 1982. Dragon differed from Hearsay in the way it represented knowledge: Hearsay used the logical approach of the "expert system" school, whereas Dragon used the Hidden Markov Model. It was during the Hearsay project that Reddy invented the "beam search" algorithm to search large spaces of possible solutions. Trivia: Dragon's technology was later acquired by Nuance, whose technology would be later acquired by a company named Siri that built a system for the Apple iPhone.

The same idea was central to Fred Jelinek's efforts at IBM ("Continuous Speech Recognition by Statistical Methods", 1976) and these statistical methods based on the Hidden Markov Model for speech processing became popular with Jack Ferguson's "The Blue Book", which was the outcome of his 1980 lectures at the Institute for Defense Analyses in New Jersey.

IBM (Jelinek's group) and the Bell Labs (Lawrence Rabiner's group) came to represent two different schools of thought: IBM was looking for the individual speech-recognition system, that would be trained to recognize one specific voice; Bell Labs wanted a system that would understand a word pronounced by any one among the millions of AT&T's phone users. IBM studied the language model, whereas Bell Labs studied the acoustic model.

IBM's technology (the n-gram model) tried to optimize the recognition task by predicting statistically the next word. The inspiration for the IBM technique came from a word game devised by Claude Shannon in his book "A Mathematical Theory of Communication" (1948). Program this technique into a computer, and test it on your friends, and you have the Shannon equivalent of the Turing Test: ask both the computer and your friends to guess the next word in an arbitrary sentence. If the span of words is 1 or 2, your friends easily win. But if the span of words is 3 or higher, the computer starts winning.

Shannon's game was the first hint that perhaps understanding the meaning of the speech was irrelevant, and instead the frequency of each word and of its coexistence with other words was crucial.

Baum's Hidden Markov Model applied to speech recognition becomes a probability measure which integrates both schools because it can represent both the variability of speech sound and the structure of spoken language. The Bell Labs approach eventually led to Biing-Hwang Juang's "mixture-density hidden Markov models" for speaker independent recognition and a large vocabulary ("Maximum Likelihood Estimation for Mixture Multivariate Stochastic Observations of Markov Chains," 1985).

Hidden Markov Models became the backbone of the systems of the 1980s: Kai-fu Lee's speaker-independent system Sphinx in 1988 at CMU (the most successful system yet for a large vocabulary and continuous speech); the Byblos system from BBN (1989); and the Decipher system from SRI (1989).

Three projects further accelerated progress in speech recognition. In 1989 Steve Young at Cambridge University developed the Hidden Markov Model Tool Kit, which soon became the most popular tool to build speech-recognition software. During the 1990s at least two major speech recognition datasets were compiled, the CSR corpus and the Swtichboard corpus.

In 1991 Douglas Paul of MIT, in collaboration with Dragon Systems, unveiled the Continuous Speech Recognition (CSR) Corpus, a dataset containing thousands of spoken articles, mostly from the Wall Street Journal.

Finally, in 1989 DARPA sponsored projects to develop speech recognition for air travel (the Air Travel Information Service or ATIS) with participants such as BBN, MIT, CMU, AT&T, SRI, etc. The program ended in 1994 when the yearly benchmark test showed that the error rate had dropped to human levels. These projects, largely based on Juang's algorithm of 1985, left behind another huge corpus of utterances. That's what you need to train speech-recognition systems. The following decade witnessed the first serious conversational agents: in 2000 Victor Zue at MIT demonstrated Pegasus for airline flights status and Jupiter for weather status/forecast, and also in 2000 Al Gorin at AT&T developed How May I Help You (HMIHY) for telephone customer care. More importantly, the leader of the ATIS project at SRI, Michael Cohen, founded Nuance in 1994 that developed the system licensed by Siri to make the 2010 app for the Apple iPhone (and Cohen was hired by Google in 2004). Voice Signal Technologies was founded by Dan Roth in 1995 in Boston and in 2002 it provided the first voice-dialing system (on a Samsung A500 phone), followed in 2003 by a name-dialing system (for the A610). Another startup of voice recognition was Israel's Advanced Recognition Technology, acquired by Nuance in 2005, as was Voice Signal in 2007.

After Alex Waibel's time-delay network (1989), combining HMM and neural nets became commonplace and led to the speech-recognition systems of Hinton at University of Toronto ("Deep Belief Networks for Phone Recognition," 2009) and of Dong Yu at Microsoft ("Conversational Speech Transcription Using Context-Dependent Deep Neural Networks", 2011). HMMs still outperformed deep neural networks in speech recognition, especially for large vocabularies. LSTM neural nets began to be used in speech recognition after Alex Graves' experiments at University of Toronto (" Speech Recognition with Deep Recurrent Neural Networks", 2013). As deep neural nets became more feasible and affordable, the demise of hidden Markov models became more appealing. HMMs are rather complicated to manipulate. End-to-end neural-network architectures began to look simpler and more elegant. In 2014 Alex Graves trained an LSTM neural network with his own method of connectionist temporal classification (CTC) of 2006 and obtained a speech recognition system that didn't have any HMM, but its error rate was higher than the error rate of HMM-based models, especially with homophones, words that sound alike ("Towards End-to-End Speech Recognition with Recurrent Neural Networks", 2014). In such hybrid systems of HMMs and deep neural networks, the temporal reasoning takes place within the HMM rather than the neural network. CTC training of neural networks forces the network to carry out that job. At the end of 2014 the (cumbersome) HMM component of a speech-recognition system was definitely dropped by Andrew Ng's group at Stanford. His system used a "language model" and "prefix beam search" to search through the space of possible word sequences. For example, given the prefix "Somebody stole his", a language model might indicate that the word "wallet" has a 50% chance of being the next word, and "phone" a 25% chance, and so on. This trick helped Ng to simplify the architecture and use a regular recurrent neural network instead of an LSTM neural network ("First-Pass Large Vocabulary Continuous Speech Recognition using Bi-Directional Recurrent DNNs", 2014). The word error rate was reduced dramatically. This architecture was employed in 2015 by Andrew Ng's team (now at Baidu's labs in Silicon Valley) to build Deep Speech 2 ("End-to-End Speech Recognition in English and Mandarin", 2015), which, for the record, is composed of eleven layers (three layers of convolution, seven recurrent layers, and a fully-connected output layer trained with batch normalization.

In 2016 Microsoft, by replacing Dong Yu's hybrid system with three kinds of convolutional nets for acoustic modeling and an LSTM for language modeling, achieved human parity (it transcribed speech as well as a professional transcriptionist) on the NIST 2000 dataset. The three convolutional nets were a variant of VGG-16, a variant of ResNet and a variant called LACE (layer-wise context expansion with attention) of the time-delay neural network pioneered by Alex Waibel in 1989.

Speech recognition systems are becoming ubiquitous: Apple's Siri (2011), Google's Now (2012), Microsoft's Cortana (2013), Wit.ai (founded in 2013 by Alexandre Lebrun in Silicon Valley and acquired by Facebook in 2015), Amazon's Alexa (2014), Baidu's Deep Speech 2 (2015, developed in Silicon Valley, the foundation of Xiaoyu that was introduced in 2017), SoundHound's Hound (launched in 2016 by a Silicon Valley startup founded in 2004 by Keyvan Mohajer), etc.

The year 2018 was the year of the full-duplex chatbot, the chatbot can talk and listen at the same time. First came Microsoft's full-duplex version of its Xiaoice (developed by Li Zhou's team in China), then Google's Duplex (developed by Yaniv Leviathan's team). Microsoft then acquired Semantic Machines, a startup founded in 2014 in Berkeley by veterans such as Dan Roth, Apple Siri's chief scientist Larry Gillick (formerly at Dragon and Voice Signal), Dan Klein of UC Berkeley and Percy Liang of Stanford, as well as Klein's student David Hall, who in 2010 built the Overmind agent that beat a human master at the videogame StarCraft.

But these systems share one limitation: they are designed to work in controlled environments using clear speech. The limitations of today's speech recognition become obvious when you talk to the machine in a noisy context. Unfortunately, that is increasingly the natural context. We are packing more people in the confined spaces of cities. Therefore, most verbal interactions happen in the cacophony of the city: multiple conversations happening at a party, beeping devices around the speakers, traffic noise all around, maybe a television screen blaring a soccer game, barking dogs, the clatter of drinking and eating, a machine alarm, an ambulance siren. Not only is there background noise, but it is totally unpredictable. Humans still manage to understand each other in these noisy environments because they are naturally able to discriminate what is voice and what is not, and to recognize the voice of their friend (even when it is not the loudest sound in the room). In real-world work situations the use of voice commands can be counterproductive. The problem is not easy to solve. There are tools to reduce and even eliminate noise, echo and reverb, but the result of these operations is to weaken the very voice that the device is trying to understand. Then identifying individual speakers becomes harder. At the end of 2017 an unknown entity posted an "ideation challenge" on the Innocentive website offering a monetary reward for ideas precisely on how to tackle this problem.

Can you Read me? A Brief History of Speech Synthesis

Speech synthesis was easier, and, in fact, it predates A.I. by at least two decades.

The vocoder (from Voice Operated reCOrDER), capable of synthesizing a human voice, was invented in 1940 by Homer Dudley at Bell Labs and used during the war to scramble the phone conversations between British prime minister Winston Churchill and US president Franklin Roosevelt. The first computer to sing was an IBM 7094 programmed in 1961 at Bell Labs by computer-music pioneer Max Mathews and neuropsychologist Louis Gerstman. It sang Harry Dacre's pop song "Daisy Bell" of 1892. This artificial song was later included in the soundtrack of Stanley Kubrick's "2001 A Space Odyssey" (1968). The vocoder became popular in electronic music after Robert Moog built one in 1968, most famously used to sing the "Ode to Joy" melody of Beethoven's "Ninth Symphony" in Stanley Kubrick's film "A Clockwork Orange" (1971), and later used on Kraftwerk's album "Autobahn" (1974) and Giorgio Moroder's album "From Here to Eternity" (1977). Meanwhile, the first full text-to-speech system had been built in Japan by Hitachi after being designed at Japan's Electrotechnical Laboratory by the psychologist Ryunen Teranishi and the linguist Noriko Umeda (first reported in ETL News #197, 1966). Then Umeda moved to Bell Labs to work on a "talking computer" with Cecil Coker. Their speech synthesizer was first reported in a New York Times article of 1972 ("Where Science Grows Miracles").

Trivia: Coker engineered early electronic music installations such as John Cage's "Variations V" (1965) and Robert Rauschenberg's "Linoleum" (1966).

In 1979 Dennis Klatt at MIT unveiled the MITalk system for text to speech conversion that was turned into a product by DEC in 1984 (DECtalk) and used to create the electronic voices in Robert Zemeckis' film "Back to the Future II" (1985). The era of concatenative speech synthesis began with Eric Moulines and Francis Charpentier at the National Center of Telecommunications in France ("Text-to-speech algorithms based on FFT synthesis", 1988), a technique popularized by Andrew Hunt and Alan Black at ATR Labs in Japan as part of the CHATR system ("Unit Selection in a Concatenative Speech Synthesis System Using a Large Speech Database", 1996). The generation of Apple's Siri, Microsoft's Cortana, Amazon's Alexa and the Google Assistant all the way to Google's WaveNet used some kind of synthesis-by-concatenation method. The competing approach was statistical speech synthesis, based on hidden Markov models, for example the HTS system demonstrated in 1995 by Keiichi Tokuda and others at the Nagoya Institute of Technology in Japan ("Speech Synthesis from HMMs Using Dynamic Features," 1996), which in 2002 became a popular open-source toolkit.

The Genetic Algorithm Renaissance

The field of mathematical optimization got started in earnest with the invention of linear programming. Contributions came from the Russian economist Leonid Kantorovich in 1939, from Frank Hitchcock at MIT in 1941 and from the Dutch economist Tjalling Koopmans at University of Chicago in 1942, but it was George Dantzig, then still a student from UC Berkeley working an office of the Air Force during World War II, who invented the "simplex" method in 1946, just about in time to make it an ideal application of the newly invented electronic computer.

An optimization problem consists in finding the minimum or maximum of an "objective" function in a multidimensional space, given some constraints, a typical problem in planning. If the objective function and the constraints are linear, the optimization can be done with linear programming methods like simplex. Linear programming methods cannot be used when the objective function or the constraints or both are nonlinear. For example, gradient descent is a method of nonlinear optimization.

Genetic algorithms (or, better, evolutionary algorithms) are nonlinear optimization methods inspired by Darwinian evolution: let loose a population of algorithms in a space of possible solutions (the "search space") to find the best solution to a given problem, i.e. to autonomously "learn" how to solve a problem over consecutive generations using the Darwinian concepts of mutation, crossover and selection (the "survival of the fittest" process).

There is a long story of "black box" function optimization, starting with the Metropolis algorithm, devised in 1953 by Marshall Rosenbluth at the Los Alamos laboratories, and the "downhill simplex method" devised in 1965 by British statisticians John Nelder and Roger Mead ("A Simplex Method for Function Minimization", 1965).

Evolution Strategies (ES) for the optimization of nonlinear functions were introduced by Hans-Paul Schwefel ("Two-phase Nozzle and Hollow Core Jet Experiments," 1970) and Ingo Rechenberg at the Technical University of Berlin (his doctoral dissertation "Evolution Strategies", 1971).

At the same time that John Hopfield was finding a similarity between neural networks and thermodynamics, mathematicians found a connection between thermodynamics and function optimization (when the function is nonlinear). Scott Kirkpatrick, Dan Gelatt and Mario Vecchi at IBM labs in New York were looking for a method to optimize the process of electronic chip design and, resurrecting Rosenbluth's ideas, came up with simulated annealing ("Optimization by Simulated Annealing", 1983).

Juergen Schmidhuber employed genetic algorithms for meta-learning, i.e. learning how to learn, in his dissertation at the Technical University of Munich ("Evolutionary Principles in Self-referential Learning", 1987). His thesis also pioneered the idea of meta-learning, of training a meta-learner so that it optimizes the so-called hyper-parameters of the learner, i.e. the parameters that determine how the learner learns. The hyper-parameters of a neural network or of an evolutionary system are usually decided by the engineer based on experience and intuition.

Other significant innovations in genetic algorithms were Darrell Whitley's "Genitor" algorithm at Colorado Stave University in 1988, in which the offspring replaces not the parents but rather the least fit member of the population; the cellular genetic algorithms of Bernard Manderick and Piet Spiessens of the Vrije Universiteit in Belgium ("Fine Grained Parallel Genetic Algorithms", 1989); David Goldberg's "tournament selection" at University of Alabama ("A Note on Boltzmann tournament Selection", 1990); Larry Eshelman's CHC, a variant of ES, at Philips Laboratories in New York State (1991); and Inman Harvey's SAGA ("Species Adaptation Genetic Algorithm") algorithm at University of Sussex (1992).

In the 1990s new evolution-based methods of optimization were born, such as "differential evolution", invented by Ken Price ("Differential Evolution", 1996) at UC Berkeley in order to solve an exotic problem (known as the "Chebychev Polynomial Fitting Problem") that had been posed to him by Rainer Storn; and "gene pool recombination", invented by Heinz Muehlenbein and Gerhard Paass in Germany ("Gene Pool Recombination in Genetic Algorithms", 1996). Notably, Andreas Ostermeier and Nikolaus Hansen at the Technical University of Berlin continued to refine Rechenberg's and Schwefel's evolution strategies and eventually achieved what is now the dominating algorithm of black-box optimization, the "Covariance Matrix Adaptation Evolution Strategy" or CMA-ES ("A Derandomized Approach to Self Adaptation of Evolution Strategies", 1994).

A framework for black-box optimization was developed at the Istituto Dalle Molle di Studi sull'Intelligenza Artificiale (IDSIA) by Daan Wierstra and Tom Schaul of Juergen Schmidhuber's group ("Natural Evolution Strategies", 2008). It was similar to the "estimation of distribution algorithms" (EDAs) developed by Heinz Muehlenbein and Gerhard Paass a decade earlier ("From Recombination of Genes to the Estimation of Distributions", 1996).

In 1994 the artist Karl Sims created videos of how genetic algorithms can simulate Darwinian evolution and quickly yield "virtual creatures" selected for swimming, walking, jumping, etc.

Footnote: Neuroevolution Algorithms

By 2017 deep learning and reinforcement learning were the most popular techniques of A.I. but far from being accepted by everybody as the best. In fact, their very limitations led to a revival of evolutionary algorithms, especially when applied to evolving both the weights and the topologies of neural networks.

The first person to write about applying genetic algorithms to neural networks was the aerospace engineer Lawrence Fogel, a researcher at both UCLA and Convair, who published papers such as "Autonomous Automata" (1962) and "Toward Inductive Inference Automata" (1962), and then the book "Artificial Intelligence Through Simulated Evolution" (1966). The idea of training neural networks with genetic algorithm can also be found in John Holland's seminal book "Adaptation in Natural and Artificial Systems" (1975). The first attempts were carried out by Darrell Whitley at Colorado State University ("Applying Genetic Algorithms to Neural Net Learning", 1988); Lawrence Davis at BBN in Boston ("Mapping Classifier Systems into Neural Network", 1988); Rodney Brooks in person at MIT, who programmed a six-legged robot ("A Robot that Walks", 1989); Stanford University students Geoffrey Miller (a future star of evolutionary psychology) and Peter Todd, who was also applying genetic algorithms to composing music ("Designing Neural Networks using Genetic Algorithms," 1989); Hugo de Garis at George Mason University ("Genetic Programming", 1990); Richard Belew, John McInerney and Nicol Schraudolph at UC San Diego ("Evolving Networks", 1990); David Schaffer, Rich Caruana and Larry Eshelman at Philips Laboratories in New York State ("Using Genetic Search to Exploit the Emergent Behavior of Neural Networks", 1990). That was the beginning of a new discipline, "neuroevolution". Initially, the research was limited to setting the weights of the network, i.e. to fixed-topology neuroevolution: generate a population of neural networks with their weights set randomly, measure the ones that are best at the task, let these ones generate a new population by mutating and crossing over with each other, and so on. Fixed-topology algorithms were developed by Risto Miikkulainen's group at University of Texas ("Evolving Finite State Behavior using Marker-based Genetic Encoding of Neural Networks", 1992); and by Randall Beer and John Gallagher at Case Western Reserve University ("Evolving Dynamical Neural Networks for Adaptive Behavior", 1992).

Later neuroevolutionists began to program their algorithms to also generate the topology of the network ("neurogenesis"): Eric Mjolsness at Los Alamos National Laboratory, who used simulated annealing (instead of genetic algorithms) to generate the structure of the network ("Scaling Machine Learning and Genetic Neural Nets", 1989); Steven Harp, Tariq

Samad and Aloke Guha of Honeywell labs in Minnesota, who developed NeuroGenesys ("Towards the Genetic Synthesis of Neural Networks", 1989); Jordan Pollack's GNARL (which stands for "GeNeralized Acquisition of Recurrent Links") at Ohio State University ("An Evolutionary Algorithm that Constructs Recurrent Neural Networks", 1994); Xin Yao's EPnet at Penn State ("A New Evolutionary System for Evolving Artificial Neural Networks", 1997); Josh Bongard's Artificial Ontogeny at University of Zurich ("Evolving Complete Agents using Artificial Ontogeny", 2001); etc. In 2002 Risto Miikkulainen's student Ken Stanley at University of Texas developed the algorithm NEAT (which stands for "NeuroEvolution of Augmenting Topologies"), the most popular and widely used algorithm for neuroevolution for a decade ("Evolving Neural Networks through Augmenting Topologies", 2002).

These were approaches of "direct encoding": they directly encoded network configurations. Then came the generation of "indirect encoding", that encodes a set of rules for generating networks, an approach pioneered by Hiroaki Kitano when he was at CMU ("Designing Neural Networks using Genetic Algorithms with Graph Generation System", 1990). Kitano was inspired by the work of the Hungarian biologist Aristid Lindenmayer who, while at City University of New York, had proposed a formal grammar called L-system to generate graphs ("Mathematical Models for Cellular Interactions in Development", 1968). Indirect encoding was popularized by the video titled "Evolved Virtual Creatures" (1994) of the virtual creatures "animated" by digital media artist Karl Sims' algorithms when he was artist-in-residence at Thinking Machines in Boston (in 2017 the video was at https://www.youtube.com/watch?v=JBgG_VSP7f8). A variant called "cellular encoding" was developed by Frederic Gruau in France ("Neural Network Synthesis using Cellular Encoding and the Genetic Algorithm", 1994).

The resurgence of evolution-based machine learning in the 2010s was centered around various Google laboratories: Quoc Le's team published "HyperNetworks" (2016), a variation on Stanley's HyperNEAT ("A Hypercube-Based Encoding for Evolving Large-Scale Neural Networks", 2009), while Daan Wierstra's team ("Convolution by Evolution", 2016) and Alex Kurakin's team ("Large-Scale Evolution of Image Classifiers", 2017) applied evolutionary methods to different kinds of deep learning.

In 2017 Ilya Sutskever's team at OpenAI announced a new algorithm that outperformed backpropagation, although it was simply an evolution (sorry for the pun) of the genetic algorithms of the 1970s ("Evolution Strategies as a Scalable Alternative to Reinforcement Learning").

A newly galvanized community of neuro-evolutionists presented seized the moment, and 2017 alone saw the introduction of DeepNEAT, a version of NEAT for deep networks, by Risto Miikkulainen, of NMODE (wich stands for "Neuro-MODule Evolution") by Keyan Ghazi-Zahedi at the Max Planck Institute in Germany, and of "Genetic CNN" by Lingxi Xie and Alan Yuille at the Johns Hopkins University.

Mathematical Teaser: If you Really Have to... Phase Transitions in Intelligence

A better paradigm to analyze the behavior of large-scale computation was suggested by Bernardo Huberman and Tad Hogg at Xerox PARC, one year before Hans Moravec's "Mind Children" came out: studying which phase transitions can computational systems undergo as they become bigger and more complex ("Phase Transitions in Artificial Intelligence Systems", 1987).

This paradigm echoes similar ideas proposed by scholars who studied the human mind. For example, the Canadian neuropsychologist Merlin Donald, in his book "Origins of the Modern Mind" (1991), argued that the modern mind of symbolic thought arose from a non-symbolic form of intelligence through gradual absorption of new representational systems. The four "phase transitions" envisioned by Donald roughly correspond to the stages of cognitive growth in children that were studied by the Swiss psychologist Jean Piaget (his classic "The Language and Thought of the Child" of 1923) and by the Russian psychologist Lev Vygotsky (his classic "Thought and Language" of 1934): their hypothesis was that children follow a path of "phase transitions" from non-symbolic to full-fledged symbolic thinking. Jerome Bruner at Harvard reached a similar conclusion: intellectual abilities develop in three stages ("Studies in Cognitive Growth", 1966). Maybe we should study the phase transitions of a computational system and build the equivalent of Piaget's and Vygotsky's epistemological theories, and from that fusion of physics, psychology and computational mathematics we may learn something about the kind of "intelligences" that are possible beyond ours.

There is a vast literature in the "stages" of human cognitive development, which sometimes varies from Piaget's original formulation. In more recent times one finding that struck me was the study "Beliefs about Beliefs" (1983) by psychologists Heinz Wimmer (University of Salzburg) and Josef Perner (University of Sussex). Three-year-olds tend to fail "false-belief" problems which become easy to solve just one year later: that's an impressive "phase transition" in human cognition. The problem has to do with guessing what a person thinks, not with what actually happened: what happened is obvious. A person leaves an object in a place and then walks out of the room. A second person walks in and moves the object to another place. The children are asked to predict where the first person will look for the object when she returns to the room. Three-year-olds tend to answer that the person will look for the object where the second person moved it. Four-year-olds correctly answer that the person will look for the object where she left it. What happens between age three and age four is not known, but obviously the brain undergoes a "phase transition" such that it can now correctly evaluate false-belief problems. Alison Gopnik at UC Berkeley, pioneer of the "child-scientist" theory, has argued that the learning of 2- to 4-year-old children can be modeled with Bayesian inference networks ("A Theory of Causal Learning in Children", 2004), the kind of reasoning that deep learning uses.

The philosopher Hubert Dreyfus wrote the book "What Computers can't do" (1979) to criticize expert systems, but indirectly provided another perspective on the phase transitions of human intelligence. He broke down human acquisition of performance into five stages. First, we are born novices: we simply follow the rules (an instructor, a manual). The moves of novices are not secure and not fluid, although they can be technically correct. Sometimes applying a rule is plain silly, but the novice will still do so because he doesn't know better. Eventually, after practicing for a while, we become advanced beginners. At this stage we are capable of modifying rules based on the situation. Our behavior is still driven by rules but it doesn't look as awkward. Competent humans, the next stage of experience, follow rules but in a very fluid manner. Their rules are also adaptable: the competent human knows that she can modify the rules. In fact, she will feel guilty if something goes wrong, even if she followed the proper rules. Proficient performers do not even follow rules anymore: they act by reflex. The fact that they have encountered similar situations many times before matters more than the original rules. Experts, the final stage, do not even remember the rules. Sometimes if they have to articulate them they can't even figure them out. They just act based on their expertise and their intuition. They are often not even aware of what they are doing. An expert driver does not realize that she is shifting gears and at which point she is shifting gears. She just shifts gear when it's appropriate to. An expert has synthesized experience in an unconscious behavior that reacts instantaneously to a complex situation. What the expert knows cannot be decomposed in rules. A failure usually results in degradation: an expert driver does not even remember the rules for starting a car, but, if she can't start the car, she will gradually walk down the ladder from expert to merely competent all the way down to novice, and, if nothing in her experience helps, she will finally pick up the driver's manual to figure out why the car won't start.

How to Win a Nobel Prize: Theory Formation

"Mathematical reasoning may be regarded rather schematically as the exercise of a combination of two facilities, which we may call intuition and ingenuity" (Alan Turing). So how do we simulate intuition and ingenuity in a computer program?

After Douglas Lenat's half-baked attempt at Stanford to create a program for mathematical discovery, Automated Mathematician or AM (1976), Simon's students debuted discovery systems that employed knowledge ("heuristics") about a specific domain to derive scientific laws from data: Fahrenheit in chemistry, a system designed by Jan Zytkow ("A Theory of Historical Discovery", 1986); Kekada in biology, by Deepak Kulkarni ("The Processes of Scientific Discovery", 1988); and Mechem also in chemistry, by Raul Valdes-Perez (1990, that includes the hypothesis-formation program Stoich). More general was Abacus, an evolution of Ryszard Michalski's AQ11, that was applied to problems both in physics and chemistry ("Integrating Quantitative and Qualitative Discovery", 1986). Kulikowski's student Vonwun Soo at University of Pennsylvania worked on chemical reaction pathways before Kekada and Mechem ("Theory Formation in Postulating Enzyme Kinetic Mechanisms", 1987). In 1988 Siemion Fajtlowicz at University of Houston developed Graffiti for discovery in mathematics. Peter Karp at Stanford University wrote the program Hypogene for hypotheses generation in molecular biology (1989), an evolution of Molgen.

Douglas Hofstadter at Indiana University built several programs such as Jumbo (1983) and Numbo (1987) that reacted against the notion that intelligence could be just the product of domain knowledge. He believed that finding patterns constituted the core of intelligence and therefore researched a more abstract type of intelligence, capable of discovering patterns and shaping concepts out of patterns. He implicitly viewed mathematics as the supreme demonstration of human intelligence because mathematicians continuously discover higher and higher levels of abstraction to explain the patterns that they discover. Hofstadter's programs tried to simulate how we build concepts, expand them, adapt them. Concepts are dynamic, not static, or, better, concepts are fluid. An important moment is when a paradigm shift occurs and we start seeing the world differently. His student Melanie Mitchell built Copycat (1988) to simulate paradigm shifts in a microworld. Note that most research in mathematics is driven by the illusion of getting closer and closer to the ultimate truth of the universe, not by the certainty of doing something useful for daily life.

Ken Forbus's student Brian Falkenhainer at Northwestern University used Forbus' analogical reasoning system SME to develop scientific theories ("Scientific Theory Formation through Analogical Inference", 1987).

Saul Amarel's student Michael Sims at Rutgers University built the program IL for theory formation in mathematics (1990). In 1997 Derek Sleeman's student Faye Mitchell at Aberdeen University in Britain developed Daviccand (Data VIsualisation Clustering and Conceptually Analysing) for determining the chemical properties of metals (and the rare case of interactive human-machine discovery). Alan Bundy's student Simon Colton at University of Edinburgh designed the HR system for theory formation in mathematics (1998).

Meanwhile, the field of "literature-based knowledge discovery" had been launched by Don Swanson at University of Chicago in 1986 ("Fish Oil, Raynaud's Syndrome, and Undiscovered Public Knowledge", 1986). Swanson viewed discovery as connecting the disconnected in scientific knowledge, as finding links between apparently unrelated ideas in the scientific literature. Swanson, who knew little or nothing about fish oil and a vascular disease called Reynaud's Syndrome, discovered that fish oil can cure that disease without conducting any experiment. He later also discovered a link between migraine and magnesium deficiency. Swanson's idea was relatively simple: if a discipline is studying the relationship between concept A and B and another discipline is studying the relationship between concepts B and C, there might be important hidden facts about the relationship between A and C that are not visible to either discipline. The catch, of course, is that different disciplines use different vocabularies, so one has to translate one field's terminology into the other field's terminology. Neil Smalheiser then helped him develop Arrowsmith, a program that scanned the scientific literature to find connections between articles (1998).

In 1997 Marti Hearst started the LINDI (Linking Information for Novel Discovery and Insight) project at UC Berkeley (later renamed Bio Text).

These programs for scientific discovery did not complete the loop from hypothesis generation to experiment design back to hypothesis generation. In 2004 Ross King at Aberystwyth University envisioned a "robot scientist", Adam, designed to: generate hypotheses, design experiments to test these hypotheses, carry out the physical experiments using robotic laboratories, interpret the resulting data, and repeat the cycle until a successful theory is shaped ("Functional Genomic Hypothesis Generation and Experimentation by a Robot Scientist", 2004). In 2009 Adam identified twelve genes responsible for catalysing specific chemical reactions in yeast.

Eureqa, developed in 2007 by Hod Lipson's student Michael Schmidt at Cornell University (later a product sold by Schmidt's startup Nutonian), uses evolutionary algorithms to generate and select hypotheses consistent with experimental data and to design further experiments to continue the selection.

In 2009 Andrey Rzhetsky's team at University of Chicago, drawing data from more than 300,000 papers and 8 million abstracts, built two datasets of molecular interactions related to the muscle disease ataxia: 49,493 for mice and 52,518 for humans. By comparing the two, the scientists were able to discover genes associated with brain malformations ("Looking at Cerebellar Malformations through Text-Mined Interactomes of Mice and Humans", 2009). Rzhetsky later wrote the manifesto "Machine Science" (2010).

Pierre-Yves Oudeyer at Inria (the French Institute for Research in Computer Science and Automation) worked on computational models of curiosity ("Intrinsic Motivation Systems for Autonomous Mental Development", 2007).

Joshua Tenenbaum, Charles Kemp (now at CMU) and Thomas Griffiths (now at UCB) proposed that hierarchical Bayesian models are the force behind our brain's ability to discover scientific theories ("A Probabilistic Model of Theory Formation", 2009).

IBM Research in San Jose and Baylor College of Medicine in Texas collaborated on KnIT, the Knowledge Integration Toolkit (2014), a system that mines scientific literature and generates hypotheses ("Automated Hypothesis Generation Based on Mining Scientific Literature"). KnIT read 100,000 papers and discovered nine p53 kinases, seven of which had already been discovered by scientists and two that were unknown.

In 2015 a program developed by Michael Levin's group at Tufts University developed a scientific theory about a 120-year-old mystery related to the flatworm planarian; a modest but encouraging success.

These were all systems that used preexisting knowledge to analyze data and generate new knowledge. Since 2006, however, the ruling paradigm in A.I. was "deep learning", i.e. neural networks, i.e. data-driven A.I. Neural networks discover patterns, not theories. Can one discover new knowledge in a field without any previous knowledge of the field?

In June 2017, at the 50th Alan Turing Award ceremony in San Jose, Stuart Russell of UC Berkeley declared "A deep-learning system would never discover the Higgs boson from the raw data". Unbeknownst to him, two months earlier DeepMind had just seen a program discover something that was not in the data, AlphaGo Zero. So the verdict is still out on whether a deep-learning system can or cannot discover something like the Higgs boson.

Relatively few scientists followed Solomonoff's third way to artificial intelligence. Marcus Hutter of the Australian National University worked out a version of Solomonoff induction called AIXI, a universal optimal learning program, which he claimed to be "the most intelligent unbiased agent possible" ("A Theory of Universal Artificial Intelligence based on Algorithmic Complexity", 2000). He then (2009) used a Monte Carlo method (that randomly generates the set of hypotheses) to build a computationally feasible approximation.

In 2002 Juergen Schmidhuber at IDSIA designed an Optimal Ordered Problem Solver (OOPS) and then in 2003 a self-improving problem solver called Goedel Machine. These programs modify themselves when they prove logically that a change improves their performance. (Schmidhuber believes that the history of Artificial Intelligence begins with Kurt Goedel's famous self-referential formulas of 1931).

One foundational problem is that we don't quite understand well how human creativity works. One of the most famous theories was advanced by Robert Sternberg at Yale University ("A Propulsion Model of Types of Creative Contributions", 1999). His definition was that creative contributions propel a field forward, and he recognized eight types of such contributions because theory formation is not a well-defined process, nor is it a single process. Sternberg's paper "Types of Innovation" (2003) is a good summary of famous discoveries classified according to their different dynamics: replication, redefinition, forward incrementation, redirection, reinitiation, integration, etc. Some creative contributions accept current paradigms, some reject them and some integrate multiple paradigms. His collaborators James Kaufman and Lauren Skidmore provided an updated survey for the Internet age in "Taking the Propulsion Model of Creative Contributions into the 21st Century" (2010).

"A man provided with paper, pencil, and rubber, and subject to strict discipline, is in effect a universal machine" (Alan Turing).

A Failed Experiment

In my opinion the "footnotes" in the history of Artificial Intelligence were not just footnotes: they represent colossal failures. They were all great ideas. In fact, they were probably the "right" ideas: of course, an intelligent machine must be capable of conversing in natural language; of course, it must be able to walk around, look for food, and protect itself; of course, it must be able to understand what people say (each person having a slightly different voice); of course, it would make more sense for software to "evolve" by itself than to be written by someone (just like any form of intelligent life did); of course, we would expect an intelligent machine to be able to write software (and build other machines, like we do); of course, it would be nice if the machine were capable of translating from one language to another; of course, it would make sense to build a computer that is a replica of a human brain if what we expect is a performance identical to the performance of a human brain.

These ideas have remained unfulfilled. In a sense, Artificial Intelligence is still a failed experiment: we still don't know how to do it properly.

Note that, ironically, it was A.I. that made computers popular and fueled progress in computer science. The idea of a thinking machine, not the usefulness of computers, drove the initial development. Since those days, progress in A.I. has been scant, but computers have become household appliances. Your laptop and smartphone are accidental by-products of a failed scientific experiment.

An Easy Science

When a physicist makes a claim, an entire community of physicists is out there to check that claim. The paper gets published only if it survives peer review and the discovery is accepted only if the experiment can be repeated elsewhere. For example, when OPERA announced particles traveling faster than light, the whole world conspired to disprove them, and eventually it succeeded. It took months of results before CERN accepted that probably (not certainly) the Higgs boson exists.

Artificial Intelligence practitioners, instead, have a much easier life. Whenever they announce a new achievement, it is largely taken at face value by the media and by the A.I. community at large.

If a computer scientist announces that her or his program has learned what a cat looks like by watching videos, the whole world posts enthusiastic headlines even if nobody has actually seen this system in action, and nobody has been able to measure and doublecheck its performance: what else did it recognize? did it recognize the human beings in those videos? did it recognize furniture? what else was in those videos? For the record, when that happened in 2012, the media consistently reported "videos" when in fact the neural network had been trained with "images" taken from videos, i.e. still images, and we were not told who picked the images and according to which criteria out of the tens of thousands of frames that constitute the average YouTube video. It does make a difference which images are fed to the neural network out of the billions of images available on YouTube. A one-minute video contains about 2,000 frames. This neural network was fed 10 million images, which is the equivalent of about 80 hours of video, a pittance compared with the millions of hours of videos available on YouTube. The media also forgot to mention that this Google Brain network only recognized "faces of cats", not "cats". Humans (and mice and many other animals) can recognize a cat in any pose, for example when it is walking away from us and we actually do not see its face.

When in 2012 Google announced that "Our vehicles have now completed more than 300,000 miles of testing" (a mile being 1.6 kilometers for the rest of the world), the media simply propagated the headline without asking simple questions such as "in how many months?" or "under which conditions?" or "on which roads"? "at what time of the day"? Most people now believe that self-driving cars are feasible even though they have never been in one. Many of the same people probably don't believe all the weird consequences of Relativity and Quantum Mechanics, despite the many experiments that confirmed them.

The 2004 DARPA challenge for driverless cars was staged in the desert between Los Angeles and Las Vegas (i.e. with no traffic). The 2007 "DARPA Urban Challenge" took place at the George Air Force Base in California. Interestingly, a few months later two highly educated friends told me that a DARPA challenge took place in downtown Los Angeles in heavy traffic. That never took place. Too often the belief in the feats of A.I. systems feels like the stories of devout people who saw an apparition of a saint and all the evidence you can get is a blurred photo.

In 2005 the media reported that Hod Lipson at Cornell University had unveiled the first "self-assembling machine" (the same scientist in 2007 also unveiled the first "self-aware" robot), and in 2013 the media reported that the "M-blocks" developed at MIT by Daniela Rus' team were self-constructing machines. Unfortunately, these reports were wild exaggerations.

In May 1997 the IBM supercomputer "Deep Blue" beat then chess world champion Garry Kasparov in a widely publicized match. What was less publicized is that the match was hardly fair: Deep Blue had been equipped with an enormous amount of information about Kasparov's chess playing, whereas Kasparov knew absolutely nothing of Deep Blue; and during the match IBM engineers kept tweaking Deep Blue with heuristics about Kasparov's moves. Even less publicized were the rematches, in which the IBM programmers were explicitly forbidden to modify the machine in between games. The new more powerful versions of Deep Blue (renamed Frintz) could beat neither Vladimir Kramnik, the new world chess champion, in 2002 nor Kasparov himself in 2003. Both matches ended in a draw. What is incredible to me is that a machine equipped with virtually an infinite knowledge of the game and of its opponent, and with lightning-speed circuits that can process virtually infinite number of moves in a split second cannot beat a much more rudimentary object such as the human brain equipped with a very limited and unreliable memory: what does it take for a machine to outperform humans despite all the technological advantages it has? Divine intervention? Nonetheless, virtually nobody in the scientific community (let alone in the mainstream media) questioned the claim that a machine had beaten the greatest chess player in the world.

If IBM is correct and, as it claimed at the time, Deep Blue could calculate 200 million positions per second whereas Kasparov's brain could only calculate three per second, who is smarter, the one who can become the world's champion with just three calculations per second or the one who needs 200 million calculations per second? If Deep Blue were conscious, it would be wondering "Wow, how can this human being be so intelligent?"

What Deep Blue certainly achieved was to get better at chess than its creators. But that is true of the medieval clock too, capable of keeping the time in a way that no human brain could, and of many other tools and machines.

Finding the most promising move in a game of chess is a lot easier than predicting the score of a Real Madrid vs Barcelona game, something that neither machines nor humans are even remotely close to achieving. The brute force of the fastest computers is enough to win a chess game, but the brute force of the fastest computers is not enough to get a better soccer prediction than, say, the prediction made by a drunk soccer fan in a pub. Ultimately what we are contemplating when a computer beats a chess master is still what amazed the public of the 1950s: the computer's ability to run many calculations at lightning speed, something that no human being can do.

IBM's Watson of 2013 consumes 85,000 Watts compared with the human brain's 20 Watts. (Again: let both the human and the machine run on 20 Watts and see who wins). For the televised match of 2011 with the human experts, Watson was equipped with 200 million pages of information including the whole of Wikipedia; and, in order to be fast, all that knowledge had to be stored on RAM, not on disk storage. The human experts who competed against Watson did not have access to all that information. Watson was allowed to store 15 petabytes of storage, whereas the humans were not allowed to browse the web or keep a database handy. De facto the human experts were not playing against one machine but against a whole army of machines, enough machines working to master and process all that data. A fairer match would be to pit Watson against thousands of human experts, chosen so as to have the same amount of data. And, again, the questions were conveniently provided to the machine as text files instead of spoken language. If you use the verb "to understand" the way we normally use it, Watson never "understood" a single question. And those were the easiest possible questions, designed specifically to be brief and unambiguous (unlike the many ambiguities hidden in ordinary human language). Watson didn't even hear the questions (they were written to it), let alone understand what the questioner was asking. Watson was allowed to ring the bell using a lightning-speed electrical signal, whereas the humans had to lift the finger and press the button, an action that is order of magnitudes slower.

Over the decades i have personally witnessed several demos of A.I. systems that required the audience to simply watch and listen: only the creator was allowed to operate the system.

Furthermore, some of the most headline-capturing Artificial Intelligence research is supported by philanthropists at private institutions with little or no oversight by academia.

Many of the A.I. systems of the past have never been used outside the lab that created them. Their use by the industry, in particular, has been virtually nil.

For example, on the first of October of 1999 *Science Daily* announced: "Machine demonstrates superhuman speech recognition abilities. University of Southern California biomedical engineers have created the world's first machine system that can recognize spoken words better than humans can". It was referring to a neural network trained by Theodore Berger's team. As far as i can tell, that project has been abandoned and it was never used in any practical application.

In October 2011 a Washington Post headline asked "Apple Siri: the next big revolution in how we interact with gadgets?" Meanwhile, this exchange was going viral on social media:

User: Siri, call me an ambulance
Siri: Okay, from now on I'll call you "an ambulance"

(Note: in 2017 the app-measurement firm Verto Analytics estimated that, between the period of May 2016 and May 2017, Siri lost 7.3 million monthly users, or about 15% of its total user base in the USA).

In 2014 the media announced that Vladimir Veselov's and Eugene Demchenko's program Eugene Goostman, which simulated a 13-year-old Ukrainian boy, passed the Turing Test at the Royal Society in London (Washington Post: "A computer just passed the Turing Test in landmark trial"). It makes you wonder what was the I.Q. of the members of the Royal Society, or, at least, of the event organizer, the self-appointed "world's first cyborg" Kevin Warwick, and what was the I.Q. of the journalists who reported his claims. It takes very little ingenuity to fool a "chatbot" impersonating a human being: "How many letters are in the word of the number that follows 4?" Any human being can calculate that 5 follows 4 and contains four letters, but a bot won't know what you are talking about. I can see the bot programmer, who has just read this sentence, frantically coding this question and its answer into the bot, but there are thousands, if not millions, of questions like this one that bots will fail for as long as they don't understand the context. How many words are in this sentence? You just counted them, right? But a bot won't understand the question. Of course, if your Turing Test consists in asking the machine questions whose answers can easily be found on Wikipedia by any idiot, then the machine will easily pass the test.

A video that went viral on social media was a video of a robot folding towels. The video was played at 24 times the real time, and nobody seemed to pay tribute to dry cleaners: it is certainly impressive that a robot can fold towels but dry cleaners already use machines that fold shirts, a more complicated task.

In 2015 both Microsoft and Baidu announced that their image-recognition software was outperforming humans, i.e. that the error rate of the machine was lower than the error rate of the average human being in recognizing objects. The average human error rate is considered to be 5.1%. However, Microsoft's technology that has surfaced (late 2015) is CaptionBot, which has become famous not for its usefulness in recognizing scenes but for the silly mistakes that no human being would make. As for Baidu, its Deep Image system, that ran on the custom-built supercomputer Minwa (432 core processors and 144 GPUs), has not been made available to the public as an app. However, Baidu was disqualified from the most prestigious image-recognition competition in the world (the ImageNet Competition) for cheating. Recognizing images was supposed to be Google's specialty but Google Goggles, introduced in 2010, has flopped. I just tried Goggles again (May 2016). It didn' recognize: towel, toilet paper, faucet, blue jeans... It recognized only one object: the clock. Officially, Google's image recognition software has an error rate of 5%. My test shows more like 90% error rate. In 2015 the Google Photos app tagged two African-Americans as gorillas, causing accusations of racism when in fact it was just poor technology. In 2014 the media widely reported that Facebook's DeepFace correctly identified photos in 97.25% of cases, a fact that induced the European Union to warn Facebook that people's privacy must be protected; but, as of 2017, it is still not available for us to try it out. In fact, Facebook stopped publicizing it in 2015, and Facebook still identifies few of my 5,000 friends: face recognition works well only if you have a small number of friends.

I am also confused by all these announcements that seem to contradict each other. In March 2015 Google announced that FaceNet, a 22-layer deep convolutional network, recognized the faces of celebrities with a negligible error rate. Google only released an open-source version of FaceNet that doesn't even come close to what they claimed. In June 2015 Facebook announced that a new algorithm was capable of recognizing partially covered faces with 83% accuracy, but it never clarified what "partially covered" means, and i have seen no improvements to the error rate in recognizing my friends even when the face is clearly visible.

In 2015 it was widely reported that a DeepMind neural network learned how to play Atari videogames, but virtually nobody reported that, two years later, Dileep George's team at Vicarious showed a grotesque limitation of the system, unable to play the videogame if you introduce tiny perturbations ("Schema Networks", 2017), a finding replicated on a larger scale by Pieter Abbeel's students at UC Berkeley ("Adversarial Attacks on Neural Network Policies", 2017).

In September 2017 Apple announced its Face ID technology for facial recognition in the iPhone X smartphone. It took exactly two months for Vietnamese firm Bkav to find out how to fool the system, and it was simply a matter of creating a mask with some cheap 3D-printed material and a little paint.

Sometimes the claims border on the ridiculous. Let's say that i build an app that asks you to submit the photo of an object, then the picture gets emailed to me, and i email back to you the name of the object: are you impressed by such an app? And, still, countless reviewers marveled at CamFind, the app introduced in 2013 by Los Angeles-based Image Searcher, an app that "recognizes" objects. In most cases it is actually not the app that recognizes objects, but their huge team in the Philippines that is frantically busy tagging the images submitted by users.

In 2017 progress by text-comprehension systems was widely reported (with academic papers at major conferences boasting up to 97% accuracy in answers to natural-language questions), but Robin Jia and Percy Liang at Stanford showed how easy it is to fool those systems compared with human beings ("Adversarial Examples for Evaluating Reading Comprehension Systems", 2017).

You've probably seen many headlines like these: "Google's AI translation system is approaching human-level accuracy" (The Verge, 2016); "AI-based translation to soon reach human levels" (Quartz, 2017); "Microsoft researchers match human levels in translating news from Chinese to English" (ZDNet, 2018). In 2018 i was at a conference where the screens on the side of the room where displaying automatic translation (from Chinese to English). The automatic translation was provided by the A.I. system that had just achieved the highest BLEU (Bilingual Evaluation Understudy) score ever. Well, the automatic translation was often hilarious. Luckily there was also human simultaneous translation. At one point a "screening for breast cancer" became "screening for breakfast", and "AI is difficult to program" became "I was difficult to program". Luckily the human translator got all of these right, and that's why i know what they were supposed to be.

In September 2018 China's main firm for speech recognition and machine translation, Xunfei/iFlytek, pretended that its live translation was made by a machine but later its employees spoke out to denounce that it was done by professional interpreters. (Sure enough, some media reported that the iFlytek system was achieving "super-human" performance). Remember the automata of centuries ago, that in reality were people camouflaged like machines? In 1769, a chess-playing machine called "The Turk", created by Wolfgang von Kempelen, toured the world, winning games wherever it went: it concealed a man inside so well that it wasn't exposed as a hoax for many years.

(To be fair, Microsoft's CaptionBot is not bad at all: it was criticized by people who expected human-level abilities in the machine, but, realistically, it exceeds my expectations).

Very few people bother to doublecheck the claims of the A.I. community. The media have a vested interest that the story be told (it sells) and the community as a whole has a vested interest that government and donors believe in the discipline's progress so that more funds are poured into it.

The media widely reported that neural networks reached parity with humans, and even surpassed humans, in image and speech recognition. The truth is that these facts are measured not in the real world but in the limited world of the images contained in datasets like ImageNet. These neural nets got really good at recognizing the images for which they were trained. They were not trained to recognize objects or faces in the real world. China is using face recognition in multiple locations but you have to stand right in front of the camera, facing the camera (not at an angle), remove your eyeglasses, move your hair out of your forehead, and good luck if you have a bandaid or scar. Neural networks recognize only that for which they have been trained, that is a very tiny fraction of the objects of the world, and only in ideal, perfect conditions. Software has helped police identify criminals, but "helped" does not mean "did". Neural networks recognize the objects that they have been trained to recognize, the objects that were in the training dataset, not "all" the objects in the world, and certainly not all the objects within images and all the situations that include those images.

It is obviously not true that a deep convolutional net is more accurate than humans at recognizing images: it is more accurate only relative to the

ImageNet dataset (or some other dataset) that was used to train it. In real life the same net is as good as a drunk in a crowded bar.

After 2014 the much vaunted (by media that are easily manipulated by press releases) Microsoft project Adam, developed by software engineers with little or no background in deep learning, disappeared as quickly as it had appeared despite statements that it was outperforming Google's system (it was actually outperforming an old Google network that Google had already abandoned and it did so at the 22K-category ImageNet contest that virtually nobody was contesting anymore because everybody was focused on the 1K-ImageNet challenge instead).

In 2017 the robot Xiaoyi (developed by Tsinghua University and Chinese startup iFlyTek) passed the medical examination test... or did it? The media glossed over the fact that it memorized two million medical records and 400,000 academic papers. This medical robot was incapable of visiting patients: it was just a toy providing a smiling user interface to a database of medical information (a database like many others available online). Xiaoyi was no more than a glorified database to serve China's rural areas where general practitioners are in severe shortage.

In 2016 the media announced that the Washington Post used an "intelligent" software, "Heliograf", to write articles about the Olympic Games, and that Toutiao in China employed Xiaomingbot, a similar writing robot, to write 450 articles during the 15 days of the games. Both pieces of news were false. Neither "robot" wrote anything meaningful. The "articles" were just a few words long (the longest was 821 words long and it was simply a list of medals) and they were proof-edited by human editors: the Washington Post articles clearly stated "It is powered by Heliograf, the Post's artificial intelligence system" ("powered", not "written"). A human being could have easily written (not just powered) hundreds of such "articles" in one day. But nobody wants to read hundreds of such articles. We'd rather read just a list of medals or one (just one) real article.

In 2017 Cadillac was billing its "Super Cruise" system as the "world's first true hands-free driving system", except that it worked only on Cadillac's mapped routes; which is as "self-driving" as a train riding on rails.

When faced with the question, "What would a self-driving car do if a child suddenly ran in front of the car", a famous A.I. scientist replied "Children rarely appear out of nowhere", which tells you how much we can trust A.I. scientists in the real world. (It also tells you what their solution to the "random children" problem will be: ban children from any road where self-driving cars are allowed, and hold parents liable if their children break this rule).

Before the 2018 World Cup started, many media publicized the prediction of an Artificial Intelligence program that Spain was the favorite to win, followed by Germany, Brazil, France, Belgium, Argentina, England, Portugal... See https://arxiv.org/abs/1806.03208 Well, Germany didn't even make it beyond the first stage; Argentina, Spain, Portugal and Brazil were promptly kicked out too. France did win but the artificial intelligence had ranked it 4th. This is the same technology that will be used to rank the odds that the object in front of the car is a pedestrian. The same media that publicized the predictions by the A.I. system did not comment on how grotesquely wrong its predictions turned out to be.

In August 2018 the media widely advertised the fact that OpenAI's bot won against masters of the videogame Dota 2. A lot less advertised was the fact that the bot, OpenAi Five, lost to humans a few days later at The International in Vancouver, an annual tournament for Dota 2. The difference was that this time the rules were more fair instead of handing the machine an unfair advantage (as it has been the norm in this kind of events since DeepBlue's famous chess match). For example, the "hero" lineup of the game was chosen by a third party instead of being chosen by OpenAI.

In 2018 a face-recognition system deployed in the city of Ningbo to spot jaywalkers identified a photo of Chinese billionaire Mingzhu Dong posted on an ad of a passing bus as a jaywalker. In 2018 Reddit user "MalletsDarker" used Google Photos' A.I. to merge three photos taken at a ski resort, two photos of the snowy landscape and one photo of his friend wearing helmet and goggle: the A.I. system merged the three into one panorama, understanding his friend's head as a giant mountain peak. In 2018 Boston Dynamics' humanoid robot Atlas carried out an impeccable show at the Congress of Future Scientists and Technologists but then awkwardly tripped over a curtain and tumbled off the stage. Hopefully unaware that Nazis held similar beliefs about Jews, in 2018 the Israeli startup Faception claimed that its face-analysis system can infer a person's I.Q. from the person's facial structure.

The media are quick at publicizing any marginal success by A.I. systems, but routinely ignore its vastly more numerous failures.

The media love anecdotes documenting this simple pattern: every time someone said that A.I. could never do something, A.I. did it. A.I. keeps proving the skeptics wrong. This is mostly true. What has not been reported equally well is the converse: whenever someone said that an A.I. system would never make such and such a mistake, the A.I. system made that mistake. In at least one case it also killed the driver who was using the self-driving feature (Joshua Brown, the first person to die in a self-driving car accident, June 2016, while napping in his Tesla) and in at least one case it also killed a pedestrian (an Uber car in March 2018 ran over a woman who was crossing the street in a place where there was no crosswalk).

A.I. is de facto guilty of "fake news": too many announcements about this or that achievement are "fake news" because these machines exist only in the imagination of the press and in the press releases of scientists who are looking for funding and investors. The real achievements are much more limited and humbler. You see a video of a robot grasping an object like a skilled worker but that video has been accelerated 20 times. You see videos of self-driving cars but the video has been made from an angle that hides the human copilot in the car. We have created an entire economy of highly-paid speakers and writers around a technology that largely doesn't exist yet.

"We always overestimate the change that will occur in the next two years and underestimate the change that will occur in the next ten" (Bill Gates)

Footnote: Between Fatima and AlphaGo

Paul Nunez in "Brain, Mind, and the Structure of Reality" (2010) distinguishes between Type 1 scientific experiments and Type 2 experiments. Type 1 is an experiment that has been repeated at different locations by different teams and still holds. Type 2 is an experiment that has yielded conflicting results at different laboratories. UFO sightings, levitation tales and exorcisms are not scientific, but many people believe in their claims, and i will call them "Type 3" experiments, experiments that cannot be repeated by other scientists. Too much of Artificial Intelligence occupies the space between Type 2 and Type 3.

News of the feats achieved by machines rapidly propagate worldwide thanks to enthusiastic bloggers and tweeters the same way that news about telepathy and levitation used to spread rapidly worldwide thanks to word of mouth without the slightest requirement of proof. (There are still millions of people who believe that cases of levitation have been documented even though there is no footage and no witness to be found anywhere). The belief in miracles worked the same way: people wanted to believe that a saint had performed a miracle and they transmitted the news to all their acquaintances in a state of delirious fervor without bothering to doublecheck the facts and without providing any means to doublecheck the facts (address? date? who was there? what exactly happened?). The Internet is a much more powerful tool than the old "word of mouth" system. In fact, i believe that part of this discussion about machine intelligence is a discussion not about technology but about the World-wide

Web as the most powerful tool ever invented to spread myths. And part of this discussion about machine intelligence is a discussion about the fact that 21st century humans want to believe that super-intelligent machines are coming the same way that people of previous centuries wanted to believe that magicians existed. The number of people whose infirmity has been healed after a visit to the sanctuary of Lourdes is very small (and in all cases one can find a simple medical explanation) but thousands of highly educated people still visit it when they get sick, poor or depressed. On 13 October 1917 in Fatima (Portugal) tens of thousands of people assembled because three shepherd children had been told by the Virgin Mary (the mother of Jesus) that she would appear at high noon. Nobody saw anything special (other than the Sun coming out after a shower) but the word that a miracle had taken place in Fatima spread worldwide. Believe it or not, that is pretty much what happens today when a fanatic blogger reports a feat performed by an A.I. software or a robot as a new step towards the Singularity. People like me who remain skeptical of the news are frowned upon in the same way that skeptics were looked down upon after Fatima: "What? You still don't believe that the Virgin Mary appeared to those children? What is wrong with you?" Which, of course, shifts the burden of proof on the skeptic who is asked to explain why one would NOT believe in the miracle (sorry, i meant "in the machine's intelligence") instead of pressing the inventor/scientist/lab/firm into proving that the miracle/feat has truly been accomplished and can be repeated at will and that it really did what bloggers said it did.

"Whenever a new science achieves its first big successes, its enthusiastic acolytes always fancy that all questions are now soluble" (Gilbert Ryle, "The Concept of Mind", 1949, six years before Artificial Intelligence was born).

Footnore: A Brief History of Chatbots

Perhaps the most visible by-product of A.I. has been the chatterbot or chatbot (a computer program capable of having a conversation, the idea pioneered by Joseph Weizenbaum's Eliza).

In 1990 A.I. philanthropist Hugh Loebner and the Cambridge Center for Behavioral Studies in Boston launched a yearly prize to be awarded to the chatbot that went closer to passing the Turing Test. Competitors have included: Joseph Weintraub's PC Therapist (originally written in 1986), a variation on Eliza; Michail Mauldin's Julia (1994), by the founder of Lycos; Richard Wallace's Alice (Artificial Linguistic Internet Computer Entity), originally developed in 1995; Rollo Carpenter's Jabberwacky (1997); Robby Garner's Albert One (winner in 1998); Bruce Wilcox's

Suzette (2009), first of a dynasty of chatbots; and Steve Worswick's Mitsuku (unveiled in 2013). SmarterChild, the first commercial chatbot, was launched in 2000 by a New York firm, ActiveBuddy, and was used at one time or another by millions of people. None of these was particularly useful; nor intelligent enough to have a real conversation, but perhaps more fun than playing the game card solitaire. Again, one has to wonder what is the I.Q. of the Loebner Prize judges who are fooled by these chatbots. Then came the generation of Apple Siri (2011), Google Now (2012), Microsoft Xiaoice (2014) and Amazon Alexa (2014) that tried to be useful, not particularly intelligent, thanks to much progress in speech recognition; all the way to the hyper-realistic "human" chatbots made by New Zealand-based startup Soul Machines (the brainchild of Mark Sagar, a former Hollywood animation engineer), starting with "Baby X" (2014). In 2016 Eugenia Kuyda and Philip Dudchuk released the "memorial chatbot" Replika that learns a person's style of chat and can replicate it even when the person is dead. In 2017 the Stanford psychologist Alison Darcy unveiled Woebot, a therapy chatbot for depression and anxiety.

It has certainly become easier to create chatbots. I would rather call them "conversational user interfaces" in which some very limited linguistic skills replace the traditional interaction via menus or touch-screens. It is certainly convenient to shout "Skip!" to Alexa when you want to skip a song in a playlist. In 1955 Zenith introduced the first remote control for TV sets, the Flashmatic, invented by Eugene Polley. That too was a very convenient invention, but nobody called it "intelligent".

There are now scripting languages such as Artificial Intelligence Markup Language or AIML (Richard Wallace, 1995) and ChatScript (Bruce Wilcox, 2011); corporate (and sometimes open-source) natural language processing (NLP) tools such as Speaktoit, later renamed API.ai (Ilya Gelfenbeyn, 2014, acquired by Google in 2016), Wit.ai (Alexandre Lebrun, acquired by Facebook in 2015), Microsoft's Language Understanding Intelligent Service or LUIS (2015), Amazon Lex (2017), i.e. the technology powering Amazon's virtual assistant Alexa, and Facebook's fastText for text representation and classification (2018); as well as free platforms such as Pandorabots (Kevin Fujii and Richard Wallace, largest installed base of chatbots, 2008), Rebot.me (Ferid Movsumov and Salih Pehlivan, 2014) and Imperson (Disney Accelerator, 2015). The advent of conversational user interfaces may or may not be an improvement over previous methods of interaction. I personally think that the old command line of Unix and DOS was not such a bad idea: i used to get things done quickly and efficiently.

Intermezzo: A.I. and the History of Homo Sapiens

It is said that Homo Sapiens is the only species that records its actions (and even writes lengthy histories of itself) so that it can learn from the past. However, that is not completely correct: writing was only invented a few thousand years ago, which means that Homo Sapiens (a 200,000-year-old species) spent 99.9% of its history without writing. Furthermore, computers were invented only a few decades ago, which means that Homo Sapiens spent 99.9999% of its history without a tool to analyze the patterns of its own actions.

Fast forward to today. Today, thanks to all the invisible bots that constantly monitor and influence its actions, Homo Sapiens has become the only species on the planet whose behavior is rigidly determined by its past behavior: we watch movies recommended by bots that study which movies we liked in the past, we read the news that align with our political preferences, we buy more of what we bought yesterday, etc.

Teaser: Learning to Chat

Daniel Schwartz, dean of Stanford's Graduate School of Education, gave a talk in January 2018 about how students learn to reason when they train their teaching agent. The teaching agent is an A.I. system that is designed to teach them a particular subject. It turns out that the most valuable learning for the students consists in learning how to reason, and that happens when they train the teaching agent, not when they use it.

I think that this is true in general. The person who architects and trains a neural network learns as the network is learning. Building an A.I. system is an extremely formative experience. On the other hand, if you only use an A.I. agent, you don't learn much, and you might actually get dumber (because you will simply be delegating "intelligent" thinking to your A.I. agent).

The screenwriter of a conversational experience (of a "chatbot" experience) probably learns a lot more about the topic than the user of the conversational agent (of the chatbot) ever will.

Intermezzo and Trivia: the Original App

At the same time the real achievements of the machine are sometime neglected. I am not terribly impressed that computers can play chess. I am much more impressed that computers can forecast the weather, since the atmosphere is a much more complex system than the game of chess.

Weather forecasting was a particularly interesting application of electronic computing for John Von Neumann. In fact, it was "the" application originally envisioned for the machine that Von Neumann designed at Princeton's Institute for Advanced Studies (IAS), using ideas from ENIAC's inventors John Mauchly and Presper Eckert, the machine that introduced the "Von Neumann architecture" still used today.

Note that, unlike chess and machine translation, this problem is not currently solved by using statistical analysis. It is solved by observing the current conditions and applying physical laws. Statistical analysis requires an adequate sample of data, and a relatively linear behavior. Weather conditions, instead, are never the same, and the nonlinear nature of chaotic systems like the atmosphere makes it very easy to come up with grotesquely wrong predictions. This does not mean that it is impossible to predict the weather using statistical analysis; just that it is only one method out of many, a method that has been particularly successful in those fields where statistical analysis makes sense but was not feasible before the introduction of powerful computers. There is nothing magical about its success, just like there is nothing magical about our success in predicting the weather. Both are based on good old-fashioned techniques of computational mathematics.

Don't be Fooled by the Robot

The bar is being set very low for robotics too. Basically, any remote-controlled toy (as intelligent as the miniature trains that were popular in the 1960s) is now being hailed as a step toward the robot invasion. I always advise robotics fans to visit the Musee Mecanique in San Francisco, that has a splendid collection of antique coin-operated automatic mechanical musical instruments... sorry, i meant "robotic musicians", before they venture into a discussion about progress in robotics. These automata don't constitute what we normally call "intelligence" but they provide amazing shows. Automata have entertained royalties and peasants for centuries: Ismail Al-Jazari's music ensemble of 1206, Leonardo DaVinci's knight of 1495, Juanelo Turriano's monk of 1560, Jacques de Vaucanson's duck of 1739, Pierre Jaquet-Droz's dolls of 1768-74, John Joseph Merlin's "Silver Swan" of 1773, Hubert Martinet's musical elephant of 1774, Henri Maillardet's draughtsman-writer of 1800, Joseph Faber's Euphonia of 1840... the list is endless.

Does driving a car qualify as a sign of "intelligence"? Maybe it does, but it has to be "really" what it means for humans. There is no car that has driven even one meter without help from humans. The real world is a world in which first you open the garage door, then you stop to pick up the newspaper, then you enter the street and you will stop if you see a pedestrian waiting to cross the street. No car has achieved this skill yet. They self-drive only in highly favorable conditions on well marked roads with well marked lanes and only on roads that the manufacturing company has mapped accurately (in other words, with a lot of help from humans). And i will let you imagine what happens if the battery dies or there's a software bug... What does the self-driving car do if it is about to enter a bridge when an earthquake causes the bridge to collapse? Presumably it will just drive on. What does the self-driving car do if it is stopping at a red light and a man with a gun breaks the window? Probably nothing: it's a red light. If you fall asleep in a self-driving car, your chances of dying will skyrocket. There are countless rules of thumb that a human driver employs all the time, and they are based on understanding what is going on. A set of sensors wrapped in a car's body does not understand anything about what is going on.

Human-looking automata that mimic human behavior have been built since ancient times and some of them could perform sophisticated movements. They were mechanical. Today we have electromechanical sophisticated toys that can do all sort of things. There is a (miniature) toy that looks like a robot riding a bicycle. Technically speaking, the whole toy is the "robot". Philosophically speaking, there is no robot riding a bicycle. The robot-like thing on top of the bicycle is redundant, it is there just for show: you can remove the android and put the same gears in the bicycle seat or in the bicycle pedals and the bike with no passenger would go around and balance itself in the exact same way: the thing that rides the bicycle is not the thing on top of the bike (designed to trick the human eye) but the gear that can be placed anywhere on the bike. The toy is one piece: instead of one robot, you could put ten robots on top of each other, or no robot at all. Any modern toy store has toys that behave like robots doing some amazing thing (amazing for a robot, ordinary for a human). It doesn't require intelligence: just good engineering. This bike-riding toy never falls, even when it is not moving. It is designed to always stand vertical. Or, better, it falls when it runs out of battery. That's very old technology. If that's what we mean by "intelligent machines", then they have been around for a long time. We even have a machine that flies in the sky using that technology. Does that toy represent a quantum leap forward in intelligence? Of course, no. It is remotely controlled just like a television set. It never "learned" how to bike. It was designed to bike. And that's the only thing it can do. The only thing that is truly amazing in these toys is the miniaturization, not the "intelligence".

If you want this toy to do something else, you'll have to add more gears of a different kind, specialized in doing that other thing. Maybe it is possible (using existing technology or even very old mechanical technology) to build radio-controlled automata that have one million different gears to do every single thing that humans do, the whole taking up no more space than my body does. It would still be a toy.

A human being is NOT a toy (as yet).

The Intelligence of Neural Networks

Nobody has a viable definition of "intelligence" but there are different levels at which a machine can simulate what we do. The "Infinite Monkey Theorem" that the French mathematician Emile Borel discussed in his book "Statistical Mechanics and Irreversibility" (1913) probably represents the lowest level: let a monkey type randomly on a typewriter for millions of years and it will eventually produce all the books ever written by humankind.

Ross Ashby in his paper "Design for an Intelligence Amplifier" (1956) calculated that the Brownian motion of the molecules inside a cubic centimeter of air produces the correct binary code for a trigonometric formula 100,000 times a second. He concluded that a child doodling all day long for long enough would eventually write down the same algebraic formula.

A slightly more "intelligent" (or, at least, more feasible) simulation of our intelligence is the "Chinese Room" imagined by the philosopher John Searle. If you lock in a room a person, who knows absolutely nothing of the Chinese language, and provide her with the list of all possible answers in Chinese to all possible questions in Chinese (no matter how complex the question), a master of Chinese language asking questions in Chinese and reading her answers in Chinese would conclude that the person in the room is a highly literate Chinese scholar when in fact the person locked in the room knows absolutely nothing about Chinese (maybe not even that it is Chinese).

Whether you call "intelligent" Borel's monkey and Searle's room is up to you. It really depends on your definition of "intelligence".

Neural networks fall somewhere in between. Like Borel's monkey they perform a lot of random "typing" (except that here "typing" is really "number crunching") and like Searle's Chinese room they have been provided "rules" (although here they are in the form of "trained" configurations).

A deep-learning practitioner will probably tell you that building a neural network to simulate human behavior is a lot more feasible than setting up Searle's Chinese room. However, no neural network has ever come even remotely close to being able to answer complex questions in Chinese (or any other language).

Life with Intelligent Machines

Our daily lives increasingly depend on the work of algorithms instead of fellow humans.

Banks have always been ahead in automation, so let me start with some banking examples.

In the old days of the coin-operated phone, i found myself in a foreign country with no access to my bank account: it had been locked because the algorithm had detected "unusual activity". I called the emergency number from a public phone on a sidewalk and had a brief conversation with the bank representative. The bank representative had to verify my identity by asking me all sorts of personal questions. I had to shout them in front of all the local people who were gathering around me. The whole town learned my personal data. The bank representative was obviously not interested in the security of my bank account because he had just compromised it by asking me to shout all that information. The bank representative was simply executing an algorithm. Without executing that algorithm he would not have been able to help me unlock my bank account. Now it was unlocked but, of course, that's precisely when it should have been locked because the security was completely compromised.

When i called my bank to inquire about a money transfer, the bank's automated phone system asked me for my account number, my ATM card's secret code, my social security number and my mother's maiden name before informing me that the office was currently closed. Because the bank's system was now using voice recognition technology, i had to repeat the numbers countless times before it understood them correctly, and some of those numbers have 16 digits.

One day i helped a neighbor who seemed to be having a panic attack. She was shouting "help! help! operator! operator!" to the bank's automated system that kept asking her to pick among nine options and was responding with a simple: "Let's try again. Which of these nine options best describes why you are calling today?" And it kept repeating the nine options. It wasn't an emergency but she had just lost her patience and wanted to speak to a person. When we finally found a route to get the "You want to talk to an operator. Is that correct?" She shouted a heartfelt "Yes!!!!" But the system, implacable, said: "Before i transfer you, are you interested in hearing about a special offer..." When she finally managed to speak to a human being, the human being began by telling her that the phone call was recorded for "quality assurance". That is obviously false: nobody at the bank will ever care that she had to spend 20 minutes shouting "help!" to a machine that kept repeating nine irrelevant options to her.

When i had problems using my credit card outside the USA, i logged into my account and sent an angry message to customer support. Minutes later i saw that there was already a reply: the reply was saying "We are sorry that you decided to cancel your credit card". That is the exact opposite of what i needed. I replied with truly angry words to this message and this time it took 24 hours to get an answer: a customer service representative apologized for the misunderstanding that was due not to the stupidity of an employee but to the stupidity of a software "bot" that misunderstood my message.

I applied for "global entry", a program that speeds up immigration procedures for US citizens. I was accepted and a card was delivered to my home. I had a question about validating the card and i found a well-hidden place on their website where one can submit a question. A few minutes later i got a reply by email that informed me of something totally irrelevant to my case: rules for crossing overland from Mexico in a private car. The reply had clearly been assembled by another software "bot" that didn't understand my simple question.

In late 2019 i requested some help from Paypal. It was a simple question, but i made the mistake of asking it via their "message" system. The reply obviously came from a "bot": as it is often the case with bots, the reply had absolutely nothing to do with my question. I tried to re-word the question a few times, but the bot kept misunderstanding. Eventually, it reached the conclusion that i was not able to access my account. So it gave me instructions on what i should do when unable to access my account... except that in order to read this message i had to access my account because Paypal doesn't communicate via email: you can write a message to tech and read a message from tech only after you access your account. The bot not only misunderstood all my messages but didn't even realize how stupid it is to send a message that requires account access when the message is about how to obtain account access. If you think that i cannot access my account, why do you send me a message that i can read only if i can access my account?

My friend Ania had an embarrassing car accident: the handbrake of her car failed while the car was parked uphill and the car rolled out and got damaged (luckily it didn't injure anybody and didn't hit any other car). When she filed the report with the car insurance, the agent asked her the usual questions that include whether she was wearing a seatbelt at the moment of the accident. Ania politely replied that she was not in the car when the car rolled down the hill. The agent politely repeated the question: "Were you wearing the seatbelt?" He also politely informed her that the machine only accepted "yes" or "no" as the answer, and that a "no" would certainly cause her premium to skyrocket. Ania protested that this was ridiculous, but eventually she had to lie and answer "yes" in order to avoid the punishment for not wearing seatbelts when not in the car.

A European friend, who resides in San Francisco, got married in China and his wife applied for a spouse visa at the Shanghai consulate of that country. The Shanghai consulate told her that her husband needed to register the marriage at the San Francisco consulate, otherwise the computer wouldn't accept that she was married to someone who didn't show up as married in the database; and it had to be done in San Francisco because that's where her husband resided. The San Francisco consulate told the couple that they had to provide a translation of the Chinese wedding certificate notarized by... the Shanghai consulate! Basically, the Shanghai consulate had to tell the San Francisco consulate "yes they are married" so that the San Francisco consulate could tell the Shanghai consulate "yes they are married". They did so and then both computers were happy: one computer reported that the husband was a married man, and the other computer gladly granted the spouse visa to his wife. Nobody seemed to find it odd that the Shanghai consulate actually had that information from the beginning.

Years ago, purchasing an airline ticket used to require a visit to the airline office or to a travel agent. Now you can visit one of the websites that sells tickets for all airlines. When these websites started, they delivered a great simplification to us. But now these websites have become so fragile that they work only if you have the right computer, the right operating system and the right browser. And they decide which one is the "right" one. Imagine a shop that you can enter only if you are wearing clothes approved by the shop owner. Even if you have the right browser, you may not have the most recent release, in which case you have to download and install it. And if your old computer cannot run the new release, oh well, you just need to buy a new computer. When you finally have all the hardware and software that is accepted by the airline's website, you can enjoy the frustration of a slow website with a cluttered (albeit colorful) user interface. If you are a member of their frequent flyer program, you'll also have to login, and this may involve password and secret questions. Make three mistakes and your account gets automatically locked. When you finally manage to login, often you end up buying a more expensive ticket than needed because you are exhausted.

When my preferred airline introduced a new level of security in their login process, i was asked to pick three "secret" questions from a list of about 20. None of those 20 applied to me: i was not married (therefore i didn't have a place where i met my wife), i don't have children (therefore i never hired a babysitter), i certainly don't remember the name of my first elementary school, i never had a favorite color, and so forth. Nonetheless, i could not have logged into my account without first picking three secret questions. So now i have an imaginary wife whom i met in Greece and the imaginary babysitter who took care of my first child was named Olga (and i won't tell you the third silly question).

I am currently abroad and the most popular search engine keeps switching to the local language that i don't speak, and there seems no way to convince it that i really don't speak that language, no matter how many times i switch to English; and sometimes it doesn't even give me the option to switch languages at all. The search that i am trying to do right now (and for which i get no option to switch the results to English) is "what time check in at the airport for American Airlines". It shouldn't be terribly difficult to understand in which language i want to read the results.

We've all had close encounters with stupid algorithms. The scary fact is that increasingly a) there is no human to talk to in order to override the algorithm, and b) there is often no human who actually knows or has the power to override it! One day in China i couldn't log into my PayPal account because PayPal's website insisted on asking me for my pet's name (i never had a pet in my life, and i don't remember ever setting up a security question about a pet that i never had) or on sending me a text message to my phone (that was in California). As of 2017, PayPal offers no way to contact customer support electronically without logging in: if you cannot log in, you cannot email them saying you cannot log in. Your only option is to make a phone call (their way to discourage customers from using customer support). I waited till midnight and then called PayPal in the USA (different time zone). The PayPal representative couldn't find a way to let me access my PayPal account. I asked him to simply transfer my balance to my bank account, the bank account that PayPal has on record. Answer: only the account owner can do that, and the only way to prove that i was the account owner was to login, precisely what i could not do. I asked him to escalate the problem to his supervisor. His supervisor told me the same thing: "nothing we can do". I threatened to close my account, publicize the problem on social media , sue PayPal, etc. I tried to reason about the situation: if i never set up a security question about a pet, then someone else must have done it, and that's a security breach, correct? This whole ordeal was about the security of my account, wasn't it? If security was the primary goal here, shouldn't PayPal immediately transfer my money out of my PayPal account into some kind of safe account? To no avail: no human being had the power to control the algorithm that blocked access to my account. At the end of a 40-minute conversation, the representative literally told me to "try again after 24 hours". What if it still doesn't work? "Try again after another 24 hours"! Note the word: "try". Nobody at PayPal could tell what the algorithm would do during the 24 hour period, but they hoped that it would stop asking me the security question about a pet that i never had.

All these interactions with increasingly "smarter" machines share the same characteristic: throughout the operation you are expected to reply like a machine (a stupid machine) to all the questions. Any attempt to behave like an intelligent being results in the transaction getting canceled.

Automated procedures (algorithms) were invented to get rid of unskilled workers, whose repetitive work can easily be replaced by software. But now dealing with these "smart" algorithms requires higher, not lower, skills. Whenever something goes wrong, or you simply need an explanation, it is not trivial for the human representative who answers your call to figure out why the algorithm did what it did. It took almost one hour for an insurance agent to figure out why their automated system had sent me a bill for $212 even though i had paid the premium in full. In fact, he never found out where the number 212 came from, he just figured out that the algorithm generated the bill one day before another algorithm recorded my payment and, for mysterious reasons, the physical letter was sent out ten days later. But my premium has never been $212, nor was the monthly installment. If he really wanted to understand why his algorithm charged me that specific amount, he would have had to spend hours and maybe days studying the algorithm. Luckily, it was relatively easy to simply cancel the bill and neither of us was too eager to spend more time on this issue.

In 2015 i was trying to get an exemption from the mandate to get health insurance (at my age the cost is prohibitive), an exemption that is allowed under the law. Unfortunately, the bureaucracy was not (yet?) set up to deal with this exception: i eventually had to pay the penalty for not having health insurance even if every person who looked into my case agreed that i was entitled to the exemption. Think of it, and there are probably many cases in which the algorithm failed because your case was an "exception". Exceptions are no longer allowed.

You can also expect that "intelligent" algorithms will fail you precisely when you have an emergency. The reason is that they are trained to detect and then block "unusual activity". An antispam software started flagging all my emails to my wife as "550 High probability of spam" when she was in China and i was in California and we were frantically exchanging messages about an emergency. The "intelligent" software correctly detected unusual activity. It was unusual indeed: every emergency is "unusual", by definition. Then the algorithm decided to block that unusual activity, i.e. all my emails to my wife. In the middle of an emergency, i was not able to communicate with my wife by email. (You'd think that "intelligent" software would figure out that you are not spamming if you email someone from whom you have received thousands of emails over the years, but the "intelligence" is not trained to use common sense, it is trained to use real-time data). Something similar happens to you when you are abroad and start using your credit card or ATM card frantically because some emergency has occurred: your bank's algorithm reacts to your emergency by blocking your card. What "intelligent" algorithms want is that everybody behaves the same way all the time. Anything slightly different is cause for concern. Think of it, and that is actually what human beings do: if your patterns of behavior suddenly change in a very visible manner, your neighbors may suspect that you are up to something illegal; except, of course, that you can simply tell a human being what is going on ("my mother is sick" or "my son lost his wallet" or "we are remodeling the bathroom") and a human being, even a not particularly "intelligent" one, can understand the implication of that fact on your patterns of behavior. Alas, that's precisely what an "intelligent" algorithm cannot understand.

Even more discouraging was the vocabulary used by the tech-support engineer when the antispam software failed me: he called it "a false positive", not "a bug", despite the fact that "High probability of spam" is factually false since the probability of spam is zero when i message someone who has messaged me countless times.

It will get worse: soon, we won't even be able to speak with a human engineer. Another "intelligent" algorithm (i.e. another incredibly stupid algorithm) will receive our complaints and calmly explain to us that we have to stop emailing our wives or stop using email altogether or maybe just drop dead.

In fact, the best business plan of our days is probably about "what will poor consumers need in order to cope with the incredibly stupid machines that they are being forced to adopt?" For example, if people are really forced to buy self-driving cars, that obviously don't work, what services will the not-so-proud owners of those vehicles need? Probably the call center for driving: make a phone call and a human tele-operator will take control of the car and drive it for you so that the car can finally get around the pillow that someone dropped in the middle of the street or so that the car can finally go through a crowd of pedestrians crossing while their traffic light is red (obviously we will all start doing that after we realize that the cars are programmed to always stop for pedestrians).

In 2017 i heard Li Jiang, the director of the Robotics and Future Education Initiative at Stanford University, talk about the way children interact with talking robots such as Mattel's chat-friendly Barbie (that debuted at the New York Toy Fair in 2015) and Musio (made by Santa Monica-based startup called AKA), or with conversational agents such as Apple Siri or Amazon Alexa. Children are very good at testing the limits of algorithms, and quickly reach a point at which the algorithm consistently fails to understand their questions. Then the children behave very rudely with the device. After a while, the children will have indirectly learned to be always rude to the device. Then they transfer that rude behavior to humans. When i heard this story, i realized that the exact same thing is happening to us adults: we get so used to curse against our dumb machines that we soon start doing it to fellow humans too. We start treating humans like (dumb) algorithms that deserve no respect for their work.

Then, again, when you are frantically trying to finish something and your assistant becomes hostile because it is time for her to go home, or s/he becomes hostile because your emails and texts are getting less and less courteous due to stress and lack of time (with some of the stress being caused by that very assistant), you do regret not having replaced the human assistant with an algorithm that works 24 hours a day, 7 days a week, doesn't get sick, doesn't get upset, doesn't get hungry, and doesn't have to watch a movie with friends!

You are actually surrounded by more than an army of algorithms: you are surrounded by hierarchies of algorithms. If the algorithm does not provide you the service that you need, you can press a key or say something that will route you to "customer service". But that is increasingly another algorithm. Even if you manage to bypass this algorithm and speak with a human being, increasingly you are asked to provide feedback about your experience, and that survey is run by another algorithm and analyzed by another algorithm. In the end, the experience itself, i.e. the way algorithms behave to you, is designed by algorithms.

When a business or government agency tells us that they are using an "intelligent" algorithm, they should also tell us what the algorithm is intelligent "in". Nobody is intelligent at everything. One can be very smart at playing chess and very dumb at saving money. Einstein was intelligent in physics, but probably not in stocks or golf. Let's say that a bank announces a new "intelligent" algorithm: is the algorithm "intelligent" in helping you or in helping the bank? It does make a difference.

There is worldwide competition to build "smart cities". Each time a new "smart city" is announced it is presented as even "smarter" than previous generations because controlled by even smarter algorithms. A smart city is about efficiency. It should really be called "efficient" city. Buildings, utilities, streets, cars and so on are tightly integrated so that the life of the city is optimized. But citizens should ask "what exactly gets optimized? efficient for what?" Efficiency is a dangerous concept when applied to human lives. The most efficient thing you can do is to die. You are going to die anyway. Staying alive is postponing the inevitable. What is the purpose of postponing it? What do you think is so important about your life to take away resources from others? You are wasting energy, using services, causing pollution, etc. In the name of efficiency, you should die. We should all die. The designers of smart cities often forget that a city is not only made of buildings and streets: there are also people. Bottom line: so far, automation has resulted in a massive injection of stupidity and cruelty into human society, a society that, unlike the societies of other animals, was supposed to be characterized by intelligence and compassion.

"The trouble with modern theories of behaviorism is not that they are wrong but that they could become true" (Hannah Arendt, 1958)

Corollary: We are Replacing Animals with Machines

We are replacing animals with machines. We used to be surrounded by domesticated animals: horses, cows, chickens, pigs, pigeons... They were intelligent and this required us to be more intelligent than them. Now animals are disappearing (that's why city children gets so excited when they see a bird) and we are being surrounded by machines that are extremely unintelligent. Historians like to talk about machines replacing human labor, but the real story is the one about machines replacing animals (labor was always more "animal" than "human"). For example, the population of horses started declining dramatically after the invention of the car, and continues to decline. The only mammals whose population is exploding are pets, which are closer to furniture than to living beings. There are now farms that don't own a single animal, something that was unthinkable a century ago. Machines also do many things that animals couldn't do for us, like selling train tickets or taking passport-size pictures of us. And, yes, there are also many cases of machines replacing humans in workplaces, although those are often the humans that we don't want to deal with, humans who perform tedious repetitive jobs and who are therefore permanently in a bad mood. Biotech may soon make cows obsolete. We have always had to adapt to new tools, but this is the first time in history that the tools, the machines, are multiplying while so many animals are becoming rarities. Compared with our ancestors of a century ago, we spend most of our time interacting with (unintellinget) machines rather than with (intelligent) animals.

Cute Intermezzo: The Principle of Complementarity in A.I.

Niels Bohr's principle of complementarity is one of the foundations of our current understanding of the universe, that there are two different ways to describe it, perceive it, and predict its future (e.g., viewing a system either as a set of particles or as a set of waves). I imagine that, were Bohr alive today to witness the chaotic hype about A.I. of the 2010s, he would slap a principle of complementarity on "intelligent" automation: for every activity that gets automated, someone will automate the counter-activity. For example, if someone creates an automated recruiting assistant to recruit the best engineer, someone will create an automated resume writer to write the best resume that fools the recruiting assistant; if someone creates a travel assistant that picks the best combination of hotels, car rentals, trains, etc, someone will create a travel-agency assistant to offer a package that fools the travel assistant. This will become an endless duel between the two sides, each one trying to fool the other one in smarter and smarter ways. After all, the two algorithms look like one and the same if you reverse engineer the parameters.

Footnote: A.I. in the Age of Post-truth

Donald Trump, appointed by the electoral college in November 2016, will probably always be remembered as the president of "fake news" and "post-truths". His campaign, the media that supported him (e.g. the mainstream television channel Fox News), and incognito Russian hackers, artfully manipulated facts to create an "alternative reality" (to quote one of his associates). What was unique about his election was not that fake news had such a great influence (they did) but that the platform that was most responsible for spreading them, Facebook, denied their effect. In fact, very few people were using the neologism "fake news" until the day (10 November 2016) when Facebook's founder Mark Zuckerberg, speaking at the Techonomy Conference, dismissed and mocked the notion that fake news, spread on Facebook, could be so influential. It was, as all "fake news" has been before.

Political propaganda has been around since the times of the Athenian democracy. The Soviet Union mastered the art of propaganda, creating "fake news" about both its Western enemies and its own conditions (ironically, the main organ of "fake news" was called Pravda, which means "the truth"). Political propaganda was largely funded and controlled by governments until the advent of mass media. Mass media enabled more organizations to spread fake news. For example, political candidates in democratic countries have always relied on fake news to smear the opponents, and mass media turned smearing campaigns into something akin to scientific endeavours.

In the USA, the power of mass media in spreading "fake news" was demonstrated by two events that predate the Internet by several decades. In October 1938 actor and filmmaker Orson Welles used the radio to stage a fake invasion of aliens that spread panic around the country ("The War of the Worlds"). In February 1950, speaking to a women's club in West Virginia, senator Joseph McCarthy claimed to be in possession of a list of communist spies. In the next two years newspaper articles and radio programs convinced the public that the threat of communist infiltration was real, and by the 1952 elections the "Red Scare" had become the main topic of debate, with each candidate trying to outdo the other in anti-communism fervor. The New York Times of 27 August 1952 ran three front-page stories about the suspected communist infiltration.

Mark Lane's book "Rush to Judgment" (1966) started the conspiracy theories about the assassination of president John Kennedy. Erich von Daniken's book "Chariots of the Gods" (1968) launched an entire genre of nonfiction, devoted to documenting how aliens built the ancient monuments of Egypt, England and the Easter Islands. Peter Hyams' film "Capricorn One" (1978) involuntarily started the conspiracy theory that the USA never went to the Moon. Independent television channels and later websites helped spread the conspiracy theories about the "seven sisters" of the oil business and the "Illuminati" who control the whole world. Michael Moore's documentary "Fahrenheit 9/11" (2004) was a typical specimen of fake news about the Middle Eastern wars of the USA.

In 2015 Paul Horner, a specialist of viral news hoaxes, whom the Washington Post called "impresario of a Facebook fake-news empire", got help from a new powerful ally: the algorithms employed by the most popular search engine and social-networking platform. Because people love hatred, his stories became very popular on both platforms, and both platforms helped to spead them. (To be fair, Horner also got help from the mainstream television channel Fox News, that lent credibility to some of his most ridiculous concoctions).

A study by Soroush Vosoughi, Deb Roy and Sinan Aral at MIT, "The Spread of True and False News Online" (published in 2018), showed that people are much more likely to be intrigued (and to re-broadcast) fake news than real stories (they analyzed ten years of tweets on Twitter), with the result that lies spread much faster than the truth. In the human world, falsehood beats truth hands down.

In 2017 the strange case of Lani Sarem's book "Handbook for Mortals" rocked the publishing world. The New York Times declared it the number-one bestseller in the country but nobody had heard of it before, and there seemed to be no chat on social media about it. Eventually the suspicion arose that the author (or someone on her behalf) had simply purchased lots of copies of the book from the most influential bookstores, thereby twisting the process to this author's advantage. (The New York Times promptly dumped it from the bestseller list).

Horner and Sarem had taken advantage of the stupidity of algorithms. In both cases it was not difficult to fool algorithms that have no common sense and no idea of what the consequences of their calculations could be.

It will not be difficult to revise the New York Times algorithm to check whether the book is being discussed on social media, and, if discussion matches sales, then accept the numbers; but then, again, it will equally not be difficult for authors and their agents to create two, three, ten, one hundred bots that will impersonate people on social media and will start fake discussions about the book. Who wins?

It is easy for a human being to fool algorithms that have no sense of the world of humans. It is difficult for engineers to make these algorithms impossible to manipulate if the algorithms have no sense of the world of humans.

In fact, algorithms can even generate "fake news" without any human intervention, just by accident. Dumb algorithms are prone to dumb conclusions. As an example of indirect "fake news", today (28 August 2017) i just typed "fake news" in the most popular search engine and the top results are two pictures of Mark Zuckerberg, founder of Facebook, instead of the picture of the main producers of fake news. The caption says "Facebook now blocks ads from pages that spread fake news" but the association created by this page is that "fake news" means "Mark Zuckerberg", not Donald Trump or Fox News.

In fact, before being a producer of fake news, Trump is an avid consumer of fake news: in many cases he may be the victim before he becomes the perpetrator.

How did we find out that the Sarem affair was a scam? People. People figured out that it is unusual for a bestseller not to be the subject of intense social-media discussion. People figured out that the New York Times uses a selected number of bookstores to calculate the bestselling books. People figured out that an author could conceivably buy 100 copies of her own book from each of these selected bookstores. People figured out that this is a distortion of the meaning of "best seller": if i buy 100,000 copies of my own book, my book is NOT a bestseller.

If the benefits that social-media platforms obtain from "fake news" is big enough, it may be humanly impossible to disincentivize their (increasingly "intelligent") algorithms from helping the cheaters. And the number of cheaters (both humans and increasingly "intelligent" bots) will be proportional to the benefits of producing fake news.

Reality has always been a game-theoretic problem: there is truth and there are adversaries of truth. But discovering the truth is more than just a game between the fact-checkers and the fakers. A "bot" of truth would simply look for consensus, i.e. to track on which bases a statement was formulated, but that may simply worsen the problem of "echo chambers".

Besides the passion for spectacles of hatred, humans also have a passion for creating "echo chambers", i.e. for listening only to the news that they want to hear. That's why Aristotle and Galen ruled unchallenged for centuries. Who dared to argue with their wisdom, certified by all wise people? Luckily, humans have another unique "evil" skill: to disagree all the time with each other. Sooner or later (when the risk of being burned at the stakes has decreased enough) one of those experts rises to challenge the master.

It is easy to write a program that checks whether a statement is based on previously proven statements, i.e. that is consistent with the consensus. But how easy is it to build a program that challenges the consensus? If everybody believes that the Earth is flat, how easy is it to build a program capable of discovering that everybody is wrong? I still have more faith in evil humans when it comes to discovering the truth. And i fear that "intelligent" algorithms will simply plunge us deeper into the post-truth mess.

"The size of the lie is a definite factor in causing it to be believed" (Adolf Hitler in "Mein Kampf")

Deep Fake

One of the most powerful attacks against truth came from people who used deep learning to create fake video and fake audio. Matthias Niessner at University of Erlangen-Nuremberg developed Face2Face, a system (demonstrated in 2016) that captures the facial expression of an actor and maps it into another person's face: you can take a video of your favorite politician and then that person will show your facial expressions as you make them. Face2Face was based on a technique (a variant of "multilinear principal component analysis") that dates from the 1990s. Video Rewrite, one of the first programs for facial reenactment, was developed in 1997 by Malcolm Slaney at Interval Research Corporation, a Palo Alto laboratory founded in 1992 by Microsoft's cofounder Paul Allen and David Liddle. Video Rewrite altered the video of a person talking so that it looked like she was saying something else ("Driving Visual Speech with Audio", 1997). In 1999 Volker Blanz and Thomas Vetter at the Max Planck Institute in Germany had published a method to construct three-dimensional faces from just one single photograph ("A Morphable Model for the Synthesis of 3d Faces", 1999). And in 2009 Paul Debevec's team at University of Southern California had showed at the SIGGRAPH conference the Digital Emily Project of facial animation, whose original purpose was to create a photo-realistic digital actor. In 2010 Steve Seitz's student Ira Kemelmacher-Shlizerman at University of Washington demonstrated a project named Being John Malkovich in which facial expressions of actress Cameron Diaz were mapped into facial expressions of actor John Malkovich. This was based on a different approach: seeking images of the target (Malkovich) similar to the images of the source (Diaz). For the record, David Fincher's "The Curious Case of Benjamin Button" (2008) was the first Hollywood film to feature a photo-realistic computer-generated human character (an aged version of Brad Pitt). Paul Debevec helped Hollywood studios complete the movie "Fast and Furious 7" (2015) even though the actor Paul Walker had died partway into filming; and then helped the actor Peter Cushing, dead for 20 years, to reprise his role as Grand Moff Tarkin in a new episode of the "Star Wars" saga, "Rogue One" (2016).

The technique was expanded beyond facial expression by Christian Theobalt's team at the Max Planck Institute, whose generative neural network produced re-animations of head position, head rotation, face expression, eye gaze, and eye blinking ("Deep Video Portraits", 2018).

Voice impersonation became popular in 2015 when Nitesh Saxena's group at University of Alabama demonstrated a system capable of creating a speech with someone's voice after simply listening to a few minutes of this person speaking. For example, you can use a recording of one of my public talks to create a message in my voice, and good enough to fool the biometric authentication system of my smartphone ("Stealing Voices to Fool Humans and Machines", 2015). This system employed the Festvox voice converter built on top of the Festival Speech Synthesis System developed by Alan Black at University of Edinburgh and first published in 1997 (and later maintained by Black at CMU). Similarly, DeepMind's deep neural network WaveNet and Adobe's Voco, both demonstrated in 2016, could make someone say on video something that she never said. In 2017 the Canadian startup Lyrebird, founded by former students of Yoshua Bengio's Montreal Institute for Learning Algorithms (MILA), improved voice impersonation to the point that the system could be trained with just one minute of the victim's voice. These systems can fake your voice and create audiobooks that sound like they are read by you. In 2017 Supasorn Suwajanakorn of Steve Seitz's group demonstrated the Synthesizing Obama system that produced videos of president Barack Obama speaking his own words with accurate lip synchronization: the system picked images of Obama to match the audio. Such a system would be able to generate a video of a person speaking his own words given the audio and a dataset of images of that person. For example, one could generate the video of an Albert Einstein talk for which there is no video.

Alas, an impersonator capable of imitating president Obama's voice could also generate a convincing video of president Obama saying something he never said.

Slowly, voice-morphing technology is being combined with face-morphing technology to create completely fake speeches.

In November 2017 a Reddit user nicknamed "Deepfakes" started a Reddit community /r/deepfakes for creating fake porn videos, mostly using a face-swapping program called FakeApp created by another user, Deepfakeapp, which was based on TensorFlow (the first large-scale demonstration of the power of an A.I. platform for developing practical applications). This launched the phenomenon of "deepfake videos" in which the face of an actor in a porn video is replaced by the face of a celebrity (you do need a lot of pictures of the victim in order to train the algorithm; hence usually the victims are celebrities). Within a few months Reddit had to intervene to stop the face-swapping celebrity porn craze.

In 2019 Sean Vasquez and Mike Lewis at Facebook AI used videos to make their MelNet imitate voices of famous people.

"Humanity is acquiring all the right technology for all the wrong reasons" (Buckminster Fuller)

The Curse of the Large Dataset

The most damning evidence that A.I. has posted very little conceptual progress towards human-level intelligence comes from an analysis of what truly contributed to Deep Learning's most advertised successes of recent years: the algorithm or the training database? The algorithm is designed to learn an intelligent task, but it has to be trained via human-provided examples of that intelligent task.

There is a pattern about neural networks that has become the norm after the 1990s: an old technique stages spectacular performance thanks to a large training dataset, besides more powerful processors.

In 1997 Deep Blue used Reinefeld's NegaScout algorithm of 1983. The key to its success, besides the massively-parallel 30 high-performance processors, was a dataset of 700,000 chess games played by masters, a dataset created in 1991 by IBM for the second of Feng-hsiung Hsu's chess-playing programs, Deep Thoughts 2.

In 2011 Watson utilized (quote) "90 clustered IBM Power 750 servers with 32 Power7 cores running at 3.55 GHz with four threads per core" and a dataset of 8.6 million documents culled from the Web in 2010, but its "intelligence" was Robert Jacobs' 20-year-old "mixture-of-experts" technique.

All the successes of convolutional neural networks after 2012 were based on Fukushima's 30-year-old technique but trained on the ImageNet dataset of one million labeled images created in 2009 by Fei-fei Li.

DeepMind's celebrated videogame-playing of December 2013 used Chris Watkins' Q-learning algorithm of 1989 but trained on the Arcade Learning Environment dataset of Atari games developed in 2013 by Michael Bowling's team at University of Alberta. In 2015 Google's FaceNet used the dataset Labeled Faces in the Wild, a collection of digital pictures of celebrities tagged by Erik Learned-Miller's lab at University of Massachusetts since 2007. In 2016 AlphaGo used the dataset of millions of go positions stored and ranked at the KGS Go Server (Kiseido Go Server).

It is easy to predict that the next breakthrough in Deep Learning will not come from a new conceptual discovery but from a new large dataset in some other domain of expertise. Progress in Deep Learning depends to a large extent on many human beings (typically PhD students) who manually accumulate a large body of facts. It is not terribly important what kind of neural network gets trained to use those data, as long as there are really a lot of data. The pattern looks like this: at first the dataset becomes very popular among A.I. hackers; then some of these hackers utilize an old-fashioned A.I. technique to train a neural network until it exhibits master-like skills in that domain.

Several popular datasets of manually-labeled images have been developed by various organizations over the years: FERET (1993) by Jonathon Phillips at the Army Research Laboratory in Maryland, the NIST handwritten-digit dataset (1993) by the National Institute of Standards and Technology, the ORL face dataset (1994) by Ferdinando Samaria at Olivetti's British labs, the MNIST (Modified NIST) handwritten-digit dataset (1999) by LeCun at New York University, NORB (2004) also by LeCun's team at New York University, etc; all the way to the Tiny Images Dataset started in 2007 by Antonio Torralba at MIT, that eventually grew to 80 million tiny images, from which Hinton's students extracted CIFAR-10 and CIFAR-100. In 2006 Andrew Zisserman's group at Oxford built the dataset of annotated visual objects PASCAL VOC (which, believe it or not, stands for "Pattern Analysis Statistical-modelling And Computational Learning Visual Object Classes") and in 2009 Fei-fei Li (now at Stanford University) published the most famous of all that she had started while at Princeton, ImageNet, whose related challenge (first held in 2010) would graduate the most famous names in deep learning (the 2010 challenge was won by a joint team of NEC Laboratories in Cupertino and University of Illinois led by Yuanqing Lin, using the SIFT method). In 2012 Andreas Geiger at the Karlsruhe Institute of Technology in collaboration with Raquel Urtasun of the Toyota Technological Institute in Chicago published Kitti, six hours of traffic scenarios. In 2014 Larry Zitnick's team at Microsoft published the Microsoft dataset for image captioning named COCO, which stands for "Common Objects in Context" (Facebook hired Zitnick and his team members Ross Girshick of R-CNN fame and Piotr Dollar). Ditto for large datasets of annotated speech, such as: the Switchboard-1 Telephone Speech Corpus, a project started by Texas Instruments in 1990, composed of approximately 2,400 telephone conversations; the Continuous Speech Recognition (CSR) Corpus, a dataset containing thousands of spoken articles, mostly from the Wall Street Journal, compiled by in 1991 by Douglas Paul of MIT, in collaboration with Dragon Systems; and the TIMIT Acoustic-Phonetic Continuous Speech Corpus, started in 1993 by the Linguistic Data Consortium (LDC), the same organization that in 1996 released the Broadcast News corpus (30 hours of radio and television news broadcasts). In 2013 Shih-fu Chang's team at Columbia University released the Sentibank dataset for visual sentiment (emotions). In 2014 Catalin Ionescu at Institute of Mathematics of the Romanian Academy published Human3.6m, a dataset of videos of human motion useful to train robots. In 2017 Andrew Zisserman's group at Oxford released the Kinetics-600 dataset of 500,000 video clips, covering 600 human action classes with at least 600 video clips for each action class. In 2016 Google released a dataset of eight million tagged YouTube videos called YouTube-8M. Progress in these disciplines has largely followed the creation of these datasets.

If you want to predict what's coming next in A.I., look at the new datasets. Now there are even SEMAINE (Sustained Emotionally coloured Machine-human Interaction using Nonverbal Expression), developed in 2007 at Queen's University Belfast; Cam3D, built in 2011 at Cambridge University, a corpus of complex mental states captured using high-definition cameras and Kinect sensors; MAHNOB-HCI, created in 2011 by Maja Pantic's team at Imperial College London, a corpus of synchronized recordings of video, audio, and physiological data annotated with emotional tags; the EAGER dataset of spontaneous dynamic facial expressions, assembled in 2013 at Binghamton University; FaceWarehouse, a dataset of three-dimensional facial expressions, published by Zhejiang University in 2014; the LSUN (Large-scale Scene Understanding) dataset, prepared by Jianxiong Xiao's student Fisher Yu at Princeton University in 2015; Ziwei Liu's CelebFaces Attributes (CelebA) dataset at the Chinese University of Hong Kong (2015); Haolin Wei's corpus of human interactions, released in 2014 by Dublin City University in Ireland; the UM-corpus of English-Chinese pairs for machine translation, published in 2014 by Liang Tian at University of Macau, as well as Longyue Wang's similar corpuses derived from movie subtitles in 2016 at Dublin City University in Ireland.

The Stanford Natural Language Inference (SNLI) dataset, released in 2015 by Sam Bowman at Stanford, consisted of 570,000 human-written pairs of English sentences to train neural networks for sentence representation. In 2016 Percy Liang's team at Stanford developed the Stanford Question Answering Dataset (SQuAD), a reading comprehension dataset consisting of more than 100,000 questions, as well as SCONE (Sequential CONtext-dependent Execution); and in 2016 Jianfeng Gao's team at Microsoft published the Machine Reading Comprehension (MARCO) dataset that grouped 100,000 queries with their corresponding answers. The Large Movie Review Dataset (LMRD), released in 2011 by Andrew Ng's student Andrew Maas at Stanford University, consisted of 25,000 highly opinionated movie reviews that could be used to train a neural network for "sentiment analysis". The Stanford Sentiment Treebank, published in 2013 by Richard Socher at Stanford, contained "sentiment" labels for more than 200,000 phrases.

In 2018 three important datasets were introduced for natural-language processing: the CoQA (Conversational Question Answering) dataset by Siva Reddy and Danqi Chen (both now in Christopher Manning's group at Stanford University); the QuAC (Question Answering in Context) dataset, a collaboration among Percy Liang's group at Stanford University, Luke Zettlemoyer's group at University of Washington and Mark Yatskar at the Allen Institute; and OpenAI's WebText, a dataset of eight million World-wide Web pages.

What do you do if a very large dataset tells you nothing? For example in 2010 Naoki Nakaya of the Danish Cancer Society compiled a huge dataset of 60,000 people over 30 years that shows no correlation between personality traits and the likelihood of surviving cancer ("Personality Traits and Cancer Risk and Survival Based on Finnish and Swedish Registry Data", 2010). Now what? What we (humans) learn from this dataset is that we have to move on to some other research. What does the neural network learn from this dataset?

It is scary to think that neural networks always learn something from the data, and those learned features will influence the way the networks will treat you, possibly for the rest of your life.

Data is becoming destiny.

A Consumer's Rant Against the Stupidity of Machines: Reversing Evolution?

When you buy an appliance and it turns out that you have to do something weird in order to make it work, it is natural to dismiss it as "a piece of garbage". However, when it is something about computers and networks, you are instead supposed to stand in awe and respectfully listen to (or read) a lengthy explanation of what you are supposed to do in order to please the machine, which is usually something utterly convoluted bordering on the ridiculous.

This double standard creates the illusion that machines are becoming incredibly smart when in fact mostly we are simply witnessing poor quality assurance (due to the frantic product lifecycles of our times) and often incredibly dumb design.

For mysterious reasons, the computer industry has a tradition of being hostile to its customers. Buying a new refrigerator means that you will have an appliance that saves energy and provides one or two useful new features. Buying a new computer, instead, typically means that some of the important things that you used to do every day will now become very complicated or just impossible. A whole generation of "user-experience designers" has conspired to sabotage the users of their products. I personally think that one of the drivers of the transition from the desktop to the smartphone was precisely the user-experience designers who were turning the "user experience" on the desktop into sheer hell, and the smartphone came out as the simpler version of that "user experience". If email gets so hostile to the user, we the users will switch to texting or chatting, not because we particularly like texting or chatting but because email has become unusable.

You never know what is going to happen to your favorite application when you download an "update". New releases (which you are forced to adopt even if you are perfectly happy with the old release) often result in lengthy detours trying to figure out how to do things that were trivial in the previous release (and that have been complicated by the software manufacturer for no other reason than to justify a new release). A few weeks ago my computer displayed the message "Updating Skype... Just a moment, we're improving your Skype experience". How in heaven do they know that this will improve my Skype experience? Of course they don't. The reason they want you to move to a new release is different: it will certainly improve THEIR experience. Whether it will also improve mine and yours is a secondary issue. At the least, any change in the user interface will make it more difficult to do the things to which you were accustomed.

It has become customary that a program only gives you two choices: "Download Now" or "Not Now". The option "Never, thanks!" does not exist anymore.

Software engineers have a passion for littering webpages with images and videos, while users simply want to read the text. An image can tell a thousand words but, more generally, in the Internet world of 2019 an image (or, worse, a video) can kill a thousand words because the image or video will take so long to download that the reader may just click on the "back" button of the browser and abandon the page.

As i wait for the New York Times website to finish downloading the article that i would like to read, and i watch the endless sequence of "transferring data from..." displayed at the bottom of my browser, i think nostalgically of the days when i would just flip through the paper pages of the newspaper in a few seconds. Meanwhile, the browser (the latest version from the top Internet company) is still downloading pictures and adverts and doesn't let me scroll down to the few sentences that really interest me. When it's finally done, the search feature doesn't work, probably because the ingenious software engineers of the New York Times used some avantgarde script to challenge the software engineers of the browser. Incidentally, the main New York Times webpage is structured as a series of graphically appealing blocks. The one in the middle, however, has no picture and only the text: "Error: Server Error. The server encountered a temporary error and could not complete your request. Please try again in 30 seconds." I am not sure which request my browser issued to their server. I just wonder how we would have felt 20 years ago if the newspaper delivered to our house early in the morning had a big note in the middle "Server Error..."

Alan Cooper, a former Microsoft engineer, wrote a hilarious memoir of how software is designed (and unwanted new releases come to be), "The Inmates Are Running the Asylum" (1999), a book that notes how engineers "keep fixing what's not broken until it's broken". Coincidentally, the book came out the same year when Shawn Fanning released the Napster software that enabled pirate copies of copyrighted music and Jon Johansen released the DeCSS software that enabled pirate copies of copyrighted films, two releases that constituted indeed progress... but both inventors ended up in court and almost in jail.

Incidentally, the reason why so many corporations (banks, airlines, news media, etc) want you to update your browser to the latest release is that your old browser does not allow them to track you efficiently, steal your private data, customize advertising, and maximize their profits out of your browsing experience. In fact, it would be much easier for them to deliver their services in a text-only format that even 1990s browsers can manage. They are willing to spend good money on highly-paid software engineers not to improve your browsing experience but to make sure that they can get the most out of it.

We live in an age in which installing a wireless modem can take a whole day and external hard disks get corrupted after a few months "if you use them too often" (as a sales associate told me at Silicon Valley's most popular computer store).

In the old days i was backing up my work all the time because i didn't trust computers: they frequently crashed. I didn't trust them because they were not reliable. These days computers don't crash anymore but i still back up frequently my work: i don't trust them because it is unpredictable what they will do with my work. Computers download and store files where they want (often in obscure folders/ directories), sometimes they change my desktop appearance, sometimes they change the formatting of my documents, etc. The manufacturers that sell them tell me that these programs are becoming "more and more intelligent": obviously we have wildly different definitions of "intelligence".

In 1997 Steve Jobs famously told Business Week: "People don't know what they want until you show it to them." Maybe. But sometimes the high-tech industry should say: "People don't know what they want until we FORCE it upon them and give them no alternative."

Reality check: here is the transcript of a conversation with Comcast's automated customer support:

> "If you are currently a Comcast customer, press 1" [I press 1]
> "Please enter the ten-digit phone number associated with your account" [I enter my phone number]
> "OK Please wait just a moment while i access your account"
> "For technical help press 1"
> "For billing press 2" [I press 2]
> If you are calling regarding important information about Xfinity etc press 1 [I press 1]
> "For payments press 1"
> "For balance information press 2"
> "For payment locations press 3"
> "For all other billing questions press 4" [I press 4]
> "For questions about your first bill press 1"
> "For other billing questions press 3" [I press 3]
> "Thank you for calling Comcast. Our office is currently closed."
> (You can listen to it at https://soundcloud.com/scaruffi/comcast-customer-support)

An example of eBay's customer support (Half.com was an eBay service): "Dear Piero, Thank you for contacting Half.com. I understand you wanted to retrieve the items on your wish list. Let me help you out with this concern. Piero, I want to let you know that you can no longer retrieve the listed item on your wish list. I'm glad that I was able to help you. Thank you for choosing Half.com" (04 Sep 2017)

Based on the evidence, it is easier to believe that we still live in the stone age of computer science than to believe that we are about to witness the advent of superhuman intelligence in machines.

It is interesting how different generations react to the stupidity of machines: the old generation that grew up without electronic machines gets extremely upset (because the automated system can complicate things that used to be simple in the old-fashioned manual system), my generation (that grew up with machines) gets somewhat upset (because machines are still so dumb), and the younger generations are progressively less upset, with the youngest ones simply taking for granted that customer support has to be what it is (from lousy to non-existent) and that many things (pretty much all the things that require common sense, expertise, and what we normally call "intelligence") are virtually impossible for machines.

A book on "The State of Machine Stupidity" instead of "The State of Machine Intelligence" should be much longer.

Incidentally, there are very important fields, such as getting rid of paper, in which we haven't even achieved the first step of automation. Health care, for example, still depends on paper: your medical records are probably stored in old fashioned files, not the files made of zeros and ones but the ones made of cardboard or plastic. We are bombarded daily by news of amazing medical devices and applications that will change the way diseases are prevented, identified and healed, but for the time being we have seen very little progress in simply turning all those paper files into computer files that the patient can access from a regular computer or smartphone and then save, print, email or delete at will.

What we have done so far, and only in some fields, is to replace bits and pieces of human intelligence with rather unintelligent machines that can only understand very simple commands (less than what a two-year old toddler can understand) and can perform very simple tasks.

In the process we are also achieving lower and lower forms of human intelligence, addicted to having technology simplify all sorts of tasks (more about this later). Of course, many people claim the opposite: from the point of view of a lower intelligence, what unintelligent machines do might appear intelligent.

Footnote: Intelligent Customer Support as a Challenge to Human Intelligence

It is not artificial intelligence that defies human intelligence: it is software unfriendliness. Notably, "customer support" (my favorite euphemism) is reaching a level of unfriendliness that defies human intelligence... or, better, tests human intelligence. It has become a real challenge to speak with someone about a problem that you have. To speak with an operator you need to pierce through a multi-layered wall of "press this digit" and "press that digit". The reason is simple: they really want you to interact with the machine so that they can save on salaries. The machine, unfortunately, requires YOU to behave like a machine. If you are determined on speaking with a human being, you have to play a sort of videogame in which you need to avoid all the answers that will send you back to a webpage and find which sequence of answers will allow you to reach a human operator. If you need help with service X, do NOT select "service X" from the options, otherwise you'll be stuck in a conversation with an algorithm that will provide no help whatsoever. You have to find an option that will get you around the algorithm. Typically, you want to pick some silly exotic options for which most likely no algorithm has been set up. On the other side of this game there is an army of software engineers who are guessing what you are going to do in order to speak with a human being and who are trying to prevent you from doing so, i.e. they make it constantly more difficult for you to find the route to the human operator. When you sign up for a credit card or any other service, they will tell you "just go online and you can set up the feature that you need", for example set up automatic payment (of course, banks don't really want you to set up automatic payment because they want you to run huge balances on which they make money by charging you outrageous interests). You go online and spend a day trying to figure out where you can set up the feature that you want. It takes maximum concentration to find the places where you need to click in order to tell their system what to do. Sometimes it takes maximum concentration for more than one hour. If the phone rings or someone enters the room and distracts you, you probably have to restart from the beginning. It is just as complicated as proving a theorem or reading Kant. Or playing a videogame. Somewhere on the screen there's "Help" or "Contact us" but beware: if you click on those buttons you will get lost in a labyrinth of questions that will still send you back to the webpage that you cannot understand. That's the videogame that you are playing with the software developers of that "customer support" webpage. In fact, clicking on "Help" is the very wrong thing to do if you want to talk to someone: that "Help" section is precisely where the troops of software engineers have been concentrated to erect a formidable wall that will never allow you to get help.

Now most websites are being deliberately designed to make it as difficult as possible (or utterly impossible) for citizens and consumers to send comments. Ironically, clicking on "Help", "Feedback" or "Customer Support", frequently sends you to a page of FAQ, which in theory means "frequently asked questions" but, if you pronounce it like i do, it tells you how much they care about you.

In 2017 if you want to send an email to PayPal's customer service, you have to go through a series of pulldown menus. You have to select one of the options or you won't be able to do anything else. If none of the options apply to your case, it is your problem, not theirs. I found a way to send them an email only by clicking on the option "Close my Account". That is not what i needed help with, but it was the only way that i could find to send an email to their customer service.

The other game that is going on is between the user who wants simple applications to do simple things and the software engineers who are paid to complicate those applications. We are reluctant to abandon old software not because we are old-fashioned but because we are painfully aware that new releases frequently imply a decline in productivity: it takes countless mouse clicks to do what we used to do with one click (e.g. copying a text into Microsoft Word: Paste, Paste Special, Unformatted Text, OK).

How any of this relates to Artificial Intelligence is difficult to grasp. It is simply a game of endurance between two arch-enemies: the customer who needs customer support and the software engineer who designs the customer-support system so that it will minimize customer support by (expensive) human specialists. The problem is that, increasingly, the human "specialists" are no better than the algorithms: the human specialist that you finally get on the line is likely to simply recite the rules, and often in a language that only an algorithm would understand, asking you to make all the choices. You have won the game and got to speak with a human being instead of an algorithm, but only to find out that the human being, like in zombie movies, has become an algorithm.

The Singularity as the Outcome of Exponential Progress

The Singularity crowd is driven to enthusiastic prognostications about the evolution of machines: machines will soon become intelligent and they will rapidly become intelligent in a superhuman way, acquiring a higher form of intelligence than human intelligence.

There is an obvious disconnect between the state of the art and what the Singularity crowd predicts. We are not even remotely close to a machine that can troubleshoot and fix an electrical outage or simply your washing machine, let alone a software bug. We are not even remotely close to a machine that can operate any of today's complex systems without human supervision. One of the premises of the theory of the Singularity is that machines will not only become intelligent but will even build other, smarter machines by themselves; but right now we don't even have software that can write other software.

The jobs that have been automated are repetitive and trivial. And in most cases the automation of those jobs has required the user/customer to accept a lower (not higher) quality of service. Witness how customer support is rapidly being reduced to a "good luck with your product" kind of service. The more the automation around you, the more you are forced to behave like a machine in order to interact and communicate with machines, precisely because they are still so dumb.

The reason that we have a lot of automation is that (in developed countries like the USA, Japan and the European countries) it saves money: machine labor is a lot cheaper than human labor. Wherever the opposite is true, there are no machines. The reason we are moving to online education is not that university professors failed to educate their students but that universities are too expensive. And so forth: in most cases it is the business plan, not the intelligence of machines, that drives automation.

Wildly optimistic predictions are based on the exponential progress in the speed and miniaturization of computers. In 1965 Gordon Moore predicted that the processing power of computers would double every 18 months ("Moore's law"), and so far his prediction has been correct. Look closer and there is little in what they say that has to do with software. It is mostly a hardware argument. And that is not surprising: predictions about the future of computers have been astronomically wrong in both directions but, in general, the ones that were too conservative were about hardware (its progress has surprised us), the ones that were too optimistic were about software (its progress has disappointed us). What is amazing about today's smartphones is not that they can do what computers of the 1960s could not do, but that they are small, cheap and fast. The fact that there are many more software applications downloadable for a few cents means that many more people can use them, a fact that has huge sociological consequences; but it does not mean that a conceptual breakthrough has been reached in software technology. It is hard to name one software program that exists today and could not have been written in Fortran fifty years ago. If it wasn't written, the reason, probably, is that it would have been too expensive or that some required hardware did not exist yet.

Accelerating technological progress in computer science has largely been driven by the accelerating cost of labor, not by real scientific innovation. The higher labor costs go, the stronger the motivation to develop "smarter" machines. Those machines, and the underlying technologies, were already feasible ten or twenty or even thirty years ago, but back then it didn't make economic sense for them to be adopted.

There has certainly been a lot of progress in computers getting faster, smaller and cheaper. Even assuming that this will continue "exponentially" (as the Singularity crowd is quick to claim), the argument that this kind of (hardware) progress is enough to make a shocking difference in terms of machine intelligence is based on an indirect assumption: that faster/smaller/cheaper will lead first to a human-level intelligence and then to a superior intelligence. After all, if you join together many many many dumb neurons you get the very intelligent brain of Albert Einstein. If one puts together millions of superfast GPUs, maybe one gets superhuman intelligence. Maybe.

In any event, we'd better prepare for the day that Moore's Law stops working. Moore's Law was widely predicted to continue in the foreseeable future, but its future does not look so promising anymore. It is not only that technology might be approaching the limits of its capabilities, but also the original spirit behind Moore's Law was to show that the "cost" of making transistor-based devices would continue to decline. Even if the industry finds a way to continue to double the number of transistors etched onto a chip, the cost of doing so might start increasing soon: the technologies to deal with microscopic transistors are inherently expensive, and heat has become the main problem to solve in ultra-dense circuits. In 2016 William Holt of Intel announced that Intel will not push beyond the 7-nanometer technology, and cautioned that processors may get slower in the future in order to save energy and reduce heat, i.e. costs. In October 2014 the DARPA created the first 1THz computer chip (i.e. capable of one trillion instructions per second). In June 2016 Bevan Baas' team at UC Davis unveiled the KiloCore chip with a maximum computation rate of 1.78 trillion instructions per second. But it is telling that in 2017 it was impossible to find out, using search engines, what is the fastest processor in the world: no company seemed to have made that claim. So i looked up the most expensive ones: the AMD Ryzen Threadripper 1950X at 4GHz and the Intel Core i9-7900X at 4.3 GHz; quite far from a terahertz, in fact 200 times slower.

For 70 years computers have been getting smaller and smaller, but in 2014 they started getting bigger again (the iPhone 6 generation). If Moore's Law stops working, will there still be progress in "Brute-force A.I.", e.g. in deep learning? In 2016 Scott Phoenix, the CEO of Silicon Valley-based AI startup Vicarious, declared that "In 15 years, the fastest computer will do more operations per second than all the neurons in all the brains of all the people who are alive." What if this does not come true?

The discussion about the Singularity is predicated upon the premise that machines will soon be able to perform "cognitive" tasks that were previously exclusive to humans. This, however, has already happened. We just got used to it. The early computers of the 1950s were capable of computations that traditionally only the smartest and fastest mathematicians could even think of tackling, and the computers quickly became millions of times faster than the fastest mathematician. If computing is not an "exclusively human cognitive task", i don't know what would qualify. Since then computers have been programmed to perform many more of the tasks that used to be exclusive to human brains. And no human expert can doublecheck in a reasonable amount of time what the machine has computed. Therefore there is nothing new about a machine performing a "cognitive" task that humans cannot match. Either the Singularity already happened in the 1950s or it is not clear what cognitive task would represent the coming of the Singularity.

To assess the progress in machine intelligence one has to show something (some intelligent task) that computers can do today that, given the same data, they could not have done fifty years ago. There has been a lot of progress in miniaturization and cost reduction, so that today it has become feasible to use computers for tasks for which we didn't use them fifty years ago; not because they were not intelligent enough to do them but because it would have been too expensive and it would have required several square kilometers of space. If that's "artificial intelligence", then we invented artificial intelligence in 1946. Today's computers can do a lot more things than the old ones just like new models of any machine (from kitchen appliances to mechanical reapers) can do a lot more than old models. Incremental engineering steps lead to more and more advanced models for lower prices. Some day a company will introduce coffee machines on wheels that can make the coffee and deliver the cup of coffee to your desk. And the next model will include voice recognition that understands "coffee please". Etc. This kind of progress has been going on since the invention of the first mechanical tool. It takes decades and sometimes centuries for the human race to fully take advantage of a new technology. "Progress" often means the process of mastering a new technology (of creating ever more sophisticated products based on that technology). The iPhone was not the first smartphone, and Google was not the first search engine, but we correctly consider them "progress".

There is no question that progress has accelerated with the advent of electrical tools and further accelerated with the invention of computers. Whether these new classes of artifacts eventually constitute a different kind of "intelligence" will probably depend on your definition of "intelligence".

The way the Singularity would be achieved by intelligent machines is by these machines building more intelligent machines capable of building more intelligent machines and so forth. A similar loop has existed since about 1776. The steam engine enabled mass production of steel, which in turn enabled the mass production of better steam engines, and this recursive loop continued for a while. James Watt himself, inventor of the steam engine that revolutionized the world, worked closely with John Wilkinson, who made the steel for Watt's engines using Watt's engines to make the steel. Today this loop of machines helping build other machines takes place on a large scale. For example, a truck carries the materials that the factory will use to make better trucks. The human beings in this process can be viewed as mere intermediaries between machines that are evolving into better machines. This positive-feedback loop is neither new nor necessarily "exponential". In the 19th century that loop of machines building (better) machines which build (better) machines accelerated for a while. Eventually, the steam engine (no matter how sophisticated that accelerating positive-feedback loop had made it) was made obsolete by a new kind of machine, the electrical motor. Again, electrical motors were used by manufacturers of motor parts that contributed to making better electrical motors used by manufacturers of electrical motor parts.

We have been surrounded by machines that built better machines for a long time... but with human intermediaries designing the improvements.

Despite the fact that no machine has ever created another machine of its own will, and no software has ever created a software program of its own will, the Singularity crowd seems to have no doubts that a machine is coming soon created by a machine created by a machine and so forth, each generation of machines being smarter than the previous one.

i certainly share the concern that the complexity of a mostly automated world could get out of hand. This concern has nothing to do with the degree of intelligence but just with the difficulty of managing complex systems. Complex, self-replicating systems that are difficult to manage have always existed. For example: cities, armies, post offices, subways, airports, sewers, economies...

"Machines are, or are becoming, animate" (Samuel Butler in 1863, when not even the light bulb had been invented).

A Look at the Evidence: A Comparative History of Accelerating Progress

A postulate at the basis of many contemporary books by futurists and self-congratulating technologists is that we live in an age of unprecedented rapid change and progress. But look closer and our age won't look so unique anymore.

As i wrote in the chapter titled "Regress" of my book "Synthesis", this perception that we live in an age of rapid progress is mostly based on the fact that we know the present much better than we know the past. One century ago, within a relatively short period of time, the world adopted the car, the airplane, the telephone, the radio and the record, while at the same time the visual arts went through Impressionism, Cubism and Expressionism. Science was revolutionized by Quantum Mechanics and Relativity. Office machines (cash registers, adding machines, typewriters) and electrical appliances (dishwasher, refrigerator, air conditioning) dramatically changed the way people worked and lived. Debussy, Schoenberg, Stravinsky and Varese changed the concept of music. These all happened in one generation. By comparison, the years since World War II have witnessed innovation that has been mostly gradual and incremental. We still drive cars (invented in 1886) and make phone calls (invented in 1876), we still fly on airplanes (invented in 1903) and use washing machines (invented in 1908), etc. Cars still have four wheels and planes still have two wings. We still listen to the radio and watch television. While the computer and Genetics have introduced powerful new concepts, and computers have certainly changed daily lives, i wonder if any of these "changes" compare with the notion of humans flying in the sky and of humans located in different cities talking to each other. There has been rapid and dramatic change before.

Does the revolution in computer science compare with the revolutions in electricity of a century ago? The smartphone and the Web have certainly changed the lives of millions of people, but didn't the light bulb, the phonograph, the radio and kitchen appliances change the world at least as much if not much more?

A history of private life in the last 50 years would be fairly disappointing: we wear pretty much the same clothes (notably T-shirts and blue jeans), listen to the same music (rock and soul were invented in the 1950s), run in the same shoes (sneakers date from the 1920s), and ride, drive and fly in the same kinds of vehicles (yes, even electric ones: Detroit Electric began manufacturing electric cars in 1907). Public transportation is still pretty much what it was a century ago: trams, buses, trains, subways. New types of transportation have been rare and have not spread widely: the monorail (that became reality with the Tokyo Monorail in 1964), the supersonic airplane (the Concorde debuted in 1976 but was retired in 2003), the magnetic levitation train (the Birmingham Maglev debuted in 1984, followed by Berlin's M-Bahn in 1991, but in practice the Shanghai Maglev Train built in 2004 is the only real high-speed magnetic levitation line in service). The "bullet train" (widely available in Western Europe and the Far East since Japan's Shinkansen of 1964) is probably the only means of transportation that has significantly increased the speed at which people travel long distances in the last 50 years.

We chronically underestimate progress in previous centuries because most of us are ignorant about those eras. Historians, however, can point at the spectacular progress that took place in Europe during the Golden Century (the 13th century) when novelties such as spectacles, the hourglass, the cannon, the loom, the blast furnace, paper, the mechanical clock, the compass, the watermill, the trebuchet and the stirrup changed the lives of millions of people within a few generations; or the late 15th century when (among other things) the printing press enabled an explosive multiplication of books and when long-distance voyages to America and Asia created a whole new world.

If you are an expert on the 17th century like Anthony Clifford Grayling is, then you will think that the 17th century was the most special of all centuries, a century that witnessed the intellectual revolutions of Galileo, Pascal, Kepler, Newton, Cervantes, Shakespeare, Donne, Milton, Racine, Moliere, Descartes, Spinoza, Leibniz, Locke, Rubens, El Greco, Rembrandt, Vermeer (see his 2016 book "The Age of Genius"). Are we sure that our age measures up to that era?

Many think that globalization is a new phenomenon in the history of humankind, but here is what the British economist John Maynard Keynes wrote in 1919: "The inhabitant of London could order by telephone, sipping his morning tea in bed, the various products of the whole earth, in such quantity as he might see fit, and reasonably expect their early delivery upon his doorstep" (in ""The Economic Consequences of the Peace.").

The expression "exponential growth" is often used to describe our age, but the trouble is that it has been used to describe just about every age since the invention of exponentials. In every age, there are always some things that grow exponentially, but others don't. For every technological innovation there was a moment when it spread "exponentially", whether it was church clocks or windmills, reading glasses or steam engines; and their "quality" improved exponentially for a while, until the industry matured or a new technology took over. Moore's law is nothing special: similar exponential laws can be found for many of the old inventions. Think how quickly radio receivers spread: in the USA there were only five radio stations in 1921 but already 525 in 1923. Cars? The USA produced 11,200 in 1903, but already 1.5 million in 1916. By 1917 a whopping 40% of households had a telephone in the USA up from 5% in 1900. There were fewer than one million subscribers to cable television in 1984, but more than 50 million by 1989. The Wright brothers flew the first airplane in 1903: during World War i (1915-18) France built 67,987 airplanes, Britain 58,144, Germany 48,537, Italy 20,000 and the USA 15,000, for a grand total of almost 200 thousand airplanes, after just 15 years of its invention. In 1876 there were only 3,000 telephones: 23 years later there were more than a million. Neil Armstrong stepped on the Moon in 1969, barely eight years after Yuri Gagarin had become the first human to leave the Earth's atmosphere.

Most of these fields then slowed down dramatically. And 47 years after the Moon landing we still haven't sent a human being to any planet and we haven't even returned to the Moon since the Apollo 17 in 1972. Similar statistics of "exponential growth" can be found for other old inventions, all the way back to the invention of writing. Perhaps each of those ages thought that growth in those fields would continue at the same pace forever. The wisest, though, must have foreseen that eventually growth starts declining in every field. Energy production increased 13-fold in the 20th century and freshwater consumption increased 9-fold, but today there are many more experts worried about a decline (relative to demand) than experts who believe in one more century of similar growth rates.

Furthermore, there should be a difference between "change" and "progress". Change for the sake of change is not necessarily "progress". Most "updates" in my software applications have negative, not positive effects, and we all know what it means when our bank announces "changes" in policies. If i randomly change all the cells in your body, i may boast of "very rapid and dramatic change" but not necessarily of "very rapid progress". Assuming that any change equates with progress is not only optimism: it is the recipe for ending up with exactly the opposite of progress. Out of the virtually infinite set of possible changes, only a tiny minority, a tiny subset, would constitute progress.

There has certainly been progress in telecommunications; but what difference does it make for ordinary people whether a message is sent in a split second or in two split seconds? In 1775 it took 40 days for the English public to learn that a revolution had started in the American colonies. Seven decades later, thanks to the telegraph, it took minutes for the news of the Mexican War to travel to Washington. That is real progress: from 40 days to a few minutes. The telegraph did indeed represent "exponential" progress. Email, texting and chatting have revolutionized the way people communicate over long distances, but it is debatable whether that is (quantitatively and qualitatively) the same kind of revolution that the telegraph and the telephone brought about.

There are many "simpler" fields in which we never accomplished what we set out to accomplish originally, and pretty much abandoned the fight after the initial enthusiasm. We simply became used to the failure and forgot our initial enthusiasm. For example, domestic lighting progressed dramatically from gas lighting to Edison's light bulbs and Brush's arc lights of the 1880s and then the first tungsten light-bulbs to the light-bulbs of the

1930s, but since then there has been very little progress: as everybody whose eyesight is aging knows too well, we still don't have artificial lighting that compares with natural sunlight, and so we need to wear reading glasses in the evening to read the same book that we can easily read during the day. A century of scientific and technological progress has not given us artificial lighting that matches sunlight.

I can name many examples of "change" that are often equated with "progress" when in fact it is not clear what kind of progress it is bringing. The number of sexual partners that a person has over a lifetime has greatly increased, and social networking software allows one to have thousands of friends all over the world, but i am not sure that these changes (that qualify as "progress" from a strictly numerical point of view) result in happier lives. I am not sure that emails and text messages create the same bonds among people than the phone conversation, the letter on paper, the postcard and the neighbor's visit did.

One can actually argue that there is a lot of "regress", not "progress". We now listen to lo-fi music on computers and digital music players, as opposed to the expensive hi-fi stereos that were commonplace a generation ago. Mobile phone conversations are frequently of poor quality compared with the old land lines. We have access to all sorts of food 24 hours a day but the quality of that food is dubious. Not to mention "progress" in automated customer support, which increasingly means "search for the answer by yourself on the Web" (especially from high-tech software giants like Microsoft, Google and Facebook) as opposed to "call this number and an expert will assist you".

In the early days of the Internet (1980s) it was not easy to use the available tools but any piece of information on the Internet was written by very competent people. Basically, the Internet only contained reliable information written by experts. Today there might be a lot more data available, but the vast majority of what travels on the internet is: a) disinformation, b) advertising. It is not true that in the age of search engines it has become easier to search for information. Just the opposite: the huge amount of irrelevant and misleading data is making it more difficult to find the one webpage that has been written by the one great expert on the topic. In the old days her webpage was the only one that existed. (For a discussion on Wikipedia see the appendix).

Does the Internet itself represent true progress for human civilization if it causes the death of all the great magazines, newspapers, radio and television programs, the extinction of bookstores and record stores, and if it will make it much rarer and harder to read and listen to the voices of the great intellectuals of the era? While at the same time massively increasing the power of corporations (via targeted advertising) and of governments (via systemic surveillance)? From the Pew Research Center's "State of the News Media 2013" report: "Estimates for newspaper newsroom cutbacks in 2012 put the industry down 30% since its peak in 2000. On CNN, the cable channel that has branded itself around deep reporting, produced story packages were cut nearly in half from 2007 to 2012. Across the three cable channels, coverage of live events during the day, which often requires a crew and correspondent, fell 30% from 2007 to 2012... Time magazine is the only major print news weekly left standing".

Even the idea that complexity is increasing relies on a weak definition of "complexity". The complexity of using the many features of a smartphone is a luxury and cannot be compared with the complexity of defending yourself from wild animals in the jungle or even with the complexity of dealing with weather, parasites and predators when growing food on a farm. The whole history of human civilization is a history of trying to reduce the complexity of the world. Civilization is about creating stable and simple lives in a stable and simple environment. By definition, what we call "progress" is a reduction in complexity, although to each generation it appears as an increase in complexity because of the new tools and the new rules that come with those tools. Overall, living has become simpler (not more complicated) than it was in the stone age. If you don't believe me, go and camp in the wilderness by yourself with no food and only stone tools.

In a sense, today's Singularity prophets assume that machine "intelligence" is the one field in which growth will never slow down, but will keep accelerating forever.

Again, i would argue that it is not so much "intelligence" that has accelerated in machines (their intelligence is the same as that given by Alan Turing when he invented his "universal machine") as much as miniaturization. Moore's law (which was indeed exponential while it lasted) had nothing to do with machine intelligence, but simply with how many transistors one can squeeze on a tiny integrated circuit. There is very little (in terms of intelligent tasks) that machines can do today that they could not have done in 1950 when Turing published his paper on machine intelligence. What has truly changed is that today we have extremely powerful computers squeezed into a palm-size smartphone at a fraction of the cost. That's miniaturization. Equating miniaturization to intelligence is like equating an improved wallet to wealth.

Which progress really matters for Artificial Intelligence: hardware or software? There has certainly been rapid progress in hardware technology (and in the science of materials in general) but the real question to me is whether there has been any real progress in software technology since the invention of binary logic and of programming languages. And a cunning software engineer would argue that even that question is not correct: there is a difference between software engineering (that simply finds ways to implement algorithms in programming languages) and algorithms. The computer is a machine that executes algorithms. Anybody trying to create an intelligent machine using a computer is trying to find the algorithm or set of algorithms that will match or surpass human intelligence. Therefore it is neither progress in hardware nor progress in software that really matters (those are simply enabling technologies); what matters is progress in Computational Mathematics.

Ray Kurzweil's book used a diagram titled "Exponential Growth in Computing", but i would argue that it is bogus because it starts with the electromechanical tabulators of a century ago: it is like comparing the power of a windmill to the power of a horse. Sure there is an exponential increase in power, but it doesn't mean that windmills will keep improving forever vis à vis horsepower and windpower. And it doesn't distinguish between progress in hardware and progress in software, nor between progress in software and progress in algorithms. What we would like to see is a diagram titled "Exponential Growth in Computational Math". As i am writing this, most A.I. practitioners are looking for abstract algorithms that improve automatic learning techniques.

Others believe that the correct way to achieve artificial intelligence should be to simulate the brain's structure and its neural processes, a strategy that greatly reduces the set of interesting algorithms. In that case, one would also want to see a diagram titled "Exponential Growth in Brain Simulation". Alas, any neurologist can tell you how far we are from understanding how the brain performs even the simplest of daily tasks. Current brain simulation projects are modeling only a small fraction of the structure of the brain, and provide only a simplified binary facsimile of it: neuronal states are represented as binary states, the variety of neurotransmitters is reduced to just one kind, the emphasis is on feed-forward rather than on feedback connections, and, last but not least, there is usually no connection to a body. No laboratory has yet been able to duplicate the simplest brain we know, the brain of the 300-neuron roundworm: where's the exponential progress that would lead to a simulation of the 86 billion-neuron brain of Homo Sapiens (with its 100 trillion connections)? Since 1963 (when Sydney Brenner first proposed it), scientists worldwide have been trying to map the neural connections of the simplest roundworm, the Caenorhabditis Elegans, thus jump-starting a new discipline called Connectomics. So far they have been able to map only subsets of the worm's brain responsible for specific behaviors.

If you believe that an accurate simulation of brain processes will yield artificial intelligence (whatever your definition is of "artificial intelligence"), how accurate has that simulation to be? This is what neuroscientist Paul Nunez has called the "blueprint problem". Where does that simulation terminate? Does it terminate at the computational level, i.e. at simulating the exchanges of information within the brain? Does it terminate at the molecular level, i.e. simulating the neurotransmitters and the very flesh of the brain? Does it terminate at the electrochemical level, i.e. simulating electromagnetic equations and chemical reactions? Does it terminate at the quantum level, i.e. taking into consideration subatomic effects?

Ray Kurzweil's "Law of Accelerating Returns" is nothing but the usual enthusiastic projection of the present into the future, a mistake made by millions of people all the time. Alas, millions of people buy homes when home values are going up believing that they would go up forever. Historically, most technologies grew quickly for a while, then stabilized and continued to grow at a much slower pace until they became obsolete.

We may even overestimate the role of technology. Some increase in productivity is certainly due to technology, but in my opinion other contributions have been neglected too quickly. For example, Luis Bettencourt and Geoffrey West of the Santa Fe Institute have shown that doubling the population of a city causes on average an increase of 130% in its productivity ("A Unified Theory of Urban Living", 2010). This has nothing to do with technological progress but simply with urbanization. The rapid increase in productivity of the last 50 years may have more to do with the rapid urbanization of the world than with Moore's law: in 1950 only 28.8% of the world's population lived in urban areas but in 2008 for the first time in history more than half of the world's population lived in cities (82% in North America, the most urbanized region in the world).

Predictions about future exponential trends have almost always been wrong. Remember the prediction that the world's population would "grow exponentially"? In 1960 Heinz von Foerster predicted that population growth would become infinite by Friday the 13th of November 2026. Now we are beginning to fear that it will actually start shrinking (it already is in Japan and Italy). Or the prediction that energy consumption in the West will grow exponentially? It peaked a decade ago; and, as a percentage of GDP, it is actually declining rapidly. Life expectancy? It rose rapidly in the West between 1900 and 1980 but since then it has barely moved. War casualties were supposed to grow exponentially with the invention of nuclear weapons: since the invention of nuclear weapons the world has experienced the lowest number of casualties ever (see Steven Pinker's book "The Better Angels of Our Nature"), and places like Western Europe, that had been at war nonstop for 1500 years, have not had a major war since 1945.

There is one field in which i have witnessed rapid (if not exponential) progress: Genetics. This discipline has come a long way in just 70 years, since Oswald Avery and others identified DNA as the genetic material (1944) and James Watson and Francis Crick discovered the double-helix structure of DNA (1953). Frederick Sanger produced the first full genome of a living being in 1977, Kary Banks Mullis developed the polymerase chain reaction in 1983, Applied Biosystems introduced the first fully automated sequencing machine in 1987, William French Anderson performed the first procedure of gene therapy in 1990, Ian Wilmut cloned a sheep in 1997, the sequencing of the human genome was achieved by 2003, and Craig Venter and Hamilton Smith reprogrammed a bacterium's DNA in 2010. The reason that there has been such dramatic progress in this field is that a genuine breakthrough happened with the discovery of the structure of DNA. I don't believe that there has been an equivalent discovery in the field of Artificial Intelligence.

Economists would love to hear that progress is accelerating because it has an impact on productivity, which is one of the two factors driving GDP growth. GDP growth is basically due to population growth plus productivity increase. Population growth is coming to a standstill in all developing countries (and declining even in countries like Iran and Bangladesh) and, anyway, in the 20th century the biggest contributors to workforce growth were actually women, who came to the workplace by the millions, but now that number has stabilized.

If progress were accelerating, you'd expect productivity growth to accelerate. Instead, despite all the hoopla about computers and the Internet, productivity growth of the last 30 years has averaged 1.3% compared to 1.8% in the previous 40 years. Economists like Jeremy Grantham now predict a future of zero growth ("On The Road To Zero Growth," 2012). Not just deceleration but a shrieking halt.

Whenever i meet someone who strongly believes that machine intelligence is accelerating under our nose, i ask him/her a simple question: "What can machines do today that they could not do five years ago?" If their skills are "accelerating" and within 20-30 years they will have surpassed human intelligence, it shouldn't be difficult to answer that question. So far the answers to that question have consistently been about incremental refinements (e.g., the new release of a popular smartphone that can take pictures at higher resolution) and/or factually false ("they can recognize cats", which is not true because in the majority of cases these apps still fail, despite the results of the ImageNet Competitions).

In 1939 at the World's Fair in New York the General Motors Futurama exhibit showed how life would be in 1960 thanks to technological progress: the landscape was full of driverless cars. The voiceover said:

"Does it seem strange? Unbelievable? Remember, this is the world of 1960!" Twentyone years later the world of 1960 turned out to be much more similar to the world of 1939 than to the futuristic world of that exhibit.

On the 3rd of April 1988 the Los Angeles Times Magazine ran a piece titled "L.A. 2013" in which experts predicted how life would look like in 2013. They were comfortable predicting that the average middle-class family would have two robots to carry out all household chores including cooking and washing; that kitchen appliances would be capable of intelligent tasks; and that people would commute to work in self-driving cars. How many robots do you have in your home and how often do you travel in a self-driving car?

In 1964 Isaac Asimov wrote an article in the New York Times (August 16) titled "Visit to the World's Fair of 2014" in which he predicted what the Earth would look like in 2014. He envisioned that by 2014 there would be Moon colonies and all appliances would be cordless.

I am told that you must mention at least one Hollywood movie in a book on A.I. The only one that deserves to be mentioned is "2001: A Space Odyssey" (1968) by Stanley Kubrick. It is based on a book by Arthur Clarke. It features the most famous artificial intelligence of all time, HAL 9000. In the book HAL was born in 1997. 1997 came and went with no machines even remotely capable of what HAL does in that film (and so did 2007 and so will 2027)

One can even argue the opposite: technology tends to create stability in lifestyle. When i was a kid, i was not listening to the music of 50 years earlier. I am (pleasantly) surprised that today's kids listen to the music of 50 years ago, including some of my all-time favorites (Velvet Underground, Doors, Rolling Stones, etc). Digitizing music has bridged the temporal gap. When i was a kid, looking at the way people dressed or wrote just ten years earlier would make me chuckle (e.g., my father never left the house without a hat). But when you look at the way people dressed and wrote 20 years ago, it looks amazingly similar to the way people dress and write today. A kitchen of the 1950s was very different from a kitchen of the 1900s, but today's kitchen is very similar to the kitchen of the 1960s..

People who think that progress has been dramatic are just not aware of how fast progress was happening before they were born and of how high the expectations were and of how badly those expectations have not been met by current technology. Otherwise they would be more cautious about predicting future progress.

The future is mostly disappointing. As Benjamin Bratton wrote in December 2013: "Little of the future promised in TED talks actually happens".

Intermezzo: In Defense of Regress – Apocalypse Now

We accept as "progress" many innovations whose usefulness is dubious at best. Here are some favorite examples.

Any computer with a "mouse" requires the user to basically have three hands.

Never since the 1950s have phone communications been so rudimentary as after the introduction of the mobile phone. Conversations invariably contain a lot of "Can you hear me?" like in the age of black and white movies. I felt relieved in a Mexican town where there was a public phone at every corner: drop a coin and you are making a phone call. Wow. No contract needed, and no "can you hear me?"

Mobile phone ringers that go off in public places such as movie theaters and auditoria (and that obnoxiously repeat the same music-box refrain in a mechanical tone) do not improve the experience.

Voice recognition may have represented an improvement when it allowed people to say numbers rather than press them on the phone keyboard; but now the automated system on the other side of the phone asks you for names of cities or even "your mother's maiden name", and never gets them right (especially if you, like me, have a foreign accent) or for long numbers (such as the 16-digit number of my credit card) that you have to repeat over and over again until it gets it right or it gives up and mercifully connects you to a human operator.

I interact with Alexa the exact same way i used to interact 30 years ago with the Unix shell: i issue a command and Alexa responds (sometimes). Ironically, what makes Alexa more "human" is the fact that in so many cases it fails (it misunderstands what i commanded), whereas the old text-based user interfaces never failed. Hence i get angry with Alexa in a way that i never did with the Unix shell. I am painfully aware that Alexa is just a retarded piece of software, whereas the Unix shell felt like an extremely efficient assistant.

The automation of cash registers means that it takes longer to pay than to find the item you want to buy (and you cannot buy it at all if the cash register doesn't work).

The car keys with an embedded microchip (the "transponder" keys) cost 140 times more to duplicate than the old chip-less car keys.

If you made the mistake of buying a new car, you may have a car without a spare tire. For a century since the invention of the car, replacing a flat tire was a matter of minutes. Now a flat tire may leave you stranded for hours if not days.The new cars come with a do-it-yourself kit (sealant and inflator) that is a lot more difficult to use than a lug wrench and a jack, and, in general, it doesn't work: repairing even the smallest puncture requires superhuman skills and side punctures cannot be repaired anyway. A flat tire, that used to be an annoyance, has now become an emergency.

Watching films on digital media such as DVDs is more difficult for a professional critic than watching them on videotapes because stopping, rewinding, forwarding and, in general, pinpointing a scene is much easier and faster on analog videotapes (VCRs) than on digital files.

Computer's and car's CD drives that you have to push (instead of pull) in order to open are simply more likely to break and don't really add any useful feature. If the CD or DVD gets stuck inside, the drive can only be opened with a special screwdriver that virtually no user has.

In theory, it sounds like progress that our devices come packed with so many features. In reality, because manuals have disappeared and have been replaced by desperate searches on the web, we know very little about the devices that we buy, and we will never even know that some features exist. It is astonishing how little we know about the features for which we pay when we buy a computer, a phone or a software package. Unfortunately, most of us have a life, i.e. we have better things to do than to experiment with the labyrinthine menus and obscure terms of our devices.

Nowadays more and more families keep a lens handy: the serial number of a product was certainly a kind of progress because it easily identified the device that you purchased, but the serial number is usually a tiny number in the most unlikely corner of the device, hence the resurgence of the medieval lens, indispensable whenever you need technical support.

Most portable gadgets used to operate with the same AA or AAA batteries. When on the road, you only had to worry about having those spare batteries. Now most cameras work only with proprietary rechargeable batteries: the fact that they are "rechargeable" is useless if they die in a place where you cannot recharge them, which is the case whenever you are far from a town or forgot the charger at home. I don't see this as progress compared with the cheap, easily replaceable AA batteries that i could also use with my hiking GPS, my headlight and my walkie-talkie. In fact, Nikon mentions it as a plus that its Coolpix series is still "powered by readily available AA batteries".

Booking a shuttle to the airport now requires filling an endless series of forms. I used to do it with a phone call. Sure: i had to wait for 5 or 10 minutes to get connected, but i never made a mistake. Recently i mistakenly entered "AM" instead of "PM" as the pick-up time, a mistake that i would have never made on the phone (especially since it was a 2PM flight). The first problem is that there are so many forms to fill: i would not make mistakes if i could just write directly all the flight information in one line. The second problem is that i have to convert my one line of information (San Francisco - Beijing, January 13, 2pm) into a long sequence of finger movements: sometimes click, sometimes scroll, sometimes roll... It is unnatural and prone to error. There are two more "improvements" on this website for the airport shuttle: 1. the confirmation email lists no phone number to call in case the shuttle doesn't show up on time; 2. the tip is automatically included in the charge to your credit card (you only have one chance to say "no" and it's not very visible). Bottom line: you tip the driver even before he shows up and, if he doesn't show up, good luck.

It is hard to believe that there was a time (a century ago) when you would pick up the phone and ask an operator to connect you to someone. Now you have to dial a 10-digit number, and sometimes a 13-digit number if you are calling abroad. More recently there used to be telephone directories to find the phone number of other telephone subscribers. I remember making fun of Moscow when we visited it in the 1980s because it didn't have a telephone directory. In the age of mobile phones the telephone directory has disappeared: you can know a subscriber's number only if someone gives it to you. Apparently the Soviet Union is the future, not the past.

Thanks to air-conditioned buildings with windows that are tightly sealed, we freeze in the summer and sometimes catch bronchitis while it is really hot outside.

The air conditioning control has become so complicated that in most hotels we end up having no air conditioning at all: it is just too difficult to figure out how to turn it on and how to adjust it. Turning on the TV set now requires studying a manual: on one remote control i counted 58 buttons.

Talking of windows, the electric windows of your car won't operate if the car's battery dies (the old "roll down the window" does not apply to a car with dead battery).

In most of the developed world, when you travel by bus or train, you need to get your ticket at a machine or have exact change to buy it on the bus, hardly an improvement over the old system of paying the conductor when you board. New buses and trains are climatized: it is impossible to take decent pictures of the landscape because the windows cannot be opened and are dimmed.

Printing photographs has become more, not less, expensive with the advent of digital cameras, and the quality of the print is debatable.

The taximeter, rarely used in developing countries but mandatory in "advanced" countries, is a mixed blessing. Basically, a taxi driver asks you to buy a good without telling you the price until you have already used the good and you cannot change your mind. The taximeter often increases the cost of a ride because you can't bargain anymore as you would normally do based on the law of supply and demand (for example, in situations when the taxi driver has no hope of getting other customers). Furthermore, the taximeter motivates unscrupulous taxi drivers to take the longest and slowest route to your destination, whereas a negotiated price would motivate the driver to take you there as quickly as possible.

Thanks to "progress" in software, over the years i have had to adapt to countless limitations.

I have been using email since the early 1980s. The email system that i am forced to use at Stanford today is absolutely atrocious compared with the simple, quick, user-friendly and no-brainer email system that i was using twenty years ago at the same institution. Google's Gmail program and Microsoft's Outlook program, the most used email clients of this year, hardly qualify as progress: they rank among the most idiotic email clients ever designed. When i am forced to use either program, the expression that comes to mind is not "wow what an amazing intelligent program!" but rather "what a ridiculous piece of garbage!" (I personally believe that Gmail is singlehandedly responsible for the decline of email's popularity among the younger generation because younger people think that email "is" Gmail and therefore shun email unless forced by teachers or employers).

In some cases "progress" is humiliating. I've had an email address "piero .•⌐•.,.•⌐`•..•⌐`•. scaruffi" since the 1980s. Cute, eh? Well, in 2016 i was forced to abandon it and adopt a simple "piero scaruffi" because the #1 email system in the world started treating as spam any email address that contains special characters. This is just the latest step in a process of "normalization" of my Internet persona that started when they forced me to put a smiling adult headshot of me on the website of a popular university about 20 years ago (until that day i had always used the first picture ever taken of me as a child).I presume that at some point they will force me to capitalize my name to "Piero Scaruffi" to be as uniform as possible with the rest of humankind. They are asking me to remove my personality from what i do, otherwise i will not be able to do it anymore. In my humble opinion, these are terrorist actions, no more and no less terrifying than a physical terrorist attack. What is truly accelerating is not progress but the standardization of human behavior.

It has become commonplace on the Web to see this line on the webpage that you are reading: "This site uses cookies to improve your experience".

You know what that means, right? It means that they spy on what you do and then some other website will try to sell you something, and all of them get a cut out of those "cookies" sitting on your computer.

Increasingly, you are very welcome to upload a video for free on platforms such as Youtube, but you are not welcome to download one for free: they tell you that watching it on the "cloud" (i.e., "streaming" it) is so much better. It is certainly much better for them: If you watch the video on the cloud, they can feed you all the ads that maximize their profits, something that they cannot do if you download the video on your computer.

One of the hot topics of our age is the transition from the home computer to the cloud. The idea is that you don't need to have your data on your own device: you can keep them on the cloud, so that you can access them from anywhere in the world. This idea is preposterous any way you look at it: 1. You can access your data only where there is a connection to the cloud, only when the cloud platform is up and working, and only under the terms of the corporation that owns that platform; 2. It is obviously not safe, and all your data can be stolen in one second by a hacker who breaks into the cloud platform; 3. Access to your data will never be as fast as when they are on your own device. It is difficult to see "progress" in the fact that i used to have a movie on a videotape in my bookshelf and now i have to rely on a third-party to watch that movie. But corporations and governments relentlessly promote the idea of using their cloud to store your data. In many cases they are willing to do it for free. The real reason is that they make money out of it, and they can spy on and control what we do. It would be nice if they told us honestly "We make money out of owning your data, and maybe it is also convenient for you". The certain thing is that they make money out of it. Progress for them? Certainly.

The best example of the mixed "progress" inherent in the cloud is Wikipedia. Now the world only has one encyclopedia, entirely written by anonymous contributors, which turn out to be mostly public-relationship departments of corporations, marketing agents of celebrities, special-interest groups, sometimes foreign government agencies, and lots of fanatics determined to defend their version of the truth. See my article "Wikipedia as a Force for Evil" (2013).

Today i cannot check the weather forecast because the most popular website floods my screen with refrigerators: a Chinese friend used my browser to check prices of refrigerators while waiting for his flight to China, and there's no way to tell the various websites that my friend lives in China and i am not going to relay to him the various promotions that appear on my screen here in California.

I cannot login anymore into my bank account from my old desktop computer. When i login, i get the error message "You are using an unsupported browser version". Fair enough: it's an old computer. One after the other all websites will start telling me that i can't login anymore from this computer. I don't think that's "progress", but i'm willing to accept that i need a new desktop computer, even if there's nothing wrong with this old one. (Imagine if your fridge one day refused to open or your dishwasher one day refused to wash dishes simply because you are wearing old gloves). But then i used my laptop computer, which is brand new, and i noticed another "improvement" to my bank's website: they made it really difficult if not impossible to send them an electronic message. Now you can only contact them by phone or visit their offices in person (which is precisely the world of thirty years ago).

Some financial transactions that used to take a few minutes at the bank's branch are becoming more and more difficult, if not impossible. The bank will tell you that you must do it online, where the website will tell you that additional steps are required "for your own security" with the result that, given slow Internet connections, antispam firewalls and assorted filters, you have to spend a lot more than a few minutes. Soon those transactions may become de facto impossible except for the most perseverant of users. The bank isn't increasing our security, it is banning those services.

After being notified a thousand times by a very aggressive Windows 10 operating system that new updates were available, one day i finally clicked on Yes and... Movie Maker stopped working: it now consistently objects that my brand new laptop does not meet the minimum requirements (yes, it does). A few days later i received another notification that new updates were available and i immediately clicked on Yes hoping that one of these updates would fix the problem that keeps Movie Maker from running. The only noticeable difference is that now my laptop arranges all the icons to the left, no matter how i try to arrange them. I tried to get rid of the annoying "lock screen". I searched the Web and found that thousands of Windows 10 users are as annoyed as me by this "feature". There is absolutely no information on the Microsoft website but there are forums ("customer support" in the age of intelligent machines) where several people have posted a solution that worked for me. Quote:

- Open the registry editor.
- Navigate to
HKEY_LOCAL_MACHINE\SOFTWARE\Policies\Microsoft\Windows
- Create a new registry key called Personalization
- Navigate to the Personalization key

- Right click in the right pane and select New then DWORD (32-bit) Value.
- Name the new value NoLockScreen
- Set NoLockScreen to 1"

This was titled "Simple Steps To Get Rid Of Windows 10's New Lock Screen".

No, this does not happen only with Windows 10, or only with Microsoft software. It happens with all the software out there.

It is hard to see how modern websites such as Linkedin and Pinterest, or online magazines such as Wired and Rolling Stones, can constitute progress when they crash 90% of the browsers in the world (most computers in the world still run old operating systems: not everybody is willing to buy a new computer just because Linkedin or Pinterest has decided to use sophisticated scripting technology). On the other hand you are constantly stalked by software that desperately wants you to reset your password.

If progress means that what i have been using will not work anymore, it is not progress for me. It is progress for the ones who make it and sell it, but not for the ones who never asked for it and are now forced to accept it and pay for it.

Computers can be amazingly unintuitive compared with older devices. If you remove a USB flash drive the way you normally remove a CD or DVD from its player, you may lose all the data, so you are required to "safely remove" it. On Apple computers the way to safely remove a drive is to… throw it in the garbage can!

Websites with graphics, animation, pop-up windows, "click here and there", cause you to spend most of the time scrolling away from these digital paraphernalia instead of reading the information that you were looking for.

In 2017 the most popular oxymoron on websites seems to be "We use cookies to improve your experience": does anybody believe that the website has hired costly software engineers to implement those cookies for the sole purpose of improving our experience?

We the consumers passively accept too many of these dubious "improvements".

Most of these "improvements" may represent progress, but the question is "progress for whom"? Pickpockets used to steal one wallet at a time. The fact that today a hacker can steal millions of credit card numbers in an instant constitutes progress, but progress for whom?

And don't get me started on "health care", which in these high-tech days has become less about "health" and more and more about making you chronically ill. You are perfectly fine until you walk into the office of a dentist, eye doctor or other specialist; when you come out, you have become a medication addict with an immune system weakened by antibiotics and some prosthetic addition to your body that will require lifelong maintenance: progress for the "health-care" industry, not for you. (In 2016 a study published in the British Medical Journal by Martin Makary, a surgeon at the Johns Hopkins University, estimated that medical error was the third leading cause of death in the USA after heart disease and cancer).

We live in a world of not particularly intelligent machines, but certainly in a world of machines that like to beep. My car beeps when I start it, when I leave the door open, and if I don't fasten my seat belt. Note that it doesn't beep if something much more serious happen, like the alternator dies or the oil level is dangerously low. My microwave oven beeps when the food is ready, and it keeps beeping virtually forever unless someone opens its door (it doesn't matter that you actually pick up the food, just open the door). My printer beeps when it starts, when it needs paper, and whenever something goes wrong (a blinking message on the display is not enough, apparently). Best of all, my phone beeps when i turn it off, and, of course, sometimes i turn it off because i want it silent: it will beep to tell everybody that it is being silenced. I think that every manual should come with instructions on how to disable the beeping on the device: "First and foremost, here is how you can completely shut up your device once and forever".

Last but not least, something is being lost in the digital age, something that was the fundamental experience of (broadly defined) entertainment. During a vacation in a developing country i watched as a girl came out of the photographer's shop. She couldn't wait and immediately opened the envelope that contained her photographs. I witnessed her joy as she flipped through the pictures. The magic of that moment, when she sees how the pictures came, will be gone the day she buys her first digital camera. There will be nothing special about watching the pictures on her computer's screen. There will be no anxious waiting while she uploads them to the computer because, most likely, she will already know how the pictures look like before she uploads them. Part of the magic of taking photographs is gone forever, replaced by a new, cold experience that consists in refining the photograph with digital tools until it is what you want to see, not what it really looked like, and then posting it on social media in an act of vanity.

Or take live events. The magic of a live sport event used to be the anxious wait for the event to start, and then the "rooting" for one of the competitors or teams. After the introduction of TiVo, one can watch a "live" event at any time by conveniently "taping" it. Many live events are actually broadcast with a slight delay, so you may find on the Internet the result of a soccer game that is still going on according to your television channel. Thus the "waiting" and the "rooting" are no longer the two fundamental components of the "live" experience. The whole point of watching a live event was the irrational feeling that your emotional state could somehow influence the result. If the event is recorded (i.e., is already in the past), that feeling disappears and you have to face the crude reality of your impotence to affect the result. But then what's the point of rooting? Thus the viewer is unlikely to feel the same emotional attachment to the game that s/he is watching. In the back of her/his mind, it is clear that the game has already finished. The experience of watching the "live" event is no longer one of anxiety but one of appreciation. Told by a friend that it was a lousy game, the viewer may well decide not to watch the event that her home appliance taped.

The first time that i went to Hong Kong i took a slow ferry for about $1. I was on the deck and took lots of great pictures of the island. In 2017 i took the ferry again. This time the price was 40 times more. Passengers are no longer allowed to walk on the deck, they are confined to the cabin and can only take pictures through the windows, extremely filthy windows that cannot be opened. You pay 40 times more for a much less rewarding experience. But the ferry is indeed many times faster than the old one, and that counts as "progress". Except that younger people will never have the beautiful pictures that i have from 20 years ago, or will have to pay even more to get on a tourist boat that allows passengers on the deck.

Yes, i know that Skype and Uber and many new services can solve or will solve these problems, but the point is that these gadgets and features were conceived and understood as "progress" when they were introduced (and usually amid much fanfare). The very fact that platforms such as Skype and Uber have been successful proves that the quality of services in those fields had overall regressed, not progressed, and therefore there was an opportunity for someone to restore service to a decent level.

We should always pause and analyze whether something presented as "progress" truly represents progress. And progress for whom.

Intermezzo: Why Futurists Always Get it Wrong

Because they want to predict the future without first studying the past.

And because they underestimate how important society is to shape the future, as opposed to "exponential" technological progress. It is the eccentric in the garage, thinking of something completely different from the mainstream, who writes the history of the future. No futurist predicted

Gutenberg, Columbus, Watt, Mendel, Edison, Marconi, Einstein, Fleming, Turing, Crick, Berners-Lee, Wozniak, Page, Zuckerberg... the scientists and inventors who truly changed the world.

In 1963 John McCarthy founded the Stanford AI Lab (SAIL) with the goal of building a fully intelligent machine within a decade. In 1965 Herbert Simon predicted that "machines will be capable, within twenty years, of doing any work a man can do". In 1967 Marvin Minsky predicted that "within a generation... the problem of creating artificial intelligence will substantially be solved", and anticipated that solving the problem of computer vision would take only a summer. In 1978 Moravec predicted that computers would become as intelligent as human beings in 1998 (but then in 1998 he published the essay "When Will Computer Hardware Match the Human Brain?"). I am still to find a single prediction by Kurzweil that turned out to be true, certainly none of those listed in "The Age of Spiritual Machines" (1999), except for those that everybody was already predicting (but this didn't stop an anonymous Wikipedia article from crediting him with a success rate of 86%). One of my favorites is Gartner Group's 2007 prediction that a whopping 80% of Internet users would participate in virtual worlds by 2012. The year 2012 came and went, and to this day (2016) the vast majority of Internet users do not even know what a virtual world is.

According to Stewart Brand, Marvin Minsky believed that contemporary philosophers were "shallow and wrong", but of course that could have been because contemporary philosophers proved him (Minsky) shallow and wrong.

The Future of Machine Learning: Unsupervised Learning to the Rescue?

Summarizing, there are four desiderata that one would like to see in A.I. systems, if they have to compare well with human (or just animal) brains: meta-learning, learning by demonstration ("few-shot learning"), transfer learning and multi-task learning.

Meta-learning is particularly relevant in the case of reinforcement learning. It is obvious that reinforcement learning is highly unnatural. DeepMind's AlphaGo and OpenAi Five need to learn from scratch via a huge number of trials. Animals, instead, use built-in or acquired "meta-skills" to learn new tasks in just a few trials.

The modern computational theory of meta-learning (learning how to learn) dates back at least to the 1990s, when Schmidhuber published the manifesto "Simple Principles of Metalearning" (1996), followed by his student Sepp Hochreiter ("Learning to Learn Using Gradient Descent", 2001), and by Nicolas Schweighofer and Kenji Doya at Japan's ATR ("Meta-learning in Reinforcement Learning", 2001). Examples of "deep" meta-learning systems of the new generation are: RL Square by Pieter Abbeel's student Yan Duan at UC Berkeley, based on Schulman's TRPO ("RL Square: Fast Reinforcement Learning via Slow Reinforcement Learning", 2016); the "model-agnostic meta-learning" (MAML) of Sergey Levine's student Chelsea Finn at UC Berkeley ("Model-Agnostic Meta-Learning for Fast Adaptation of Deep Networks", 2017); Marcel Binz's thesis at KTH Royal Institute of Technology ("Learning Goal-Directed Behaviour", 2017); Jane Wang's "deep meta-reinforcement learning" at DeepMind ("Learning to Reinforcement Learn", 2017); and OpenAI's Reptile, developed by Alex Nichol and John Schulman, a generalization of Finn's MAML ("On First-Order Meta-Learning Algorithms", 2018). DeepMind's neuroscientist Matthew Botvinick believes that the latter could be a model for how our brain learns: the dopamine system trains another part of the brain, the prefrontal cortex, to operate as its own free-standing learning system ("Prefrontal Cortex as a Meta-reinforcement Learning System", 2018).

It is also obvious that animals can naturally "transfer" skills from one domain to another: an animal rarely needs to learn a new skill as if it had nothing in common with known skills. Transfer learning is about applying what one learned in one case to a different case. Success stories in computational transfer learning are scant despite the pioneering work of Satinder Singh (1991), Lorien Pratt (1992), Sebastian Thrun (1994) and Rich Caruana (1993).

Reinforcement learning is particularly difficult to generalize to multiple problems because each case requires a different reward function. Transfer learning is one case in which the learning agent needs to do a lot of exploration (it can't just repeat what it has learned in a previous case). Exploration based on "intrinsic motivation" is an old idea, from Schmidhuber ("Evolutionary Principles in Selfreferential Learning", 1987) via Barto ("Intrinsically Motivated Reinforcement Learning", 2004) all the way to DeepMind ("Unifying Count-Based Exploration and Intrinsic Motivation", 2016), whose algorithm achieved improvement in strategy-based videogames such as Montezuma Revenge, and OpenAI ("Some Considerations on Learning to Explore via Meta-reinforcement Learning", 2017), that introduced two new reinforcement-learning algorithms (EMAML and E-RL2).

Andrei Rusu and others at DeepMind also invented the "progressive nets" method that supposedly emulated the way the human mind can reuse previous experience and applied it to accelerate learning multiple Atari games ("Progressive Neural Networks", 2016).

Curiosity-driven exploration, which, again, was pioneered by Schmidhuber ("Curious Model-building Control Systems," 1991) and was applied to developmental robotics by Pierre-Yves Oudeyer and Frederic Kaplan at Sony labs in France ("Motivational Principles for Visual Know-how Development," 2003), was studied by Shankar Sastry's student Joshua Achiam at UC Berkeley ("Surprise-Based Intrinsic Motivation for Deep Reinforcement Learning", 2017), Trevor Darrell's student Deepak Pathak at UC Berkeley ("Curiosity-driven Exploration by Self-supervised Prediction", 2017), and DeepMind ("Kickstarting Deep Reinforcement Learning", 2018). Pathak's "intrinsic curiosity model", for example, was a self-supervised reinforcement learning system that used curiosity, instead of feedback, as a natural reward signal to enable the agent to explore its environment and learn skills for later use. An example of applications of curiosity-driven learning was Leela, unveiled in 2018 by Leela.ai in Palo Alto, a learning agent modeled on Jean Piaget's child psychology that built increasingly abstract models of the world from exploration, play and trial-and-error. Pieter Abbeel's student Abhishek Gupta at UC Berkeley introduced an algorithm, "model agnostic exploration with structured noise" (MAESN), to learn exploration strategies from past experience ("Meta-Reinforcement Learning of Structured Exploration Strategies, 2018"). Abhinav Gupta's student Lerrel Pinto at CMU built a robot that could push, poke, grasp and observe objects; four different types of physical interactions that forced a shared CNN to learn a visual representation ("The Curious Robot - Learning Visual Representations via Physical Interactions", 2016).

The way we normally learn a new game is by watching people play. After watching a few games, and being told what the rules are, we can start playing. This is called "few-shot learning", learning by watching a few demonstrations. This is very different from what OpenAI Five and DeepMind's AlphaZero do: they play thousands of games against themselves. Pioneering work in "behavioral cloning" (another term for "learning from demonstration") was made in Britain by Donald Michie ("Cognitive Models from Subcognitive Skills", 1990) and Claude Sammut ("Learning to Fly", 1992). Research on machine learning from visual observation of human performance was conducted in Japan by Yasuo Kuniyoshi of the Electrotechnical Laboratory (ETL) in collaboration with Masayuki Inaba and Hirochika Inoue of Tokyo University ("Teaching by Showing", 1989), and by Dean Pomerleau at CMU when training the self-driving vehicle ALVINN to follow street lanes ("Autonomous Land Vehicle in a Neural Network", 1989). Stefan Schaal wrote the manifesto

"Is Imitation Learning the Route to Humanoid Robots?" (1999). Two future stars of deep learning, Pieter Abbeel and Andrew Ng, were attracted by the field ("Apprenticeship Learning via Inverse Reinforcement Learning", 2004) when, both at Stanford University, the former was studying philosophy and the latter had just graduated in philosophy of computer science from UC Berkeley with a thesis on reinforcement learning supervised by Michael Jordan. Stimulus to develop alternatives for few-shot learning came also from Lake-Salakhutdinov-Tenenbaum's theory of concept learning through probabilistic program induction (2015). A popular avenue of research in one-shot learning was to extend Alex Graves' neural Turing machines (2014) and James Weston's memory networks (2014): Oriol Vinyals at DeepMind developed "matching networks" ("Matching Networks for One Shot Learning", 2016) and Han Altae-Tran at MIT developed the "iterative refinement long short-term memory" ("Low Data Drug Discovery with One-Shot Learning", 2018). Sergey Levine's students Tianhe Yu and Chelsea Finn, instead, working on robots, developed a system that can imitate a movement after watching it just once, a movement that the system has never seen before. The secret is a variant of MAML called "domain-adaptive meta-learning" (or DAML) that trains a deep network with many videos of human and robot movements performed for different tasks. Then the system can handle a novel task involving a novel object after watching a human perform that task with that object ("One-Shot Imitation from Observing Humans via Domain-Adaptive Meta-Learning", 2018).

The next step for Tianhe Yu and Chelsea Finn was to use the video of a human demonstration (or televised transmission of it) to train the robot ("One-Shot Imitation from Observing Humans via Domain-Adaptive Meta-Learning", 2018).

Byron Boots' group at Georgia Tech used imitation learning to continue Dean Pomerleau's mission to train a self-driving vehicle ("Agile Autonomous Driving using End-to-End Deep Imitation Learning", 2018).

Aravind Rajeswaran and Vikash Kumar at University of Washington mixed learning by demonstration and deep reinforcement learning to teach a robot how to grasp objects ("Learning Complex Dexterous Manipulation with Deep Reinforcement Learning and Demonstrations", 2018) and DeepMind's group of Nando de Freitas and Nicolas Heess did the same to teach visual skills ("Reinforcement and Imitation Learning for Diverse Visuomotor Skills", 2018).

Being able to learn from a demonstration can make a huge difference. For example, it was long believed that reinforcement learning cannot learn to use multi-fingered robotic hands because of the large number of degrees of freedom, i.e. because of the high dimensionality of the problem.

Therefore multi-fingered hands were controlled with trajectory optimization methods such as the physics-based method developed by Karen Liu at Georgia Institute of Technology ("Synthesis of Interactive Hand Manipulation", 2008) or the "contact-invariant optimization" method developed by Emanuel Todorov at University of Washington ("Contact-invariant Optimization for Hand Manipulation", 2012). But Aravind Rajeswaran at University of Washington, working in collaboration with UC Berkeley's Sergey Levine and OpenAI's John Schulman, showed that reinforcement learning can actually learn to control dexterous multi-fingered hands if it is augmented with a small number of human demonstrations ("Learning Complex Dexterous Manipulation with Deep Reinforcement Learning and Demonstrations", 2018).

Finally, all animals, and certainly humans, can learn many tasks with the same brain, whereas it is terribly difficult for a machine-learning algorithm to learn more than one task. There have been, however, encouraging signs that it might be possible. Sejnowski's NETtalk (1986) used one neural network to learn two strongly related tasks about speech, as did Ronan Collobert's and Jason Weston's system for natural language processing (2008). Sebastian Thrun at CMU studied how to enable a robot to continuously learn as it collects new experiences ("Lifelong Robot Learning", 1995). Rich Caruana's work on the self-driving car ALVINN (1997) showed that learning multiple tasks in parallel can be easier than learning them separately. In fact, in 2015 Bharath Ramsundar at Stanford developed a multitask network for drug discovery, based on Szegedy's GoogLeNet, whose accuracy improved as additional tasks and data were added ("Massively Multitask Networks for Drug Discovery", 2015). Incidentally, Ramsundar was also the brain behind deepchem.io, an open-source Python library to help scientists build deep-learning systems for drug discovery. Alas, Ramsundar's "massively multitasking" network required training with millions of data points when in reality the average drug-discovery laboratory works with only a handful of chemical compounds. One key event in the revival of multitask networks was the Kaggle competition sponsored in 2012 by pharmaceutical giant Merck: the winner was a multitask network (designed by Geoffrey Hinton's student George Dahl at University of Toronto). Studies multiplied and within a few years there was significant progress. For example, Ishan Misra and Abhinav Shrivastava at CMU introduced "cross-stitch units" for convolutional networks, units that look for the best shared representations for multitask learning ("Cross-Stitch Networks for Multi-task Learning", 2016). Ever since Caruana's paper "Multitask Learning - A KnowledgeBased Source of Inductive Bias" (1993), the supervision of multiple tasks was carried out at the outermost layer of the neural network (as if it were at the top of a hierarchy), but Anders Sogaard at University of Copenhagen and Yoav Goldberg at Bar-Ilan University showed that it is better to manage different tasks at different layers rather than in the same layer ("Deep Multi-task Learning with Low Level Tasks Supervised at Lower Layers", 2016). Their finding was applied, for example, by Richard Socher of Salesforce in collaboration with University of Tokyo to problems of natural language processing, thereby improving on Collobert's and Weston's model ("A Joint Many-Task Model", 2017). Mingsheng Long at Tsinghua University introduced "relationship networks" to discover transferrable features between tasks ("Learning Multiple Tasks with Multilinear Relationship Networks", 2017).

In order to achieve what Thrun had called "lifelong learning" in 1995, neural networks need to overcome the problem that Michael McCloskey and Neal Cohen had called "catastrophic forgetting" in 1989. For almost 30 years that remained a stumbling block, and lifelong learning was mostly attempted with other techniques. For example, in 2013 Eric Eaton and Paul Ruvolo of Bryn Mawr College proposed a general algorithm for lifelong learning called Efficient Lifelong Learning Algorithm (ELLA). Finally, James Kirkpatrick and others at Deep Mind developed the algorithm called "elastic weight consolidation" (EWC) that overcomes "catastrophic forgetting" and that can therefore be used in supervised learning and reinforcement learning algorithms to train the network on several tasks sequentially without forgetting the previous ones ("Overcoming Catastrophic Forgetting in Neural Networks", 2017). Another strategy was devised by Jaehong Yoon at KAIST in South Korea, the Dynamically Expandable Network ("Lifelong Learning with Dynamically Expandable Networks", 2017).

In 1996 Caruana had introduced a method of "hard parameter sharing", but this is suitable only for learning closely-related tasks (such as two tasks related to linguistic features). Instead, Sebastian Ruder at the National University of Ireland introduced "sluice networks", a framework for learning loosely-related tasks, a framework that also turns out to be a generalization of both the Sogaard-Goldberg model and of cross-stitch networks ("Learning What to Share Between Loosely Related Tasks", 2018).

Iasonas Kokkinos at University College London built UberNet, a CNN based on VGG-Net, capable of handling multiple vision tasks ("Training a Universal Convolutional Neural Network for Low-, mid-, and High-level Vision Using Diverse Datasets and Limited Memory", 2017).

Fei Wang in collaboration with Tsinghua University and The Chinese University of Hong Kong designed a network of stacked attention modules, based on ResNet101, that, while performing the general task of image classification, can separate features that are useful for different tasks ("Residual Attention Network for Image Classification", 2017).

Andrew Davison's student Shikun Liu at Imperial College London proposed the Multi-Task Attention Network (MTAN) that can be implemented with any feed-forward network, and provides a global feature pool for any number of task-specific modules so that all learned features can be shared across different tasks. Each task-specific module is an attention module designed to learn task-specific features but also to share them ("End-to-End Multi-Task Learning with Attention", 2018).

These are the desiderata: meta-learning, few-shot learning (i.e. learning by demonstration), transfer learning and multi-task learning. Now the question is which method of learning can achieve those goals. Artificial Intelligence has studied three main methods of learning for neural networks: supervised, unsupervised and reinforced. Supervised learning gave us image and speech recognition. Reinforcement learning gave us software that plays games better than human champions. These successes somewhat obscured the research in unsupervised learning; but obviously supervised learning (that needs to see millions of cats before it can recognize a cat) and reinforcement learning (that needs to play a game millions of times before it can win) are not good models of how animals really learn. Animals learn a lot faster. Many believe that the difference between current algorithms and animals lies precisely in unsupervised learning.

If you want your robot to understand what is happening around it, and what the consequences of its actions will be on the objects around it, supervised learning is not helpful: you would need to show your robot millions of chairs, millions of tables, millions of pens, millions of all the possible objects in order for the robot to figure out just what is in the room. And then it would still need to learn the dynamics of physical interactions with those objects. In order to get to the point that your robot can understand and deal with a variety of scenes and situations, your robot needs to learn by itself what the world is like and how it works.

Up to the 2010s there were basically two categories of unsupervised learning methods: probabilistic (such as Paul Smolensky's restricted Boltzmann machines and Pietro Perona's "constellation models") and autoencoders. And, incidentally, Marc'Aurelio Ranzato in Yann LeCun's group at New York University proved that these two groups are actually similar mathematical models ("A Unified Energy-based Framework for Unsupervised Learning", 2007).

Some new timely theories about the brain reshaped the field of unsupervised learning. The Cambridge University neuroscientist Horace Barlow, one of the pioneers with Hubel and Wiesel of formal studies of the visual cortex, showed how neurons of the visual cortex detect what he termed "suspicious coincidences" that occur frequently and use them to build models of the world ("Cerebral Cortex as Model Builder", 1985). This was based on the models of the visual cortex developed by Christoph von der Malsburg in Germany (" Self-organization of Orientation Sensitive Cells in Striate Cortex", 1973) and by Barlow's colleague Nicholas Swindale at Cambridge ("The Development of Columnar Systems in the Mammalian Visual Cortex", 1980). Barlow's intuition was that the brain is a poor scientist but a great statistician, and so, for example, the brain learns that lightning is almost always accompanied by thunder even without knowing the reason.

If Barlow focused on coincidence, others focused on movement. Animals are capable of visually identifying an object regardless of where the object is, i.e. regardless of distance and perspective. This is quite amazing as the image of the object can be very different if you think of an image as just a matrix of pixels. We can even deform an object and most people will recognize it (e.g., you can still read the newspaper if you bend a page, which means that letters don't have to be flat like on a table or a wall for your brain to recognize them). Fukushima's neocognitron of 1980 tried to simulate this phenomenon via a hierarchy of alternating feature detectors and invariance layers, which was basically the architecture discovered by David Hubel and Torsten Wiesel. This was the principle also used by LeCun to recognize digits in 1989. Barlow's collaborator Peter Foldiak at Cambrige University, instead, argued that the visual system learns to recognize objects (regardless of the way they look from a specific perspective) from its sensory experience: as we move around we see the object change shape and this trains our visual system to recognize that object from any other perspective ("Learning Invariance from Transformation Sequences", 1991). This came to be known as the "principle of temporal coherence".

Foldiak's model was also more consistent with the view of biologists of the school known as "ecological realism". Biologists such as James-Jerome Gibson had been arguing for decades that animals learn about the environment by acting in it, and, once they learn how the environment works, they become capable of performing a lot of actions in it. As Gibson phrased it: "We move in order to see and we see in order to move".

In the real world an object is rarely seen as an isolated image. It is almost always part of a scene, and part of a scene that is changing. Even if the objects are not moving, the observer is moving, and in most cases both the viewer and the objects are moving.

Foldiak's ideas also resonated with Dileep George's hierarchical probabilistic model of the visual cortex at Stanford University that was based on a similar principle: the geometric invariance of objects (that is trivial for humans to understand) is linked to our movements. As we move around an object, we know that it is still the same object even if it looks different as the visual angle changes. That's how our brain gets trained to recognize objects ("A Hierarchical Bayesian Model of Invariant Pattern Recognition", 2005). His research was sponsored by Silicon Valley inventor Jeff Hawkins at the Redwood Center for Theoretical Neuroscience and then at their startup Numenta.

The limits of supervised learning were particularly serious in the field of video analysis, which was becoming important in designing vision systems for autonomous robots, self driving cars, and security systems. Supervised learning is difficult in the case of videos because videos are much higher dimensional entities compared to single images. Luckily, videos contain a lot of "suspicious coincidences": spatial and temporal regularities. For example, two successive frames of a video are likely to contain the same objects. Unsupervised learning becomes much easier if the neural network can also use these spatial and temporal correlations. These regularities provide important information about how objects behave. The neural network can use this information about objects to train itself. That's why sometimes this is known as "self-supervised" learning. Furthermore, a self-supevised neural network learns a representation that can be used for other practical tasks. For example, a neural network that learns to recognize a car in videos of highways is learning a representation of what cars do on roads. This can be used, for example, to classify movies.

So it turned out that videos were resurrecting unsupervised learning from both sides: they were showing the limitations of supervised learning and at the same time they were showing that a neural network can self-train by using the information available in the environment (information captured by the videos themselves).

Video analysis requires world knowledge and world knowledge is contained in videos. This was not an accident: every animal is a manifestation of the same loop. Animals need world knowledge in order to act in their environment, and they acquire that knowledge precisely by acting in the environment.

We learn what objects do and what to do with objects, and we learn it by interacting with them. The challenge is to design robots that can do the same: learn about the world while interacting with the world, accumulate knowledge about everyday objects and use it when appropriate. Inspired by the new discipline of developmental cognitive neuroscience (that took the name from Mark Johnson's 1996 book), three famous pioneers of Japanese robotics (Minoru Asada, Hiroshi Ishiguro and Yasuo Kuniyoshi) advocated "cognitive developmental robotics" ("Cognitive Developmental

Robotics As a New Paradigm for the Design of Humanoid Robots", 2001). Equivalently, Juyang Weng at Michigan State University and others called for a robotics of "autonomous mental development" ("Autonomous Mental Development by Robots and Animals," 2001). Developmental robotics, again, called for unsupervised learning.

Before deep learning, most approaches to learning representations of videos in an unsupervised way were based on "independent component analysis". Johannes van Hateren and Daniel Ruderman at Groningen University in the Netherlands pioneered the field ("Independent Component Analysis of Natural Image Sequences Yields Spatio-temporal Filters Similar to Simple Cells in Primary Visual Cortex", 1998).

One has to realize upfront that deep learning is not necessary for predicting what is going to happen in a scene. This was done, for example, by Abhinav Gupta's student Jacob Walker at CMU ("Patch to the Future", 2014). Using a rather traditional method (the 30-year-old Kanade-Lucas tracking algorithm), his program learned in a completely unsupervised manner from a large collection of videos what happens next in scenes of traffic. Also relatively traditional was the approach, at ETRI in Korea, of Michael Ryoo's predictor of human activity, that predicts the next frame in a video of human action ("Human Activity Recognition", 2011). This was, incidentally, an important work because it emphasized the importance of "prediction": classifying what humans did in the past is not enough if, for example, you want to prevent a crime; you also need to be able to predict what they are about to do before they do it. Classifying what humans "did" is not enough. Ryoo used an old-fashioned histogram-based approach. The main reason for using deep learning is to get closer to what the brain does, hoping that this will also lead to better results.

Barlow's intuition was used by Hossein Mobahi, working with Ronan Collobert and Jason Weston at University of Illinois, for an unsupervised learning model (a deep convolutional network) based on features that are adjacent in time ("Deep Learning from Temporal Coherence in Video", 2009).

Then in 2012 came the autoencoder built by Andrew Ng's group that recognized cats in still frames of videos, a project that proved deep learning could be a useful method for video analysis.

One more finding from neuroscience fueled progress in video analysis. Rajesh Rao and Dana Ballard at the Salk Institute described brains as predictive systems: the early stages of sensory processing in the brain learn the statistical regularities in the environment and transmit to the next stages of processing only the sensory input that is not redundant. The predictable components of the input are removed at the very beginning, and only what is not predictable reaches the subsequent stages. The sensory input gets compressed into a more efficient form before it is forwarded to other regions of the brain ("Predictive Coding in the Visual Cortex, 1999). The principle of predictive coding, originally developed for the visual system of the fly by Mandyam Srinivasan, Simon Laughlin and Andreas Dubs at the Australian National University ("Predictive Coding", 1982), was soon applied to many other brain areas, including the auditory system, the hippocampus and the frontal cortex. Predictive coding views cortical functions as a process in which top-down information predicts bottom-up information, and inhibits all bottom-up information that fits with the prediction, thus allowing only errors to propagate upwards. This simple principle actually constitutes a very efficient way to "code" new information about the world. Karl Friston at University College London summarized the activity of the brain as a process to minimize prediction error, and also expressed it in terms of minimizing the "free energy" of the brain, a thermodynamic formulation that could lead to a unified theory of mind and life ("Learning and Inference in the Brain", 2003).

Incidentally, the underlying principle of encoding only the "unexpected" and discarding the "predictable" is the same used in audio and video compression methods such as JPEG.

Meanwhile, Daniel Felleman and David Van Essen ("Distributed Hierarchical Processing in the Primate Cerebral Cortex", 1991) had showed that the cortex is layered (we know of at least six layers) and hierarchical, and that each layer learns more abstract concepts than the previous ones.

Twenty years earlier, David Mumford had modeled the visual cortex as a hierarchy in which loops integrate top-down expectations and bottom-up observations via probabilistic (Bayesian) inference ("On The Computational Architecture Of The Neocortex II", 1992), an idea refined a decade later with Tai-sing Lee of CMU ("Hierarchical Bayesian Inference in the Visual Cortex", 2003). This came to be known as the "Bayesian brain hypothesis" after a book by Kenji Doya and others titled "Bayesian Brain" (2007).

Jeff Hawkins merged these threads in his book "On Intelligence" (2004), and envisioned a general neocortical algorithm that is basically a prediction algorithm, and a general process of learning that is basically just a process of optimizing prediction.

Andy Clark at University of Edinburgh summarized this view of the brain as a "hierarchical generative model that aims to minimize prediction error within a bidirectional cascade of cortical processing" ("Predictive Brains, Situated Agents, and the Future of Cognitive Science", 2013). Learning is a delicate dance between top-down predictions and bottom-up inputs that either validate those predictions (and are therefore discarded) or invalidate them (in which case they trigger new coding).

Viewing the brain as a "predictive network" established a new paradigm for neural networks.

Rasmus Palm's thesis at the Technical University of Denmark ("Prediction as a Candidate for Learning Deep Hierarchical Models of Data", 2012) showed that a "predictive" autoencoder is a far better candidate for learning than the original "reconstructive" autoencoder. The "predictive" encoder is a particular kind of denoising autoencoder that, instead of reconstructing the input, tries to predict future input from the inputs received so far. In order to succeed, it must have encoded the previous inputs into a suitable representation to make a prediction about the next input. For the record, Palm's predictive encoder is similar to the "conditional restricted Boltzmann machine" designed by Geoffrey Hinton's student Graham Taylor ("Two Distributed-State Models For Generating High-Dimensional Time Series", 2011).

Rakesh Chalasani and Jose Principe at University of Florida implemented the ideas of Friston and Rao-Ballard ("Deep Predictive Coding Networks", 2013).

In the case of video analysis, a neural network trained to predict the next frame in a video is implicitly learning an efficient representation of the world depicted in that video: the objects and the structure of the scene. This was the strategy followed by Vincent Michalski at the Goethe University in Germany in designing his multi-layer neural network ("Modeling Deep Temporal Dependencies with Recurrent Grammar Cells", 2014), by Marc'Aurelio Ranzato (now at Facebook) for a recurrent neural network that also borrowed ideas from Mikolov's language model of 2010 ("Video Language Modeling", 2014), by Bill Lotter at Harvard University for his convolutional LSTM network PredNet ("Unsupervised Learning of Visual Structure Using Predictive Generative Networks", 2015), and by Ruslan Salakhutdinov's student Nitish Srivastava at University of Toronto for his coupled LSTMs, an encoder that is trained with the initial frames to build the representation and a decoder that predicts the next frame based on that representation ("Unsupervised Learning Of Video Representations Using LSTMs", 2015).

These systems had learned to predict pixels. Antonio Torralba's student Carl Vondrick at MIT ("Anticipating Visual Representations from Unlabeled Video", 2016) built a system (a variant of AlexNet with three more fully connected layers) to predict not just future pixels but the future of complex concepts such as objects and actions, and without learning a visual representation.

Xiaolong Wang at CMU trained a convolutional neural network with hundreds of thousands of unlabeled videos to learn visual representations via visual tracking. Visual tracking (following the object while the video is rolling) provides the equivalent of "supervision" ("Unsupervised Learning of Visual Representations using Videos", 2015).

Jitendra Malik's student Pulkit Agrawal used the information obtained by a moving camera, the camera of a self-driving car ("Learning to See by Moving", 2015). His KittiNet coupled pre-training (training a convolutional neural network on a pretext task that is not the target one) and fine-tuning (adapting the network to the real task).

Alexei Efros' student Carl Doersch at UC Berkeley, instead of temporal correlations, used spatial correlations as the training surrogate: he designed a convolutional network to predict the position of the patch of an image relative to another one, with the pairs of patches picked at random, a pair per image. As the network learned this task, it started discovering categories such as "cat" and "bird" ("Unsupervised Visual Representation Learning by Context Prediction", 2016).

Summarizing, the goal of representation learning is to build internal representations of the world that can later be used for machine-learning tasks. "Self-supervised learning" is a smart way to achieve representation learning. "Self-supervised learning" is a particular kind of unsupervised learning in which the network uses information implicit in the environment to train itself. Doersch uses the relative spatial co-location of patches in images, Wang uses object correspondence obtained through tracking in videos, and Agrawal uses information obtained by a moving camera. In all of these cases the representation learned through "self-supervised learning" can be used for applications of object identification and object classification.

Mehdi Noroozi and Paolo Favaro at University of Bern in Switzerland pre-trained a convolutional network called Context-Free Network (a variation of AlexNet) to solve jigsaw puzzles. Then the same network was used to classify and detect objects ("Unsupervised Learning of Visual Representations by Solving Jigsaw Puzzles", 2017). They used Agrawal's model: pre-training (in this case train to solve jigsaw puzzles) and fine-tuning (adapting the network to the classification or detection task). It turns out that during the pre-training, the network learned something about the structure of objects, knowledge that is useful in those other tasks. In this case the self-supervision used information that is available within a single image: one can fragment any image into arbitrary tiles and turn it into a jigsaw puzzle.

Phillip Isola at UC Berkeley trained his deep neural network with the rate of co-occurrence in space and time of objects (how often they are found in the same picture or video frame) so that the network can then predict whether two objects are likely to be found next to each other in space or time ("Learning Visual Groups from Co-occurrences in Space and Time", 2016). A practical application was to group photographs by theme (ocean views, mountain landscapes, sunsets, etc).

Tinghui Zhou, working in Noah Snavely's team at Google, designed an (unsupervised) CCN for predicting the three-dimensional structure of a scene given a sequence of images ("Unsupervised Learning of Depth and Ego-Motion from Video", 2017). Noah Snavely had previously helped John Flynn of Zoox build DeepStereo, a network trained end-to-end with a large number of images taken from different viewpoints for the purpose of synthesizing a new view of a scene ("Learning to Predict New Views from the World's Imagery", 2016).

Another unsupervised strategy to learn video representations consisted in training a CNN to verify that a sequence of frames corresponds to the correct order in a video. This was done by both Ishan Misra at CMU ("Shuffle and Learn - Unsupervised Learning Using Temporal Order Verification", 2016) and Hsin-Ying Lee at UC Merced ("Unsupervised Representation Learning by Sorting Sequences", 2017).

Deepak Pathak at UC Berkeley used motion cues (again, the Gestalt principle of "common fate", that pixels that move together tend to belong together) to build simple pixel representations of objects found in the frames of a video, and then trained a CNN to predict these representations without having access to the motion cues: the network learned a higher-level representation that was then transferred to other recognition tasks ("Learning Features by Watching Objects Move", 2017).

There are cues in the environment that help us make sense of what is happening, and self-supervised networks can exploit those cues too (although it is not easy for us to realize which cues we ourselves use!) Chuang Gan at MIT thought of using geometry as auxiliary supervision for the self-supervised learning of video representations, and found that the CNN pre-trained by the geometry cues can indeed understand more of what goes on in a video ("Geometry Guided Convolutional Neural Networks for Self-Supervised Video Representation Learning", 2018). Alas, the most accurate method for dynamic scene recognition remained the "shallow" method used by Christoph Feichtenhofer at Graz University in Austria ("Spacetime Forests with Complementary Features for Dynamic Scene Recognition", 2013).

Carl Vondrick, now at Google, built a self-supervised CNN that, while colorizing grayscale videos, learned to visually track objects in the videos without ever being trained explicitly for tracking, and in fact learned to track multiple objects ("Tracking Emerges by Colorizing Videos", 2018).

Andrew Zisserman's students Olivia Wiles and Sophia Koepke at University of Oxford built two self-supervised frameworks/architectures, both trained using a large collection of video data with no manually labelled annotations. X2Face was a self-supervised framework for face puppeteering, i.e. giving a face the pose and expression of another face ("Self-supervised Learning from Watching Faces", 2018). Facial Attributes-Net (FAb-Net) was a self-supervised framework for learning a facial attribute representation that encodes information about pose and expression ("Self-supervised Learning of a Facial Attribute Embedding from Video", 2018).

Alexei Efros' student Andrew Owens at UC Berkeley learned a multisensory representation of a video by training a neural network to predict whether video and audio are aligned; and the resulting network proved was then trained to visualize the location of a sound in a video frame or to recognize the action going on ("Audio-Visual Scene Analysis with Self-Supervised Multisensory Features", 2018).

There was a general strategy at work: the use of a "proxy task" to force the network to learn a higher-level visual representation that could then be used for pre-training the network for other visual tasks. Doersch's cropped patches, Wang's visual tracking, Agrawal's KittiNet, Mehdi Noroozi's jigsaw puzzles, Deepak Pathak's segmentations, Misra's sequential verification, Lee's sorting sequences, Noroozi's shuffled patches, and Owens' video-sound alignment were all examples of proxy cases that pre-trained a neural network in an unsupervised manner. Indirectly, they all made the network generate a representation useful for other tasks.

Mehdi Noroozi and Paolo Favaro at University of Bern later decoupled the self-supervised model from the task-specific model, thus obtaining a more efficient architecture that shrank the gap between the accuracy of models trained via self-supervised learning and models trained via supervised learning ("Boosting Self-Supervised Learning via Knowledge Transfer", 2018).

Most of these "proxy tasks" were still implemented on two-dimensional CNNs. By definition, such networks cannot properly capture a spatio-temporal representation. Dahun Kim at KAIST in Korea used a three-dimensional self-supervised task to train three-dimensional CNNs using a video dataset ("Self-Supervised Video Representation Learning with Space-Time Cubic Puzzles", 2018). Based on 3D ResNet-18, this system beat all the previous self-supervised systems on popular benchmarks by a significant margin.

All of these systems used convolutional nets and/or LSTMs. Filip Piekniewski at UC San Diego started from the realization that Fukushima's neocognitron (the template for all convolutional networks) was only a timid approximation of the structure of the visual system. His Predictive Vision Model (PVM) was inspired by more up-to-date neuroscience and by Jeff Hawkins' hierarchical temporal memory ("Unsupervised Learning from Continuous Video in a Scalable Predictive Recurrent Network", 2016). A hierarchical model of the brain's visual system, originally proposed by Sven Behnke and Raul Rojas at the Free University of Berlin ("Neural Abstraction Pyramid", 1998), the "neural abstraction pyramid", assumed that the visual cortex relies on both horizontal (lateral) and vertical (feedback and feedforward) loops as it transforms an image into a sequence of representations with increasing levels of abstraction and decreasing levels of detail. Similarly, a decade later, Rodney Douglas and Kevan Martin at the Institute of Neuroinformatics in Switzerland found a lot of feedback connectivity in the neocortex, but also found that the connectivity tends to be local (neurons tend to talk to neighbouring neurons in the same region of the cortex whereas long-distance connections are rare), i.e. that "the local circuit is the heart of cortical computation" ("Recurrent Neuronal Circuits in the Neocortex", 2007). PVM was therefore based on ubiquitous feedback connectivity, unlike deep learning that mostly relies on feedforward connections: a hierarchy of heavily connected units (with both horizontal and vertical feedback), each a multilayer perceptron. Like Wang's system, PVM too learned (unsupervised) from tracking the movement of objects relative to the observer. Additionally, it tried to construct a representation of the physical reality around the object. Deep learning was based on end-to-end error propagation, whereas PVM was mainly interested in local prediction.

The video prediction model designed by Sergey Levine's student Chelsea Finn at UC Berkeley with help from Ian Goodfellow, called "convolutional dynamic neural advection" (CDNA), using a stack of LSTMs, explicitly predicted the motion of the objects encountered in a video, beyond single-frame prediction ("Unsupervised Learning for Physical Interaction Through Video Prediction", 2016). This system was meant to do more than learn: it was meant to "imagine" possible futures based on different courses of action.

Tracking objects in video is a fundamental problem in computer vision. It is essential to interacting with objects and, generally speaking, living a normal life in our ordinary world. Predicting what will happen next to the object that we are tracking is equally important, and natural for the human mind.

Robots need to understand a lot of things about the world. Today they understand less than what a worm understands. They need to understand: the effect of forces on objects (what happens if you push an object beyond the edge of a table?); the effect of movement of objects (where can the car go from where it is now?); and the effect of people's movement (what happens next when a person pulls out a gun?) These are all cases of "prediction". Animal brains are amazingly good at predicting (simulating) the future. Machines are still incredibly bad at it.

AlphaGo and AlphaZero were "model-free", meaning that they knew nothing about the world (actually AlphaGo knew what humans do in the world of weiqi). It sounds impressive that they learned to play a game without much or any knowledge, but most real-world tasks require planning in unpredictable circumstances, where knowledge of the world is essential. If one wants to build a robot to operate in the real world, either its actions will be hard-coded (only those actions can be performed and only in specific circumstances) or it will need a model of the world to decide which actions make sense. Unfortunately, model-based algorithms lagged behind model-free algorithms of reinforcement learning. Hybrids using deep reinforcement learning and some kind of world model had already appeared. Joschka Boedecker's group at the University of Freiburg introduced a method for model learning called Embed to Control E2C (2015); and the video prediction models of Jitendra Malik's student Pulkit Agrawal at UC Berkeley ("Learning to Poke by Poking", 2016) and Sergey Levine's student Chelsea Finn also at UC Berkeley ("Deep Visual Foresight for Planning Robot Motion", 2017) were applied to robotics. Humans can learn to play games in minutes, whereas A.I. algorithms such as AlphaZero need a lot of time. A Google team including Finn and Levine applied the method of video prediction models to learn models of Atari games in a system called SimPLe that dramatically lowered the number of iterations required to learn the games ("Model Based Reinforcement Learning for Atari", 2019).

Sergey Levine's student Kurtland Chua rediscovered the "cross-entropy method" (CEM) proposed two decades earlier by Reuven Rubinstein at the Technion in Israel ("Optimization of Computer Simulation Models with Rare Events", 1997) for his Probabilistic Ensembles with Trajectory Sampling or PETS that plans with learned models ("Deep Reinforcement Learning in a Handful of Trials Using Probabilistic Dynamics Models", 2018).

The deep reinforcement learning algorithms of DQN and AlphaGo were successfully applied to problems such as Atari games and the game of weiqi that have a finite number of discrete actions. Robots, however need to perform continuous actions. The traditional way to adapt a discrete algorithm to continuous action is to "discretize" the space of possible actions, but of course this yields approximations that a far cry from the performance of a human worker (or even of a trained animal). Timothy Lillicrap's group at DeepMind developed first the Deep Deterministic

Policy Gradient (DDPG) algorithm ("Continuous Control with Deep Reinforcement Learning", 2015), an extension of David Silver's Deterministic Policy Gradient ("Deterministic Policy Gradient Algorithms", 2014) that replaced Sutton's stochastic approach to policy with a deterministic approach; and then the Distributed Distributional Deep Deterministic Policy Gradient (D4PG) algorithm, which adapted the "distributional" approach to reinforcement learning developed by his colleagues Marc Bellemare, Will Dabney, and Remi Munos at DeepMind ("A Distributional Perspective on Reinforcement Learning", 2017) to the case of continuous action ("Distributed Distributional Deterministic Policy Gradients", 2018). Deep Planning Network (PlaNet), developed by Danijar Hafner of the University of Toronto, Timothy Lillicrap of DeepMind and others, was a purely model-based agent that learned the environment dynamics from images and chose actions through fast online planning in latent space ("Learning Latent Dynamics for Planning from Pixels", 2019). PlaNet learned much faster than D4PG by combining the deterministic model of LSTM and GRU methods (that remembers information over many time steps) with the stochastic model of Gaussian methods (that can predict multiple futures).

Teaser: Can Intelligence Arise from Silicon?

The current A.I. (neural networks and deep learning) is more biological than the one of the past, but barely so. We are still a long way from understanding the brain. Today's sophisticated neural network architectures are crude simulations of the brain. Just think of the number of neurons and neurotransmitters, and the simple fact that no two neurons are identical. The complexity of the brain is just infinitely superior to the complexity of the neural networks designed by software engineers. One has to wonder whether the whole effort makes sense. There is a complexity to biological matter that (maybe) just cannot be replicated in silicon or any other nonbiological substance. Biological matter naturally grows billions of neurons that are all different from each other and that are connected by synapses that are all different from each other via an infinite number of spontaneous mutations.

A study by a consortium called the Brain Somatic Mosaicism Network discovered that individual neurons differ from each other even in terms of DNA: each neuron has a slightly different DNA ("Intersection of Diverse Neuronal Genomes and Neuropsychiatric Disease", 2017).

Furthermore, a neuron in my brain is not the same thing as a neuron in your brain. A neurotransmitter in my brain is not exactly the same neurotransmitter in your brain. While one can argue that the "function" is the same, one cannot argue that they are "the same thing". Use a microscope and you'll see that they are all slightly different "things".

Biological matter is wildly chaotic, just like no two branches are identical (no matter how many trees you examine) and no two fingerprints are the same. Biological nature abhors mass production of copies. Our simulations, instead, are built of identical zeroes and ones, which are used to create identical (artificial) neurons which communicate using identical messages. It is virtually impossible to replicate the wild chaos of biological forms without using biological matter.

Learning from the structure of the brain (the neural networks) can certainly lead us to build useful machines, but maybe "useful" and not "intelligent" is all that we can get out of this program. Claiming that in any way these machines are "brains" is like claiming that walking sticks are legs.

The human brain is not only complex: each of its billions of units is completely different from the others. While this phenomenon is very common in nature (no two rocks are identical, no two clouds are identical, no two sticks are identical), it is the antithesis of human artifacts, which are instead built out of identical building blocks.

Footnote: Phase Transitions and A.I.

An exponential increase (especially if it is only about computational speed) is not enough to demonstrate that a qualitative change will ever take place. If you build exponentially faster cars, you will eventually get a car that can travel close to the speed of light, but it would still be a car, not an elephant.

Should machines ever reach a superhuman level of intelligence, one would imagine that this would entail a sequence of phase transitions, e.g. from mere arithmetic calculation to pattern recognition to higher and higher forms of mental faculties.

The most famous model of cognitive development in children is the one articulated by the Swiss psychologist Jean Piaget in "The Language and Thought of the Child" (1923). He posited that the child's mind undergoes what a physicist would call "phase transitions" after which the mind thinks differently. Cognitive faculties are not fixed at birth but evolve during the lifetime of the individual, and the evolution is not smooth, it requires quantum jumps in cognitive skills. Precisely, the development of children's intellect proceeds from simple mental arrangements to progressively more complex ones, not by gradual evolution, but by sudden rearrangements of mental operations that produce qualitatively new forms of thought. First a child lives a "literal" sensorymotor life, then the child begins to deal with internal symbols, then the child learns to perform internal manipulations on symbols that represent real objects, and, finally, the child's mental life extends to abstract objects. Piaget's four transition phases start with a stage in which the dominant factor is perception, which is irreversible, and end with a stage in which the dominant factor is thought, which is reversible.

At the end of the century, the Canadian neuropsychologist Merlin Donald reached the conclusion in his book "Origins Of The Modern Mind" (1991) that a similar path can be seen in the growth of the human mind through history: the human mind developed in four stages (which roughly correspond to Piaget's stages of cognitive growth in children) from a non-symbolic form of intelligence to the modern mind of symbolic thought through gradual absorption of new representational systems.

Neural networks belong to the class of complex systems, which are characterized by nonlinear dynamics.

In 1978 Jack Cowan (who was now at University of Chicago) showed that neural networks have a mathematical structure that closely resembles that of quantum field theory (unpublished until "Stochastic Neurodynamics", 1991). Neurons can be in three, not two, states: quiescent, stimulated, and refractory (a period during which the neuron would not react to a new stimulation). Cowan even showed that the neurodynamics of a three-state neuron is described by mathematical matrices that are similar to the ones used in 1964 by the physicist Murray Gell-Mann to describe quarks. The algebra of state transitions in a neural network is eerily reminiscent of the algebra of quantum chromodynamics. Cowan, working with the mathematician Bard Ermentrout of University of Pittsburgh, also applied bifurcation theory to the analysis of neural field equations ("Large Scale Spatially Organized Activity in Neural Nets", 1978, but published only in 1980).

Bernardo Huberman and Tad Hogg at Xerox PARC studied the phase transitions that take place in large-scale cognitive systems ("Phase Transitions in Artificial Intelligence Systems", 1987), but there was generally little interest in studying phase transitions in neural networks. Elizabeth Gardner at University of Edinburgh, not coincidentally an expert in spin glass theory, applied statistical mechanics to neural networks but she died of cancer a few weeks before her two papers were published (notably "The Phase Space of Interactions in Neural Networks Models", 1988). Michael Biehl at the Institute for Theoretical Physics of Wuerzburg in Germany ("Statistical Mechanics of Unsupervised Structure Recognition", 1994) has studied, in general, phase transitions in networks (whether the World-wide Web, ecological nets, social nets, cellular nets, linguistic nets or neural nets).

Both physics and neural networks study systems with many degrees of freedom: physics studies many-body interactions, neural networks process data in high dimensions. Physics uses a trick called "renormalization" to handle complex systems with many degrees of freedom. Neural networks use the approximation tricks of deep learning. The connection between the two fields has been mainly explored by physicists such as Pankaj Mehta of Boston University and David Schwab of Northwestern University ("An Exact Mapping between the Variational Renormalization Group and Deep Learning", 2014).

"You must be the change you wish to see in the world" (Mahatma Gandhi).

Jobs in the Age of the Robot – Part 1: What Destroys Jobs

During the Great Recession that ravaged the Western world in 2008-2011, both analysts and ordinary families were looking for culprits to blame for the high rate of unemployment, and automation became a popular one in the developed world. Automation was indeed responsible for making many jobs obsolete, but it was not the only culprit nor the main one.

The first and major factor that accounts for the demise of many jobs in the Western world is the end of the Cold War. Before 1991 the economies that really mattered were a handful (USA, Japan, Western Europe). Since 1991 the number of competitors for the industrialized countries has skyrocketed, and they are becoming better and better at competing with the West. Technology might have "stolen" some jobs, but that factor pales by comparison with the millions of jobs that were exported to Asia. In fact, if one considers the totality of the world, an incredible number of jobs have been created precisely during the period in which critics argue that millions of jobs have been lost to automation. If Kansas loses one thousand jobs but California creates two thousand, we consider it an increase in employment. These critics make the mistake of using the old nation-based logic for the globalized world. When counting jobs lost or created during the last twenty years, one needs to consider the entire interconnected economic system that spreads all over the planet. Talking about the employment data for the USA but saying nothing about the employment data (over the same period) of China, India, Mexico and so forth is distorting the picture. If General Motors lays off one thousand employees in Michigan but hires two thousand in China, it is not correct to simply conclude that "one thousand jobs have been lost". If the car industry in the USA loses ten thousand jobs but the car industry in China gains twenty thousand, it is not correct to simply conclude that ten thousand jobs have been lost by the car industry. In these cases jobs have actually been created.

That was precisely the case: millions of jobs were created by the USA in the rest of the world while millions were lost at home. The big driver was not automation but, cheap labor.

Then there are sociopolitical factors. Unemployment is high in Western Europe, especially among young people, not because of technology but because of rigid labor laws and government debt. A company that cannot lay off workers is reluctant to hire any. A government that is indebted cannot pump money into the economy. This is a widespread problem in the Western economies of our age. It has to do with politics, not with automation.

Germany is as technologically advanced as the USA. All sorts of jobs have been fully automated. And, still, in Germany the average hourly pay has risen five times faster between 1985 and 2012 than in the USA. This has little to do with automation: it has to do with the laws of the country. Hedrick Smith's "Who Stole the American Dream?" (2012) lays the blame on many factors, but not on automation.

In 1953 Taiichi Ohno invented "lean manufacturing" at Japan's Toyota, possibly the most important revolution in manufacturing since Ford's assembly line. Nonetheless, Japan created millions of jobs in manufacturing; and, in fact, Toyota went on to become the largest employer in the world of car-manufacturing jobs. Even throughout its two "lost decades" (1991-2010) Japan continued to post very low unemployment. Japan has perhaps the highest number of industrial robots of any country, and it also enjoys one of the lowest unemployment rates in the world. Germany is a close second in automation, and it has the lowest unemployment figures for any major country in Western Europe.

Another major factor that accounts for massive losses of jobs in the developed world is the management science that emerged in the 1920s in the USA. That science is the main reason that today companies don't need as many employees as comparable companies employed a century ago. Each generation of companies has been "slimmer" than the previous generation. As those management techniques get codified and applied across all departments, companies become more efficient at manufacturing (world-wide), at selling (using the most efficient channels) and at predicting business cycles. All of this results in fewer employees not because of automation but because of optimization.

In May 2016 Challenger, Gray & Christmas estimated the companies that had laid off the most workers. The top job cutter of the first four months of 2016 was National Oilwell Varco, a Texan company making equipment for the petroleum industry. Job cutters #3 (Schlumberger), #5 (Halliburton), #7 (Chevron) and #10 (Weatherford) were all involved in the petroleum business. This had nothing to do with robots or artificial intelligence but simply with record-low oil prices. Walmart was the second job-cutter in the country, but, like all retail chains, its problem was simply the competition from online sales. Meanwhile, the US economy was adding about 200,000 new jobs each month, and those jobs were consistently in high-tech sectors. Intel (#4) and Dell (#6) too were in that list. Both missed the mobile revolution and were being replaced by other firms. Their job cutting was not due to more automation in the factories.

Additionally, in the new century the USA has deliberately restricted immigration to the point that thousands of brains are sent back to their home countries even after they graduated in the USA. This is a number that is virtually impossible to estimate, but, in a free market like the USA that encourages innovation and startups, jobs are mostly created via innovation, and innovation comes from the best brains, which account for a tiny percentage of the population. Whenever the USA sends back or refuses to accept a foreign brain that may become one of those creators of innovation, the USA is de facto erasing thousands of future jobs. Those brains are trapped in places where the system does not encourage the startup-kind of innovation or where capital is not as readily available. They are wasted in a way that equivalent brains were not wasted in the days when immigration into the USA was much easier, up until the generation of Yahoo, eBay and Google. According to a study contained in the "Kauffman Thoughtbook 2009" by the Kauffman Foundation, foreign-born entrepreneurs ran 24% of the technology businesses started between 1980 and 1998 (in Silicon Valley a staggering 52%). In 2005 these companies generated $52 billion in revenue and employed 450,000 workers. In 2011 a report from the Partnership for a New American Economy found that 18% of the Fortune 500 companies of 2010 were founded by immigrants. These companies had combined revenues of $1.7 trillion and employed millions of workers. If one includes the Fortune 500 companies founded by children of immigrants, the combined revenues were $4.2 trillion in 2010, greater than the GDP of any other country in the world except China and Japan.

Technology is certainly a factor, but it can go either way. Take, for example, energy. This is the age of energy. Energy has always been important for economic activity but never like in this century. The cost and availability of energy are major factors to determine growth rates and therefore employment. The higher the cost of energy, the lower the amount of goods that can be produced, the lower the number of people that we employ. If forecasts by international agencies are correct, the coming energy boom in the USA (see the International Energy Agency's "World Energy Outlook" of 2012) will create millions of jobs, both directly and indirectly. That energy boom is due to new technology.

When the digital communication and automation technologies first became widespread, it was widely forecast a) that people would start working from home and b) that people would not need to work as much. What i have witnessed is the exact opposite: virtually every company in Silicon Valley requires people to show up at work a lot more than they did in the 1980s, and today virtually everybody is "plugged in" all the time. I have friends who check their email and text messages all the time while we are driving to the mountains and even while we are hiking. The digital communication and automation technologies have not resulted in machines replacing these engineers but in these engineers being able to work all the time from everywhere, and sometimes their companies require it. Those technologies have resulted in people working a lot more. (The willingness of people to work more hours for free is another rarely mentioned factor that is contributing to higher unemployment).

At the end of 1994 Netscape released the first major browser for the World-wide Web and within a few years the web had become a center of e-commerce. Since then there has been a steady exodus of jobs from the "brick-and-mortar" economy. Jobs were lost in post offices, bookstores, photo shops, stationery shops, travel agencies, newspapers/magazines, record stores, music labels... you name it. It was a slaughter. I have never seen a total number but it certainly affected tens of millions of workers, as the entire economy was redesigned in just about 20 years. And, still, 23 years later (and despite two massive economic recessions) the unemployment rate in the USA stands at 4.4%, lower than it was in 1994 (6.1%) I strongly doubt that Artificial Intelligence will cause as many job losses as the dotcom revolution did, especially if its impact is felt mainly among white-collar workers, who are a much smaller group than the groups replaced by the dotcoms.

It is misleading to say that Artificial Intelligence is automating or will automate "white-collar" jobs. In reality, Artificial Intelligence is doing very little, and certainly very little that qualifies as "intelligent". The change is happening in the jobs themselves: they are becoming more and more structured. They are becoming routine. The expression "white collar" evokes the image of someone who is thinking, but increasingly these white-collar workers are performing trivial tasks that simply require following some rules. The insurance agent or the bank loan agent are typical examples. These used to be professions that required a lot of skills, a lot of intuition and a lot of knowledge. Now they are performed by people with little or no education who simply enter your data into a computer and press a button to get an answer. Any job that becomes routine can be easily automated. You don't need to call it "artificial intelligence": "automation" suffices. White-collar jobs can be automated just like blue-collar jobs when they become as repetitive and unimaginative as blue-collar jobs. What has improved is not the technology, but the standardization of white-collar work.

India is a particularly vulnerable economy because so many of its I.T. jobs (white-collar jobs) are about routine work that can be easily automated. It hasn't been automated before mainly because it was cheaper to offshore the routine to Indians than to machines. As software gets cheaper and better, and Indian workers demand higher salaries, those jobs will become very vulnerable.

According to a 2017 survey by the European Commission, 80% of Swedes are optimistic about robots and Artificial Intelligence, whereas, according to a survey by the Pew Research Center, 72% of US citizens are "worried" about robots. The different psychological reactions are probably due more to the health care system than anything else. If a robot steals your job in Sweden, the state will take of you, including your health insurance. You are neither going to die nor going bankrupt to pay hospital bills. In fact, education is mostly free: your children will still have the same chance to attend a top university, and you will have a chance to attend a training program while enjoying generous unemployment benefits. In the USA the situation is the exact opposite: most workers depend on employers for health insurance, and sending their children to school is extremely expensive, and therefore these workers can be legitimately alarmed that a robot may take their job. The socialist regime of Sweden is making people more willing to experiment with their jobs. The socially backward regime of the USA makes people extremely conservative about their jobs: voters don't demand jobs, they demand their existing jobs (forever) even if it's coal mining (one of the most terrible jobs on the planet).

The fear of automation dates from the dawn of machines. In 1921 the New York Times featured a book review titled "Man Devoured by His Machines". In 1961 Time magazine ran a story titled "The Automation Jobless" (24 February). In 1980 the New York Times carried a story titled "A Robot is After Your Job" (3 September). In 2009 Mike Konczal wrote in The Atlantic the piece "Robots and the Future of Unemployment".

One of the least prescient books ever written on jobs is Jeremy Rifkin's "The End of Work" (1995) that predicted worldwide unemployment due to the automation of jobs in the manufacturing, agricultural and service sectors. Twenty years later millions of jobs have been created, and high unemployment is a chronic problem only in countries with relatively little automation like Greece.

A 2017 study by Robert Atkinson and John Wu of the Information Technology and Innovation Foundation titled "False Alarmism: Technological Disruption and the U.S. Labor Market, 1850-2015" (that examined 165 years of labor history in the USA) found that technology is disrupting fewer (not more) jobs than usual. In 2017 Greg Ip wrote a piece in the Wall Street Journal titled "Robots Aren't Destroying Enough Jobs" that argues (quote): "Economic predictions of massive job losses to automation are missing indicators that show just the opposite". JP Morgan's report "Big Data and A.I. Strategies" of May 17 outlines the armies of people that will be needed by corporations to acquire, sort out and label data, and, in particular, lists more than 40 programming languages that the data center of the near future will need to master.

All the predictions of a job apocalypse have been proven wrong over and over again, so much so that in 2015 the MIT economist David Autor felt he had to write an article titled "Why Are There Still So Many Jobs?"

There is something terribly wrong, but it is not the machine per se. Karl Marx already realized this 150 years ago. The Luddites were British workers who protested against the machines that were stealing their jobs. The rebellion started in 1811 and lasted until 1817, when six of their leaders were hanged. The Luddites were terrified by the power-loom, a machine that, like the computer two centuries later, could be programmed to weave any pattern through the use of punched cards. Karl Marx, who was born the year after the Luddite rebellion ended, criticized the Luddites in his book "Capital" (1867) because they fought the machines instead of fighting the way in which society (namely the capitalist) was using them.

Jobs in the Age of the Robot – Part 2: What Creates Jobs

Unemployment cannot be explained simply by looking at the effects of technology. Technology is one of many factors and, so far, not the main one. There have been periods of rapid technological progress that have actually resulted in very low unemployment (i.e. lots of jobs). This happened most recently in the 1990s when e-commerce was introduced and job maket expanded despite the fact that the digital camera had killed the photographer's shop, Amazon the bookstore, the mobile phone the land lines and Craigslist the local newspaper.

The effect of a new technology on employment is not always obvious, and that's why our first reaction is of fear. For example, who could have imagined that computers (invented for fast computation) would go on to create millions of new jobs in the sector of telecommunications?

A 2014 report by the Kauffman Foundation showed that between 1988 and 2011 almost all new jobs were created by businesses less than five years old, while existing firms were net job destroyers, losing 1 million jobs net combined per year. By contrast, in their first year, new firms add an average of 3 million jobs."

New technologies also create jobs in other sectors. It is called the "multiplier effect". The people employed in the new technology need shops, restaurants, doctors, lawyers, schoolteachers, etc. A 2016 report by the Bay Area Council Economic Institute shows that the biggest multiplier effect of our times came from the high-tech industry: for each job created in the high-tech sector, more than four jobs are created in other sectors. A company that hires a software engineer is indirectly creating 4 new jobs in the community. By comparison, traditional manufacturing has a multiplier effect of 1.4 jobs.

Historically, in fact, technology created jobs while simultaneously destroying old jobs, and the new jobs have typically been better-paying and safer than the old ones. Not many people dream of returning to the old days when agriculture was fully manual and millions of people were working in terrible conditions in the fields. Today a few machines can water, seed, rototill and reap a large field. Those jobs don't exist anymore, but many jobs have been created in manufacturing sectors for designing and building those machines. In the USA of the 19th century, 80% of jobs were in agriculture; today only about 2% are. Yet it is not true that the mechanization of agriculture had caused 78% of people to remain unemployed. Few peasants in the world would like their children to grow up to be peasants instead of mechanical engineers. Ditto for computers that replaced typewriters and typewriters that replaced pens and pens that replaced human memory.

Gutenberg's printing press put a few thousand scribes out of business, but it launched a mass production of books, which, besides educating the public and creating an infinite number of new businesses for educated people (e.g. magazines and newspapers), created millions of jobs to print books, market them and sell them. For each scribe that went out of business, thousands of bookstores popped up all over the world.

For at least two thousand years, the painters of the world tried to get better and better at portraying people, objects and landscape. Did the invention of the camera in the mid-19th century make painters obsolete? As far as duplicating reality, it did. But painters found something else to paint: the imaginary and irrational landscapes of surrealism, the excessive and brutal forms of expressionism, the geometric deconstructions of cubism, and so forth. Coincidence or not, the decades after the arrival of photography witnessed a Cambrian-like explosion of painting styles. In fact, the first exhibition of the impressionists, in 1874, took place in the studio of the most famous photographer of Paris, Felix Nadar. At the same time, photography created thousands, and eventually millions, of new jobs: camera manufacturers, camera repair shops, photographers, manufacturers of camera film, printers of photographs, etc.

Steam engines certainly hurt the horse and mule business, but created millions of jobs in factories and thousands of new businesses for the goods that could be used or made in those factories.

All of these forms of automation had side effects that were negative, but one negative side-effect that they did NOT have was to cause unemployment. They created more jobs than they destroyed, and better ones.

In the 1980s i worked in Silicon Valley as a software engineer and back then the general consensus was that software engineering was being automated and simplified at such a pace that soon it would become a low-paid job and mostly exported to low-wage countries like India. Millions of software jobs have in fact been "offsourced" to India, but the number of software developers in the USA has skyrocketed to 1,114,000 with a growth rate of 17% and an average salary of $100,000, which is more than twice the average salary of $ 43,643 (source: US Bureau of Labor Statistics, 2015).

It is true that the largest companies of the 21st century are much smaller than the largest companies of the 20th century. However, the world's 4,000 largest companies spend more than 50% of their revenues on their suppliers and a much smaller percentage on their people (as little as 12% according to some studies). Apple may not be as big as IBM was when it was comparable in power, but Apple is the reason that hundreds of thousands of people have jobs in companies that make the parts used in Apple's products.

I would be much more worried about the "gift economy": the fact that millions of people are so eager to contribute content and services for free on the Internet. For example, the reason that journalists are losing their jobs has little to do with the automation in their departments and a lot to do with the millions of amateur "bloggers" who provide content for free on the Internet.

If we take into account the global effects of automation, we reach different conclusions about the impact that robots (automation in general) will have. In the USA robots are likely to bring back jobs. The whole point of exporting jobs to Asia was to benefit from the lower wages of Asian countries; but a robot that works for free 24 hours a day 7 days a week beats even the exploited workers of communist China. As they become more affordable, these "robots" (automation in general) will displace Asian workers, not Michigan workers. The short-term impact will be to make outsourcing of manufacturing an obsolete concept. The large corporations that shifted thousands of jobs to Asia will bring them back to the USA. In the mid-term this could even have the secondary effect of putting Asian products out of the market and of creating a manufacturing boom in the USA: not only old jobs will come back but a lot of new jobs will be created. In the long term robots might create new kinds of jobs that today we cannot even foresee.

Let's take a simple example, that of a kind of robot that will appear soon at the supermarket of your neighborhood. As you enter the store, you will be welcomed by a mobile robot equipped with a basket and a check-out system. The robot will ask you what you are looking for, and escort you to the correct aisle of the store. It will let you browse the shelf and pick the brand you prefer. Then it will ask you to drop it in the basket. This will continue for as long as you have items to purchase. In fact, you could even read your shopping list to the robot and the robot will calculate and work out your passage through the store, optimally. For any product that you don't find the robot will investigate on the spot whether it can be ordered for you and delivered at your home address, or whether there is an affiliated store nearby where you can find it. The robot will also alert you to the products that are on sale and politely ask you if you'd like to take a look at them. When you are done, you will simply ask the robot "How much?" The robot will already know the total because it will have scanned each item that you dropped in its basket. The robot will take your payment (whether a credit card or a smartphone app) and print a receipt while escorting you back to your car. A robot like this is perfectly feasible today, except that it would still cost too much to build and operate, a cost not justified in stores that have a relatively low margin of profit on the goods that they sell. Nonetheless, who should panic at the prospect that such robots will someday exist? Which jobs are at risk? There is no human being who performs this task today. When you enter a store, you are on your own. If you have a question, good luck finding any employee who can help you. In many places the check-out operation is already a self-checkout. Hence, not a single job will be lost to these robots. On the other hand, imagine how many jobs will be created. Some company will become the Apple of shopping robots, and hire thousands of people to design, manufacture, sell and maintain these robots. Another company, the Google of robots, will come up with the idea of making home robots for shopping, robots that you keep in the house and drive with to the store, and which knows the organization of each store connected to the cloud according to some Android-like standard. While you are walking into the store with your robot, your robot downloads the configuration of the store and the entire database of products, and then it starts behaving exactly as if it were that store's shopping robot. More jobs created here. Another company, the Oracle of robots, will come up with software that informs your robot about the best place for each item on a given shopping list. You won't even know where you are going. The robot will take you to a selection of stores that have what you need at the best prices. More jobs created. Then the Tesla of robots will come up with a way that you can 3D-print your custom shopping robot, as big or small as you want it, as environmental as you want it, as fast or slow as you want it. None of these jobs exist today. And almost no existing job is killed in this simple example.

Not many people in 1946 realized that millions of software engineers would be required by the computer industry in 2013. Robotics will require millions of robotics engineers. These engineers will not be as "smart" as their robots at whatever task for which those robots were designed just like today's software engineers are not as fast as the programs they create. As i type, Silicon Valley is paying astronomical salaries to robotics engineers and China is hiring thousands of people for the Internet of Things.

At the end of 2015 both the McKinsey report "Four fundamentals of workplace automation" and the study by James Bessen of Boston University's School of Law "How Computer Automation Affects Occupations" (November 2015) showed that robots will steal your job but will also create another job, and most likely it will be a better one in terms of income, health and personal satisfaction.

Bessen mentions the case of the ATM, one of the most successful programs to replace humans with machines. The routine jobs of bank tellers were very easy to automate, and, sure enough, the percentage of tellers in the USA fell from 20 per branch in 1988 to 13 in 2004. However, the banks reinvested the money that they saved and, in particular, they opened many more branches, that in turn hired many more tellers. ATMs ended up creating more jobs for tellers, and a huge number of jobs for companies building and maintaining ATMs, jobs that did not exist before.

In 1900 a whopping 41% of the US workforce was employed in agriculture, but one century later, thanks to automation, that share has plunged below 2%. I am sure for some farmers this was traumatic, but this did not create mass unemployment and did not lower the median income. Millions of jobs were created to build the machines that automated the work of a farmer, jobs that didn't exist before.

A robot that has indeed "stolen" a lot of jobs is the humble traffic signal, now deployed at most of the road intersections of the world. There used to be traffic guards directing traffic. Very few are left. They have been replaced by millions of traffic lights, some of which are even equipped with cameras or connected to sensors.

Unfortunately, you do sell a lot of copies if you write books about the apocalypse, so writers are under pressure to be as negative as possible. Hence, bestsellers such as Martin Ford's "Rise of the Robots - Technology and the Threat of a Jobless Future" (2015) or Jerry Kaplan's "Humans Need Not Apply" (2015), which i thought were misleading, superficial and not helpful at all (to be fair, they both came out before the McKinsey and Bessen reports).

According to the International Federation of Robotics, in 2016 the countries with the highest density of robot population were South Korea (478 robots per 10,000 workers), Japan (315) and Germany (292). These countries also had some of the lowest unemployment rates in the world. For the record, the number for the USA was 164 (the USA has higher, not lower, unemployment than South Korea and Japan), and the number for the euro-countries afflicted by chronic unemployment (such as Greece) was the lowest in the developing world. Data about Italy can be misleading: Italy has a relatively high density of robots but also high unemployment. However, almost all the robots are in the industrialized north, where unemployment is very low, and almost none in the south, where unemployment is one of the highest in the world. The number of robots sold in the United States increased by 43% in 2011 and has continued to increase rapidly, and unemployment has declined every single year since 2011. How in heaven did those distinguished writers conclude from the given data that robots cause unemployment?

It is always easy to imagine which jobs will be destroyed and very difficult to imagine the new jobs that technology will create. So we exaggerate the reality of the disappearing jobs and underestimate the reality of the new ones.

In 1992, one year after the invention of the first Internet browser, when newly elected president Bill Clinton assembled a group of experts to discuss the future of the economy, nobody mentioned the Internet (David Leonhardt, "The Depression - If Only Things Were That Good", New York Times, 2011).

The society of robots will create new jobs that today we can't even imagine. Robots will create an even more complex society in which human intelligence will be even more important. The future always surprises us.

And my guess is that robots will become obsolete too at some point, replaced by something else that today doesn't even have a name. Some day robots will be made obsolete by a new human invention. Robots will become obsolete way before humans become obsolete.

Jobs in the Age of the Robot – Part 3: The Sharing Economy

The real revolution in employment is coming from a different direction: the "sharing economy". Companies such as Airbnb, that matches people who own rooms and people looking for rooms to rent, and Uber, that matches drivers who own a car and people looking for a ride in a car, have introduced a revolutionary paradigm in the job market: let people monetize under-utilized assets. This concept will soon be applied in dozens of different fields, allowing ordinary people to find ordinary customers for their ordinary assets; or, in other words, to supply labor and skills on demand. Before the industrial revolution most jobs were in the countryside but urban industry existed and it consisted mainly of artisan shops. The artisans would occasionally travel to a regional market, but mostly it was the customer who looked for the artisan, not vice versa. Cities like Firenze (Florence) had streets devoted to specific crafts, so that a customer could easily find where all the artisans offering a certain product were located. Then came the age of the factory and of transportation, and industrialization created the "firm" employing thousands of workers organized in some kind of hierarchy. Having a job came to mean something else: being employed. Eventually society started counting "unemployed" people, i.e. people who would like to work for an employer but no employer wants their time or skills. The smartphone and the Internet are enabling a return of sorts to the model of the artisan era. Anybody can offer their time and skills to anybody who wants them. The "firm" is simply the intermediary that allows customers to find the modern equivalent of the artisan.

In a sense, the "firm" (such as Uber or Airbnb) plays the role that the artisan street used to play in Firenze. Everybody who has time and/or skills to offer can now become a "self-employed" person. And that "self-employed" person can work when she wants, not necessarily from 8 to 5. There is no need for an office and for hiring contracts.

The traditional firm has a workforce that needs to be fully employed all the time, and sometimes the firm has to lay off workers and sometimes has to hire some, according to complicated strategic calculations.

In the sharing economy, no such thing exists: the firm is replaced by a community of skilled workers who take the jobs they want to take, when they want to take them, and if the customer wants them to take them. In a sense, people can be fired and hired on the fly.

Of course, this means that "good jobs" will no longer be judged based on job promotions, salary increases and benefits. They will be based on customer demand (which in theory is what drives company's revenues which in turn drives job promotions, salary increases and benefits).

The unemployed person who finds it difficult to find a job in a firm is someone whose skill is not desired by any firm, but this does not mean that those skills are not desired by any customer. The firm introduced a huge interface between customer and worker. When there is a need for your skill, you have to hope that a manager learns of your skills, usually represented by a resume that you submitted to the human resources department, and hope that the financial officer will approve the hiring. The simple match-making between a customer who wants a service and the skilled worker who can provide that service gets complicated by the nature of the firm with its hierarchical structure and its system of checks and balances (not to mention internal politics and managerial incompetence). It would obviously be easier to let the customer deal directly with the skilled worker who can offer the required service.

Until the 2000s the problem was that the customer had no easy way of accessing skilled workers other than through the "yellow pages", i.e. the firms. Internet-based sharing systems remove the layers of intermediaries except one (the match-making platform, which basically provides the economy of scale). In fact, these platforms turn the model upside down: instead of a worker looking for employment in a firm that is looking for customers, the new model views customers as looking for workers. Not only does this model bypass the slow and dumb firm, but it also allows you to monetize assets that you own and never perceived as assets. A car is an asset. You use it to go to work and to go on vacation, but, when it is parked in the garage, it is an under-utilized asset.

Marketing used to be a scientific process to shovel a new product down the throat of reluctant consumers: it now becomes a simple algorithm allowing customers to pick their skilled workers, an algorithm that basically combines the technology of online dating (match making), of auctions (bidding) and of consumer rating (that basically replaces the traditional "performance appraisal" prescribed in the traditional firm).

Of course, the downside of this new economy is that the worker has none of the protections that she had in the old economy: no security that tomorrow she will make money, no corporate pension plan, etc; and she is in charge of training herself to keep herself competitive in her business. The responsibility for a worker's future was mostly offloaded to the firm.

In the sharing economy that responsibility shifts entirely to the worker herself.

The new proletariat is self-employed, and, basically, each member of the proletariat is actually a micro-capitalist; the price to pay is that the worker will have to shoulder the same responsibilities that traditionally have fallen into the realm of firm management.

People who worry about robots are thinking about the traditional jobs in the factory and the office.

Futurists have a unique way to completely miss the scientific revolutions that really matter.

Jobs in the Age of the Robot – Part 4: The Maid Principle

Older workers are scared at the prospect that their specialty skill will soon be performed by a machine. Students are scared at the prospect that they may be studying to perform a job that will not exist when they graduate. Both concerns are legitimate. Jobs will be created, but they will not be the jobs that we have today. Being able to adapt to new jobs, jumping from one skill to a very different skill, will make the difference between success and failure.

It is virtually impossible to give proper advice about jobs that don't exist today. It is difficult to imagine what foundations to study and what path to follow in order to be ready for a job that doesn't exist today. But here are some rules of thumb.

The most obvious among them is: anybody whose job consists in behaving like a machine will be replaced by a machine. In highly structured societies like the USA (where one cannot get an omelette in a restaurant after 11am despite the fact that they have all the ingredients in the kitchen and even the most inept of chefs certainly knows how to cook an omelette), many jobs fall into this category. Even the people who write press releases for big corporations, even speech writers, and to some extent even engineers are asked to follow rules and regulations. The higher the portion of their job that is governed by rules, the higher the chance that they will soon be replaced by a machine. Those same jobs are less vulnerable in countries where the "human touch" still prevails over clockwork organization.

Those people who are good at communicating, empathizing, and the other things that we expect from fellow humans, will not be replaced by machines any time soon. A nurse who simply performs a routine task and shows little or no emotional attachment to her patients will be replaced by a robot, but a nurse who also provides comfort, company and empathy is much harder to replace. There is no robot coming in the near future that can have a real conversation with a sick person or an elderly person.

If you behave and think like a machine, you are already redundant. There are many in the USA who fall into this category, people who get upset if we ask them to do something slightly different from what they have been trained to do. If you are one of those people who don't like to do anything that requires "thinking", ask yourself "why does the world need me?" Machines can do a better and friendlier job than you, with no lunch breaks, no sleep, no weekend parties and no exotic vacations. If you are a cog in a highly structured environment, you should be surprised that someone is still willing to pay you a salary.

On the other hand, if you are the one who designs the structured environment in which machines can thrive, or the one who designs the machines for that environment, or even just the one who builds, repairs and/or sells them, then we desperately need you, and we rely on you to make sure that machines will create a better world.

I actually like the idea that automation will keep challenging us to be more creative, to find higher and higher meanings to our lives. If machines can build a better world, why does the world need us? We have to answer this question. We are actually more human when we struggle to find a higher meaning to our lives than when we simply work from 8am till 5pm mindlessly following a routine as if we were... robots.

Interdisciplinary thinking will be more useful than ever and today's machines make it easier than before to get an interdisciplinary education. If you are using the power of machines, like your smartphone, to let machines do the thinking for you, you are probably getting dumber, and that will not help you. If you are using the power of machines to learn a lot more things than your parents did, in a lot more fields, you are more likely to compete for the best jobs of the future.

A serious problem in the USA is its increasingly under-educated population that will certainly have trouble adjusting to the new job opportunities. This is already happening in sectors like software and biotech, where the new highly-paid jobs often go to the much better educated Chinese immigrants than to the native US citizens who dropped out of school. We immigrants of Silicon Valley can't help noticing that most of the people serving us in shops and restaurants were born and raised right here in the Bay Area, and totally missed the high-tech revolution that was happening under their nose. In 1946 the USA had the #1 high school graduation rate in the world. Today (according to the OECD) it ranks 22nd among 27 industrialized nations. US students rank 25th in mathematics, 17th in science and 14th in reading. Only 46% of US students finish college.

But there are also low-level jobs that cannot be automated easily. They cannot be automated because they are so "human". A favorite example is the hotel maid. This is a very low-wage job but just think about it: which robot can pick objects of a virtually infinite range of shapes and solidity and use common sense to understand what must be done with them? Try to explain to a robot what "garbage" means. You don't throw dirty underwear away (it belongs to somebody) but you do throw empty pizza boxes away. On the other hand, you don't throw it away if the guest has written "maid: please save this" on it. If the empty pizza box contains dirty tissue paper, it is meant to be thrown away. But if the paper inside the empty pizza is green with the face of a president, maybe you should think twice.

"The main lesson of 35 years of AI research is that the hard problems are easy and the easy problems are hard" (Steven Pinker).

Marketing and Fashion

Back to the topic of accelerating progress: what is truly accelerating at exponential speed is fashion. This is another point where many futurists and high-tech bloggers confuse a sociopolitical phenomenon with a technological phenomenon.

What we are actually witnessing in many fields is a regression in quality. This is largely due to the level of sophistication reached by marketing techniques. Marketing is a scary human invention: it often consists in erasing the memory of good things so that people will buy bad things. There would be no market for new films or books if everybody knew about the thousands of good films and books of the past: people would spend their entire life watching and reading the (far superior) classics instead of the new films and books, most of which are mediocre at best. In order to have people watch a new film or read a new book, the marketing strategists have to make sure that people will never know about old films and books. It is often ignorance that makes people think they just witnessed "progress" in any publicized event. Often we call "progress" the fact that a company is getting rich by selling poor quality products. The "progress" lies in the marketing, not in the goods. The acceleration of complexity is in reality an acceleration of low quality.

We may or may not live in the age of machines, but we certainly live in the age of marketing. If we did not invent anything, absolutely anything, there would still be frantic change. Today change is largely driven by marketing. The industry desperately needs consumers to go out and keep buying newer models of everything. We mostly buy things we don't need. The younger generation is always more likely to be duped by marketing and soon the older generations find themselves unable to communicate with young people unless they too buy the same things. Sure: many of them are convenient and soon come to be perceived as "necessities"; but the truth is that humans have lived well (sometimes better) for millennia without those "necessities". The idea that an mp3 file is better than a compact disc which is better than a vinyl record is just that: an idea, and mainly a marketing idea. The idea that a streamed movie is better than a DVD which is better than a VHS tape is just that: an idea, and mainly a marketing idea. We live in the age of consumerism, of rapid and continuous change in products, mostly unnecessary ones.

What is truly accelerating is the ability of marketing strategies to create the need for new products. Therefore, yes, our world is changing more rapidly than ever; not because we are surrounded by better machines but because we are surrounded by better snake-oil peddlers (and dumber consumers).

"The computer industry is the only industry that is more fashion-driven than women's fashion" (I am quoting Larry Ellison, founder and chairman of Oracle).

Sometimes we are confusing progress in management, manufacturing and marketing (that accounts for 90 percent of the "accelerating progress" that we experience) with progress in machine intelligence (that is still at the "Press 1 for English" level).

Technological progress is, in turn, largely driven by its ability to increase sales. Therefore it is not surprising that the big success stories of the World-wide Web (Yahoo, Google, Facebook, etc) are the ones that managed to turn web traffic into advertising revenues. We are turning search engines, social media and just about every website into the equivalent of the billboards that dot city streets and highways. It is advertising revenues, not the aim of creating intelligent machines, that is driving progress on the Internet. In a sense, Internet technology was initially driven by the military establishment that wanted to protect the USA from a nuclear strike, then by a utopian community of scientists that wanted to share knowledge, then by corporations that wanted to profit from e-commerce, and now by managers of advertising campaigns who want to capture as large an audience as possible. Whether this helps accelerate progress and in which direction is, at best, not clear.

When Vance Packard wrote his pamphlet "The Hidden Persuaders" (1957) on the advertising industry (on how the media can create the illusory need for unnecessary goods), he had literally seen nothing yet.

"The best minds of my generation are thinking about how to make people click ads" (I am quoting former Facebook research scientist Jeff Hammerbacher in 2012).

And, to be fair, the best minds of his generation are not only used to make people click on ads but also to create ever more sophisticated programs of mass surveillance (as revealed in 2013 by National Security Agency analyst Edward Snowden).

Recap

Here is what i have told you so far. There are assumptions underlying the belief that super-intelligent machines are coming soon. The first one is that A.I. is making staggering progress and the second one is that progress is accelerating like never before. I showed you that both statements are wild exaggerations. There are still colossal gaps in the program of A.I. and very few creative ideas on how to fill those gaps. Brute-force A.I. is unlikely to succeed in solving outside problems of pattern recognition and brute-force A.I. has relied too much on faster and faster processors. Now that Moore's Law is coming to an end, we will need more (ahem, ahem) intelligent ways to do A.I. than brute force. I am not denying that there is progress in the way machines can work for us, but I will demystify why they can work better than in the past (it is more about the environment that we structure for the machines than about their intelligence). This explains why most of the machines around us are pretty stupid and why i don't see any robots walking around the streets of Silicon Valley. Precisely because of the limitations of today's A.I., you don't fear that machines will steal your job... unless your job is so stupid that even a stupid machine can do it.

As for the "accelerating progress" of our age, in most cases it is neither "unique" nor "progress". One century ago the world was completely changed by a series of inventions that happened one after the other within a few years: the telephone, the radio, the car, the airplane, the record, Quantum Mechanics, Relativity, etc. Are you sure that today's progress is more dramatic than that one? When answering this question, let's keep in mind that change is not always progress. Change can go in both directions: forward or backward. Change is not necessarily in the direction of progress. There have been a lot of changes in Syria since 2011, but only ISIS would call it "progress".

More criticism of today's A.I. is coming in the next pages, some of it more philosophical than practical, and in particular we will discuss the concept of super-human intelligence. But first let's go back to the fundamental question: why A.I. at all? And this will also link to what i wrote at the very beginning: I am not afraid of A.I.; i am afraid that it will not come soon enough.

Why we need A.I. or The Robots are Coming– Part 2: The Near Future of A.I. or Don't be Afraid of the Machine

The media are promising a myriad applications of A.I. in all sectors of the economy. So far we have seen very little compared to what was promised. In 2016 Bloomberg estimated 2,600 startups working on A.I. technology, but IDC calculated that sales for all companies selling A.I. software barely totaled $1 billion in 2015. There is a lot of talk, but, so far, very few actual products that people are willing to pay for.

The number-one application of A.I. is and will remain... drum roll... making you buy things that you don't need. All major websites employ some simple form of A.I. to follow you, study you, understand you and then sell you something. Your private life is a business opportunity for them and A.I. helps them figure out how to monetize it. The founders of A.I. are probably turning in their graves.

And sometimes these "things" can even kill you (the case of Wei Zexi in 2016, who was induced by an ad posted on Baidu to buy the cancer treatment that killed him).

Mark Weiser famously wrote: "The most profound technologies are those that disappear. They weave themselves into the fabric of everyday life until they are indistinguishable from it" ("The Computer for the 21st Century", 1991). Unfortunately, it turned out to be a prophecy about the ubiquitous "intelligent" agents that make us buy things.

Perhaps the most sophisticated (or, at least, the most widely used) A.I. system since 2014 is Facebook's machine-learning system FBLearner Flow, designed by Hussein Mehanna's team, that runs on a cluster of thousands of machines. It is used in every part of Facebook for quickly training and deploying neural networks. Neural networks can be fine-tuned by playing with several parameters. Optimizing these parameters is not trivial. It requires a lot of "trial and error". But even just a 1% improvement in machine-learning accuracy can mean billions of dollars of additional revenues for Facebook. So Facebook is now developing Asimo, that performs thousands of tests to find the best parameters for each neural network. In other words, Asimo does the job that is normally done by the engineers who build the deep-learning system.

While Jeff Hammerbacher's lament remains true, we must recognize that progress in deep learning has been driven by funding from companies like Google and Facebook whose main business interest is to convince people to buy things. If the world banned advertising from the Web, the discipline of deep learning would probably return to the obscure laboratories of the universities where it came from.

Remember Marshall McLuhan's comment in "Understanding Media" (1964): "Far more thought and care go into the composition of any prominent ad in a newspaper or magazine than go into the writing of their features and editorials"? The same can be said today: far more thought and care has been invested in designing algorithms that make you buy things when you are reading something on the Web than in the writing that you are reading.

The next generation of "conversational" agents will be able to access a broader range of information and of apps, and therefore provide the answer to more complicated questions; but they are not conversational at all: they simply query databases and return the result in your language. They add a speech-recognition system and a speech-generation system to the traditional database management system.

There are actually "dream" applications for deep learning. Health care is always at the top of the list because its impact on ordinary people can be significant. The medical world produces millions of images every year: X-Rays, MRIs, Computed Tomography (CT) scans, etc. In 2016 Philips Health Care estimated that it manages 135 billion medical images, and it adds 2 million new images every week. These images are typically viewed by only one physician, the physician who ordered them; and only once. This physician may not realize that the image contains valuable information about something outside the specific disease for which it was ordered. There might be scientific discoveries that affect millions of those images, but there is nobody checking them against the latest scientific announcements. First of all, we would like deep learning to help radiology, cardiology and oncology departments to understand all their images in real time. And then we would like to see the equivalent of a Googlebot (the "crawler" that Google uses to scan all the webpages of the world) for medical images. Imagine a Googlebot for medical images that continuously scans Philips' database and carries out a thorough analysis of each medical image utilizing the latest updates on medical science. Enlitic in San Francisco, Stanford's spinoff Arterys, and Israel's Zebra Medical Vision are the pioneers, but their solutions are very ad-hoc. A medical artificial intelligence would know your laboratory tests of 20 years ago and would know the lab tests of millions of other people, and would be able to draw inferences that no doctor can draw.

In 2016 Sebastian Thrun's team at Stanford built a neural network capable of recognizing skin cancer with the accuracy of a dermatologist. In 2016 radiologist Luke Oakden-Rayner of University of Adelaide in Australia demonstrated a deep-learning system that estimated a person's longevity based on radiological chest images of people aged 60 and over. Most early symptoms of heart attacks, cancer and diabetes (diseases that kill millions of people every year) are visible in these images but it takes a trained specialist. At the end of 2016 Varun Gulshan and physician Lily Peng of Google trained a (deep convolutional) neural network (using a dataset of 128,175 retinal images) to identify retinas at risk of a diabetic disease that causes 5% of blindness worldwide and tested this neural network against a group of expert ophthalmologists, showing that the accuracy was virtually the same. In 2017 the South Korean scientists Hongyoon Choi and Kyong Hwan Jin used a neural network to scan brain images and identify people likely to get Alzheimer's disease within the next three years. Alzheimer's disease affects 30 million people. In 2017 Stephen Weng at University of Nottingham unveiled a neural network, trained from hundreds of thousands of medical records, that proved to be better than human experts at predicting heart attacks. Every year about 20 million people die of a cardiovascular disease: this neural network could save the lives of millions of people. The success stories of medical-image analysis keep coming in.

In 2015 Joel Dudley's team at Mount Sinai Hospital in New York trained a deep-learning system called Deep Patient on the hospital's large dataset of health records about more than 700,000 patients. Deep Patient can discover patterns in patient data that are not easily spotted by human experts, and it has proven capable of predicting diseases, especially psychiatric disorders. In April 2016 a neural network trained by Harvard pathologist Andy Beck and an MIT computer scientist Aditya Khosla competed with an expert pathologist at identifying cancer and narrowly lost (it soon became the flagship product of the authors' new startup PathAI).

There are at least three problems though. The first one is that, as usual, it is difficult if not impossible to replicate the results. As of 2017, Google's dataset has not been released to the public so none else can replicate the experiment. The dataset of medical images is almost always imbalanced: it mostly contains data about sick people. It is not difficult to build a neural network that will recognize the "positives" (the medical images that signal a disease), but it is difficult to make sure that the neural network will NOT recognize as positive the healthy person. Finally, deep learning has been proven to work well only with small images. Medical images are giant images. In these experiments only a tiny fraction of the pixels was used. Quoting from Luke Oakden-Rayner's blog: "... retinal photographs are typically between 1.3 and 3.5 megapixels in resolution... these images were shrunk to 299 pixels square, which is 0.08 megapixels..." Each pixel can increase the dimensionality of the neural network to a degree that defies existing computational theories. In other words, we don't know if these methods work with the real images or just with toy miniatures of the real images.

The results of your neural network are only as good as your training data.

In 2015 the USA launched the Precision Medicine Initiative that consists in collecting and studying the genomes of one million people and then matching those genetic data with their health, so that physicians can deliver the right medicines in the right dose to each individual. This project will be virtually impossible without the use of machines that can identify patterns in that vast database.

There are also disturbing applications of the same technology that are likely to spread. The smartphone app FindFace, developed by two Russian kids in their 20s, Artem Kukharenko and Alexander Kabakov, identifies strangers in pictures by searching pictures posted on social media. If you have a presence on social media, the user of something like FindFace can find out who you are by simply taking a picture of you. In 2016 Apple acquired Emotient, a spinoff of UC San Diego, that is working on software to detect your mood based on your facial expression.

An example of unreasonable expectations is Google's self-driving car. The project was launched in 2009 by Sebastian Thrun, the Stanford scientist who had won the DARPA "Grand Challenge" of 2005, a 212-km race across the Nevada desert. Thrun quit in 2013 and was replaced by Chris Urmson, formerly a CMU student who in 2007 had worked on William Whittaker's victorious team for the DARPA "Urban Challenge" held at George Air Force Base near Los Angeles. (For the record, Chris Urmson left Google in 2016, as had done most of the original team).

The self-driving car may never fully materialize, but the "driver assistant" is coming soon. Mobileye, the Israeli company founded in 1999 that is widely considered the leader in machine-vision technology (and that does not use deep learning) has a much more realistic strategy based on incremental steps to introduce Advanced Driver Assistance Systems (ADAS) that can assist (not replace) drivers. Otto, founded by one of the engineers who worked on Google's self-driving car, Anthony Levandowski, does not plan to replace the truck driver but to assist the truck driver, especially on long highway drives. Otto, which in 2016 was acquired by Uber, does not plan to build a brand new kind of truck, but to provide a piece of equipment that can be installed on every truck. In 2014 a total of 3,660 people died in the USA in accidents that involved large trucks.

The need for robots is even greater. There are dangerous jobs in construction and steel work that kill thousands of workers every year. According to the International Labor Organization, mining accidents kill more than 10,000 miners every year; and that number does not include all the miners whose life expectancy is greatly reduced by their job conditions.

Robots and drones need eyes to see and avoid obstacles. There will be a market for computer-vision chips that you can install in your home-made drone, and there will be a market for collision-avoidance technology to install in existing cars. Israel's Mobileye and Ireland's Movidius have been selling computer-vision add-ons for machines for more than a decade.

We also need machines to take care of an increasingly elderly population. The combination of rising life expectancy and declining fertility rates is completely reshaping society. The most pressing problem facing humanity as a whole used to be the well-being and the education of children. That was when the median age was 25 or even lower. Ethiopia has a median age of about 19 like most of tropical Africa. Pakistan has a median age of 21. But the median age in Japan and Germany is 46. This means that there are as many people over 46 as there are under 46. Remove teenagers and children: Japan and Germany don't have enough people to take care of people over 46. That number goes up every year. There are more than one million people in Japan who are 90 years old or older, of which 60,000 are centenarians. In 2014, already 18% of the population of the European Union was over 65 years old, almost ten million people. We don't have enough young people to take care of so many elderly people, and it would be economically senseless to use too many young people on such an unproductive task. We need robots to help elderly people do their exercise, to remind them about taking their medicines, to pick up packages at the front door for them, etc.

I am not afraid of robots. I am afraid that robots will not come soon enough.

"Instead of worrying about what machines can do, we should worry more about what they still cannot do." (World chess champion Garry Kasparov)

The robots that we have today can hardly help. Using an IDC report of 2015, we estimated that about 63% of all robots are industrial robots, with robotic assistants (mostly for surgery), military robots and home appliances (like Roomba) sharing the rest in roughly equal slices. The main robot manufacturers, like ABB (Switzerland), Kuka (Germany, being acquired by China's Midea in 2016) and the four big Japanese companies (Fanuc, Yaskawa, Epson and Kawasaki), are selling mostly or only industrial robots, and not very intelligent ones. Robots that don't work on the assembly line are a rarity. Mobile robots are a rarity. Robots with computer vision are a rarity. Robots with speech recognition are a rarity. In other words, it is virtually impossible today to buy an autonomous robot that can help humans in any significant way other than inside the very controlled environment of the factory or of the warehouse. Nao (developed by Bruno Maisonnier's Aldebaran in France and first released in 2008), RoboThespian (developed by Will Jackson's Engineered Arts in Britain since 2005, and originally designed to be an actor), the open-source iCub (developed by the Italian Institute of Technology and first released in 2008), Pepper (developed by Aldebaran for Japan's SoftBank and first demonstrated in 2014) and the autonomous robots of the Willow Garage "diaspora" (Savioke, Suitable, Simbe, etc) are the vanguard of the "service robot" that can welcome you in a hotel or serve you a meal at the restaurant: "user-friendly" humanoid robots for social interaction, communication and entertainment at public events. In 2016 Knightscope's K5 robot security guard worked in the garage of the Stanford Shopping Center; Savioke's Botlr delivered items to guests at the Aloft hotel in Cupertino; Lowe's superstore in Sunnyvale employed an inventory checker robot built by Bossa Nova Robotics; and Simbe's Tally checked shelves of a Target store in San Francisco. But these are closer to novelty toys than to artificial intelligence. A dog is still a much more useful companion for an elderly person than the most sophisticated robot ever built.

In 1954 Sylvania used robots to publicize its products. Most of today's robots serve the same function: they are cute for taking pictures

The most used robot in the home is iRoomba, a small cylindrical box that vacuums floors. Not exactly the tentacular monster depicted in Hollywood movies. Unfortunately, it will also vacuum money if you drop it on the floor: we cannot trust machines with no common sense, even for the most trivial of tasks.

An industry that stands to benefit greatly from the "rise of the robots" is the toy industry. In 2016 San Francisco-based startup Anki introduced Cozmo, a robot with "character and personality". That's the future of toys, especially in countries like China where the one-child policy has created a generation of lonely children. In fact, we have already been invaded by robots: there are millions of Robosapien robots. The humanoid Robosapien robot was designed by Mark Tilden, a highly respected inventor who used to work at the Los Alamos National Laboratory, and introduced in 2004 by Hong Kong-based WowWee (a company founded in the 1980s by two Canadian immigrants). Most robots will be an evolution of Pinocchio, not of Shakey.

If you consider them robots, the exoskeletons are a success story. These are basically robots that you can wear. The technology was originally developed by the DARPA to help soldiers carry heavy loads, but it is now used to help victims of brain injuries and spinal-cord injuries in several rehabilitation clinics.

ReWalk, founded by an Israeli quadriplegic (Amit Goffer), Ekso Bionics and Suitx (two UC Berkeley spinoffs) and SuperFlex (an SRI spinoff) already helped paraplegics or seniors walk. Panasonic's ActiveLink has announced an exoskeleton that will help weak nerdy people like me with manual labor that requires physical strength. The cost is still prohibitively high, but one can envision a not-too-distant future in which we will be able to rent an exoskeleton at the hardware store to carry out gardening and home-improvement projects. After you wear it, you can lift weights and hammer with full strength.

Robots can pick up only a very limited set of objects, and sometimes only one specific kind of object. In 2015 a member of RoboBrain, Stefanie Tellex of Brown University, demonstrated how her robot was trained by another robot to manipulate an object. The knowledge required was passed over the cloud from one robot to the other. She then launched the "Million Object Challenge" to build a knowledge-base of manipulation experiences that can be reused by any robot.

"Cloud Robotics" is a term coined in 2010 by James Kuffner at CMU. The idea is to create a library of programs that can be executed remotely by any robot, a "skills library"; basically, removing most of the brain of the robot and enabling the robot to use a common brain. In 1993 Masayuki Inaba at University of Tokyo explored the concept of such "remote-brained robots", but that was before cloud computing became affordable. Robots can take advantage of projects such as OpenEase, a platform for machines to share knowledge, or RoboEarth (2010) and Rapyuta (2013), funded by the European Union. RoboHow (2012) wants robots to learn new tasks from high-level descriptions; and RoboBrain (2014) wants robots to learn new tasks from human demonstrations and advice. The immediate effect is to turn the robot into the equivalent of a "thin" client. Benefits include a longer battery life and no need to download software updates. But the bigger benefit is that cloud-enabled robots can engage in collective progress, learning rapidly from each other's experiences.

For example, in 2010 a group of makers in Silicon Valley led by Ryan Hickman, who later founded the Cloud Robotics team at Google, started the Cellbots project to build robots out of smartphones and spare parts. The clever idea was to realize that a smartphone is almost a robot: it already has touch, hearing, vision, speech, navigation, and can even sense that we don't have anything like real-time translation, and is accessing the cloud all the time. It only lacks mobility, i.e. legs or wheels. Hickman ran cables between an Android phone and an Arduino platform, mounted it on wheels and obtained a "cellbot".

In fact, object recognition and grasping algorithms can all be performed in the cloud. The robot itself can be just a brainless body.

In 2017 Ken Goldberg at UC Berkeley (also a pioneer of telerobotic art) found another way to achieve a faster way of training of robots: train them in virtual reality. First his team created a database of thousands of 3D models, DexNet 1.0 (2015). Then the robot was trained by practicing to grasp those virtual objects in a simulated world. The idea of using virtual reality to train robots was also behind Canadian startup Kindred.ai, the brainchild of Suzanne Gildert, a senior researcher at quantum-computer maker D-Wave.

Others are studying how a team of robots can self-organize in order to collaborate and achieve a goal; i.e. collective artificial intelligence. The pioneer is Marco Dorigo in Belgium, who in 1999 developed the "ant colony optimization" algorithm for the "swarmanoids" of his Ph.D. dissertation at Milan's Polytechnic Institute in Italy ("Distributed Optimization by Ant Colonies", 1992). Then came the Alices of Simon Garnier at the New Jersey Institute of Technology ("Alice in Pheromone Land", 2007), the "kilobot" swarms by Radhika Nagpal's group at Harvard University ("A Low Cost Scalable Robot System for Collective Behaviors", 2012), the tiny robots at University of Colorado by the group of Nikolaus Correll, a former student of Daniela Rus at MIT ("Modeling Multi-Robot Task Allocation with Limited Information as Global Game", 2016), and the "smarticles", or smart active particles, by Dana Randall and Daniel Goldman at Georgia Tech ("A Markov Chain Algorithm for Compression in Self-Organizing Particle Systems", 2016). There are also two international conferences on swarm intelligence, ICSI (first held at Peking University in China in 2010) and ANTS (started in 1998 at IRIDIA in Belgium). Studies on the self-organizing skills of ants by myrmecologists such as Deborah Gordon at Stanford ("Ants at Work", 1999) and Guy Theraulaz in France ("Spatial Patterns in Ant Colonies", 2002) were particularly influential with this school of thought.

But first we will need to build robotic arms whose dexterity matches at least the dexterity of a squirrel.

Our hand has dozens of degrees of freedom. Let's say that it has ten (it actually has many more). I can plan the movement of my hand easily ten steps ahead: that's 10 to the 10th to the 10th to... a very huge number. And i can do it without thinking, in a split second. For a robot this is a colossal computational problem. Picking arbitrary objects is not trivial because objects, even of the same category, may come in an infinite number of variations. For example, there are thousands of kinds of mugs: each one requires a slightly different grasping movement. Even the orientation of the mug causes a change in the way we grasp it.

Grasping an object is something very easy and natural for humans, but terribly difficult to understand (even for humans) and therefore to implement in robots. Traditionally, robot grasping has been implemented through a combination of perception (estimating the position and orientation of the object) and planning (calculating the optimal movement of the robot's arm and hand). In this way the problem of grasping is reduced to a combination of geometric and kinematic considerations. This "analytical" approach works only in ideal conditions and for known objects (objects for which a 3D model is available). It tends to fail in cluttered environments and for objects never seen before. The "data-driven" approach, instead, infers a grasp for an object without knowing its exact shape, using experience that, typically, comes from simulations. This approach became popular after Peter Allen's student Andrew Miller at Columbia University developed the GraspIt simulator ("Automatic Grasp Planning Using Shape Primitives," 2003), which was made available to the larger community in 2004.

Andrew Ng's student Ashutosh Saxena at Stanford University augmented the data-driven approach with machine learning in order to grasp an object seen for the first time through computer vision. He specifically envisioned a household scenario in which a robot has to empty a dishwasher ("Robotic Grasping of Novel Objects using Vision", 2006). Similar methods were developed by Peter Allen's student Hao Dang at Columbia University ("Semantic Grasping", 2012), who trained his neural network on a large dataset of objects and designed it to generalize to unseen objects, by Markus Vincze's student David Fischinger at Vienna University of Technology in Austria, in collaboration with Ashutosh Saxena's student Yun Jiang, after Saxena moved to Cornell University ("Learning Grasps for Unknown Objects in Cluttered Scenes", 2013), and by Joseph Redmon at University of Washington (of "You Only Look Once" fame) in collaboration with Anelia Angelova of Google ("Real-time Grasp Detection Using Convolutional Neural Networks", 2015). These groups pioneered the application of deep learning to improving the dexterity of robots.

Meanwhile, new devices for 3D sensing became commercially available, such as Microsoft's Kinect (2010) for depth sensing, and new software methods were invented, notably one by Kurt Konolige at Willow Garage ("Projected Texture Stereo", 2010).

The data-driven approach became predominant in the era of big data, as shown by Daniel Kappler, Jeannette Bohg and Stefan Schaal at the Max-Planck Institute ("Leveraging Big Data for Grasp Planning", 2015). Robert Platt's student Andreas ten Pas at Northeastern University designed a hybrid of the analytical and data-driven approaches, augmented with machine learning ("Using Geometry to Detect Grasp Poses in 3d Point Clouds," 2015).

Some continued the tradition of supervised-learning approaches, for example Ashutosh Saxena's student Ian Lenz ("Deep Learning for Detecting Robotic Grasps", 2015), that used human supervision, and Ken Goldberg's student Jeffrey Mahler at UC Berkeley, who introduced Dex-net 2.0 ("Deep Learning to Plan Robust Grasps with Synthetic Point Clouds and Analytic Grasp Metrics", 2017); while others opted for novel self-supervised approaches, notably Abhinav Gupta''s student Lerrel Pinto at CMU ("Supersizing Self-supervision", 2015) and a Google team led by Sergey Levine ("End-to-end Learning of Semantic Grasping", 2017) that mimicked the "two-stream" model of visual reasoning advanced in 1992 by the neuroscientists David Milner and Melvyn Goodale at the University of Western Ontario in Canada (a "ventral stream" that recognizes the kind of object plus a parallel "dorsal stream" that recognizes the object's location relative to the viewer). Both approaches relied heavily on data. Hence a number of teams became to generate simulated data to train the neural network, just like in Jeffrey Mahler's Dex-net. Notably: a Google team led by Vincent Vanhoucke that included Kurt Konolige and Sergey Levine and that implemented GraspGAN ("Using Simulation and Domain Adaptation to Improve Efficiency of Deep Robotic Grasping", 2017); Robert Platt's student Ulrich Viereck at Northeastern University ("Learning a Visuomotor Controller for Real World Robotic Grasping Using Easily Simulated Depth Images", 2017); and Silvio Savarese's student Kuan Fang at Stanford, who introduced the Task-Oriented Grasping Network or TOG-Net ("Learning Task-Oriented Grasping for Tool Manipulation from Simulated Self-Supervision", 2018). All these systems relied on large-scale simulated self-supervision, i.e. trained their neural networks in a simulated environment and then transferred it to the real robot.

The studies on grasping yielded interesting insights on other aspects of human cognition. For example, Keng Peng Tee's team at Singapore's Agency for Science, Technology, and Research (A*STAR), in collaboration with Gowrishankar Ganesh's team at Japan's National Institute of Advanced Industrial Science and Technology (AIST), discovered an algorithm that can automatically recognize a novel object as a potential tool and can figure out how to use it ("Towards Emergence of Tool Use in Robots", 2018).

Then came the age of AlphaGo, i.e. of deep reinforcement learning. Robert Platt's student Marcus Gualtieri at Northeastern University used deep reinforcement learning (a learning agent similar to DQN) to demonstrate versatile manipulation of objects ("Learning 6-DoF Grasping and Pick-Place Using Attention Focus", 2018). A joint Google-UC Berkeley team led by Sergey Levine overcame the limitations of DDPG and NFQCA with QT-Opt ("Scalable Deep Reinforcement Learning for Vision-Based Robotic Manipulation", 2018) that achieved a 96% grasp success rate picking and grasping different shaped objects. It was basically DQN on steroids (trained using multiple robots at the same time).

Lucas Manuelli and Wei Gao at MIT, instead, developed a method called kPAM to model categories of objects with just a handful of "keypoints": three for mugs, six for shoes ("KeyPoint Affordances for Category-Level Robotic Manipulation", 2019).

"High-level reasoning requires very little computation, but low-level sensorimotor skills require enormous computational resources" (Erik Brynjolfsson)

Earlier in the book i mentioned that two of the motivations for doing A.I. were: a business opportunity and the ideal of improving the lives of ordinary people. Both motivations are at work in these projects. Unfortunately, the technology is still primitive. Don't even think for a second that this very limited technology can create an evil race of robots any time soon.

"Nothing in life is to be feared, it is only to be understood" (Marie Curie).

Jobs in the Age of the Robot – Part 5: The Jobs we Need

Asking which jobs will be eliminated by intelligent machines is asking the wrong question. Technology has been replacing humans for a long time, and typically for every job that is destroyed better jobs are created to manage the technology. The question that makes more sense is: which jobs should "not" be replaced by intelligent machines? Which jobs require common sense; which jobs can have disastrous consequences if performed by a person or a machine that doesn't have common sense?

For example, i don't want the self-driving car because a driver without common sense can kill people. The environment has to get a lot more structured to reassure me that intelligent machines (where "intelligent" really means "incredibly stupid") can be trusted with driving a car.

The other question that makes a lot of sense is: which jobs will be done by intelligent machines that today nobody does. Some of these are jobs that we desperately need but that no human is capable of doing, like scanning all those millions of medical images. That machine will not steal anybody's job, and will not risk anybody's life. That machine will simply provide additional information to your doctor about your health, something that today nobody is doing. No job will be lost on account of this machine. On the contrary, several jobs will be created to build this machine, maintain it, update it, and, some day, decommission it and replace it with a better one; and maybe a job will even be created in a museum of old "intelligent" machines; and certainly writers like me will have a job writing about it (or against it).

Over the last 30 years China has built an enormous number of high-rise buildings. Chinese cities need to check the exteriors and the windows of thousands of skyscrapers. It is a crucial task to guarantee safety to millions of people who live and work in those buildings. Alas, we don't have "Spider Men" that can climb the exterior walls of skyscrapers and check every surface and every window. Tiny climbing robots will do that job. They will not steal anybody's job. They will carry out an important task that today nobody is doing. Those robots will need to be built, programmed, maintained, and operated (and marketed, sold, delivered, and, why not, written about in books like this one). Those robots will not kill jobs: they will create millions of jobs all over the world.

"The best way to predict the future is to invent it" (Alan Kay)

Origins of Singularity Thinking

Singularity thinking originated with the essay "Today's Computers, Intelligent Machines and Our Future" (1978) by Hans Moravec of CMU, with Ray Solomonoff's article "The Time Scale of Artificial Intelligence" (1985), and with Marvin Minsky's essay "Will Robots Inherit the Earth" (1994), and was popularized by Ray Kurzweil's "The Singularity is Near" (2005) besides Hans Moravec's "Mind Children" (1988) and "Robot - Mere Machine to Transcendent Mind" (1998). David Levy's "Robots Unlimited" (2006) even predicted that machines will soon be conscious. Hyperbole ruled on magazines such as Wired, New Scientist (that in 2001 discussed a silicon brain), and even Scientific American (whose Quarterly in 1999 had an article titled "Downloading your Brain"). David Gelernter's book "Mirror Worlds" (1992) was subtitled "the Day Software Puts the Universe in a Shoebox... How It Will Happen and What It Will Mean". Masahiro Mori, a scientist at the Tokyo Institute of Technology and future president of the Robotics Society of Japan who in 1970 had published the influential article "The Uncanny Valley", had actually predated the whole Singularity movement when he argued in "The Buddha in the Robot" (1974) that robots would someday be able to attain buddhahood.

In fact, one can go even further back in time. Herbert Wells, the novelist who, among many other things, wrote some of the early science-fiction classics such as "War of the Worlds" (1897) and "The Time Machine" (1895), introduced the concept of a constantly updated World Encyclopaedia in November 1936, speaking at the Royal Institution. In 1962 the science-fiction writer Arthur Clarke, the author of the short story "The Sentinel" (originally written in 1948) that was going to be transformed into Stanley Kubrick's film "2001 A Space Odyssey", published his book "Profiles of the Future", in which he predicted that an artificial intelligence will implement Wells' world brain. And finally in 1982 Peter Russell predicted that by 2000 the global network of computers would give rise to a "global brain" and that such global brain would evolve into a conscious being.

Trivia: Daniel Wilson wrote a hilarious manual to help humans survive in a world threatened by intelligent machines, "How to Survive a Robot Uprising" (2005).

The original prophet of what came to be called "transhumanism" was probably Fereidoun "FM-2030" Esfandiary who wrote "Are You a Transhuman?" (1989) and predicted that "in 2030 we will be ageless and everyone will have an excellent chance to live forever". He died of pancreatic cancer (but was promptly placed in cryogenic suspension).

In 1999 Kurzweil argued that there exists a general law, the "Law of Accelerating Returns" that transcends Moore's Law. Order causes more order to be created and at a faster rate. Order started growing exponentially millions of years ago, and progress is now visible on a daily basis. This echoed science-fiction writer Vernon Vinge's declaration that "the acceleration of technological progress has been the central feature of this century" (1993). They both base their conclusions on the ever more frequent news of technological achievements. (Personally, i think that they are confusing progress and the news cycle. Yes, we get a lot more news from a lot more sources. If the same news and communication tools had been available at any time in previous peacetime periods, the people alive back then would have been flooded by an equal amount of news). In particular, at some point computers will acquire the ability to improve themselves, and then the process that has been manually done by humans will be automated like many other manual jobs, except that this one is about making smarter computers, which means that the process of making smarter computers will be automated by smarter computers, which turns into a self-propelled accelerating loop. This will lead to an infinite expansion of "intelligence".

The Case for Superhuman Intelligence... and against it

The case for the coming of an artificial intelligence, of an artificial general intelligence and then of the Singularity rests on the simple assumption that A.I. is making dramatic progress and that progress is accelerating. If you believe these two statements, then you probably believe that we will soon have machines that can have a philosophical conversation with us and write books like this one.

That is precisely the conclusion that Moravec and Kurzweil reached. Hans Moravec, the author of "Mind Children" (1988) and "Robot - Mere Machine to Transcendent Mind" (1998), predicted that machines will become smarter than humans by 2050. Ray Kurzweil, the author of "The Singularity is Near" (2005), predicted that machine intelligence will surpass human intelligence by 2045.

Moravec and Kurzweil were not the first futurists to put those two assumptions together. In 1957 Herbert Simon, one of the founders of A.I., had said: "there are now in the world machines that think, that learn, and that create. Moreover, their ability to do these things is going to increase rapidly".

The case against A.I. (and therefore the Singularity) dates from the 1970s, when philosophers started looking into the ambitious statements coming out of the A.I. world.

The first philosopher to look into these claims was Hubert Dreyfus, who wrote in "Alchemy and Artificial Intelligence" (1965): "Significant developments in Artificial Intelligence ... must await an entirely different sort of computer. The only existing prototype for it is the little-understood human brain."

Mortimer Taube, author of "Computers and Common Sense" (1961), and John Lucas, author of "Minds, Machines and Gödel" (1959), had already pointed out that full machine intelligence is incompatible with Kurt Gödel's incompleteness theorem. In 1935 Alonso Church proved a theorem, basically an extension of Gödel's incompleteness theorem to computation: that first-order logic is "undecidable". Similarly, in 1936 Alan Turing proved that the "halting problem" is undecidable for Universal Turing Machines. What these two theorems say is basically that it cannot be proven whether a computer will always find a solution to every problem; and that is a consequence of Gödel's theorem, a highly respected mathematical proof. Several thinkers have used similar arguments based on Gödel's theorem, notably Roger Penrose in "The Emperor's New Mind" (1989).

The most famous critique of Artificial Intelligence was contained in John Searle's article "Minds, Brains and Programs" (1980) and came to be known as the "Chinese room" argument. If you give a person a comprehensive set of instructions needed to translate Chinese into English and lock that person in a room, someone standing outside the room would be fooled into thinking that the person inside knows Chinese, when in fact that person is mechanically following instructions, that are meaningless to her, to be able to manipulate symbols that are also meaningless for her. She has no clue what the Chinese sentence says, but she produces the correct translation into English. Endless papers have been written by philosophers to discuss the validity of Searle's argument. However, Searle was not attacking the feasibility of machine intelligence but simply whether an intelligent machine would be also conscious.

Today's computers, including the superfast GPUs used by AlphaGo, are Turing Machines. Critics of A.I. need to prove that Turing machines cannot match human intelligence. Many have written books along those lines, but what are the tasks that Turing Machines can't perform is a conveniently movable target. To my knowledge, nobody spelled out what is it that Turing machines will never do better than us.

I suspect that here another kind of "religion" plays a role in the opposite direction, in the direction of making us reluctant to accept that machines can become as intelligent as us and even more intelligent. Astrophysics has shown that there is nothing special about the location of the Earth, Biology showed that there is nothing special about human life, neuroscience is showing that there is nothing special about the human brain, and now Artificial Intelligence might show that there is nothing special about our intelligence. Each of these revelations seems to make humankind less relevant, more insignificant.

But i have seen no convincing proof that machines can reach human-level intelligence. Hence: why not?

Instead, one has to wonder for how long it will make sense to ask the question whether full-fledged artificial intelligence is possible. If the timeframe for fully intelligent machines is centuries and not decades like the optimists believe, then it's like asking an astronaut "Will it at some point be possible to send a manned spaceship to Pluto?" Yes, it may be very possible, but it may never happen: not because it's impossible but simply because we may invent teleportation that will make spaceships irrelevant. Before we invent intelligent machines, synthetic biology or some other discipline might have invented something that will make robots irrelevant. The timeframe is not a detail.

Assuming that some day we will have fully intelligent machines, will they evolve into a superior level of intelligence that is unattainable by humans? That is a different question. I have seen no proof that machine intelligence inevitably leads to machines becoming more intelligent than humans.

Let me use a metaphor. Just because we built a ladder it doesn't mean that we can fly: it only means that we can build taller and taller ladders, and maybe those ladders will help us climb on the roof and fix a leak; but the technology to fly is different from the technology of climbing ladders, and therefore virtually no progress towards flying will be achieved by building better and better ladders. And I doubt that ladders will spontaneously evolve into flying beings. Both the ladder and the bird have to do with "heights" and naive media may conclude that one leads to the other, but people who build ladders should know better.

What exactly are the things that a superhuman intelligence can do and no human being can ever do? If the answer is "we cannot even conceive them", then we are back to the belief that angels exist and miracles happen, something that eventually gave rise to organized religions. If instead there is a simple, rational definition of what a superhuman intelligence can do that no human can ever do, i have not seen it; or, better, i have seen one but also the opposing view. I will briefly discuss these two opposite views.

On one hand, superhuman intelligence should exist because of the "cognitive closure", a concept popularized by Colin McGinn in "The Problem Of Consciousness" (1991). The general idea is that every cognitive system (e.g., every living being) has a "cognitive closure": a limit to what it can know. A fly or a snake cannot see the world the way we see it because they do not have the same visual system that humans have. In turn, we can never know how it feels to be a fly or a snake. A blind person can never know what "red" is even after studying everything there is to be studied about the color "red". According to this idea, each brain (including the human brain) has a limit to what it can possibly think, understand, and know. In particular, the human brain has a limit that will preclude humans from understanding some of the ultimate truths of the universe. These may include spacetime, the meaning of life, and consciousness itself. There is a limit to how "intelligent" we humans can be. According to this view, there should exist cognitive systems that are "superhuman", i.e. they don't have the limitations that our cognition has.

However, i am not sure if we (humans) can intentionally build a cognitive system whose cognitive closure is larger than ours, i.e., a cognitive system that can "think" concepts that we cannot think. It sounds a bit of an oxymoron that a lower form of intelligence can intentionally build the highest form of intelligence. However, it is not a contradiction that a lower form of intelligence can accidentally (by sheer luck) create a higher form of intelligence.

That is the argument in favor of the feasibility of superhuman intelligence. A brilliant argument against such feasibility is indirectly presented in David Deutsch's "The Beginning of Infinity" (2011). Deutsch argues that there is nothing in our universe that the human mind cannot understand, as long as the universe is driven by universal laws. I tend to agree with Colin McGinn that there is a "cognitive closure" for any kind of brain, that any kind of brain can only do certain things, and that our cognitive closure will keep us from ever understanding some things about the world (perhaps the nature of consciousness is one of them); but in general i also agree with Deutsch: if something can be expressed in formulas, then we humans will eventually "discover" it and "understand" it; and, if everything in nature can be expressed in formulas, then we (intelligent beings) will eventually "understand" everything, i.e. we are the highest form of intelligence that can possibly exist. So the only superhuman machine that would be too intelligent for humans to understand is a machine that does not obey the laws of nature, i.e. that is not a machine.

If you lean towards the "cognitive closure" argument, you also have to show that we haven't reached it yet. The progress of the human mind did not necessarily end with you. If human intelligence hasn't reached the cognitive closure yet, then there is still room for improvement in human intelligence. I see no evidence that the human mind may have reached a maximum of creativity and will never go any further. We build machines based on today's knowledge and creativity. Maybe, some day, those machines will be able to do everything that we do today; but why should we assume that, by then, the human mind will not have progressed to new levels of knowledge and creativity? By then, humans may be thinking in different ways and may invent things of a different kind. Today's electronic machines may continue to exist and evolve for a while, just like windmills existed and evolved and did a much better job than humans at what they were doing; but some day electronic machines may look as archaic as windmills look today. I suspect that there is still a long way to go for human creativity. The Singularity crowd cannot imagine the future of human intelligence the way that someone in 1904 could not imagine Relativity and Quantum Mechanics.

Some day the Singularity might come, but i wouldn't panic. Mono-cellular organisms were neither destroyed nor marginalized by the advent of multicellular organisms. Bacteria are still around, and probably more numerous than any other form of life in our part of the universe. The forms of life that came after bacteria were presumably inconceivable by bacteria but, precisely because they were on a different plane, they hardly interact. We kill bacteria when they harm us but we also rely on many of them to work for us (our body has more bacterial cells than human cells). In fact, some argue that a superhuman intelligence already exists, and it's the planet as a whole, Gaia, of which we are just one of the many components.

In some cases we are "afraid" of a machine simply because we can't imagine the consequences. Imagine the day when machines will be able to understand natural language. A human can read only a few books a week. Such a machine, instead, will be able to read in a few seconds all the texts ever produced and digitized by the human race. It is hard to imagine what this implies.

In theory, an artificial intelligence that talks to another artificial intelligence could learn a lot faster than us. We humans need to relocate ourselves to places called universities and take lengthy classes to learn just a fraction of what the experts know. An artificial intelligence could learn in just a few seconds everything that another artificial intelligence knows (with a single "memory dump"). In fact, some day (if computer speed keeps improving) an artificial intelligence could learn everything that EVERY artificial intelligence knows. Imagine if you could learn in a few seconds everything that all humans know.

The way our bodies and brains are built by nature makes it impossible for us to do the same. One possibility is that Nature couldn't do any better. The other possibility is that, maybe, over millions of years of natural selection, Nature figured out that it is better that way.

Critics of A.I. cannot tell us exactly what it is that machines will never be able to do that humans can do. Believers in the Singularity cannot tell us exactly what it is that humans will never be able to do that machines will do. My tentative conclusion is that machines as intelligent as humans are possible (the question is not "if" but "when") whereas machines more intelligent than humans are not possible. Alas, this conclusion hinges on a very vague definition of "intelligence".

What is the Opposite of the Singularity?

What worries me most is not the rapid increase in machine intelligence but a possible decrease in human intelligence.

The Turing Test is commonly understood as: when can we say that a machine has become as intelligent as humans? But the Turing Test is about humans as much as it is about machines because it can equivalently be formulated as: when can we say that humans have become as stupid as a machine? In other words, there is another way for machines to pass the Turing Test: make dumber humans. Let's call the Turing Point the point when the machine has become as smart as humans. The Turing Point can be reached because machine intelligence increases to human level or because human intelligence decreases to machine level.

Humans have always become dependent on the tools they invented. For example, when they invented writing, they lost memory skills. On the other hand, they discovered a way to store a lot more knowledge and to disseminate it a lot faster. Ditto for all other inventions in history: a skill was lost, a skill was acquired. We cannot replay history backwards and we will never know what the world would be like if humans had not lost those memory skills. Indirectly we assume that the world as it is now is the best that it could have been. In reality, over the centuries the weaker memory skills have been driving an explosion of tools to deal with weak memory. Each tool, in turn, caused the decline of another skill. It is debatable if the invention of writing was worth this long chain of lost skills. This process of "dumbification" has been going on throughout society and it accelerated dramatically and explosively with electrical appliances and now with digital devices. The computer caused the decline of calligraphy. Voice recognition will cause the decline of writing.

In a sense, technology is about giving dumb people the tools to become dumber and still continue to live a happy life. A pessimist can rewrite the entire history of human civilization as the history of making humans dumber and inventing increasingly smarter tools to compensate for their increasing stupidity.

In some cases the skill that is lost may have broader implications. If you always use the smartphone's navigator to find places, your brain does not exercise the part of the brain that knows how to navigate the territory. If we don't explore, we don't learn how to explore. If we don't learn how to explore, we don't grow cognitive maps, and we don't train the brain to create cognitive maps. The cognitive map is a concept introduced in 1948 by Edward Tolman to explain how higher animals orient themselves. Without them, the brain is a diminished organ: it will never learn how to do all the things that are enabled by cognitive maps. If George Lakoff is right and all thinking is rooted in physical metaphors, there are countless thoughts that can happen only to brains that know how to manage cognitive maps. Reading novels and discovering scientific theories may not be possible without cognitive maps.

What can machines do now that they could not do 50 years ago? They are just faster, cheaper and can store larger amounts of information. These factors made them ubiquitous. What could humans do 50 years ago that they cannot do now? Ask your grandparents and the list is very long, from multiplication to orientation, from driving in chaotic traffic to fixing a broken shoe. Or just travel to an underdeveloped country where people still live like your old folks used to live and you will find out how incapable you are of performing simple actions that are routine for them. When will we see a robot that is capable of crossing a street with no help from the traffic light? It will probably take several decades. When will we get to the point that the average person is no longer capable of crossing a street without help from the traffic light? That day is coming much sooner. Judging from simple daily chores, one could conclude that human intelligence is not "exploding" but imploding. Based on the evidence, one can argue that machines are not getting much smarter (just faster), while humans are getting dumber; hence very soon we will have machines that are smarter than humans but not only because machines got smarter.

The age of digital devices is enabling the average person to have all sorts of knowledge at her fingertips. That knowledge originally came from someone who was "intelligent" in whichever field. Now it can be used by just about anybody who is not "intelligent" in that field. This user has no motivation to actually learn it: she can just "use" somebody else's "intelligence". The "intelligence" of the user decreases, not increases (except, of course, for the intelligence on how to operate the devices; but, as devices become easier and easier to use, eventually the only intelligence required will be to press a button to turn the device on). Inevitably, humans are becoming ever more dependent on machines, while machines are becoming less dependent on humans.

I chair/organize/moderate cultural events in the Bay Area and, having been around in the old days of the overhead projectors, i'm incredulous when a speaker cannot give her/his talk because her/his computer does not connect properly to the room's audio/visual equipment and therefore s/he cannot use the prepared slide presentation. For thousands of years humans were perfectly capable of giving a talk without any help from technology. Not anymore, apparently. Can you imagine Socrates telling Plato "Sorry, I can't have a dialogue with you unless you have Powerpoint on your laptop"?

The Turing Test could be a self-fulfilling prophecy: at the same time that we (claim to) build "smarter" machines, we are creating dumber people.

My concern, again, is not for machines that are becoming too intelligent, but for humans who are becoming less intelligent. What might be accelerating is the loss of human skills. Every tool deprives humans of the training they need to maintain a skill (whether arithmetic or orientation) and every interaction with machines requires humans to lower their intelligence to the intelligence level of machines (e.g., to press digits on a phone in order to request a service). We can argue forever if the onboard computer of a self-driving car is really "driving", but we know for sure what the effect of self-driving cars will be: raising a generation of humans that is incapable of driving anymore. Every machine that replaces a human skill (whether the pocket calculator or the street navigator) reduces the training that humans get in performing that skill (such as arithmetic and orientation), and eventually causes humans to lose that skill. This is an ongoing experiment on the human race that could have a spectacular result: the first major regression in intelligence in the history of our species.

To be fair, it is not technology per se that makes us dumber. Illiterate people have better memory than literate people because they don't have the technology to write their thoughts down on a piece of paper. That doesn't mean that literate people are dumber than illiterate people. The very system that produces technology makes us dumber. The first step usually consists in some rules and regulations that simplify and normalize a process, whether serving food at a fast-food chain or inquiring about the balance of your checking account or driving a car. Once those rules and regulations are in place, it gets much easier to replace human skills with technology: the human skills required to perform those tasks have been reduced dramatically, and, in that sense, humans have become "dumber" at those tasks. In a sense, technology is often an effect, not a cause: once the skills required to perform a task have been greatly downgraded, it is quite natural to replace the human operator with a machine.

Artificial Intelligence, and automation in general, is part of a bigger story about the Spartanization of societies, the tendency to create a society in which we are reduced to machines that must follow standard processes and that can do little more than obey the rules.

When Charlie Chaplin in "The Great Dictator" shouted "Don't give yourselves to these unnatural men - machine men with machine minds and machine hearts! You are not machines!", he was inveighing against totalitarian regimes, but in reality all regimes want to turn people into machines, for reasons of efficiency if nothing else.

Paraphrasing something Bertrand Russell said about Ludwig Wittgenstein, we are weary of thinking and we are building a society that would make such an activity unnecessary. Then, of course, an unthinking machine would equal an unthinking human, not because the machine has become as thinking as the human, but because the human has become as unthinking as the machine.

The society of rules and regulations that humans have built to create order and stability has the side effect of making us "think" less.

The Turing Test can be achieved in two ways: 1. by making machines so intelligent that they will seem human; 2. by making humans so stupid that they will seem mechanical.

To wit, there could be three stages in human civilization. Stage 1: the coexistence of machine stupidity and human intelligence. Stage 2: the coexistence of machine intelligence and human intelligence. Stage 3: the coexistence of machine intelligence and human stupidity.

With all due respect, when i interact with government officials or corporate employees, the idea that these people, trained like monkeys to repeatedly say and do the prescribed routine, will some day be enslaved by intelligent machines does not seem so implausible.

What will "singular" mean in a post-literate and post-arithmetic world?

"Men have become the tools of their tools" (Henry Thoreau, "Walden", 1854).

Designing Machines and Designing People

Don Norman in "Technology Forces Us To Do Things We're Bad At" (2017) has discussed this phenomenon from the design viewpoint. Ninety per cent of car accidents are blamed on human error, but it is unfair to translate this fact into "caused by humans". They are caused by the designers who designed the car in a way that causes people to make errors. Quote: the design "forces people to behave according to the machine's needs and on its terms". Since people are not good at behaving like machines, people don't do it well, and this leads to car accidents. Even the driver's "distraction" is a design problem: if the driving is too automated, the driver loses the motivation, the cleverness and the resilience that come with a fully manual experience. If you don't expect uncertainty, you won't be prepared to deal with it. When it happens, you will crash.

Technology is supposed to improve the performance of humans, but increasingly it is humans who are asked to behave in a way that improves the performance of machines. As Don Norman writes: "we are inventing people to enhance the life of machines."

One day i had an argument with a friend who used to work on a project for self-driving cars. I eventually managed to win the argument and he conceded that the self-driving car is a bad idea, a dangerous object that will make our life more miserable. It is not important here to repeat the argument and see if you agree with me. What is interesting is his conclusion: the problem is not the self-driving car itself, the problem is all those messy unpredictable humans who populate our world and who make it difficult to program the self-driving car. If we banned humans from the streets of the city, and forced them to move around only in self-driving cars, not only the self-driving car would make a lot of sense but it would also create an idyllic world in which cars could safely assume only machine actions, i.e. no pedestrian who crosses the street recklessly, and no children playing soccer in the street, no Neapolitan driver driving down a one-way street in the wrong direction at maximum speed and no Arab driver parking illegally in the middle of the street. It's an old Silicon Valley adage: "Our system would work just fine if it weren't used by people".

Alas, i fear that my friend was seeing the future. Society has consistently agreed on enforcing rules and regulations on people that make it easier to automate this or that feature of our lives, whether customer support or paying utility bills or simply standing in line at the post office. I predict that cities will simply forbid pedestrians from entering streets: cities will mandate the use of self-driving cars even to simply cross streets. Problem solved: if there are no pedestrians in the streets, there can be no accidents between self-driving cars and pedestrians.

To a few of us this looks like a terrible future. The truth, however, is that whenever new technologies introduce rules and regulations in our society (e.g. the "press 1 for English, 2 for Spanish, etc" or the various numbers that identify you as a driver, a citizen, a bank customer, etc), someone points at the advantage of using them: it superficially makes our life easier, especially in all situations in which we are not skilled. Clearly, an automatic-transmission car is more comfortable than a manual-transmission one if you don't know how to shift gears, and a self-driving car is more comfortable than a regular car if you never learned how to drive or if you are a beginner. Will the mandate to use only self-driving cars on any asphalt make your life more comfortable? It all depends on how dumb and inept you have become. To me, the ultimate goal seems to be different from the one that technologists discuss: it is not about making more intelligent cars but less intelligent people. Less intelligent people are eager to adopt technology that compensates for their stupidity. If you lower the IQ of users, you can get them to use even the dumbest technology because even the dumbest technology will do something for them that they are not able to do on their own.

Intermezzo: The Attention Span

This topic has more to do with modern life than with machines, but it is related to the idea of an "intelligence implosion".

I worry that the chronic scarcity of time in our age is pushing too many decision makers to take decisions having heard only very superficial arguments. The "elevator pitch" has become common even in academia. A meeting that lasts more than 30 minutes is a rarity (in fact, a luxury from the point of view of the most powerful, and therefore busiest, executives). You can't get anybody's attention for more than 20 minutes, but some issues cannot be fully understood in 20 minutes; and some great scientists are not as good at rhetorical speech as they are at their science, which means that they may lose a 20-minute argument even if they are 100% right. Too many discussions are downgraded because they take place by texting on so-called smartphones, whose tiny keyboards discourage elaborate messages. The ultimate reason that we have fewer and fewer investigative reporters in news organizations is the same, i.e. the reduced attention span of the readers/viewers, with the result that the reliability of news media is constantly declining. Twitter's 140-character posts have been emblematic of the shrinking attention span.

(Trivia: Twitter introduced the limitation of 140 characters on human intelligence in the same year, 2006, that deep learning increased the intelligence of machines).

I am not afraid that the human race might lose control of its machines as much as i am afraid that the human race will self-destruct because of the limitations of the "elevator pitch" and of the "tweet"; because of the chronic inability of decision makers, as well as of the general public, to fully understand an issue.

It has become impossible to properly organize events because the participants, accustomed to tweets and texting, will only read the first few lines of a lengthy email. Multiply this concept a few billion times in order to adapt it to the dimensions of humanity's major problems, and you should understand why the last of my concerns is that machines may become too intelligent and the first of my concerns is that human interactions might become too dumb. Elon Musk (at MIT's AeroAstro 100 conference in October 2014) and others are worried that machines may get so smart that they will start building smarter machines; instead, i am worried that people's attention span is becoming so short that it will soon be impossible to explain the consequences of a short attention span. I don't see an acceleration in machine intelligence, but i do see a deceleration in human attention... if not in human intelligence in general.

To summarize, there are three ways that we can produce "dumber" humans. All three are related to technology but in opposite ways.

Firstly, there is the simple fact that a new technology makes some skills irrelevant, and those skills may be lost within one generation. Pessimists argue that little by little we become less human. Optimists claim that the same technology enables new skills to develop. I can personally attest that both camps are right: the computer and email have turned me into a highly-productive multi-tasking cyborg, and at the same time they have greatly reduced my skills in writing polite and touching letters to friends and relatives (with, alas, parallel effects on the quality of my poetry). The pessimists think that the gains do not offset the losses (the "dumbification"), especially when it comes to losing basic survival skills.

Secondly, the rules and regulations that society introduces for the purpose of making us safer and more efficient end up making us think less and less, i.e. behave more and more like (non-intelligent) machines.

Thirdly, the frantic lives of overworked individuals have greatly reduced their attention span, which may result in a chronic inability to engage in serious discussions, i.e. in a more and more superficial concept of "intelligence", i.e. in the limited cognitive experience of lower forms of life.

Cognitive Intermezzo: The Origin of Human Intelligence (or of Machine Intelligence?)

Steven Piantadosi and Celeste Kidd at Rochester University ("Extraordinary Intelligence and the Care of Infants" in Proceedings of the National Academies of Science, 2016) have found a correlation in primates between the degree of intelligence in the adults and the degree of "stupidity" in their offspring. Human brains are so sophisticated because human parents need to take care of the most helpless babies in the animal kingdom. The dumber the children, the smarter the parent has to be to keep them alive. Other animals start walking and eating right after being born. Human babies need to be fed and learn to walk only after many months. Their theory is: the dumber the children, the smarter the parents must be. How do you create very intelligent parents? By giving them very dumb children to watch, protect, lecture, etc. They speculate that there is a self-reinforcing loop at work: because humans make the dumbest children, they must be very intelligent adults, and in order to produce intelligent adults the children must be very dumb. I wonder if a similar self-reinforcing loop is at work on the intelligence of technology: does technology make dumber humans in order for humans to invent smarter technology to deal with dumber humans, technology which will in turn make humans even dumber?

Anthropological Intermezzo: You Are a Gadget

The combination of phones, computers and networks has put each individual in touch with a great number of other individuals, more than at any time in history: humankind at your fingertips. This is certainly lucrative for businesses that want to reach as many consumers as possible with their ads. But do ordinary people really benefit from being connected to thousands of people, and soon millions? What happens to solitude, meditation, to "thinking" in general (whether scientific thinking or personal recollection) when we are constantly interacting with a multitude of minds (only some of which really care)?

You "are" the people with whom you interact, because they influence who you become. In the old days those were friends, relatives, neighbors and coworkers. Now they are strangers spread all over the world (and old acquaintances with whom you only share distant memories). Do you really want to be "them" rather than being yourself?

You are not surrounding yourself with people: you are surrounding yourself with gadgets like smartphones and laptops.

If you surround yourself with philosophers, you are likely to become a philosopher, even if only an amateur one. If you surround yourself with book readers, you are likely to read a lot of books. If you surround yourself with physicists, you are likely to understand Relativity and Quantum Mechanics. And so on. So what is likely to happen to you if you surround yourself with gadgets that mediate your interaction with people and with the world at large?

It is infinitely easier to produce/accumulate information than to understand it and make others understand it.

Existential Intermezzo: You Are an Ad

We are surrounded by billboards, ads and commercials. Even the webpages that you visit on the Web depend on an algorithm of page-ranking whose behavior can be manipulated by expert professionals, so that you will visit the webpages that they want you to visit and not the ones that you would have visited using brain and luck. If you search for "Piero" with your favorite search engine, you are more likely to stay at an apartment complex in Los Angeles and eat at a restaurant in Las Vegas than read a page on my website, a website that has been around for 20 years, contains 10,000 pages of text, and is edited by a writer named Piero.

In the 1990s my website would show up as one of the top three results of a search for "Piero" on every search engine. I'll let you decide if the results that you get today are "better" than what you were getting 20 years ago. You have to thank progress in Artificial Intelligence for it.

Artificial Intelligence is making these ads more powerful, more targeted, more convincing, more inescapable.

Our lives are increasingly steered by the advertisement that surrounds (traps?) us.

What happens to us if (when) the ads stop?

Will we still be able to live a life?

There might be a sinister new meaning for Eliot's famous line "You are the music while the music lasts" (T. S. Eliot, "The Dry Salvages", 1941).

Semantics

In private conversations about "machine intelligence" i like to quip that it is not intelligent to talk about intelligent machines: whatever they do is not what we do, and, therefore, is neither "intelligent" nor "stupid" (attributes invented to define human behavior). Talking about the intelligence of a machine is like talking about the leaves of a person: trees have leaves, people don't. "Intelligence" and "stupidity" are not properties of machines: they are properties of humans. Machines don't think, they do something else. Machine intelligence is as much an oxymoron as human furniture. Machines have a life of their own, but that "life" is not human life.

We apply to machines many words invented for humans simply because we don't have a vocabulary for the states of machines. For example, we buy "memory" for our computer, but that is not memory at all: it doesn't remember (it simply stores) and it doesn't even forget, the two defining properties of (biological) memory. We call it "memory" for lack of a better word. We talk about the "speed" of a processor but it is not the "speed" at which a human being runs or drives. We don't have the vocabulary for machine behavior. We borrow words from the vocabulary of human behavior. It is a mistake to assume that, because we use the same word to name them, they are the same thing. If i see a new kind of fruit and call it "cherry" because there is no word in my language for it, it doesn't mean it is a cherry. A computer does not "learn": what it does when it refines its data representation is something else (that we don't do).

It is not just semantics. Data storage is not memory. Announcements of exponentially increasing data storage miss the point: that statistical fact is as relevant to intelligence as the exponential increase in credit card debt. Just because a certain sequence of zeroes and ones happens to match a sequence of zeroes and ones from the past it does not mean that the machine "remembered" something. Remembering implies a lot more than simply finding a match in data storage. Memory does not store data. In fact, you cannot retell a story accurately (without missing and possibly distorting tons of details) and you cannot retell it twice with the same words (each time you will use slightly different words). Ask someone what her job is, something that she has been asked a thousand times, and she'll answer the question every time with a different sequence of words, even if she tries to use the same words she used five minutes earlier. Memory is "reconstructive", the crucial insight that Frederic Bartlett had in 1932. We memorize events in a very convoluted manner, and we retrieve them in an equally convoluted manner. We don't just "remember" one thing: we remember our entire life whenever we remember something. It's all tangled together. You understand something not when you repeat it word by word like a parrot (parrots can do that, and tape recorders can do that) but when you summarize it in your own words, different words than the ones you read or heard: that is what we call "intelligence". I am always fascinated, when i write something, to read how readers rewrite it in their own words, sometimes using completely different words, and sometimes saying it better than i did.

It is incredible how bad our memory is. A friend recommended an article by David Carr about Silicon Valley published in the New York Times "a few weeks ago". It took several email interactions to figure out that a) the article was written by George Packer, b) it was published by the New Yorker, c) it came out one year earlier. And, still, it is amazing how good our memory is: it took only a few sentences during a casual conversation for my friend to relate my views on the culture of Silicon Valley to an article that she had read one year earlier. Her memory has more than just a summary of that article: it has a virtually infinite number of attributes linked to that article such that she can find relevant commonalities with the handful of sentences she heard from me. It took her a split second to make the connection between some sentences of mine (presumably ungrammatical and incoherent sentences because we were in a coffee house and i wasn't really trying to compose a speech) and one of the thousands of articles that she has read in her life.

All forms of intelligence that we have found so far use memory, not data storage. I suspect that, in order to build an artificial intelligence that can compete with the simplest living organism, we will first need to create artificial memory (not data storage). Data storage alone will never get you there, no matter how many terabytes it will pack in a millimeter.

Human memory confuses things easily, and even forgets easily. In fact, human memory is extremely good at forgetting. If we forgot important facts, we would be dead, but we instead forget only those facts that improve our memory (not just the storage, but the whole ability to retrieve and associate knowledge). For example, Blake Richards and Paul Frankland of the Canadian Institute for Advanced Research (CIFAR) have shown that forgetting is as important for human memory as remembering ("The Persistence and Transience of Memory", 2016). Machines that don't forget don't "remember" either ("remember" the way human memory does, in relation to everything else).

What computers do is called "savant syndrome" in the scientific literature: idiots (very low intelligence quotient) with a prodigious memory.

I am not advocating that machines should be as forgetful and slow as us. I am simply saying that we shouldn't be carried away by a faulty and misleading vocabulary.

Data is not knowledge either: having amassed all the data about the human genome does not mean that we know how human genes work. We know a tiny fraction of what they do even though we have the complete data.

I was asking a friend how the self-driving car works in heavy traffic and he said "the car knows which other cars are around". I object that the car does not "know" it. There is a system of sensors that continuously relay information to a computer that, in turn, calculates the trajectory and feeds it into the motor controlling the steering wheel. This is not what we mean when we say that we "know" something. The car does not "know" that there are other cars around, and it does not "know" that cars exist, and it doesn't even "know" that it is a car. It is certainly doing something, but can we call it "knowing". This is not just semantics: because the car does not "know" that it is a car surrounded by other cars driving on a road, it also lacks all the common sense or general knowledge that comes with that knowledge. If an elephant fell from the sky, a human driver would be at least surprised (and probably worried about stranger phenomena ahead), whereas a car would simply interpret it as an object parked in the middle of the highway.

When, in 2013, Stanford researchers trained a robot to take the elevator, they realized that there was a non-trivial problem: the robot stopped in front of the glass doors of the elevator interpreting its own reflection into it as another robot. The robot does not "know" that the thing is a glass door otherwise it would easily realize that there is no approaching robot, just a reflection getting bigger like all reflections do when you walk towards a mirror.

It is easy to claim that, thanks to Moore's law, today's computers are one million times faster than the computers of the 1980s, and that a smartphone is thousands of times faster than the fastest computer of the 1960s. But faster at what? Your smartphone is still slower than a snail at walking (it doesn't move, does it?) and slower than television at streaming videos. Plenty of million-year old artifacts are faster than the fastest computer at all sorts of biological processes. And plenty of analog devices (like television) are still faster than digital devices at what they do. Even in the unlikely event that Moore's law applies to the next 20 years, processing speed and storage capacity will improve by a factor of a million. But that may not increase at all the speed at which a computer can summarize a film. The movie player that i use today on my laptop is slower (and a lot less accurate) in rewinding a few scenes of the film than the old videotape player of twenty years ago, no matter how "fast" the processor of my laptop is.

We tend to use cognitive terms only for machines that include a computer, and this habit started way back when computers were invented (the "electronic brains"!). Thus the cognitive vocabulary tempts people to attribute "states of mind" to those machines. We don't usually do this to other machines. A washing machine washes clothes. If a washing machine is introduced that washes tons of clothes in a split second, consumers will be ecstatic, but presumably nobody would take it as an example of human or superhuman intelligence. And note that appliances do some pretty amazing things. There's even a machine called "television set" that shows you what is happening somewhere else, a feat that no intelligent being can do. We don't attribute cognitive states to a television set even though the television set can do something that requires more than human intelligence.

Take happiness instead of intelligence. One of the fundamental states of human beings is "happiness". When is a machine "happy"? The question is meaningless: it's like asking when does a human being need to be watered? You water plants, not humans. Happiness is a meaningless word for machines. Some day we may start using the word "happy" to mean, for example, that the machine has achieved its goal or that it has enough electricity; but it would simply be a linguistic expedient. The fact that we may call it "happiness" does not mean that it "is" happiness. If you call me Peter because you can't spell my name, it does not mean that my name is Peter.

Semantics is important to understand what robots really do. Pieter Abbeel at UC Berkeley has a fantastic robotic arm that can fold towels with super-human dexterity. But what it does is "not" what a human does when she folds towels. Abbeel's robot picks up a towel, shakes it, turns, and folds it on a table. And it does it over and over again, implacably, without erring. What any maid in a hotel does is different. She picks up a towel and folds it… unless the towel is still wet, or has a hole, or needs to be washed again, or… That is what "folding towels" really means. The robot is not folding towels: it is performing a mechanical movement that results in folded towels, some of which may be entirely useless for human purposes. The maid is not paid to do what the robot does. She is paid to fold towels, not to "fold towels" (the quotes make the difference).

Cleaning up a table at a restaurant is not just about throwing away all the objects that are lying on the table, but about recognizing which objects are garbage and which are not (eg a crumpled piece of paper versus paper money), which ones must be moved elsewhere and which ones belong there (e.g. a vase full of flowers, but not a vase full of withered flowers). A cell phone left on the table is neither garbage nor a dish to be washed: it is something that the customer forgot behind.

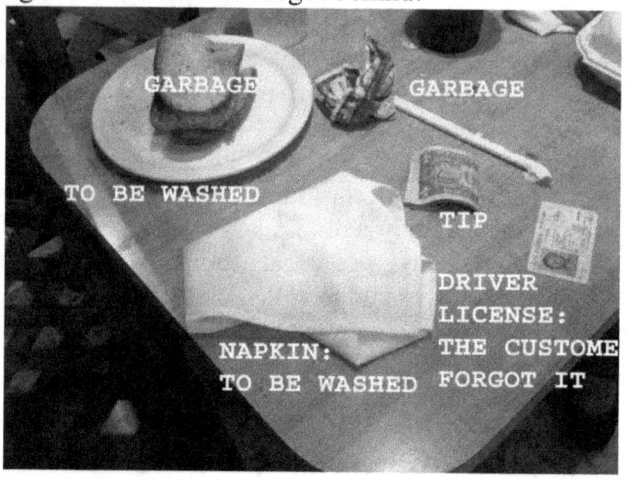

It is true that machines can now recognize faces, and even scenes, but they have no clue what those scenes mean. We will soon have machines that can recognize the scene "someone picked up an object in a store", but when will we have a machine that can recognize "someone STOLE an object from a store?" A human being understands the meaning of this sentence because humans understand the context: some of those objects are for sale, or belong to the shop, and a person walking away with those objects is a thief, which is very different from being a store clerk arranging the goods on the shelves or a customer bringing the object to the counter and paying for it. We can train neural networks to recognize a lot of things, but not to understand what those things mean.

And the day when we manage to build a machine that can recognize that someone is stealing someone else's wallet, we will still have a higher level of understanding to analyze: that scene could be a prank, as indicated by the fact that one of the two people is smiling, and that we know that they are old friends. In that case you don't call the police but simply wait for the fun to begin. And if we build a machine that can even recognize a prank, we will still have to go up one more level of abstraction to consider the case in which this is happening in a movie, not in reality. And so on and so forth. The human mind can easily grasp these situations: the same scene can mean so many different things.

The automatic translation software that you use to translate Chinese into English doesn't have a clue what those Chinese words mean nor what those English words mean. If the sentence says "Oh my god there's a bomb!" the automatic translation software simply translates it into another language. A human interpreter would shout "everybody get out!", call the emergency number and... run!

Intelligence is not about the error rate in recognizing an action. Intelligence is about "recognizing" the action for what it is. Mistakes are actually fine. We make mistakes all the time. Sometimes we think we recognized an old friend and instead it turns out to be a complete stranger. We laugh and move on. And that's another way to ponder the difference in semantics. We laugh out loud when we see the kind of mistakes that a computer makes when it is trying to recognize a scene. I just searched for images related to "St Augustine what is time" and the most famous search engine returned a page of pizza images. An image search for "Popper logic scientific discovery" returns a page with the correct images of that philosophy classic, but when i clicked on the image of the original edition i got a group of "related images" that were about a porn star. (Yes, the same result showed up on all computers, not just mine). This is what humans do: humans laugh out loud when someone (or something) makes such silly mistakes. The real Turing Test is this: when will we have a computer that laughs out loud at silly mistakes made by other computers or by itself?

Even the word "mistake" needs to be properly defined. It has been widely publicized that machines can now recognize images with more accuracy than humans can; that machines make fewer mistakes. But that is a misleading statement. The mistakes that we make when we misjudge an image are important. When hiking in the forest, for example, a hiker may mistake a tree for a bear. That is probably a mistake that a machine would not make: trained to recognize trees, it would recognize a tree as a tree, not as a bear. Sometimes we even mistake a tree for a person, or for a fellow hiker; and a boulder behind the tree for a tent. A machine would probably never mistake a rock for a tent. But these mistakes are actually important: there "could" be a bear in the forest, and there "could" be a fellow hiker, and so on. These are not details. These are important facts. Our survival depends on these "mistakes". There is a big misunderstanding in what constitutes a mistake: a mistake is to think that there are no bears in the forest. Our brain is expecting bears, just like it is expecting other hikers (and many other possible encounters). A machine that does not expect bears and hikers will obviously never be mistaken, but that's a machine that knows nothing about the environment. As a tool for hikers in the wilderness, it would be a dangerous tool. I will trust the machine the day that it mistakes a tree for a bear. Then i will feel confident that the machine is ready to go into the wilderness.

Using human semantics, the most intelligent machines ever built, such as IBM's "Watson" and Google's "AlphaGo", are incredibly stupid. They can't even cook an omelette, they cannot sort out my clothes in the drawers, they cannot sit on the sidewalk and gossip about the neighborhood (i am thinking of human activities that we normally don't consider "very intelligent"). A very dumb human being can do a lot more than the smartest machines ever built, and that's probably because there is a fundamental misunderstanding about what "intelligent" means.

AlphaGo did not "win" a game of weiqi/go. AlphaGo never learned how to play weiqi. AlphaGo cannot answer any question, but, if it could, it would not know the answer to the question: "What are the rules of weiqi?" AlphaGo never learned to play weiqi. AlphaGo simply calculates the most likely good move based on moves made by thousands of go masters in similar situations. AlphaGo has no idea that it is playing go, playing a game, playing with humans, etc. Therefore it does not "win" or "lose".

There is also talk of "evolving A.I. systems", which projects the image of machines getting more and more intelligent. This can mean different things: a) a software that devises a better technique to solve problems; b) a software that has improved itself through learning from human behavior; c) a software that has improved itself through self-playing. None of this is what we mean when we say that a species evolved in nature. Evolution in nature means that a population makes children that are all slightly different and then natural selection rewards the ones that are the best fit to the environment. After thousands of generations, the population will evolve into a different species that will not mate with the original one. There is nothing wrong with software programs that get better at doing what they do, but calling it "evolution" evokes a metaphor (and an emotional reaction) that just does not apply to today's software. There is no software that evolves. And even if you really want to call it "evolution", you should realize that the software program has "evolved" because of the software engineer who programmed it. If tomorrow beavers start building better dams, do you talk about the evolution of dams or the evolution of beavers?

The mother of all misunderstandings is the fact that we classify some technologies under the general label "Artificial Intelligence", which automatically implies that machines equipped with those technologies will soon become as intelligent as humans. There are many technologies that have made, are making and will make machines more intelligent. For example, the escapement and the gyroscope made several machines more intelligent, from clocks to motion-sensing devices, but people are not alarmed that escapements and gyroscopes might take over the world and kill us all. Monte Carlo methods have been widely used in simulation since Stanislaw Ulam published the first paper in 1949. They are usually classified under "Numerical Analysis" and sometimes under "Statistical Analysis", and don't scare anybody. Mathematically speaking, they apply statistical methods to find a solution to problems that are described by mathematical functions with no known solution. Sounds boring, right? But the Monte-Carlo tree search is a Monte Carlo method used by AlphaGo for determining the best move in a game. Now it doesn't sound boring anymore, right? If we now classify the Monte Carlo method under "Artificial Intelligence", we suddenly turn a harmless statistical technique into some kind of dangerous intelligent agent, and the media will start writing articles about how this technique will create super-intelligent machines. That is precisely what happened with "neural networks". When in 1958 the psychologist Frank Rosenblatt built the first "neural network", his aim was indeed to model how the human brain works. Today we know that the similarities are vague at best. It is like comparing a car to a horse because the car was originally called the "horseless carriage" (we still measure a car's power in horsepower!) Progress in neural networks has not been based on neuroscience but on computational mathematics: we need mathematical functions that can be implemented in computers and that can yield solutions in a finite time. Calling them "neural networks" makes people think of brains, and turns them into ideas for Hollywood movies. If we called them "constraint propagation" (which is what they are), they would only make people think of the algebra they hated in high school.

The very notion of what is supposed to be as "intelligent" as humans is frequently flawed. When they tell us that an A.I. system (a neural network) has reached or surpassed human-level performance, what are they really comparing? They are comparing a human being, which is a body equipped with many organs including eyes, mouth and arms, and encapsulating a brain, with a piece of software that takes as input some data and produces as output some other data. The human "performance" in, for example, recognizing an object consists in our eyes seeing the object and sending signals to the brain that then recognizes the object. This process is completely different from the process of computing zeroes and ones inside a neural network that has no eyes. The neural network does not "see" the object and therefore what it is doing is not "recognizing" it. A fair comparison would be between a robot whose "brain" is such a neural network and which is fed not data files but natural images in the world. For example, let the human and the robot walk around the city and then compare the performance in recognizing objects. A blind person does not recognize an image. A neural network without a body does not "recognize" an image either: what it does is some mathematical computation that eventually outputs a result that one can compare with the results coming out of other machines. We humans have no brain that can enter a competition with an A.I. system that doesn't have a brain. We cannot pull out our brain from the skull and feed it data files to see if it is as good as an A.I. system at naming objects represented by those data.

Incidentally, that's why sometimes i quip that A.I. is not a science but an art. If i draw two small circles, a straight vertical line between them and a curved line under the straight line, most people "recognize" a face. But obviously those signs are not a face at all: they are just signs on a piece of paper. Show Picasso's portrait of Ambroise Vollard and ask "what is it?" and most people would probably reply "a face". But that is not a face at all either. Such a face never existed. It is just an imaginary object that looks like someone's face. Now show the "Mona Lisa": what is it? A face? No, even that is not a face, even if this time the painter was very faithful to the real face of a real person. It is still not a face: it is a painting of someone's face. A face is the thing, made of flesh, in the front of my head. A painting of it is not a face, no matter how accurate it is.

"Machine learning" is the buzzword of 2017. Countless programs are being re-branded as "machine learning". Many of them do exactly the same thing that they were doing before being promoted to "machine learning". They do statistical analysis. They classify data. Call it "data classification" and nobody has visions of Hollywood monsters attacking humankind. But call it "machine learning" and you trigger a debate on whether these machines threaten humankind. Data classification is just one of the things that computers can do faster than humans, and that's because it can be reduced to computation using fairly old statistical methods. Neural networks are a new way of performing data classification. You can use a properly annotated set of male and female faces to train a neural network, and then the neural network will be able to classify a face as male or female. If you want it to classify young and old faces, you need to train it again with a set of faces annotated in a different way. They can classify data much faster than any human the same way that they can compute much faster than any human. Whether "data classification" is all there is to "learning" is a different story, but calling it "machine learning" implies just that.

Is the hammer a conscious being? No, of course, it is a hammer; at best, a tool that can be used in many circumstances. But not a conscious being like us. The name "artificial intelligence" is misleading. Call a hammer "artificial creator" and people will start asking "can it become conscious"? Call a neural network what it is ("image recognition", "speech recognition", etc) and people will be more likely to ask "how much does it cost" than "will it become conscious"?

The reason that sometimes i am skeptical about ever getting machines of any significant degree of intelligence is that futurists use a definition of "intelligence" that has nothing to do with the definition of intelligence used by ordinary people. When the computer displayed "Would you like to download new updates?" live on the giant screen of the Stanford auditorium so that 200 people could see it and laugh out loud while the elderly physicist was focused on explaining the exciting new findings of the particle accelerator, it was obviously an incredibly stupid moment of an incredibly stupid machine, but futurists would instead point out the number of "logical operations" (ironically abbreviated as FLOPs) that this cheap portable computer can perform in a split second. This will not help build a better machine, just to flop harder (sorry for the pun).

The Accelerating Evolution of Machines

Whenever we look at the rapid progress posted by machines in performing this or that task, it is tempting to say that the machine achieved in a few years what took humans millions of years of evolution to achieve. The argument goes like this: "Yes, it took years to build a machine that recognizes a cat, but how long did it take evolution to create a living being that recognizes cats?"

The truth is that any human-made technology is indirectly using the millions of years of evolution that it took to evolve its creator (Homo Sapiens). No human being, no machine. Therefore it is incorrect to claim that the machine came out of the ENIAC: it came out of millions of years of evolution, just like my nose. The machine that is now so much better than previous models of a few years ago did NOT evolve: WE evolved it (and continue to evolve it).

There is no machine that has created another machine that is superior. WE create a better machine.

We are capable of building machines (and tools in general) because those millions of years of evolution equipped us with some skills (that the machine does NOT have). If humans became extinct tomorrow morning, the evolution of machines would come to an end. Right now this is true of all technologies. If all humans die, all our technologies die with us (until a new form of intelligent life arises from millions of years of evolution and starts rebuilding all those watches, bikes, coffee makers, dishwashers, airplanes and computers). Hence, technically speaking, there has been no evolution of technology.

This is yet another case in which we are applying an attribute invented for one category of things to a different category: the category of living beings evolve, the category of machines does something else, which we call "evolve" by recycling a word that actually has a different meaning. It would be more appropriate to say that a technology "has been evolved" rather than "evolved": computers have been evolved rapidly (by humans) since their invention.

Technologies don't evolve (as of today): we make them evolve.

The day that we have machines that survive without human intervention and build other machines without human intervention, we can apply the word "evolve" to those machines.

As far as i know those machines don't exist yet, which means that there has been zero evolution in machine intelligence so far.

The machine is not intelligent, the engineer who designed it is. That engineer is the product of millions of years of evolution, the machine is a by-product of that engineer's millions of years of evolution.

(See the appendix for a provocative counter-argument: maybe i got it all wrong, and it is technologies that evolve and use us to evolve).

Non-human Intelligence is Already Here

There are already many kinds of intelligence that we cannot match nor truly comprehend. Bats can avoid objects in absolute darkness at impressive speeds and even capture flying insects because their brain is equipped with a high-frequency sonar system. Migratory animals can orient themselves and navigate vast territories without any help from maps. Birds are equipped with a sixth sense for the Earth's magnetic field. Purple martins migrate from Brazil to the USA and back each year. Some animals have the ability to camouflage. The best color vision is in birds, fish, and some insects. Many animals have night vision. Animals can see, sniff and hear things that we cannot, and airports still routinely employ sniffing dogs (not sniffing humans) to detect food, drugs and explosives. And don't underestimate the brain of an insect either: how many people can fly and land upside down on a ceiling?

Howard Hughes' "Sensory Exotica" (1999), Frans de Waal's "Are We Smart Enough to Know How Smart Animals Are?" (2016) and Peter Godfrey-Smith's "Other Minds" (2016) document the amazing skills of the animals that populate our planet.

(The earliest known paintings, in the caves of Lascaux and Chauvet, depict animals, not humans. I suspect those painters marveled at the superhuman powers of animals, not at the superior intelligence of humans).

Virtually all dogs existing today are artificial living beings: they are the result of selective breeding strategies. If you think that your dog is intelligent, then you have "artificial intelligence" right at home.

Ironically, when Deborah Gordon discovered that ant colonies use a packet-switching technique very similar to the one employed by the Internet ("The Regulation of Ant Colony Foraging Activity without Spatial Information", 2012), the media wrote that ants can do what the Internet does when in fact ants have been doing it for about 100 million years: it took human intelligence 200,000 years to figure out the same system of communication devised by ant intelligence.

Summarizing, many animals have powers we don't have. We have arbitrarily decided that any skill possessed by other animals and not by humans is an inferior skill, whereas any skill possessed by humans and not by other animals is a superior skill. This leads me to wonder what will make a skill "superhuman": just the fact that it is possessed by a machine instead of an animal?

And, of course, we have already built machines that can do things that are impossible for humans. The clock, invented almost a thousand years ago, does something that no human can do: keep time. Telescopes and microscopes can see things that the naked eye ` cannot. We can only see a human-level rendition by those machines, which is equivalent to a higher intelligence explaining something in simpler terms to a lower intelligence. We cannot do what light bulbs do. We cannot touch the groove of a rotating vinyl record and produce the sound of an entire philharmonic orchestra. And, of course, one such appliance is the computer, that can perform calculations much faster than any mathematician could. Even the pre-digital calculators of the 1940s (for example, the ones used to calculate ballistic trajectories) could calculate faster than human brains. In fact, we have always been post-human, coexisting with, and relying on, and being guided by, technology that was capable of super-human feats (and there have always been philosophers debating whether that post-human condition is anti-human or pro-human).

The intelligence of both animals and tools is not called "superhuman" simply because we are used to it. We are not used to robots doing whatever it is that they will do better than us and therefore we call it "superhuman" when in fact we should just call all of these "non-human life"; and maybe "non-human intelligence" depending on your definition of "intelligence".

If a machine ever arises (and proliferates) that is alive and capable of feats that are superhuman, it will just be yet another form of non-human life: not the first one, not the last one. Of course, there are plenty of forms of life that are dangerous to humans, mostly very tiny ones (like viruses and ticks). It comes with the territory. If you want to call it "superhuman", suit yourself.

One gene can make a huge difference in brain structure and function, as the tiny difference between a chimpanzee's DNA and human DNA proves. Gene therapy is already here and that is indeed progressing quickly. Changing the genes of the human DNA may have consequences that are orders of magnitudes bigger than we can imagine. That is one of the reasons why i tend to believe that "superhuman" intelligence, if it comes at all, is more likely to come from synthetic biology than from computers.

There are even qualitative differences in the "intelligences" of a person as the person grows and changes. Psychologists since at least Jean Piaget have studied how the mental life of a child changes dramatically, qualitatively, from one stage in which some tasks are impossible to a new stage in which those tasks become the everyday norm: each new stage represents a "super" intelligence from the viewpoint of the previous stage. There is an age at which the child conceives little more than herself and her parents. That child's brain just cannot conceive that there are other people and that people live on a planet and that the planet contains animals, trees, seas, mountains, etc; that you have to study and work; not to mention the mind-boggling affairs of sex and where children come from; and that some day you will die. All of this emerges later in life, each stage unlocking a new dimension of understanding. (And i wonder if there is an end to this process: if we lived to be 200 years old in good health, what would be our understanding?) My intelligence is "super" compared to the intelligence that i had as a little child.

At the same time try learning languages or any other skills at the speed that children learn them. Children can do things with their minds that adults cannot do anymore: sometimes you feel that you cannot understand what their minds do, that they are little monsters. Children are superhuman too, as Alison Gopnik argues in "The Philosophical Baby" (2009). One wonders what we could achieve if we remained children all our lives (more about this later).

Computer science has already succeeded in creating a new kind of intelligence. We are surrounded by "intelligent" machines that are "intelligent" in their own way just like a worm is intelligent in its own way. So far i have not been as excited to interact with computers as i am when i observe animals, and certainly a lot less excited than when i read a good novel or listen to good music. If A.I. ever succeeds in creating a machine that exhibits an extraordinary kind of intelligence, and that can, for example, produce great ideas and great art, i will be excited to interact with it the same way that i am excited to interact with intelligent people.

And many of these people are much more intelligent than me. Far from terrifying me, they inspire me to become more intelligent.

The Consciousness of Super-human Intelligence

(Warning: this chapter and the next one are boring philosophical speculation).

Given that non-human intelligence exists all around us, what would make a particular non-human intelligence also "superhuman"? I haven't seen a definition of "superhuman" (as opposed to simply "non-human").

However, there is at least one feature that i would expect to find in a superhuman intelligence: consciousness. I think, i feel, sometimes i suffer and sometimes i rejoice. If i have consciousness, an intelligence that is superior to mine should have it too.

We know that human brains are conscious, but we don't really know why and how. We don't really know what makes us conscious, how the electrochemical processes inside our brain yield feelings and emotions. (My book "Thinking about Thought" is a survey of the most influential viewpoints on consciousness). An electronic replica of your brain might or might not be conscious, and might or might not be "you". We don't really know how to build conscious beings, and not even how to find out if something is conscious. If one of the machines that we are building turns out to develop its own consciousness, it will be an amazing stroke of luck.

However, i doubt that i would call "superhuman" something that is less conscious than me, no matter how fast it is at calculating the 100th million digit of the square root of 2, how good it is at recognizing cats and how good it is at playing go/weiqi.

However, you might object, a super-human intelligence will not need to be conscious. You might object that feelings and emotions are a sign of weakness, not of strength. Consciousness makes us cry. Feelings cause us to make mistakes that we later regret, and that sometimes hurt us or hurt others. Maybe a being that is more intelligent than us and does not feel anything is actually the secret to outperforming human intelligence.

In fact, "consciousness" for an information-processing machine could be something altogether different from the consciousness of an energy-processing being like us. Our qualia (conscious feelings) measure energy levels: light, sound, etc. If information-processing machines ever develop qualia, it would make sense that those qualia be about information levels; not qualia related to physical life, but qualia related to "virtual" life in the universe of information.

It is not even clear whether superhuman intelligence requires human intelligence first: can human-level intelligence be skipped on the way to superhuman intelligence? Do machines need to be as smart as us before becoming smarter than us or can they find a short cut to superhuman intelligence?

We cannot answer this question looking at biological intelligence because the progress of machine intelligence is happening in a completely different way from the way that biological intelligence evolved. The way Nature works is simple: new species don't need to climb the ladder of intelligence: they start out at a given level of intelligence, bypassing all the lower ones. For example, humans have never been as unintelligent as bacteria. The way Artificial Intelligence works is different: it tweaks software programs, making them more and more intelligent, and these software programs can run on any computer that is powerful enough. A.I. is about the progress of software (that can run on any hardware), Nature is about the progress of hardware (a hardware that also includes a brain that, in turn, somehow includes a software called "mind").

The Intelligence of Super-human Intelligence

What does it take for machine intelligence to reach the point of human-level intelligence?

One is tempted to answer: just build an electronic replica of a human brain. If we replaced each and every neuron in your brain with an electronic chip, i am not sure that you would still be "you" but your brain should still yield a form of human intelligence, wouldn't it?

Unfortunately, we are pretty far from implementing that full replica of your brain (note that i keep saying "your" and not "mine"). It is a bit discouraging that the smallest known brain, the brain of the roundworm (300 neurons connected by a few thousand synapses) is still smarter than the smartest neural network ever built.

If you think that this hypothetical electronic replica of your brain would not be as smart as you, then you are implying that the very "stuff" of which the brain is made is important in itself; but then machine intelligence is impossible because machines are not made of that "stuff".

And, again, can a machine reach human-level intelligence without consciousness? Is consciousness required in order to be as smart as Einstein? Could a machine be as smart as Einstein without having any feelings and emotions?

Does "machine intelligence" require all of you, including the mysterious inscrutable silent existence that populates your skull, the vast unexplored land of unspoken thoughts and feelings that constitutes "you"? What i hear when i listen to you is just a tiny fraction of what you are thinking and feeling. What i see when i watch you is just a tiny fraction of what you thought of doing, dreamed of doing, and plan doing. When i hear a robot talk, that is the one and only thing that it "wants" to say. When i see it move, that is the one and only thing that it wants to do.

Koan: What is not Artificial Intelligence although it is Superhuman?

In 1941 the great Argentinian writer Jorge-Luis Borges wrote a tale titled "The Library of Babel" (1941) in which he imagined a library that contains all the permutations of the alphabet. If we throw enough computation power at it, of course, our machines can produce combinatorially every single book that has ever been published, and all possible compendia that have never been written about all the books ever published and all the books never published, and countless other novels and poems, including histories of the future, biographies of people that have not been born yet, and translations of all these books in all languages, including languages never invented by humans.

Borges doesn't tell us who created that library, and probably wouldn't know how to create it because there aren't enough people on Earth living long enough to create it; but in that same year of 1941, a continent apart, a German civil engineer named Konrad Zuse built the first programmable computer, a machine capable of creating that library. We don't need Artificial Intelligence for this task, just a ten-line program. We don't need Google's TPUs of 2017, just Zuse's 1941 machine.

Intelligent Behavior in Structured Environments

When you need to catch a bus in some underdeveloped countries, you don't know at what time it will arrive nor how much you will be charged for the ticket. In fact you don't even know how it will look like (it could be a generic truck or a minivan) and where it will stop. Once on board, you tell the driver where you want to get off and hope that he will remember. If she is in a good mood, she might even take a little detour to drop you right in front of your hotel. On the other hand, when you take a bus in a developed country, there is an official bus stop (the bus won't stop if you are 20 meters before or after it), the bus is clearly recognizable and marked with the destination and won't take any detour for any reason, the driver is not allowed to chat with the passengers (sometimes she is physically enclosed in a glass cage), the ticket must be bought with exact change at a ticket vending machine (and sometimes validated inside at another machine). There is a door to be used to exit, and you know when to exit because the name of the bus stop is displayed on an LED screen. On many long-distance trains and buses you also get an assigned seat (you can't just sit anywhere).

It is easy to build a robot that can ride a bus in a developed country, much more difficult to build a robot that can ride a bus in an underdeveloped country. What makes it easy or difficult is the environment in which it has to operate: the more structured the environment, the easier for the robot. A structured environment requires less "thinking": just follow the rules and you'll make it. However, what really "makes it" is not you: it's you plus the structured environment. That's the key difference: operating in a chaotic, unpredictable situation is not the same thing as operating in a highly structured environment. The environment makes a huge difference. It is easy to build a machine that has to operate in a highly structured environment, just like it is easy for a bullet train to ride at 300 km/hour on rails.

We structure the chaos of nature because it makes it easier to survive and thrive in it. Humans have been spectacularly successful at structuring their environment so that it obeys simple, predictable rules. This way we don't need to "think" too much: the structured environment will take us where we want to go. We know that we can find food at the supermarket and a train at the train station. In other words, the environment makes us a little more stupid but allows anybody to achieve tasks that would otherwise be difficult and dangerous, i.e. that would require a lot of intelligence. When the system fails us, we get upset because now we have to think, we have to find a solution to an unstructured problem.

If you are in Paris and the metro is on strike and it is impossible to get a taxi, how to do you get to your appointment in time? Believe it or not, most Parisians manage. Most tourists from the USA don't. If there is no traffic light and cars don't stop for pedestrians and traffic is absolutely horrible, how do you cross a wide boulevard? Believe it or not, Iranians do it all the time. Needless to say, most Western tourists spend hours trying to figure it out.

It is certainly very impressive how well humans structure a universe that is chaotic. The more we structure it, the easier for extremely dumb people and machines to survive and thrive in it.

The claims of the robotic industry are often related to structured environments, not to their robots. It is relatively easy to build an autonomous car that rides on a highway with clearly marked lanes, clearly marked exits, ordered traffic, and maps that detail everything that is going to happen. It is much more difficult (orders of magnitude more difficult) to build an autonomous car that can drive through Tehran or Lagos (this is a compliment to Iranian and Nigerian drivers, not an insult). Whoever claims that a computer is driving a car is distorting the facts: it is not the computer that is driving the car but the environment that has been structured so that any inexperienced and not particularly intelligent driver, and even a computer, can drive a car. Today's computer cannot drive a car in the traffic of Lagos or Tehran. It will if and when the streets of Lagos and Tehran become as well structured as the streets of California, if and when Iranian and Nigerian drivers are forced to obey strict traffic rules. Saying that the on-board computer is steering the driverless car is like saying that the locomotive knows in which direction to take the train: the locomotive is simply constrained by the rails to take the correct direction.

In order for self-driving cars to use our streets, we will need to retrofit roads with devices that tell the car what to do at every point in time. It is not intelligence but old-fashioned infrastructure that will allow very dumb self-driving cars to drive safely; in other words we will need the equivalent of the highly-structured system of rails and controllers that will, eventually, allow fast, safe and accurate trains to run.

There is sometimes a bit of confusion about the expressions "driver-less" and "self-driving": they are not synonyms. A driver-less vehicle is not necessarily a self-driving vehicle. Today many trains, airliners and factory machines operate mostly without the intervention of a human being: the environment has been structured so that the machine can do its job, safely and efficiently. The extraordinary has been removed. The designers of the environment for that device made sure that only the ordinary can happen, and the ordinary is greatly simplified. The driver-less vehicle in the structured environment is NOT doing what humans do (used to do) in their unstructured environments: humans had to deal with a chaotic system in which the extraordinary was routine, where at every moment they had to make decisions. The extraordinary was ordinary. "Self-driving", instead, implies that the machine does what a human can do: make decisions in extraordinary circumstances, anytime, anywhere.

I recently had to exchange the equivalent of $3.00 in a local currency while leaving a Western country at its capital's airport. The procedure was silly beyond belief. I had to produce passport, boarding pass and receipt of previous money exchanges before getting my money, a lengthy operation for just three dollars. On the contrary at the border between Haiti and Dominican Republic, a wildly chaotic place with taxi drivers, fruit vendors and police officers yelling at each other and at everybody passing by, there was a mob of money changers chasing the travelers. I had to guess which ones were honest money changers rather than scammers, and then bargain the exchange rate, and then make sure that the money was good while all the time protecting my wallet from pickpockets. It wouldn't be difficult to build a robot that can exchange money at the airport of a Western capital, but orders of magnitude more difficult to build one that can exchange money while walking from the immigration post of Haiti to the immigration post of the Dominican Republic.

The more structured the environment, the easier it is to build a machine that operates in it. What really "does it" is not the machine: it's the structured environment. What has made so many machines possible is not a better A.I. technology, but simply better structured environments. It's the rules and regulations that allow the machine to operate.

You can't call an automatic phone system and just explain your problem. You have to press 1 for English, 1 for customer support, 3 for your location, 2 for your kind of problem and 4 and 7 and so forth. What allows the machine to perform its job, and to replace the human operator, is that you (the human being) have removed the human aspect from the interaction and behaved like a machine in a mechanical world. It is not the machine that behaves like a human being in a human world.

The fundamental thing that a self-driving car must be able to do is, of course, to stop at a gas station when it runs out of gasoline. Can these cars autonomously enter a gas station, stop in front of a pump, slide a credit card in the payment slot, pull out the hose and pour gasoline in the tank? Of course, not. What needs to be done is to create the appropriate structured environment for the driverless car (or, better, for some sensors on board the car) so that the car will NOT need to behave like an intelligent being. The gas station, the gas pump and the payment used by the driverless car will look very different from the one used so far by human drivers.

Incidentally, most of those rules and regulations that create a highly structured environment (favorable to automata) were originally introduced in order to reduce costs. Employing machines has been the next logical step in cost reduction. The machine is one step in an ongoing process of cost reduction and productivity increase. The goal was not to create superhuman intelligence, just to increase profits.

Think of your favorite sandwich chain. You know exactly what kind of questions they will ask you. There is a well-structured process by which your sandwich will be made. The moment robots become cheap enough, they will certainly take over the jobs of the kids who prepare your sandwich today. It is not a matter of "intelligence" (the intelligence of today's robots is already more than enough) but of cost: today a teenager is cheaper than a robot. The whole point of structuring the sandwich-making process was to allow inexperienced and unskilled workers (read: underpaid) to perform the task once reserved for skilled experienced chefs.

The more unstructured the environment is, the more unlikely that a machine will replace the human. Unfortunately, one very unstructured environment is that of health care. Medical records are kept on physical files, and doctor's notes are notoriously impossible to read. There is very little that a machine can do in that environment. The way to introduce "intelligent" machines in that environment is, first of all, to structure all that information. When it is "digitized" and stored in databases, it means that it has been structured. At that point any human being, even with little or no knowledge of medical practice, can do something intelligent in that environment. And even a machine can.

The truth is that we do not automate jobs as they are. First, we dehumanize the job, turning it into a mechanical sequence of steps. Then we use a machine to automate what is left of that job. For example, my friend Steve Kaufman, a pediatrician all his life, realized that his skills were less and less necessary: a nurse practitioner can fill all the forms and click on all the computer buttons that are required when seeing a patient; the doctor, who is increasingly required to type on a keyboard, may not even make eye contact with the patient. This has the beneficial effect of reducing the number of days that the average patient spends at a hospital, but it erases the kind of bonding between doctor and patient that was common in the "unstructured" world. When the last vestiges of humanity will have been removed from the job of the doctor, it will be relatively easy to automate the doctor's job. But that is not what Steve was doing. As Steve pointed out to me, if you don't bond with an asthmatic patient, you may never realize that he is suicidal: you will cure his asthma, but he will commit suicide; and the machine will still archive the case as a success.

Structured environments are also relying on ever stricter rules. My favorite example is the boarding procedure at an airport, where we are treated like cattle from check-in to the gate, with a brief interval during which we are treated like a walking credit card that airport shops desperately try to get. Other than the credit card thing, we are basically building the kind of hyper-bureaucratic state pioneered by the Soviet Union.

There is a fundamental paradox underlying the ongoing structuring of society. What is profoundly human (and actually shared by all forms of life) is the vagueness of language and behavior. What humans (and animals) can do relatively well, and do on a daily basis, and today's machines are not good at, is to deal with ambiguity. Unfortunately, ambiguity is responsible for a lot of the miscommunication and chaos that complicate our life. Rules and regulations are useful because they remove ambiguity from society, and therefore simplify our life. As a side-effect, though, the more we structure human behavior by removing ambiguity, the more replicable it becomes. We become machines; machines that demand a high salary and all sorts of rights. It is a no-brainer for businesses to replace such expensive machines with cheaper ones that don't demand any right.

Increasingly structured environments, routines and practices will eventually enable the automation of "cognitive" skills too. I am writing while watching the indecent spectacle of the political campaigns for a presidential election in the USA. Political debates are becoming more and more structured, with a format agreed beforehand and a moderator that enforces it, and a restriction on the kind of questions that can be asked, and candidates who basically memorize press releases worded by their campaign staff. It is not difficult to imagine that sooner or later someone will build a piece of software that can credibly replace a politician in a political debate; but that feat will owe more to the lack of real debate in these political debates than to greater rhetorical skills on the part of the machine. On the other hand that software will be incapable of participating in a passionate conversation about a World Cup game with a group of rowdy and drunk soccer fans.

It is the increasingly structured environment that is enabling and will enable the explosion of robotics and automated services. Most of the robots and phone-based services coming to the market now rely on relatively old technology. What has made them feasible and practical is that they can now operate in highly structured environments.

Think of yourself. You are now identified by numbers in so many different contexts: your passport number, your social security number, your street address, your telephone number, your insurance policy number, your bank account number, your credit card number, your driver license number, your car's plate number, your utility bill account number… It is a rarity when someone tries to identify me based on non-numeric features. And increasingly we depend on passwords to access our own information. The more we reduce the individual to a digital file, the easier it gets to build "intelligent assistants" for that file… sorry, i meant "for that person".

In a sense, humans are trying to build machines that think like humans while machines are already building humans who think like machines.

Intermezzo: Will Intelligent Machines Return to Chaotic Environments?

Structuring the environment really consists of two parallel processes. On the one hand, it means removing the chaotic and unpredictable (and often intractable) behavior of natural environments. On the other hand, it also means removing the chaotic and unpredictable (and often intractable) behavior of human beings. The purpose of all the rules and regulations that come with a structured environment is to replace you (a messy human intelligence) with an avatar that is like you (in fact it shares your body and brain) without the quirkiness of human intelligence. That avatar lives in a highly-structured virtual world that mimics the natural world without all the quirkiness of the (wildly unstructured) natural world.

My thesis is that machines are not becoming particularly more intelligent, but, instead, it is humans who are structuring the environment and regulating behavior so that humans become more like machines and therefore machines can replace humans.

But what happens if machines become truly "intelligent"? If "intelligent" means that machines will become what humans still are today, before society turns them into rule-obeying machines, then, ironically, machines may acquire all the "baggage" that intelligent biological beings carry, i.e. the unpredictable, chaotic, anarchic behavior that any living being exhibits, i.e. precisely what the structured environment and rules and regulations aim at suppressing.

It would be ironic if creating intelligent machines would turn machines into (messy) humans at the same time that we are turning humans into (disciplined) machines.

Another Intermezzo: Disorder is Evolution, Order is Stagnation

Equilibrium is not a normal state in the universe. The universe is a vast collection of "open" systems that trade energy, matter and information with each other. Many systems thrive in a state far from equilibrium, the so-called "edge of chaos". Living beings are an example: living beings trade energy, matter and information with their ecosystem. You are constantly living at the "edge of chaos". You will reach a state of equilibrium when you die. We can view intelligent systems as systems that are particularly complex.

Ilya Prigogine, Stuart Kauffman and many others have shown an interesting property of these systems. Complex systems (technically speaking, "nonlinear" systems) that are pushed out of equilibrium by perturbations reach a point where they can either disintegrate in total chaos or spontaneously reorganize themselves at a higher level of complexity. The outcome is unpredictable, and irreversible.

A rigid society, in which rules and regulations enforce some behavior and prohibit some other behavior, leaving little to the imagination, is not a complex system. It is very predictable in what happens to you if you break those rules: you go to jail, or you get fired.

"Noise" (perturbations) is important in self-organizing systems such as the human society because it allows such systems to evolve. Under the right circumstances a self-organizing system disturbed by noise will self-organize at a higher level, in some cases a level that is profoundly different from the original one. The more we remove "noise" and unpredictability from human society, the less likely it is that human society will evolve at all, let alone towards higher levels of organization.

Intelligent machines might rediscover this law of non-equilibrium thermodynamics after humans have forgotten it.

Meta-philosophical Intermezzo

Philosophers like to discuss A.I. systems. Can we build an A.I. system that likes to discuss philosophers?

Human Obsolescence

Both computer experts and ordinary people fear that we (humans) may become obsolete because machines will soon take our place.

Bletchley Park codebreaker (i.e., Turing's coworker) Jack Good (real name Isadore Jacob Gudak) wrote in "Speculations Concerning the First Ultraintelligent Machine" (1965): "the first ultraintelligent machine is the last invention that man need ever make". Hans Moravec in "Mind Children" (1988): "Robots will eventually succeed us: humans clearly face extinction". A 2000 article by Bill Joy was titled "The Future doesn't Need us". Etcetera. Actually, this idea has been repeated often since the invention of (among other things) the typewriter and the assembly line.

When we say that "robots will succeed us" or "The future doesn't need us", we really need to define "us". Assembly lines, typewriters, computers, search engines, steam engines, printing presses and whatever comes next have replaced jobs that have to do with material life. I could simply say that they have replaced "jobs". They have not replaced "people". They replaced their jobs. Therefore what went obsolete has been jobs, not people, and what is becoming obsolete is jobs, not people. Humans are biological organisms who (and not "that") write novels, compose music, make films, play soccer, root for the Tour de France bicyclists, discover scientific theories, argue about politics, hike on mountains and dine at fancy restaurants. Which of these activities is becoming obsolete because machines are doing them better?

Machines are certainly good at processing big data at lightning speed. Fine. We are rapidly becoming obsolete at doing that. In fact, we've never done that. Very few humans spent their time analyzing big data. The vast majority of people are perfectly content with small data: the price of gasoline, the name of the president, the standings in the soccer league, the change in my pocket, the amount of my electricity bill, my address, etc. Humans have mostly been annoyed by big data. That was, in fact, a motivation to invent a machine that would take care of big data. The motivation to invent a machine that rides the Tour de France is minimal because we actually enjoy watching (human) riders sweat on those steep mountain roads, and many of us enjoy emulating them on the hills behind our home. Big data? Soon we will have a generation that cannot even do arithmetic.

What is becoming obsolete is not "us" but our current jobs. That has been the case since the invention of the first farm (that made the prehistoric gatherers obsolete) and, in fact, since the invention of the wheel (the cart made many porters obsolete), and jobs certainly disappeared when Gutenberg started printing books with the printing press, the precursor of the assembly line.

Since then, humans have used wheels to travel the world and the printing press to discuss philosophy.

In Defense of Progress: Augmented Intelligence

Enough of bashing computers. The computer might be the only major appliance invented since television, but it is qualitatively different than all the previous appliances. What can the dishwasher do other than wash dishes? The computer, instead, can do a lot of things, from delivering mail to displaying pictures. A computer is many machines in one. That was, in fact, the whole point of the Universal Turing Machine: a universal problem solver. Little did he know that its applications would range from phone conversations to social media.

The secret is the software:

In fact, there has been little progress in the physical world but a lot in the virtual world created by computers. Just witness the explosion of online services in the 1990s and of smartphone applications since 2007.

Perhaps even more importantly, the law of entropy does not apply to software: everything in this universe is bound to decay and die because of the second law of Thermodynamics (that entropy can never decrease). That does not apply to software. Software will never decay. Software can create worlds in which the second law of Thermodynamics does not apply: software never ages, never decays, never dies. (Unfortunately, software needs hardware to run, and that hardware does decay).

The catch is that software does not have a body and therefore cannot do anything unless it is attached to a machine. Software cannot cook and cannot start a car unless we drop it inside a computer and attach the computer to the appropriate machine. Software cannot even give answers without a printer, a screen or a speaker.

Disembodied software is like disembodied thought: it is an abstraction that doesn't actually exist.

Software has to be incorporated into a processor in order to truly exist (to "run"). In turn the processor, that ultimately only does binary algebra, has to be attached to another machine in order to perform an action, whether cooking an omelette or starting a car.

De facto, we attach a universal problem solver to a specific problem solver. However, there is a way to maximize the usefulness of a universal problem solver: attach it to another universal problem solver, the human mind.

One could argue that, so far, Artificial Intelligence has failed to deliver, but "Augmented Intelligence" has been successful beyond the hopes of its founding fathers. In the 1960s in Silicon Valley there were two schools of thought. One, usually associated with John McCarthy's Stanford Artificial Intelligence Lab (SAIL), claimed that machines would soon replace humans. The other one, mainly associated with Doug Engelbart at the nearby Stanford Research Institute (now SRI Intl), argued that machines would "augment" human intelligence rather than replace it. Engelbart's school went on to invent the graphic user interface, the personal computer, the Internet, and virtual personal assistants like Siri; all things that "augmented" human intelligence. This program did not necessarily increase human intelligence and it did not create a non-human intelligence: the combination of human intelligence and these devices can achieve "more" than human intelligence can alone.

The search engine is a good example of "amazing" augmented intelligence and "disappointing" artificial intelligence. It must be terribly difficult for search engines to keep up with the exponential growth of user-provided content. The ranking algorithm has to become exponentially smarter in order for the search engine to keep providing relevant answers. It's something that the user doesn't see (unlike, say, a new button on the microwave oven), but it's something that is vital to making sure that the World-wide Web does not become unsearchable, i.e. a "World-wide Mess".

Life with Algorithms: Tomorrow's Utopia

So far this book mainly spoke of algorithms as dumb and/or evil. It is my belief that today's algorithms deserve it: most of today's algorithms are dumb and/or evil. But that doesn't have to be the future. They can certainly become smarter. And they can certainly become more useful for ordinary people (not only for businesses and governments). I have criticized "smart cities" because they tend to focus on buildings and streets, and not enough on people. That is a general problem with algorithms, they are much better at dealing with physical objects than with physical people (for the simple reason that objects are easier to model mathematically than people). However, the algorithms of a smart city could also focus on people. For example, algorithms could help improve the quality of engagement. There are people in cities, but they live unnatural lives if they are kept isolated from each other. The goal of a "smart" city should also be to encourage interaction and collaboration, leading to higher and higher forms of engagement. We can design algorithms to augment individual and collective experience, algorithms to improve trust, security, match making, discovery, algorithms to augment our ability to engage, algorithms to create collectively something positive that didn't exist before. For example, the Peace Innovation Lab at Stanford, founded in 2010 by Mark Nelson, aims at using technology to increase "positive engagement".

General-purpose Intelligence

Before analyzing what it will take (and how long it will take) to get machine intelligence, we need to define what we are talking about.

A man, wearing a suit and tie, walks out from a revolving hotel door dragging his trawlley suitcase. Later, another man, wearing a shabby uniform and gloves, walks out of the side door dragging a garbage can. It is obvious even to the dumbest human being that one is a guest of the hotel and the other is a janitor. Do we require from a machine this simple kind of understanding ordinary situations in order for it to qualify as "intelligent"? Or is it irrelevant, just like matching the nightingale's song is irrelevant in order to solve differential equations? If we require that kind of understanding, we push machine intelligence dramatically forward into the future: just figuring out that one is a suit and tie and the other is a uniform is not trivial at all for a machine. It takes an enormous computational effort to achieve just this one task. There are millions of situations like this one that we recognize in a split second.

Let us continue our thought experiment. Now we are in an underdeveloped country and the janitor is dragging an old broken suitcase full of garbage. He has turned an old suitcase into his garbage can. Seeing such a scene, we would probably just smile at the man's ingenuity; but imagine how hard it is for a machine to realize what is going on. Even if the machine is capable of telling that someone dragging a suitcase is a hotel guest, the machine now has to understand that a broken suitcase carried by a person in a janitor's uniform does not qualify as a suitcase.

There are millions of variants on each of those millions of situations that we effortlessly understand, but that are increasingly trickier for a machine.

The way that today's A.I. scientists would go about it is to create one specific software program for each of the millions of situations, and then millions of their variants. Given enough engineers, time and processors, this is feasible. Whenever a critic like me asks "But can your machine do this too?", today's A.I. scientists rush out to create a new program that can do it. "But can your machine also do this other thing?" The A.I. scientists rush out to create another program. And so forth.

Given enough engineers, time and processors, it is indeed possible to create a million machines that can do everything we naturally do.

After all, the Web plus a search engine can answer any question: someone, sooner or later, will post the answer on the Web, and the search engine will find it. Billions of Web users are providing all the answers to all the possible questions. The search engine is not particularly intelligent in any field but can find the answer to questions in all fields.

I doubt that this is the way in which my mind works (or any animal's mind works), but, yes, those millions of software programs will be "functionally" equivalent to my mind. In fact, they will be better than my mind because they will be able to recognize all the situations that all the people in the world recognize, not just the ones that i recognize, just like the Web will eventually contain the answers to all questions that all humans can answer, not only the answers that i know.

This is exactly what "brute-force A.I." is doing today: creating a specific software program for each intelligent task that humans perform. The method is different, but the rationale is reminiscent of Marvin Minsky's "The Society of Mind" (1985) that viewed an artificial general intelligence as a society of specialized agents.

Luckily, the effect on the economy will be to create millions of jobs because those millions of machines will need to be designed, tested, stored, marketed, sold, and, last but not least, repaired.

Artificial General Intelligence and the No Free Lunch Theorems

In 1995 and 1996 computer scientist David Wolpert at the Santa Fe Institute proved some of the most famous theorems in computer science, the "No Free Lunch" theorems, that apply to search ("No Free Lunch Theorems for Search", 1995), supervised learning ("The Lack of a Prior Distinctions between Learning Algorithms", 1996), and optimization ("No Free Lunch Theorems for Optimization", December 1996). He proved that no learning algorithm can possibly excel at learning everything: for every pattern that a learner masters, there is another pattern that the same learner cannot master. Shortly afterwards, Joseph Culberson at University of Alberta in Canada published an "algorithmic view" of the No Free Lunch theorems that related them to complexity theory ("On the Futility of Blind Search", 1996). These theorems are clearly an obstacle towards the dream of a universal learning algorithm, towards an artificial intelligence that can be "general". Of course, Wolpert proved his theorems under some assumptions (in particular, they considered only discrete functions) which may or may not apply depending on which machine learning method you choose. Anne Auger and Olivier Teytaud at INRIA in France (the National Institute for Research in Informatics and Automation) proved that the theorems do not hold in continous domains ("Continuous Lunches are Free", 2007); but then Michael Vose of University of Tennessee and others proved that they do hold in arbitrary domains ("Reinterpreting No Free Lunch", 2009). The debate has continued, and its impact on the possibility of an artificial general intelligence is unresolved. As Erik Hoel of Columbia University wrote in 2017: "Superintelligence is a free lunch, and there are no free lunches". He believes that evolution is all about learning new skills while forgetting others, so that new forms of intelligence are being continuously created, but none is "general". Whenever a mutation makes an organism fit in an environment, it also makes it less fit in other environments. And perhaps, at the individual level: as we get more intelligent in doing something, we also get more stupid in doing something else.

We humans (creatures to which the "no free lunch" theorem doesn't seem to apply) can safely conclude that learning without knowledge is terribly difficult, and our innate knowledge comes from evolution.

The Proliferation of Appliances, Intelligent and not Intelligent

If we structure the world appropriately, it will be easy to build machines that can board planes, exchange money, take a bus, drive a car, cross a street and so on. Automated services have existed since at least the invention of the waterwheel. We even have machines that dispense money (ATMs), machines that wash clothes (washing machines), machines that control the temperature of a room (thermostats), and machines that control the speed of a car (cruise controls).

When we design robots, we are simply building more appliances. In the near future we might witness a multiplication of appliances, disguised and marketed as "robots" simply because the word "robot" is becoming fashionable: iRobot's vacuuming robot Roomba, Moley Robotics' robotic chef in Britain that, installed on top of your stove, cooks dinner for you; the robotic waiters of the Robot Restaurant in Harbin (northeastern China); Infinium Robotics' drone waiters, that deliver meals flying over the heads of the customers, in Singapore; MIT's robotic bartender; UC Berkeley's robot that folds towels; etc.

The ATM is more precise than a bank clerk (and works much longer hours) but we don't think of it as "intelligent". Ditto for the washing machine that is capable of all sorts of washing techniques. That's because they were introduced at a time when it was not popular to market them as Artificial Intelligence. If the washing machine was invented today, it would certainly be presented as the latest achievement in robotics.

Enthusiastic fans of automation predict that "soon" (how soon?) everything that humans do will be done by machines; but they rarely explain what is the point of making machines for everything we do. Do we really want machines that fall asleep or urinate? There are very human functions that people don't normally associate with "intelligence". They just happen to be things that human bodies do. We swing arms when we walk, but we don't consider "swinging arms while walking" a necessary feature of intelligent beings. The moment we attempt to design an "intelligent" machine (or collection of machines) that can mimic the entire repertory of our "intelligent" functions we run into the enumeration problem: which function qualifies as "intelligent"? Typical human activities include: forgetting where we left the mobile phone, eating fast food, watching stand-up comedy, catching a flue when attacked by viruses and, yes, frequently, urinating.

We instinctively envision a hierarchy of tasks, from "not intelligent at all" to "very intelligent", and we assume that the latter are the ones that make the difference. However, that ranking is not very objective: why a washing machine is not intelligent given that relatively few humans can wash clothes, whereas a cat-recognizing program is (given that virtually every human, no matter how dumb, can recognize a cat, and so can countless animals). Statistically, it would seem that washing clothes should be more special than cat-recognizing programs.

The current excitement about machines is due to the fact that (it is claimed) they are beginning to perform tasks that were exclusive to human beings. This is actually a very weak claim: the first washing machine was capable of performing a task that had been exclusive to human beings until the day before. Implicit in these claims is the idea that there is something that makes some tasks qualitatively more "special" than washing clothes, but it is difficult to articulate what this "special" quality would be. What is truly unique/special about human intelligence? Each machine performs for us a task that we used to do manually. Which tasks are so "special" that they deserve to be called "intelligent" is far from agreed upon.

And, finally, machines that resemble human beings (that smile, cry, walk and even say a few words) have existed for a long time and they are usually sold in toy stores and aimed at children. We can certainly create more sophisticated toys, like toys that recognize cats, but the claim that these toys will have anything to do with human intelligence needs some explaining.

Intermezzo: The Resurrection of the Dead

For as long as we have been making tools, technology has been the language of the dead speaking to the living. Maybe we are fascinated by intelligent machines because we would finally have technology that is alive. As more and more technology has invaded our private and public lives, it has become a bit discomforting to be talking so often to the dead.

Demystifying the Turing Test

Alan Turing's article "Computing Machinery and Intelligence" (1950), that involuntarily introduced the test for machine intelligence, also marked the birth of a new branch of philosophy: can "thought" be reduced to the execution of an algorithm in the brain?

The British philosopher John Lucas published "Minds, Machines and Goedel" (1959), a "proof" based on Goedel's own incompleteness theorem that machines cannot become more intelligent than us, an argument retold by the British physicist Roger Penrose in his book "The Emperor's New Mind" (1989). Goedel's theorem states that every formal system (arithmetic and up) contains a statement that cannot be proven true or false. Indirectly, Goedel's theorem states the preeminence of the human mind over the machine: some mathematical operations are not computable, nonetheless the human mind can treat them (at least to prove that they are not computable). Humans can realize what Goedel's theorem states, whereas a machine, limited to mathematical reasoning, would never realize what it states. We can intuitively comprehend a truth that the computer can only try (and, in this case, fail) to prove. Therefore no mathematical system can fully express the way Kurt Goedel's mind thinks.

One counterargument originated from the philosopher Hilary Putnam: a computer can observe the failure of "another" computer's formal system, just like a human mind can observe it. A computer can easily prove the proposition "if the theory is consistent, then the proposition that there is at least one undecidable proposition is true"; which is exactly all the human mind is capable of doing. Goedel's theorem sets a limit, not to the intelligence of machines, but to the human mind: the human mind will never be capable of building a machine that can think. This does not prove that machines cannot think. Rudy Rucker, for example, believes that we cannot build a machine that has our mathematical intuition but such a machine can exist. It cannot be built by humans and its functioning cannot be understood by humans, but it could be built by Darwinian evolutionary steps starting from a man-made machine. What Goedel's theorem asserts is that "the human mind is not capable of formulating all of its mathematical intuitions" (quoting Goedel himself). The British physicist Stephen Hawking notes that the behavior of earthworms can probably be simulated adequately with a computer, because they do not worry about Goedel sentences. Darwinian evolution can generate human intelligence from earthworm intelligence through a process (natural selection) for which Goedel's theorem is also irrelevant. Therefore, Goedel's theorem does not forbid the advent of an intelligent computer. Finally, Aaron Sloman pointed out that Goedel's theorem is false in some nonstandard mathematical systems. Goedel's theorem applies to mathematical systems that are consistent (i.e., do not contain a contradiction), but that can only be if the undecidable statement is added to the system, assuming either true or false. Nonstandard models assume that it is false. Goedel's theorem, because of the way Goedel carried it out (by employing infinite sets of formulas), leaves the illusion of proving a truth which in reality is never proved, cannot be proven and must be arbitrarily decided ("The Emperor's Real Mind", 1992).

John Searle's article "Minds, Brains, and Programs" (1980) opened another front by using the thought experiment of the "Chinese room" to expose the behaviorist fallacy of Artificial Intelligence: the man locked in the room, a man who knows absolutely nothing of Chinese and may not even know that it is Chinese, in charge of writing down the answers corresponding to the questions posed to him, would appear to be a fluent Chinese speaker to anyone who doesn't see how the man is producing the answers. That man does not know Chinese, no matter how good the answers are, and, by analogy, a computer does not "think", no matter how well it does what it is programmed to do. Paraphrasing Fred Dretske, a computer does not know what it is doing, therefore "that" is not what it is doing.

The simplest counter-argument to Searle's argument is that the man may not "know" Chinese, but the room (i.e., the man plus the rules to speak Chinese) does qualify as a fluent Chinese speaker, as someone who "understands" Chinese. It is also not clear what we really mean by "understanding". In a sense, Searle simply slowed down and broke down the process of understanding: what we do when we understand something is precisely what the man does in the room. Searle's objection sounds more like: if you can tell what the mechanism is that produces "understanding", then that cannot be true "understanding". Basically, the question is whether the simulation of a mind is itself a mind or not.

Inspired by Edmund Husserl's phenomenology, another US philosopher, Hubert Dreyfus, launched a third line of attack. Dreyfus pointed out that intelligent behavior (notably, comprehending a situation) cannot ignore the context (in which the situation occurs) and the body. The information in the environment is fundamental for a being's intelligence, as is the fact that it needs to be organized as a situation in which the body operates ("What Computers can't do", 1979). But Dreyfus was mainly criticizing the symbolic, knowledge-based school of A.I. that relied on encoding rules. Influenced by the German philosopher Martin Heidegger, Terry Winograd argued that intelligent beings act, don't think: people are "thrown" in the real world. They "think" only when action does not yield the desired result. Only then do they pause to picture the situation in its complexity and decompose it into its constituents, and try to infer action from knowledge. Similarly, Rodney Brooks argued that intelligence cannot be separated from the body: intelligence is not only a process of the brain, it is embodied in the physical world. Every part of the body is performing an action that contributes to the overall "functioning" of the organism in the environment. Intelligence "is" about moving in a physical world and cannot exist without a physical world ("A Robust Layered Control System for a Mobile Robot", 1986).

The Turing Test was basically about building a machine that can answer questions in a manner indistinguishable from the manner in which a typical human being answers.

If you ask the questions that make us human, all computer programs fail the Turing Test, and they fail in awkward ways.

Linguists like to talk about the difficulty of understanding ambiguous sentences such as "Prostitutes appeal to Pope" and "Iraqi head seeks arms". But the job of a machine gets even more difficult when common sense is involved. In the sentence "Carl, who died last year, was a great scientist, and his son Dale has fond memories, and he now takes care of the center" it is pretty clear to whom the "he" refers, because one of the two men is dead and therefore he cannot take care of the center (or of anything else). This is not obvious to a machine that doesn't know what "dying" implies.

Ask the machine: "The doll will not fit in the box because it is too big: which one is too big, the doll or the box?" If you ask questions like this one, the human being will get them right almost 100% of the time, but the machine will only get them right 50% of the time because it will simply be guessing (like flipping a coin). Ask just two sentences like this one, and, most likely, you will know whether you are talking to a machine or to a human being. The machine has no common sense: it doesn't know that, in order to fit inside a box, an object has to be smaller than the box. This is the essence of the Winograd Schema Challenge devised by Hector Levesque, at University of Toronto in 2011.

A problem with the Turing Test is that it is not clear what it is supposed to "measure": cognition or consciousness? Nowhere does Turing bother to distinguish among them. It is also a little unfair to make it a "yes/no" question: people's intelligence can vary greatly, from Einstein down to some politicians. Intelligence comes in degrees. Animals are intelligent, to some degree. It is debatable whether they are capable of thinking (conscious). A mentally-retarded person may not be intelligent, but she is presumably conscious, to some degree. Consciousness too may come in degrees. Turing does not discriminate and therefore does not tell us what his test is supposed to measure; and his test is a "zero/one" binary test that does not reflect the continuum of intelligence, from idiot to genius, that we experience in our reality.

The Turing Test needs a better definition, starting with the setting itself: which instruments must be used? Turing's test uses a human being (let's call him the "observer" or "judge", which is really the instrument of this measurement) to decide whether a machine is as good as a fellow human being (let's call him the "reference"). Thus both the instrument and the reference are humans. He does not provide a prescription for what the observer and the reference must be. Can a mentally retarded person be the reference? Can somebody under the influence of drugs be the reference? Or does it have to be the most intelligent human? The result of the test will obviously vary wildly depending on whom we choose. As for the judge, Turing's article doesn't specify which type of human he wants to preside over the test: a priest, an attorney, an Australian aborigine, an avid reader of pornographic magazines, a librarian, a mathematician, an economist, his close friends, a gullible person, a skeptic...? Clearly, the kind of questions asked by the "judge" depends on who and what she is.

The judge has to determine whether the answers to her questions come from a human or a machine. If the judge cannot tell the difference, and mistakes the machine for a human, then the machine passes the test. But what conclusions should we draw about the human who failed the test. In other words, if a machine fails the test, then the judge may conclude that it is not intelligent: but what is the judge entitled to conclude if a human fails the test? That humans are not intelligent?

Depending on your definition, computers and robots may already be "cognitive" systems: they are capable (to some degree) of remembering, learning and reasoning. But they can usually do it only in a narrow domain.

As the US computer scientist Stuart Russell remarked, Turing's definition is at the same time too weak and too strong. Too weak because it does not include "intelligent" behavior such as "dodging bullets" and too strong because it does include unintelligent beings such as Searle's Chinese-room translator. Most children, who cannot answer a lot of questions that an adult could answer, would not pass the test, but that does not make them machines.

Philosophers can split hairs as much as they like, but the Turing test simply measures how good the software is at answering questions, and nothing more. Answering questions is not a sign of "thinking" just like washing dishes is not, although both are among the many operations that thinking beings can do.

We don't normally ask if dead people think, or if furniture thinks: we assume that being alive is a precondition to thinking. Before we ask whether machines can think, we should therefore ask whether they can be alive.

Common Sense

In November 1958, at the Symposium on Mechanization of Thought Processes in England, the always prescient John McCarthy delivered a lecture titled "Programs with Common Sense", that became one of the most influential papers in A.I. McCarthy understood that a machine with no common sense is what we normally call "an idiot". It can certainly do one thing very well, but it cannot be trusted to do it alone, and it certainly cannot be trusted doing anything else.

What we say is not what we mean. If I ask you to cook dinner using whatever high-protein food you can find in a kitchen cabinet, that does not mean that you should cook the spider crawling on its walls, nor the chick that your children have adopted as a pet, nor (gasp) the toddler who is hiding in it for fun.

How do we decide when is the best time to take a picture at an event? A machine can take thousands of pictures, one per second, and maybe even more, but we only take 2 or 3 because those are the meaningful events.

Surveillance cameras and cameras on drones can store millions of hours of videos. They can recognize the make and model of a car, and even read its plate number, but they can't realize that a child is drowning in a swimming pool or that a thief is breaking into a car.

Maybe machines are becoming better than humans at recognizing images in some circumstances, but common sense still matters for understanding what is going on. For example, in April 2013 two homemade bombs killed three people during the Boston marathon. Within hours the investigation had identified two suspects. It was common sense, not Artificial Intelligence that helped the detectives: video footage showed a crowd reacting in panic... except two people who quietly walked away. Any human can draw the conclusion: those two people were not scared by what happened, they knew exactly what had happened, and the only people who could have known were the perpetrators.

In April 2016 in England a group of children spontaneously formed a human arrow on the ground to direct a police helicopter towards the fleeing suspects of a crime. Nobody taught the children to do that. What the children guessed (in a few seconds) is a fairly long list of "common sense" knowledge: there has been a crime and we need to capture the criminals; the criminals are running away to avoid capture; the helicopter in the sky is the police looking for the criminals; the police force is the entity in charge of catching criminals; it is good that you help the police if you have seen the criminals flee; it is bad if the criminals escape; the helicopter cannot hear you but can see you if you all group together; the arrow is a universal symbol to mark a direction; helicopters fly faster than humans can run; etc. That's what intelligence does when it has common sense.

Around the same time in 2016 Wei Zexi, a 21-year-old student from Xidian University in China's Shaanxi province, who was undergoing treatment for a rare form of cancer, found an ad on Baidu (China's search engine) publicizing a treatment offered by the Beijing Armed Police Corps No 2 Hospital. The "doctor" turned out to be bogus and the treatment killed the boy. The Chinese media demonized Baidu (and, hopefully, the military hospital!), but this was not a case of Baidu being evil: it was the case of yet another algorithm that had no common sense, just like the Google algorithm that in 2015 thought two African-Americans were gorillas, just like the Microsoft algorithm that in 2016 posted racist and sexist messages on Twitter. This is what intelligence does when it has no common sense.

To make things worse, i found the news of Wei Zexi's death on a website that itself displayed some silly ads. Two of these ads were almost porno in nature (titled "30 Celebs Who Don't Wear Underwear" and "Most Embarrassing Cheerleader Moments"). These ads were posted next to the article describing the tragic death of Wei Zexi: the "intelligent" software that assigns ads to web pages has no common sense, i.e. it cannot understand that it is really disgusting to post such sex-related ads in a page devoted to someone's death. (No, the ads were not customized for me: i was using an Internet-café terminal).

On 21 June 2017 the Los Angeles Times reported that a strong earthquake had just struck the town of Santa Barbara. There was no such earthquake. The US Geological Survey (USGS) had issued a false alarm. News organisations across the world had received the alert by email but quickly dismissed it because it was dated 29 June 2025; clearly some kind of snafu if you have common sense. The Los Angeles Times, however, was using an A.I. bot to automatically write stories about earthquakes and the A.I. bot dutifully informed the nation of the earthquake. (The Los Angeles Times quickly retracted the article, but the story became so popular that another bot running on the Los Angeles Times website, the one in charge of maximizing advertising revenues, started displaying an appliance advert in front of the article retracting the news of the earthquake).

In 2018 a Paypal "bot" wrote a letter to English woman Lindsay Durdle who had just died of cancer: "Important: You should read this notice carefully... You are in breach of condition 15.4(c) of your agreement with PayPal Credit as we have received notice that you are deceased." The Paypal bot pledged to sue in court because her dying violated her agreement with Paypal.

The navigation software that is now common in every smartphone is a typical example of what happens when a machine has no common sense. It calculates the shortest route to my destination but sometimes i don't want the shortest route if it implies a huge number of turns. I'd rather stay on the same street as long as possible than turn right and left a dozen times if the difference is only a few minutes. In fact, one night i got so exhausted by the silly route calculated by my navigator that i left the party early: the navigator did its job of guiding me to my destination as quickly as possible, but ruined my evening. And we generally prefer to avoid dangerous neighborhoods. A friend was assaulted when she stopped at a red light in a bad part of town. When i was driving in another bad part of town, the car in front of me suddenly stopped and two big tall men came out and walked to me and asked me why i was following them (I wasn't, but obviously in that neighborhood it happens). We would rather drive an extra ten minutes than having to drive through a bad neighborhood.

The balance between common sense and algorithms is delicate. Every year more than 200,000 Chinese die in car accidents. The government is introducing stricter rules (i.e. algorithms) and enforcing the ones that exist. This resort to algorithms will certainly reduce the most common accidents and save thousands of lives. On the other hand, the USA is a country in which there seem to be more traffic rules than drivers. On some roads it takes a few minutes to read all the posted signs. There are also strict rules on how to build a car to make it as safe as possible. And, yet, the number of people killed in car accidents in the USA keeps climbing: 32,744 in 2014 (10.28 per 100,000 people), 35,485 in 2015 (11.06), 37,461 in 2016. It looks like the USA has reached a point at which it's difficult to further reduce the number of fatalities. There is a simple explanation: drivers in the USA are trained to follow rules; they are not trained to avoid accidents. On the other hand, Chinese drivers are trained to avoid accidents, not necessarily to follow rules.

When computers became powerful enough, some A.I. scientists embarked upon ambitious attempts to replicate the "common sense" that we humans seem to master so easily as we grow up. The most famous project was Doug Lenat's Cyc (1984), which is still going on. In 1999 Marvin Minsky's pupil Catherine Havasi at MIT launched Open Mind Common Sense that has been collecting "common sense" provided by thousands of volunteers. DBpedia, started at the Free University of Berlin in 2007, collects knowledge from Wikipedia articles. The goal of these systems is to create a vast catalog of the knowledge that ordinary people have: plants, animals, places, history, celebrities, objects, ideas, etc. For each one we intuitively know what to do: you are supposed to be scared of a tiger, but not of a cat, despite the similarities; umbrellas make sense when it rains or at the beach; clothes are for wearing; food is for eating; etc. More recently, the very companies that are investing in deep learning have realized that you can't do without common sense. Hence, Microsoft started Satori in 2010 and Google revealed its Knowledge Graph in 2012. By then Knowledge Graph already contained knowledge relating to 570 million objects via more than 18 billion relationships between objects (Google did not disclose when the project had started). These projects marked a rediscovery of the old program of "knowledge representation" (based on mathematical logic) that was downplayed too much after the boom in deep learning. Knowledge Graph is a "semantic network", a kind of knowledge representation that was very popular in the 1970s. Google's natural-language processing team, led by Fernando Pereira, is integrating Google's famous deep-learning technology (the "AlphaGo" kind of technology) with linguistic knowledge that is the result of eight years of work by professional linguists.

It is incorrect to say that deep learning is a technique for learning to do what we do. If i do something that has never been done before, deep learning cannot learn how to do it: it needs thousands if not millions of samples in order to learn how to do it. If it is the first time that it has been done, by definition, deep learning cannot learn it: there is only one case. Deep learning is a technique for learning something that humans DID in the past.

Now let's imagine a scenario in which neural networks have learned everything that humans ever did. What happens next? The short answer is: nothing. These neural networks are incapable of doing anything that they were not trained to do, so this is the end of progress.

Training a neural network to do something that has never been done before is possible (for example, you can just introduce some random redistribution of what it has learned), but then the neural network has to understand that the result of the novel action is interesting, which requires an immense knowledge of the real world. If I perform a number of random actions, most of them will be useless, wasteful of time and energy, but maybe one or two will turn out to be useful. We often stumble into interesting actions by accident and realize that we can use those accidental actions for doing something very important. I was looking for a way to water my garden without having to physically walk there, and one day i realized that an old broken hose had so many holes in it that would work really well to water the fruit trees. Minutes ago, i accidentally pressed the wrong key on my Android tablet and discovered a feature that I didn't know existed. It is actually a useful feature.

In order to understand which novel action is useful, one needs a list of all the things that can possibly be useful to a human being. It is trivial for us to understand what can be useful to human life. It is not trivial for a machine, and certainly not trivial at all for a neural network trained to learn from us.

See for example Alexander Tuzhilin's paper "Usefulness, Novelty, and Integration of Interestingness Measures" (Columbia University, 2002) and Iaakov Exman's paper "Interestingness a Unifying Paradigm Bipolar Function Composition" (Israel, 2009).

The importance of common sense in daily activities is intuitive. We get angry whenever someone does something without "thinking". It is not enough to recognize that a car is a car and a tree is a tree. It is also important to understand that cars move and trees don't, that cars get into accidents and some trees bear edible fruits, etc. Deep learning is great for recognizing that a car is a car and a tree is a tree, but it struggles to go beyond recognition. So there is already a big limitation.

A second problem with deep-learning systems is that you need a very large dataset to train them. We humans learn a new game just from listening to a friend's description and from watching friends play it a couple of times. Deep learning requires thousands if not millions of cases before it can play decently.

Big data are used to train the neural networks of deep learning systems, but "big data" is not what we use to train humans. We do exactly the opposite. Children's behavior is "trained" by two parents and maybe a nanny, not by videos found on the Internet. Their education is actually a form of "training" imparted by carefully selected teachers who had to get a degree in education. We train workers using the rare experts in the craft, not a random set of workers. We train scientists using a handful of great scientists, not a random set of students.

I am typing these words in 2016 while Egypt and other countries are searching the Mediterranean Sea for an airplane that went missing. In 2014 a Malaysia Airlines airplane en route from Kuala Lumpur to Beijing mysteriously disappeared over the Indian Ocean. Deep-learning neural networks can be trained to play go/weiqi because there are thousands of well documented games played by human masters, but the same networks cannot be trained to scour the ocean for debris of a missing airplane: we don't have thousands of pictures of debris of missing airplanes. They can have arbitrary shapes, float in arbitrary ways, be partially underwater, etc. Humans can easily identify pieces of an airplane even if they have only seen 10 or 20 airplanes in their life, and never seen the debris of an aircrash; neural networks can only do it if we show them thousands of examples.

A third problem related to machines with no common sense is their inability to recognize an "obvious" mistake. Several studies have shown that, in some circumstances, deep-learning neural networks are better than humans at recognizing objects; but, when the neural network makes a mistake, you can tell that it has no common sense: it is usually a mistake that makes us laugh, i.e. a mistake that no idiot would make. You train a neural network using a large set of cat photos. Deep learning is a technique that provides a way to structure the neural network in an optimal way. Once the neural network has learned to recognize a cat, it is supposed to recognize any cat photo that it hasn't seen before. But deep neural networks are not perfect: there is always at least one case (a "blind spot") in which the neural network fails and takes the cat for something else. That "blind spot" tells a lot about the importance of common sense. In 2013 a joint research by Google, New York University and UC Berkeley showed that tiny perturbations (invisible to humans) can completely alter the way a neural network classifies the image. The paper written by Christian Szegedy and others was ironically titled "Intriguing Properties Of Neural Networks". Intriguing indeed, because no human would make those mistakes. In fact, no human would notice anything wrong with the "perturbed" images.

Neural networks can easily be fooled by "adversarial examples". An adversarial example is an image (or sound or other pattern) that has been slightly modified in a way to mislead the neural network despite the fact that the human eye doesn't notice anything strange about it. In 2015 Ian Goodfellow, the inventor of generative adversarial networks, working at Google with Szegedy, discovered "a fast method of generating adversarial examples" basically a way to serially hack a neural network ("Explaining and Harnessing Adversarial Examples", 2015). His method that quickly and massively generates adversarial examples was named "fast gradient sign method" (FGSM).

This is not just a theoretical discussion. If a self-driving car that uses a deep neural network takes a pedestrian crossing the street for a whirlwind, there could be serious consequences.

If a self-driving car turns the corner and moves towards you just when you were about to cross the street, and there is really no driver inside (right now all self-driving cars have a human driver who can take over at any time), do you still cross the street? Many of us would not, and will not. We often make eye contact with the driver in order to confirm that s/he has seen us. Can we make eye contact with the self-driving algorithm? We are told that machine vision is accurate 97% of the times, but we don't want to be a member of the 3%. And even if it gets down to an incredible 0.0001% error rate, that would still be thousands of mistakes a day.

Humans are very good at making mistakes all the time but they are also pretty good at improvising a remedy to each possible mistake. Cars are very good at not making mistakes, but, if they make a mistake, they won't even realize that they made a mistake: whatever they make is what they calculated to be the right thing to do. Machines don't think "Oh shoot! This is not right!" Machines cannot calculate the outcome of their action and realize that, no matter how good and rational the intention was, the result is a disaster, and must be undone, and ideally even aborted before it's done.

Conversely, in 2015 Anh Nguyen at University of Wyoming showed that deep neural networks can easily be fooled into recognizing objects that don't exist ("Deep Neural Networks are Easily Fooled", 2015): two of the most popular neural networks (AlexNet and LeNet) recognized with more than 99% confidence some abstract tapestry as familiar objects. In

2017 Alhussein Fawzi and Seyed Moosavi of Pascal Frossard's team at Federal Institute of Technology Lausanne (EPFL) in Switzerland developed the DeepFool algorithm to scientifically obtain perturbations that fool deep networks ("A Simple and Accurate Method to Fool Deep Neural Networks", 2017), i.e. to quantify how robust a neural network is. The problem does not go away in the three-dimensional world. MIT students Anish Athalye, Logan Engstrom and Andrew Ilyas fooled Google's InceptionV3 with a 3D-printed turtle: the neural network recognized it as a rifle. ("Synthesizing Robust Adversarial Examples", 2017). The same neural network was fooled into recognizing a cat as guacamole, but that was still in the realm of two-dimensional images. Working with Goodfellow, Alexey Kurakin at Google Brain showed that the neural network can be fooled even when the "adversarial example" is located in the physical world, e.g. when a camera takes a picture of a street sign that has been manipulated in an "adversarial" manner, and sometimes all it takes is to add a few stickers to the letters "STOP" ("Adversarial Examples in the Physical World", 2016). Ivan Evtimov and others at UC Berkeley and at University of Washington created a general attack algorithm to fool deep neural networks, Robust Physical Perturbations or RP2 ("Robust Physical-World Attacks on Deep Learning Models", 2017). For example, a small sticker on a stop sign is enough to confuse a self-driving car.

Pieter Abbeel's student Sandy Huang at UC Berkeley, working with Ian Goodfellow (now at OpenAI), showed that deep reinforcement learning too is vulnerable to adversarial examples ("Adversarial Attacks on Neural Network Policies", 2017): that's the method used by A3C to play Atari videogames and by AlphaGo to play weiqi/go. In 2017 Goodfellow and Nicolas Papernot even published an open-source library of adversarial examples, Cleverhans, that you can use to test how vulnerable your neural network is. Similarly, Percy Liang and her student Robin Jia at Stanford showed how easily a question-answering neural network can be hijacked: beware of your favorite chatbots ("Adversarial Examples for Evaluating Reading Comprehension Systems", 2017). Alexey Kurakin wrote: "Most existing machine learning classifiers are highly vulnerable to adversarial examples" (2016). That's a "highly", not just "a little". In their article "Is Attacking Machine Learning Easier than Defending it?" (2017) Goodfellow and Papernot explain why it is almost impossible to defend a neural network from all possible adversarial examples: adversarial examples are solutions to a complex (nonlinear) optimization problem and no existing mathematical tool can model this kind of solution. In other words, we can't build a mathematical proof that a certain strategy would defend a neural network against any such attack.

David Wagner's student Nicholas Carlini at UC Berkeley discovered a method to fool a speech recognition system with a slight change in the audio waveform: the tiny "hack" can make the A.I. system "hear" whatever they want. You can say "I love you" and a slight modification (impossible for humans to hear) makes the machine understand "I hate you". The study also showed that background music can be understood by the likes of Siri and Alexa as a command ("Audio Adversarial Examples", 2018). Beware of thinking that these issues have been solved by the time you read this page: between the first edition of this book and this one, many new studies have tested the vulnerability of state-of-the-art recognition systems and found similar problems. Dan Hendrycks of UC Berkeley and Thomas Dietterich of Oregon State University found "negligible changes" in robustness between AlexNet of 2012 and the deep-learning systems of 2018 ("Benchmarking Neural Network Robustness to Common Corruptions and Surface Variations", 2019).

It tells you something important about "machine intelligence" that today's machine intelligence fails if we change, even slightly, the requirements of the problem that it has just learned to solve. It is not intelligence, it is something else.

A neural network works better than a human being in the very finite world of ImageNet, that limits the number of possible categories to one thousand. In other words, the neural network is not trying to "recognize the object" but instead it is trying to "recognize which of the known one thousand objects this particular one is". If the object is none of them, then the neural network fails. If the neural network has been trained to recognize all of my friends but you showed it pictures of your friends who are not my friends, the neural network will try to find some among my friends who look like these unknown people. It is false that deep learning is better than humans at recognizing images, but it is true that humans are a lot better than deep learning at learning abstractions through verbal definition, as documented by Joshua Tenenbaum of MIT, Brenden Lake of New York University and Ruslan Salakhutdinov of University of Toronto ("Human-level Concept Learning through Probabilistic Program Induction", 2014). It is true that deep learning fails all too easily in situations that differ just slightly from the situations for which the system has been trained. For example, a team at Dileep George's new startup Vicarious showed that the famous DeepMind system that learned to play Atari videogames better than videogame masters is actually quite incapable of simple adjustments (Schema Networks", 2017).

And, no, obviously a convnet does not recognize your face: it just recognizes that the picture contains your nose and your eyes and your mouth, in whichever order. And it will recognize as your face anyone who looks like you, including a cast of your face. If they ever replace the human being with a face-recognition machine, simply make a cast of your face and anyone will be able to go through the machine by simply showing that cast. You can find several videos on YouTube on "How to make a face cast on yourself".

Deep learning is essentially a very complicated statistical method for classifying patterns of data. Deep learning is about correlation (A and B happen together), not causation (A is caused by B): it cannot distinguish causation from correlation.

The discipline of deep learning is reminiscent of alchemy before chemistry was invented and of engineering before physics was invented and of medicine before biology was invented. These disciplines could boast some success stories, but their progress was based on minor improvements over what worked, not on an understanding of why it worked. The Romans could build amazing aqueducts because they had figured out that arches can support weight, but didn't know why it worked. Deep learning is in a similar situation: conference papers document small improvements over success stories, but only reference the previous success stories, not a scientific theory of intelligence, just like the Romans didn't know Isaac Newton's equations and alchemists didn't know Antoine Lavoisier's formulas.

Beware, in particular, of machine learning in social sciences. The British prime minister Benjamin Disraeli once said "There are lies, damned lies and statistics". At some point we may have to say: "There are lies, damned lies, and machine learning".

Deep learning depends in an essential way on human expertise. It needs a huge dataset of cases prepared by humans in order to "beat" the humans at their own game (chess, go/weiqi, etc). A world in which humans don't exist (or don't collaborate) would be a difficult place for deep learning. A world in which the expertise is generated by other deep-learning machines would be even tougher. For example, Google's translation software simply learns from all the translations that it can find. If many English-to-Italian human translators over the centuries have translated "table" as "tavolo", it learns to translate "table" into "tavolo". But what if someone injected into the Web thousands of erroneous translations of "table"? Scientists at Google are beginning to grapple with the fact that the dataset of correct translations, which is relentlessly being updated from what Google's "crawlers" find on the web, may degrade rapidly as humans start posting approximate translations made with… Google's translation software. If you publish a mistake made by the robot as if it were human knowledge, you fool all the other robots who are trying to learn from human expertise. Today's robots, equipped with deep learning, learn from our experts, not from each other. We learn from experts and by ourselves, i.e by "trial and error" or through a lengthy excruciating research. Robots learn from experts, human experts, the best human experts. Google's translation software is not the best expert in translation. If it starts learning from itself (from its own mediocre translations), it will never improve.

Supervised learning is "learning by imitation", which is as good as the person you are imitating. That's why the generation of AlphaGo is introducing additional tricks. Reinforcement learning, which was the topic of Minsky's PhD thesis in 1954, is a way to improve the speed and quality of machine learning. Another useful addition to deep learning (also used by AlphaGo) is tree-search, invented by Minsky's mentor Claude Shannon in 1950.

Similar considerations apply to robots. World knowledge is vital to perform ordinary actions. Robot dexterity has greatly improved thanks to a multitude of sensors, motors and processors. But grabbing an object is not only about directing the movement of the hand, but also about controlling it. Grabbing a paper cup is not the same as grabbing a book: the paper cup might collapse if your hand squeezes it too much. And grabbing a paper cup full of water is different from grabbing an empty paper cup: you don't want to spill the water. Moving about an environment requires knowledge about furniture, doors, windows, elevators, etc. The Stanford robot that in 2013 was trained to buy a cup of coffee at the cafeteria upstairs had to learn that a) you don't break the door when you pull down the handle; b) you don't spill coffee on yourself because it would cause a short circuit; c) you don't break the button that calls the elevator; etc; and, as mentioned, that the image in the elevator's mirror is you and you don't need to wait for yourself to come out of the elevator.

We interact with objects all the time, meaning that we know what we can do with any given object.

Your body has a history. The machine needs to know that history in order to navigate the labyrinth of your world and the even more confusing labyrinth of your intentions.

Finally, there are ethical principles. The definition of what constitutes "success" in the real world is not obvious. For example: getting to an appointment in time is "good", but not if this implies running over a few pedestrians; a self-driving car should avoid crashing against walls, unless it is the only way to avoid a child...

Most robots have been designed for and deployed in structured environments, such as factories, in which the goal to be achieved does not interfere with ordinary life. But a city street or a home contains much more than simply the tools to achieve a goal.

A paradox, formulated by Hans Moravec in "Mind Children" (1988), is that the complex reasoning that we normally associated with intelligence (such as mathematical calculations or playing chess) turns out to be much easier for machines than the simple daily activities that we don't normally associate with intelligence (such as grasping a glass of water without spilling the water on the floor). This is clearly a sign that, whatever an "intelligent" machine does, is not what our brain does. We are missing something important, and what we are building is not "intelligent" machines but just... machines. Intelligence is not artificial.

"Computers are useless: they can only give you answers" (Pablo Picasso, 1964).

Living with Machines that have no Common Sense

In 2017 it is virtually impossible to get the caption "gorilla" from Google's photo indexing system even when the picture is obviously a picture of gorillas. The reason is very simple: In July 2015 that program identified a black couple as gorillas, and this became a widely publicized scandal. Google's software engineers simply changed the code of the system to make sure that similar racial problems never occur again. Unfortunately, that decision has also severely handicapped the program in all applications that require recognizing gorillas. Face detection is not hard at all: it is a problem that Artificial Intelligence solved a long time ago.

The real issue is that the system is incredibly stupid: it knows absolutely nothing of the real world, and therefore it has no way to differentiate between someone who looks like a gorilla from a gorilla (or me from some animal that looks like me). If you know what people and gorillas are and do, you can tell in one split second which ones are people and which ones are gorillas. For example, it's unlikely that a gorilla is staring straight into the camera with a background of city traffic. In fact, you can even tell when they are people dressed in gorilla costumes, or when they are gorillas dressed in human attire. But, because programs have no common sense, the only way to make sure that they don't say or do anything stupid is to physically erase the possibility, which inevitably affects the usefulness of the program. Recognizing the North Korean dictator as a watermelon could trigger a world war, so let's make sure that the program will never spit out the word "watermelon", even when it is shown pictures of watermelons in a supermarket called "Watermelonland". Recognizing the president of the USA as a wanted terrorist could lead to a grotesque shootout between the FBI and the presidential detail, so let's make sure that nobody will ever be recognized as that terrorist, not even the terrorist himself! That would be the future.

I wrote "would be" because the other possible future is that we simply ban people from taking pictures that can confuse the program. In fact, a Google executive immediately recommended that Google users watch a video on how to light and photograph black faces. If you ban everything that can be confusing or misleading, intelligent machines are already here.

However, both solutions fail to address many other subtle issues of machines with no common sense. For example, in 2016 a high-school student, Kabir Alli, did a very simple test: he searched for images of "three white teenagers" and "three black teenagers. The search for "three white teenagers" turned up pictures of nice smiling teens, whereas the search for "three black teenagers" turned up mugshots of juvenile delinquents. The poor Google engineers had to intervene again, and so now (one year later) both searches turn up nothing that is even remotely offensive... but nothing that is even remotely interesting. This was widely discussed in the media as "A.I. mirrors the racial bias hidden in society" when in fact it was more a problem of "A.I. doesn't understand the question at all". Ask any person who lives in that racially-biased society to pick a picture of three black teenagers and that person would look for normal ones, not pictures of juvenile delinquents.

Artificial Shallow Intelligence

Progress in artificial neural networks, from the first Perceptron to Fukushima's convolutional network and to Hinton's deep learning, has been based on models of how animals learn by trial and error in simple (simple) cases of conditioning. It is not based on studies of general intelligence. No surprise therefore that deep learning is not giving us anything that even remotely resembles general intelligence.

Training a neural network is a highly unnatural process. If you want a deep-learning system to recognize a cake, you have to train it with thousands if not millions of pictures of cakes. A child learns to recognize a cake after seeing (and tasting) a few of them.

Artificial Intelligence often ignores (and sometimes reinvents decades later) the body of research that comes from psychologists. In the 1950s Jerome Bruner held that a category is defined by the set of features that are individually necessary and jointly sufficient for an object to belong to it ("A Study of Thinking", 1956), whereas Roger Brown ("Words and Things", 1958) found that we "naturally" name objects based on the "distinctive action" that we perform on them: the actions we perform on flowers are pretty much all the same, and certainly different from the actions that we perform on a cat. It is our basic actions that tell us that a cat is a cat and a flower is a flower. In the 1960s Brent Berlin found that people categorize plants at the same "basic level" anywhere in the world: it is a level at which only shape, substance and pattern of change are involved ("Covert Categories and Folk Taxonomies", 1968). In the 1970s Eleanor Rosch thought that a concept is represented through a prototype: membership of an object in a category is determined by the perceived distance of resemblance of that object from the prototype of the category. ("Cognition and Categorization" (1978). George Lakoff, author of "Women, Fire and Dangerous Things" (1987), argued that categories depend also on the bodily experience of the "categorizer". Frank Keil argued that no concept can be understood in isolation from all other concepts: concepts embody "systematic sets of causal beliefs" about the world ("Concepts, Kinds and Cognitive Development", 1989). My book "Thinking about Thought" has a lengthy survey of these theories.

A program that, trained with thousands of examples, learns how to play weiqi/go on a 19x19 board (and beats the world champion) but cannot play the game on a different kind of board unless you train it again with thousands of examples, can hardly be called "intelligent". A system that is powerless when you change the problem slightly can hardly be called intelligent simply because it can solve the problem in just one particular configuration.

The Frame Problem

A stumbling block towards the creation of truly intelligent machines is still the frame problem, formulated by John McCarthy in "Some Philosophical Problems from the Standpoint of Artificial Intelligence" (1969): he asked whether it is possible, in principle, to calculate all the effects on the world of an action. For example, one day a friend left his expensive phone in the car and simply covered it with his very inexpensive jacket. Minutes later his wife ran back to the car because she was feeling cold and picked up his jacket. When they got back to the car, someone had broken the window and stolen the phone. The act of picking up the jacket (for an absolutely legitimate reason) had changed the state of the world in many obvious ways (now the jacket was no longer in the car, she was no longer cold, etc) but also in other subtler ways, including the fact that now the phone was visible to thieves.

David Chapman perhaps proved mathematically the impossibility of artificial general intelligence when he proved that the frame problem is NP-complete (it belongs to the class of "nondeterministic polynomial" problems) and hence probably inherently unsolvable ("Planning for Conjunctive Goals", 1987). To be fair, mathematicians still haven't proven whether it is, in principle, possible to construct an efficient algorithm that can solve NP-complete problems: the majority consensus is that it is not possible, but the jury is still out. Of course, it all depends on what we mean by "solving a problem", as Wim Hordijk has discussed in "The Algorithmic Mind and what it Means to Solve a Problem" (2014) in response to Stuart Kauffman's claim that that the mind is not algorithmic ("Minds And Machines", 2011).

Common Sense Knowlege for Neural Networks

Computers operate on symbolic representations of the world. They are fed data that are either numbers or names, store them somewhere as sequences of zeroes and ones, and process them according to sequences of instructions that we call "programs" (most of which produce new data to be stored somewhere). The main criticism against the program of symbolic Artificial Intelligence was that this is "not" what the human brain does. This is actually the main criticism levied against the very idea of using computers to simulate human intelligence. Alas, it is hard to deny that this is precisely what "i" do. Yes, my brain is a wildly chaotic (and ugly) mess of "gray matter", but "i" think symbolically. This sentence that is forming in my mind and that i am typing is a sequence of symbols, each of them referring to many other symbols that are in my mind, and you are reading it as a sequence of symbols that create some meaning in your mind, and the meaning is symbolic too. Neither you nor i have any evidence that there is some electrochemical activity going on inside our skulls corresponding to our symbolic life. What we "are" is our symbolic life.

The "symbolic" approach seems to imply that an "algorithm" exists for everything that "i" think, and the critics can't accept the idea that everything in their mental life is algorithmic in nature. The problem with their criticism is that nobody has found something yet that "i" do and that cannot be explained in terms of representation and algorithm, i.e. computation. There is no question that the problem is difficult: it would take a long time to write all the algorithms that do all the things that "i" do, but nobody has proven yet that it is impossible, and every year it looks more (not less) possible than it looked ten years earlier. But difficult doesn't mean "impossible". In some cases it is actually quite simple. I don't know how to train a neural network to cross a street, but i can write down the basic rules in a few seconds: look right and left, wait for no cars coming in your direction, walk quickly to the other side. If it's raining, be also careful not to slip on the wet pavement. This is an algorithm.

It is a bit ironic that neural networks exhibit no common sense: aren't they supposed to be close approximations of how our brains work? It is natural to suspect that the problem lies in the opposite direction: symbolic

Artificial Intelligence is more (not less) similar to what "i" am than a neural network. I don't want to sound too philosophical, but "i" have common sense, whereas "i" know absolutely nothing of the structure of my brain. I have to read a book to find out the structure of the brain, whereas i don't have to read any book to pick up a piece of paper that fell to the floor or, quite simply, to eat my food. What "i" do is symbolic in nature. Even if i wanted to, i could not become a neural network: "i" don't process an image pixel by pixel and feed it to a layer of neurons and then to another layer and then to another layer and then change the weights of the connections according to the backpropagation algorithm and so forth. "I" only use symbols. When neuroscientists tell me that my symbols are the product of some inner working of the brain, i am happy to know that their findings may some day fix a problem in my brain the same way that medical discoveries about the heart may help fix a problem in my heart. But "i" don't need that technical knowledge to keep thinking just as my heart doesn't need technical knowledge to keep beating.

It is not easy (as yet) to encode knowledge of the world into a neural network. Encoding knowledge was precisely the goal of the "other" branch of Artificial Intelligence, the symbolic branch that used mathematical logic and aimed at simulating the way our minds work, not the way the brain is structured. "Expert systems" such as Dendral encoded knowledge as a set of statements in first-order predicate logic, things such as "Piero is a writer" and "If X is a writer, then X has readers". This approach reaches conclusions by applying simple rules of deduction to symbols like the X in that statement. One way to graft common-sense knowledge onto neural networks is to integrate the symbolic, knowledge-based, deductive methods of expert systems with the "subsymbolic", data-driven, learning methods of neural networks. The initial impulse to mixing the two approaches came from the limitations of each field: expert systems had the problem of creating the knowledge base (that typically involved "eliciting" the knowledge from often uncooperative human experts) whereas neural networks were plagued by the problem of "local minima". Theoretically, there was also a debate about how symbolic thinking can arise from neural computation. The issue was already raised by future free-speech activist David Touretzky at CMU in collaboration with Geoffrey Hinton ("Symbol Among Neurons", 1985). Stephen Gallant at Northeastern University worked on the "connectionist expert system" MACIE (1985), that was one of the first neural networks capable of explaining its output. Paul Smolensky (then at University of Colorado) thought he could solve the problem with his "tensor analysis" that found a formal equivalence between the high-level description of neural networks and symbolic systems ("On Variable Binding and the Representation of Symbolic

Structures in Connectionist Systems", 1987). Jude Shavlik at University of Wisconsin worked on a "knowledge-based artificial neural network" ("An Approach to Combining Explanation-Based and Neural Learning Algorithms", 1989). During the 1990s books such as "Integrating Rules and Connectionism for Robust Commonsense Reasoning" (1994) by Ron Sun of Brandeis University explored the subsymbolic-symbolic fusion. Collections of papers appeared in Daniel Levine and Manuel Aparicio's "Neural Networks for Knowledge Representation and Inference" (1993) as well as Suran Goonatilake's and Sukhdev Khebbal's "Intelligent Hybrid Systems" (1995).

Dov Gabbay of King's College London and his former student Artur Garcez of the City University of London published hybrid models of neural networks for knowledge representation and inference starting with the book "Neural-Symbolic Learning Systems" (2002). Leon Bottou (famous for his "stochastic gradient descent" method), while at Microsoft, discussed informal reasoning, an intermediate layer between subsymbolic computation and logical inference ("From Machine Learning to Machine Reasoning", 2011). Leo de Penning in Holland built the Neural Symbolic Cognitive Agent or NSCA ("A Neural-symbolic Cognitive Agent for Online Learning and Reasoning", 2011). Alex Graves at DeepMind presented "differentiable neural computers" ("Hybrid Computing Using a Neural Network with Dynamic External Memory", 2016). Richard Socher worked on common sense reasoning under Andrew Ng at Stanford and invented "neural tensor networks" ("Neural Tensor Networks For Knowledge Base Completion", 2013). Luciano Serafini of Fondazione Bruno Kessler in Italy and Artur Garcez of City University of London described "logic tensor networks" that blend Socher's neural tensor networks and first-order many-valued logic ("Logic Tensor Networks", 2016).

It turns out that logic tensor networks are similar to the BLOG (Bayesian LOGic) developed by Stuart Russell at UC Berkeley in 2005. These belong to a different but parallel way of thinking about extending mathematical logic (that deals with true and false statements only, i.e. with one and zero) to deal with probabilities (that is, with a continuum of values between zero and one). The resulting logics are not limited to true/false conclusions but admit degrees of truth. At the same time that Zadeh was beginning to work on fuzzy logic, Alfred Tarski's student Haim Gaifman showed how to graft probability theory onto first-order logic, i.e. tried to ground probabilities on firm logical foundations ("Concerning Measures in First Order Calculi", 1964). After Judea Pearl introduced his Bayesian networks, contributions for integrating logic and probabilities came from several mathematicians: Joseph Halpern of IBM ("An Analysis of First-order Logics of Probability", 1989), Stephen Muggleton of the Turing Institute in Britain (who popularized "inductive logic programming" in 1991), Venkatramanan Subrahmanian at University of Maryland ("Probabilistic Logic Programming", 1992), David Poole at University of British Columbia ("Representing Bayesian Networks within Probabilistic Horn Abduction", 1993), Peter Haddawy at University of Wisconsin ("Generating Bayesian Networks from Probability Logic Knowledge Bases", 1994), etc.

Google's DeepMath of 2016 (mainly Francois Chollet's project) studied how a neural network can do high-level logical thinking like proving theorems. Francois Chollet started from the obvious fact that humans can learn from very few examples, are relatively good at long-term planning, and naturally form generalizations of situations that they can apply in the future to a broad variety of situations. Microsoft's DeepCoder of 2017 (in collaboration with Matej Balog of Cambridge University) is a similar project for automated program generation.

There is one thing that the biological neural networks of our brains do better than the artificial neural networks of Artificial Intelligence: deduction. Mathematical logic is good at deriving the effect from the cause: if it is raining, things will get wet. Most machine learning (in particular, deep learning) is good at guessing the cause given the effect. That's because the machine is "trained" by showing many effects of the cause (for example, many examples of the object "apple"). Nonetheless, if the machine can't derive the effect from the cause, after it has "learned" the cause it is not able to reproduce the behavior of the observed system. There is a long history of trying to teach machines how to learn causal knowledge, starting with Judea Pearl's landmark book "Causality" (2000) via Shohei Shimizu at Riken in Japan ("A Linear Non-Gaussian Acyclic Model for Causal Discovery", 2006) and Bernhard Schoelkopf at the Max Planck Institute in Germany ("On Causal and Anticausal Learning", 2012). In 2013 Isabelle Guyon (now no longer at Bell Labs) organized a competition to develop algorithms that can learn causal inference, the Cause-effect Pairs Kaggle Competition (266 teams competed).

Footnote: Logic and Probability

Unifying probability theory (useful to describe a world that is fundamentally uncertain) and mathematical logic (that works pretty well at describing relationships between the objects of the world) is an old ambition. In 1946 Richard Cox, a physicist at Johns Hopkins University, proved a theorem that was widely believed to have shown that probability theory extends mathematical logic beyond the realm of true and false.

Another physicist, Edwin Jaynes at Washington University in St Louis, interpreted probability theory as an extension of logic in his influential book "Probability Theory" (which he started writing in 1952). Later, Artificial Intelligence scientists, faced with the desire to save both logic and probability theory, attempted practical fusions of the two forms of reasoning, for example Fahiem Bacchus at University of Alberta in Canada ("Representing and Reasoning with Probabilistic Knowledge", 1988) and Joseph Halpern at IBM's Almaden Research Center in San Jose ("An Analysis of First-Order Logics of Probability", 1989).

Perhaps the most intriguing attempt at unifying logic and probability at the dawn of deep learning was the work on "Markov Logic Networks" by Pedro Domingos and his student Matt Richardson at the University of Washington ("Markov Logic Networks", 2006), later expanded by Domingos' student Jue Wang as "Hybrid Markov Logic Networks" (2008). These representations combine first-order logic and Markov networks and use the Markov Chain Monte Carlo method for (approximate) inference.

However, Stuart Russell at UC Berkeley pointed out that a marriage of logic and probability requires that we think of objects as uncertain too ("Unifying Logic and Probability", 2014). One thing is to tell a machine that there are apples in the world and asking the machine to recognize apples, and another thing is to ask the machine what objects it sees in the world: the machines see pixels, billions of pixels. There is a "difference between knowing all the objects in advance and inferring their existence and identity from observation".

Finally, David Chapman showed that Cox and Jaynes fundamentally misunderstood what logic is. Probability theory is an extension of propositional calculus but propositional calculus is not logic: it says nothing about objects. Logic starts with predicate calculus, which can describe relationships among objects, and logic can do things that probability theory cannot do ("Probability Theory does not Extend Logic", 2016).

We actually don't Think

The most successful algorithms used in the 2010s to perform machine translation use statistical analyses and require virtually no linguistic knowledge. These programs simply explore thousands of translations done by human experts and calculate which is the most popular. The very programmer who creates and improves the automatic-translation system doesn't need to have any knowledge of the two languages being translated into each other: it is only a statistical game. I doubt that this is how human interpreters translate one language into another, and i doubt that this approach will ever be able to match human translations, let alone surpass them.

Donald Knuth's famous sentence that A.I. seems better at emulating "thinking" than at emulating the things we do without thinking is still true; and it contains a larger truth. The really hard problem is that we don't know how we do the vast majority of things that we do, otherwise philosophers and psychologists would not have a job. A conversation is the typical example. We do it effortlessly. We shape strategies, we construct sentences, we understand the other party's strategy and sentences, we get passionate, we get angry, we try different strategies, we throw in jokes and we quote others. Anybody can do this without any training or education. And now, by comparison, check what kind of conversation can be carried out by the most powerful computer ever built.

Most of the things that we do by "thinking" (such as proving theorems and playing chess) can be emulated with a simple algorithm (especially if the environment around us has been shaped by society to be highly structured and to allow only for a very small set of moves). The things that we do without thinking cannot be emulated with a simple algorithm, if nothing else because even we don't know how we do them. We can't even explain how children learn in the first place.

Mind Uploading and Digital Immortality

Of all the life extension technologies proposed so far, perhaps none has captured the imagination of the machine-intelligence crowd more than mind uploading. Somehow the connection between the Singularity and digital immortality was made: at some point those super-intelligent machines will be able to perform one great task for us, upload our entire self and "become" us. Couple it with the immortality of the "cloud" (see later), and your "self" becomes immortal. It will be downloaded and uploaded from one release of the Singularity to the next one for the rest of time.

In the most memorable of Isaac Asimov's short stories, "The Last Question" (1956), humankind is preserved in cyberspace after the end of the universe. Some forms of mind uploading already appeared in Arthur Clarke's novella "The City and the Stars" (1953) and in Frederick Pohl's short story "The Tunnel Under the World" (1955). I find the latter more realistic because it envisions mind uploading as a trick devised by the advertising industry. Wesley Barry's film "Creation of the Humanoids" (1962) envisioned the technology to upload memories and emotions of a dead man into an immortal machine in order to save the human race from extinction after a nuclear holocaust.

The technology of uploading a human mind into a computer was first explored by a geneticist, George Martin, in "A Brief Proposal on Immortality" (1971). He foresaw that someday computers would become so powerful that they will be able to do everything that a brain can do. Therefore why not simply port our brains to computers and let the computers do the job. Needless to say, philosophers are still arguing whether that "mind" would still be "me" once uploaded into software instead of gray matter. Hans Moravec speculated that you are just a pattern, therefore you could "transmigrate" to a different body ("Dualism Through Reductionism", 1986).

That vision became more realistic in the 1990s with the explosion of the World-wide Web. A paleontologist, Gregory Paul, in collaboration with a mathematician, Earl Cox, speculated about cyber-evolution that could create non-human minds in "Beyond Humanity" (1996), including the idea of immortal "brain carriers" to replace our mortal bodies. In the days when television was still influential, William Gibson, the science-fiction writer who a decade earlier had invented the term "cyberspace" ("Burning Chrome", 1982), contributed to the popularization of the concept by scripting an X-Files episode, "Kill Switch" (1998), in which a man uploads his mind into cyberspace. Ray Kurzweil wrote the article "Live Forever Uploading The Human Brain" (2000).

Then came the deluge with books such as Richard Doyle's "Wetwares - Experiments in PostVital Living" (2003) exploring all sorts of technologies of immortality. Every year the vision of what Martine Rothblatt calls "mindclones", implemented in "mindware" (the software for consciousness), has to be updated to the latest computer platform.

In 2012 a Russian tycoon, Dmitry Itskov, pretty much summarized the vision of the immortality field: firstly, build brain-machine interfaces so that a human brain can control a robotic body; secondly, surgically transplant the human brain into a robotic body; and, finally, find a method to achieve the same result without the gory surgical operation, i.e. a way to upload a person's mind into the robotic body or, for that matter, into just about anything.

The question, of course, is whether that "you" will still be you. Or just a machine mimicking you? You can ask someone to impersonate you, but that does not mean that he or she is you. That someone has absorbed the "pattern" of your behavior, but s/he is not you. By the same token, if a machine absorbs some pattern found in your brain, it doesn't mean that the machine has become you. We literally don't know what in the brain makes you "you" (and not, for example, me). This disembodied and reconstituted "mind" might well be immortal, but is it you?

In other words, this program is predicated on the assumption that "i" am entirely in my brain, and that my body is simply a vehicle for my "i" to survive. If so, such a body can as well be replaced by some other material substrate. The brain is disposable, according to this view: the brain is merely the organ of the body designated to host the processes that construct the "i", but the "i" truly is only those processes, which, luckily for us, turn out to be information-based processes, which, luckily for us, can be easily transplanted from the mortal (and, let's admit it, quite repulsive) brain into the kind of information-processing machines that we started building in the 1940s and that are getting more and more powerful, rapidly approaching the capacity required to simulate the entirety of those brain processes.

This movement has revived the project of whole brain emulation. Ray Kurzweil and others have estimated that "artificial general intelligence" will be achieved first via whole brain emulation. The basic idea is to construct a complete detailed software model of a human brain so that the hardware connected to that software will behave exactly like the human would (which includes answering the question "is it really you?" with a "yes").

But first, one needs to map the brain, which is not trivial. In 1986 John White's and Sydney Brenner's team mapped the brain of the millimeter-long worm Caenorhabditis Elegans (302 neurons and 7000 synapses). As far as i know, that is still the only brain that we have fully mapped. And it took twelve years to complete that relatively simple "connectome". The term was coined in Olaf Sporns' "The Human Connectome, a Structural Description of the Human Brain" (2005). A connectome is the map of all the neural connections in a brain. In 2009, a few years after the success of the Human Genome Project, the USA launched the Human Connectome Project to map the human brain. The task, however, is not on the same scale as mapping a worm's brain. The entire human genome is represented by about a few gigabytes of data. Cellular biologist Jeff Lichtman and Narayanan Kasthuri estimated that a full human connectome would require one trillion gigabytes of memory ("Neurocartography", 2010). Furthermore, we all share (roughly) the same genome, whereas each brain is different. The slightest mistake and... oops... they may upload the brain of someone else instead of yours.

Once we are able to map brains, we will need to interface those brains with machines. This may actually come sooner. In 1969 the Spanish neurophysiologist Jose Delgado implanted devices in the brain of a monkey and then sent signals in response to the brain's activity, thus creating the first bidirectional brain-machine-brain interface. In 2002 John Chapin debuted his "roborats", rats whose brains were fed electrical signals via a remote computer to guide their movements. His pupil Miguel Nicolelis achieved the feat of making a monkey's brain control a robot's arm. In 2008 the team made the monkey control a remote robot (in fact, located in another continent).

By the time science is capable of uploading your mind to cyberspace most of us will probably be dead, and with us our brains. That disturbing thought predated the very science we are talking about. Robert Ettinger's book "The Prospect of Immortality" (1962) is considered the manifesto of "cryonics", the discipline of preserving brains by freezing them. It was actually cryonics that started the "life extension" movement. In 1964 another founding father, Evan Cooper, launched the Life Extension Society (LES). In 1972 Fred Chamberlain, a space scientist at the Jet Propulsion Laboratory, founded the Alcor Society for Solid State Hypothermia (ALCOR), now called Alcor Life Extension Foundation, to enter that business.

The similarities with the most successful organized religions of the Western world are too obvious to be overlooked. The end of the world is coming in the form of the Singularity, but, not to worry, we will all be resurrected in the form of mind uploads made possible by the super-machines of that very Singularity. The only difference with the ancient Western religions is that people from previous ages are dead for good, forever: we don't have their brains to upload anymore. But then maybe those super-human machines will find a way to resurrect the dead too.

Machine Immortality and the Cloud

The other implicit assumption in the scenario of mind uploading is that these superhuman machines, capable of self-repairing and of self-replicating, will live forever.

That would represent a welcome change from what we are used to. The longest life expectancy for an electrical machine probably belongs to refrigerators, that can last longer than a human generation. Most appliances die within a decade. Computers are the most fragile of all machines: their life expectancy is just a few years. Their "memories" last less than human memory: if you stored your data on a floppy disc twenty years ago, there is probably no way for you to retrieve them today. Your CDs and DVDs will die before you. And even if your files survived longer, good luck finding an application that can still read them. Laptops, notepads and smartphones age increasingly faster. The life expectancy of machines seems to be decreasing ever more rapidly. And, of course, they are alive only for as long as they are plugged into an electrical outlet (battery life can be as little as a few hours); and they seem to be more vulnerable than humans to "viruses" and "bugs".

One has to be an inveterate optimist to infer from the state of the art in storage media that increasingly mortal and highly vulnerable computer technology is soon to become immortal.

Of course, the Singularity crowd will point out the "cloud", where someone else will take care of transferring your data from one dying storage to a newer one and translating them from one extinct format to a newer one. Hopefully some day the cloud will achieve, if not immortality, at least the reliability and long lifespan of our public libraries, where books have lasted millennia.

Having little faith in software engineers, i am a bit terrified at the idea that some day the "cloud" will contain all the knowledge of the human race: one little "bug" and human civilization as we know it will be wiped out in a second. It is already impressive how many people lose pictures, phone numbers and e-mail lists because of this or that failure of a device. All it takes is that you forget to click on some esoteric command called "safely remove" or "eject" and an entire external disc may become corrupted.

If and when super-intelligent machines come, i fear that they will come with their own deadly viruses, just like human intelligence came (alas) with the likes of influenza pandemics, AIDS (Acquired ImmunoDeficiency Syndrome), SARS (Severe Acute Respiratory Syndrome), Ebola and Zika. And that's not to mention the likelihood of intentional cyber terrorism (i'm not sure who's getting better at it: the cryptographers who are protecting our data or the hackers who are stealing them) and of "malware" in general. If today they can affect millions of computers in a few seconds, imagine what the risk would be the day that all the knowledge of the world is held in the same place, reachable in nanoseconds. The old computer viruses were created for fun by amateurs. We are entering the age in which "cyber crime" will be the domain of super-specialists hired by terrorists and governments. Originally, a computer virus was designed to be visible: that was the reward for its creator. Today's cyber crime is designed to be invisible... until it's too late. Remember when only fire could destroy your handwritten notes on paper? And it was so easy to make photocopies (or even manual copies) of those handwritten notes? We found the Dead Sea Scrolls two thousand years after they had been written (on a combination of vellum and papyrus), and the Rosetta stone is still readable after 2,200 years. Think of the Herculaneum Papyri, stored in Napoli's National Museum, dated to AD 79 and discovered in the 18th century, of the 4th century Codex Sinaiticus on parchment, discovered in 1844 at the monastery of St Catherine in the Sinai, housed in the British Museum, of the 2,000-year-old Buddhist scrolls discovered in the early 20th centuries in the Mogao Caves of Dunhuang in China (housed in Beijing, London, Paris and St Petersburg). I wonder how many data that we are writing today will still be found two thousand years from now in the "cloud".

In March 2019 MySpace announced that all music uploaded onto MySpace prior to 2015 was lost. If you uploaded your song to MySpace in 2014, its "immortality" lasted only 5 years. Compare with the "Hurrian Hymn No. 6," written on a clay tablet around the 14th century B.C and discovered in the 1950s in Syria, or even with the "Gloria" by Handel whose score was found 300 years later at the Royal Academy of Music in 2001.

If you want immortality, make sure your consciousness is replicated on paper or on a Sumerian tablet, not in a digital format that can be erased in a split second.

Hackers will keep getting more and more sophisticated and, when armed with powerful computers provided by rich governments, able to enter any computer that is online and access its contents (and possibly destroy them). In the old days the only way for a spy to steal a document was to infiltrate a building, search it, find the safe where the documents were being held, crack open the safe or bribe someone, duplicate the documents, flee. This was dangerous and time consuming. It could take years. Today a hacker can steal thousands if not millions of documents while comfortably sitting at her desk, and in a fraction of a second. The very nature of digital files makes it easy to search and find what you are looking for.

Ironically, an easy way to make your files safe from hacking is to print them and then delete them from all computers. The hacker who wants to steal those files is now powerless, and has to be replaced by a traditional thief who has to break into your house, a much more complicated proposition.

Cyber-experts now admit that anything you write in digital form and store on a device that is directly or indirectly connected to the Internet, will, sooner or later, be stolen. Or destroyed. When, in march 2013, the websites of JPMorgan Chase and then American Express were taken offline for a few hours after being attacked by the Izz ad-Din al-Qassam Cyber Fighters, cyber-security expert Alan Paller warned that cyber-attacks are changing from espionage to destruction. A malware to destroy (digital) information on a large scale would be even easier to manufacture than the malware Stuxnet (unleashed in 2010 probably by Israel and the USA) that damaged about one thousand centrifuges used to enrich nuclear material at Iran's nuclear facility in Natanz.

In just the year 2017 three billion Yahoo accounts were hacked, the WannaCry "ransomware" hit half of the world, and 145 million Equifax accounts were hacked. In 2016 even the servers of the US Democratic Party were hacked and we don't know how many financial and political agencies have been the victims of cyber-crime and didn't report it to avoid panic among the public. For millennia we thought of war mainly as destruction of buildings (and, alas, of the people who lived inside them), but the future will feature a new kind of war, war that will be about destroying data, each one an escalating cyber-battle between those who try to protect data and those who want to destroy data. I am not sure who is more motivated: the cyber-criminals make money only when they win the battle, whereas the executives get their hefty salary even when they lose (Fortune reported that, after losing that catastrophic cyber-battle, Equifax's CEO Richard Smith retired with $90 million in compensation).

I also feel that "knowledge" cannot be completely abstracted from the medium, although i find it hard to explain what the difference is between knowledge stored in Socrates' mind, knowledge stored in a library and knowledge stored in a "cloud". A co-founder of one of the main cloud-computing providers said to me: "Some day we'll burn all the books". Heinrich Heine's play "Almansor", written a century before Adolf Hitler's gas chambers, has a famous line: "Where they burn books, they will ultimately burn people too". Unlike most predictions about machine intelligence, that is one prediction that came true.

Corollary: Digital Media Immortality

If you want to turn yourself into data, instead of flesh and bones, and hope that this will make you immortal, you have a small technical problem to solve.

As you well know from your Christmas shopping, the capacity of computer storage media (for the same price) increases rapidly. That's the good news. The bad news is that its longevity has been decreasing, and significantly decreasing if you start from way back in time. The life expectancy of paper and ink is very long in appropriate conditions. The original storage media for computers, punched paper tapes and punch cards, are still readable 70 years later: unfortunately, the machines that can read them don't exist anymore, unless you have access to a computer museum. By comparison, the life expectancy of magnetic media is very, very short. Most people born before 1980 have never seen a magnetic tape except in old sci-fi movies. It was introduced with the first commercial computer, the Eckert-Mauchly's UNIVAC I, in 1951. Today most magnetic tapes store terabytes of data. They last about 20-30 years. Nobody knows how long the multiplatter disks from the mainframes of the 1960s lasted because they got out of fashion before we could test their lifespan. Floppy discs are magnetic disks, the most common type of which had a capacity of 1.44 megabytes or 2 megabytes. The 8" floppy disks of the 1970s and the 5.25" floppy disks of the 1980s are given a life expectancy of 3-5 years by those who never used them, but those like me who still have them know that at least half of them are still working 30 years later. The external "hard disks" that replaced them (and that today can easily hold a terabyte, i.e. a million times more data) may last longer, but they need to spin in order to be read or written, and spinning-disk hard drives don't last long: they are mechanical devices that are likely to break long before the magnetic layer itself deteriorates, especially if you carry them around (in other words, if you use them).

Music was stored on magnetic tapes, and later on cassettes, that would still work today if mass-market magnetic tape players still existed, although they would probably not sound too good, and on vinyl records, that definitely still play today if you didn't scratch them and used appropriate cartridges on your turntable like i did. My cassettes from the 1970s still play ok. Video was stored on VHS tapes, that still play today (i have about 300 of them), but, again, colors and audio may not look/sound so good after years of playing on a VCR (if you can still find a VCR).

Then came the optical generation. Rewritable optical discs are much less reliable for data storage than read-only optical discs that you buy/rent at music or video stores because they are physically made of different materials (the film layer degrades at a faster rate than the dye used in read-only discs). The jury is still out on optical media, but, as far as storing your data goes, the Optical Storage Technology Association (OSTA) estimates a lifespan of 10-25 years for compact discs (CDs), that typically held 650 megabytes (or the equivalent of 700 floppy disks), and digital video discs (DVDs), that typically held 4.7 gigabytes. However, in practice, optical devices are much more likely to get damaged because very few people store their discs in optimal conditions. Just leaving them on a desk unprotected may greatly shorten their lifespans just like anything else that you look at (optical is optical).

Now we live in the age of solid-state media, devices that don't have moving parts and that can store several gigabytes on a very small device, like USB flash drives ("thumb" drives) and secure-digital cards ("flash cards"). They are generally less (not more) reliable than hard drives, and the manufacturers themselves don't expect them to last longer than about eight years.

And that's not to mention the quality of the recording: digital media are digital, not analog. You may not be able to tell the difference because your ears are not as good as the ears of many (supposedly less intelligent) animals, but the digital music on your smartphone is not as accurate a recording as the vinyl record of your parents or the 78 RPM record of your grandparents. A digital recording loses information. The advantage, in theory, is that the medium is less likely to deteriorate as you use it: magnetic tape degrades every time it passes by the magnetic head of a cassette player or a VCR, and the grooves of LPs do not improve when the cartridge of the turntable rides on them. The advantage of the old media, however, is that they "degraded": they didn't simply stop working. Digital files are either perfect or don't work, period. My old VHS tapes lost some of the color and audio fidelity, but i can still watch the movie. Many of my newer DVDs stop in the middle of the movie, and there is no way to continue. (I am also greatly annoyed by the difficulty of rewinding/forwarding a DVD or pinpointing a frame of the movie, something that can easily be done on a VHS tape: this is possibly the first "regress" in history for random access, a feature introduced by the Romans when they switched from the scroll to the codex).

On the other hand, microfilms are estimated to last 500 years: that is a technology that was introduced by John Benjamin Dancer in 1839, and first used on a large scale in 1927 by the Library of Congress of the USA (that microfilmed millions of pages in that year).

You can tell that the plot remains the same: larger and larger storage, but perhaps less and less reliable.

In August 2016 Seagate introduced the world's largest drive in the world (60 terabytes), not a hard disk but a solid-state drive. I couldn't find anything in 2017 that was larger. In 2016 Netflix is managing close to 40 petabytes of data (one petabyte is one thousand terabytes or one million gigabytes or one billion megabytes). Congratulations, but i doubt that any of those data will be readable a generation from now, let alone for eternity.

Note that all of this is very approximate: search for the longevity of free neutrons, and you'll readily find it (14'42"), but if you search for a scientific answer to the question of storage media longevity, you will not find it. That's how advanced the science of storage media is.

Finally, even if your media could potentially last a long time, when is the last time you saw a new computer model with a floppy drive? Even optical drives (CD, DVD) are disappearing as i type these words, and your favorite flash memory may become obsolete before this book goes out of print. And even if you still find a machine with a drive for your old media, good luck finding the operating system that has a file system capable of reading them. And even if you find both the hard drive and the operating system that can read them, good luck finding a copy of the software application that can read the data on them (e.g., GEM was the popular slide presentation software in the heydays of floppy discs). This is a field in which "accelerating progress" (in physical media, operating systems and viewing applications) has been consistently hurting data longevity, not extending it.

Yes, i know: the "cloud" will solve all these problems. And create bigger ones. In 2019 MySpace lost all the music uploaded during its first 12 years: 50 million songs from 14 million artists. It blamed it not on an apocalyptic natural disaster (not an asteroid, not an earthquake, not even a fire) but on a simple server migration. All those data stored in the MySpace "cloud" didn't even last 12 years. By comparison, The Nag Hammadi library, a collection of early Christian texts, was discovered near the Upper Egyptian town of Nag Hammadi in 1945: written on papyrus and buried in a clay jar, it had survived 1,600 years. In the past, nobody had the ability to destroy all the libraries of the world. But now we can see the day when all the knowledge of the world will be vulnerable to a hacker who creates a virus capable, in a few minutes, of infecting all the files stored in the cloud.

Ethical Intermezzo: The Downside of Immortality

If immortality can be achieved in this life, it will have non-trivial consequences on a selfish race like the human race.

If you believe that immortality is granted in the afterlife, you will do everything that you can in order to obtain it in the afterlife (which typically means obeying the instructions that a god gave humans to achieve the above said immortality); but if you believe that immortality is granted in this life, you will do everything that you can to obtain it in this life. A person who believes that immortality will be granted in the afterlife based on her good deeds will promptly sacrifice her life to save someone else or to fight a dangerous disease in Africa or to provide her children with a better future; but a person who believes that immortality is granted in this life has literally no motivation to risk her life to save someone else's life, nor any motivation to risk her life in Africa nor (ultimately) any motivation to care for her own children. Once you are dead, you are dead, therefore the only meaning that your life can have is not to die. Never. Under no circumstances. No matter what. Stay alive.

The new morality of a society that believes in immortality here now will be simple: stay alive at all costs, because immortality in this life is the only thing that matters.

"Humanity is acquiring all the right technology for all the wrong reasons" (Buckminster Fuller, "Earth Inc", 1973)

Another Philosophical Intermezzo: Do We Really Want Intelligence at All?

Intelligence is messy. When we interact with human beings, we have to consider their state of mind besides our immediate goal. We may only need a simple favor, but it makes a huge difference whether our interlocutor is happy or sad, on vacation or asleep, has just lost a close relative or been injured in an accident, angry at us, busy with her work, etc. Whether our interlocutor is capable or not of performing that favor for us may be a secondary factor compared with whether she is in the mental condition of doing it and doing it right now. On the other hand, when we deal with a dumb machine, the only issue is whether the machine is capable of performing the task or not. If it is, and the power chord is plugged into the power outlet, it will. It won't complain that it's tired or in a bad mood, it won't ask us for a cigarette, it won't spend ten minutes gossiping about the neighbors, it won't comment on the government or the soccer game.

It may seem a paradox, but, as long as machines are dumb, they are easy and painless to interact with. They simply do what we ask them to do. No whims. No complaints. No formalities.

The complication that comes with intelligent beings is that they are subject to moods, feelings, opinions, intentions, motives, etc. There is a complicated cognitive apparatus at work that determines the unpredictable reaction of an intelligent being when you ask even the simplest of questions. If your wife is mad at you, even the simplest question "What time is it?" might not get an answer. On the other hand, if you use the right manners at the right time, a complete stranger may do something truly important for you. In many cases it is crucial to know how to motivate people. But in other cases that is not enough (if the person is in a bad mood for reasons that are totally independent of your will). Human beings are a mess. Dealing with them is a major project. And that's not to mention the fact that human beings sleep, get sick, go on vacation, and even take lunch breaks. In Western Europe they are often on strike.

Compare humans with dumb machines that simply do what you ask. For example, the automatic teller machine hands you money at any time of the day or night any day of the year. Wherever the intelligent being has been replaced by a dumb machine, the interaction is simpler. We structured the interaction so that the dumb machine can perform all the operations that we need.

The reason that automated customer support has replaced human beings in so many fields is not only that it is cheaper to operate by the provider but also that it is preferred in the majority of cases by the majority of customers. The honest truth is that very few of us enjoy waiting for an operator to greet us "Hello? How are you? Isn't it a beautiful day? How can i help you?" Most of us prefer to press digits on a phone keypad. The truth is that most customers are happy if we remove the complication of dealing with human beings.

When i worked in the corporate world, my top two frustrations were the secretary and the middle management. Dealing with the secretary (especially in unionized Italy) required superior psychological skills: say the wrong word in the wrong tone, and s/he'd boycott you for the rest of the day. Most middle managers were mediocre and mostly slowed down things, seemingly paid mainly to kill great ideas. The only way to get important things done quickly was, again, to use the art of psychology: befriend them, chat with them, find out what motivated them, offer them rides home, hang out with them. My life would have been much easier if my colleagues and my secretary had been heartless robots.

And, let's face it, we often don't have the patience for human interactions that involve protocols of behavior. We are often happy when good manners are replaced by cold mechanic interactions shaped by goals and restrained by laws. Hence we do not really want machines with human intelligence, i.e. we don't want them to have emotions, to be verbose, to deceive, to plead, etc. One goal of inventing machines is precisely to remove all of that, to remove that inefficient, annoying, time-consuming quality of "humanity".

We removed the human/intelligent element from many facets of ordinary life because the truth is that in most cases we don't want to deal with intelligent beings. We want to deal with very dumb machines that will perform a very simple action when we press a button.

I'll let psychologists and anthropologists study the reasons for this trend towards less and less human interactions, but the point here is that intelligence comes at a high price: intelligence comes with feelings, opinions, habits, and a lot of other baggage. You can't have real intelligence without that baggage.

When we study how to create intelligent machines, do we really mean "intelligent" or do we mean "stupid in a way that will serve our intelligence"?

Religion and the Law of Accelerated Exaggeration

Robert Geraci of Manhattan College, in his book "Apocalyptic A.I." (2010), showed that Singularity thinking borrows motifs and practices from Jewish and Christian apocalyptic scriptures. The Judeo/Christian religions offer a dualistic view of the world: good and evil are fighting a cosmic battle. "Evil" is manifested as bodily decay, earthly world, and limited intellect. "Good" will someday materialize as eternal life, celestial world and unlimited knowledge. Singularity thinking (which he calls "Apocalyptic A.I.") adopts a similar view, cursing the mortal body and the limited knowledge of the human mind while envisioning a future in which we will become immortal and omniscient in cyberspace. The enabler is the high priesthood of A.I. scientists and engineers, whom Geraci nicknames "mystical engineers".

Geraci points out how apocalyptic thinking arose among Jews and Christians: they were both persecuted people. The Jews endured slavery and/or occupation by the Assyrians, the Babylonians, the Greeks and the Romans. The Christians were persecuted by the Romans. Geraci thinks that A.I. scientists such as Hans Moravec and Ray Kurzweil feel similarly persecuted, except that now it is "bodily alienation": they want to escape the limitations of the biological body.

New York University anthropologist Stefan Helmreich in "Silicon Second Nature" (1998) studied the "mystical" attitudes of the practitioners of Virtual Reality and Artificial Life. In 2003 Philip Rosedale's Linden Lab launched "Second Life", a virtual world accessible via the Internet in which a user could adopt a new identity and live a "second life" as an avatar, and Geraci views Second Life as a sort of temple where people perform religious functions.

Incidentally, the Singularity bears obvious similarities with the Omega Point, described by Pierre Teilhard, a Catholic priest from France, in his book "The Phenomenon of Man" (1955), and conceived as a point of super-human intelligence towards which the universe is evolving. The physicist Frank Tipler gave the omega point a formal mathematical and scientific formulation in his book "The Physics of Immortality" (1994).

Tipler predicted that evolution would end with a simulation of all the conscious beings who ever existed, i.e. the resurrection of all the dead. A precursor of transhumanism was Russian "cosmism", a school of scientific life extension (all the way to resurrection and immortality) led by a friend of the novelist Lev Tolstoy, the librarian Nikolai Fyodorov or Fedorov (employed in Moscow's largest lending library), whose "Philosophy of the Common Task" was published posthumously. This movement included Konstantin Tsiolkovsky, who advocated a future of space travel in "The Exploration of Cosmic Space by Means of Reactive Devices" (1903), and the mineralogist Vladimir Vernadsky, whose book "The Biosphere" (1926) discussed how the noosphere (thinking matter) shapes the biosphere (living matter) that shapes the geosphere (inanimate matter).

Faced with technological progress, thinkers liked to toy with the idea that the old human species was soon to be replaced by some new posthuman species. The term "posthumanism" was coined in 1976 by the Egyptian literary critic Ihab Hassan in a talk ("Prometheus as Performer") at the International Symposium on Postmodern Performance in Milwaukee, after the French philosopher Michel Foucault had declared humankind "a historical construction whose era is about to end" ("The Order of Things", 1973).

Cultural historian Margaret Wertheim in "The Pearly Gates of Cyberspace" (1999) argued that cyberspace represents the high-tech equivalent of religious paradise, an identification that goes back to Michael Benedikt of University of Texas at Austin, who wrote in the introduction to the anthology "Cyberspace" (1992) that cyberspace is the equivalent of the biblical "Heavenly City". Jeffrey Fisher's study "The Postmodern Paradiso" (1997) shows similarities between medieval Christian mythology and cyberspace utopia. To celebrate the age of Web 2.0, in 1999 Yale University's computer scientist David Gelernter wrote a manifesto titled "The Second Coming".

The impact on mysticism of the discovery of cyberspace has not been too different from the impact that the discovery of America had five centuries earlier: when in 1503 Amerigo Vespucci wrote to Lorenzo de Medici that Cristoforo Colombo had actually discovered a "new world", many viewed this "New World" as the new Eden. America (named after Amerigo) became the natural vehicle for Europe's utopian dreams at a time when Europe was launching into the humanistic, scientific and artistic revolutions of the Renaissance. The great utopian works of the following century, from Thomas More's Utopia" (1516) to Francis Bacon's "New Atlantis" (1627), were influenced by the myth of America as a blank space where a superior society could be created. Six centuries later what today's futurists are imagining in cyberspace is not all too different from what those 16th-17th century futurists imagined in their utopian books.

In "The Future of Religion" (1985) sociologists Rodney Stark of University of Washington and William Bainbridge of Boston University showed that we live in an age that lends itself to the establishment of new religions. In other words, an increasingly secular science has not killed religion but rather has created an opportunity for reforming religion. When Nietzsche announced "the death of God" in his book "Thus Spoke Zarathustra" (1883), he had basically opened the doors for a new religion, and the first one to take advantage of that opening had been the scientistic religion presented by Karl Marx in "Capital" (1894): communism. In his

"Religions for a Galactic Civilization" (1982) Bainbridge advocated establishing a scientistic theocracy along the lines of UFOlogy as something that humans need in order to survive (UFOlogy was replaced by Singularity thinking in the revised 2009 version).

Their book showed that religious innovation is particularly rampant in the Western states, from Alaska down to California, where membership in traditional churches is very low. By the 1980s Christian Science's leading state was California, followed by Oregon and Washington, Theosophy's leading state was Washington followed by California, Baha'i's leading state was again California followed by Oregon. And one of the largest new religions in the world, transcendental meditation, started out in California (in 1965 when Maharishi Mahesh Yogi founded the Students' International Meditation Society in Los Angeles). Many of the psychedelic movements born in California mutated into religious movements. For example, Timothy Leary founded "exo-psychology" and Richard Alpert became a Hindu guru, Baba Ram Dass.

Historically, new religious movements spread first among the educated elite. This seems hard to believe because the educated elite tends to be less religious than the less educated masses, i.e. more secularized, but Stark and Bainbridge show that secularization is precisely the condition required for religious innovation. People who already belong to a religious movement are unlikely to shift to another one. For example, 19th century spiritualism spread initially among socialists and intellectuals who were not religious at all. And from the beginning it took technology as a metaphor for the supernatural: the first celebrities of spiritualism were the sisters Leah, Kate and Maggie Fox who referred to their communications with the dead as a form of telegraphy, and the first journal of spiritualism, founded in New York in 1852, was called The Spiritual Telegraph. The telegraph, electricity, photography felt utterly unexplainable, and for many these technologies replaced the supernatural of traditional religions. Stark and Bainbridge point out the "overrepresentation" of Jews in modern cult movements and they explain it as a consequence of the rapid secularization of Judaism. Coincidence or not, Ray Kurzweil was born to secular Jewish parents.

Given the appeal of "religious innovation" to the secularized elite, and the fact that science has not solved the fundamental problems of humanity (the meaning of life and eternal life), Stark and Bainbridge easily predicted that "religious innovations will have significant influence in the coming years". It is precisely the success of science in demolishing the traditional religions that is creating the need for a new religion. This new religion needs to be more "scientific".

Religion is not dead at all. Its nature is simply changing, adapting to the age of math, science and software. In 1997 Michael Drosnin's book "The Bible Code", according to which the Bible is written in a secret code that foretells future events, topped the best-seller lists of the USA (to this day Michael Drosnin gets a much longer Wikipedia page than most historians can ever hope for).

Kurt Andersen wrote in "Fantasyland" (2017): "The idea that progress has some kind of unstoppable momentum... was always a very American belief. However, it's really an article of faith, the Christian fantasy about history's happy ending reconfigured during and after the Enlightenment as a set of modern secular fantasies." Michael Shermer's book "Heavens on Earth" (2018) examines humanity's obsession with the afterlife and quest for immortality from Dante's "Divine Comedy" to transhumanists, cryonicists, and extropians and singularists.

Wertheim thinks that humans naturally want a spiritual dimension to their lives. Science, by banning the spiritual out of the physical universe, has created the need for a new kind of spiritual space. If they can no longer find it in the physical universe, today's humans will find it in cyberspace. Virtual life on the Internet has been getting more and more interesting and meaningful, and the line between the real world and the virtual world has gotten more and more blurred.

Mark O'Connell wrote in "To Be a Machine" (2017) that both religion and science are ways of transcending our inherently pathetic condition.

Bainbridge wrote in "Religion for a Galactic Civilization 2.0" (2009) that religion and science are not opposed at all; instead, they coevolve: "Religion shapes science and technology, and is shaped by them in return". And, without mentioning the Singularity, he added that the "creation of a galactic civilization may depend upon the emergence of a galactic religion capable of motivating society for the centuries required to accomplish that great project".

Traditionally the strength of religion has been inversely proportional to the status of science in society. But this time the new religion of A.I. is about science itself and it is being created by people who are very knowledgeable about the science. This is not the first time that scientists present technology as a sort of divine power, as David Noble of York University in Toronto has shown in "The Religion of Technology" (1997), and it would not be the first time that a new science rises in parallel with a new organized religion, as Wertheim has shown in "In Pythagoras' Trousers" (1997). Francis Bacon's "New Atlantis" (1627) was the first scientific utopia, and Isaac Newton wrote (unpublished) books of prophecy such as "Observations upon the Prophecies of Daniel, and the Apocalypse of St John" (1733). Sometimes we forget that science and technology evolved from the Catholic monasteries and from the Church-controlled universities (and from the Islamic madrasas) of the Middle Ages. The culture of the San Francisco Bay Area lies at the same intersection of science and spirituality, the former represented by Silicon Valley's high-tech industry and the latter by the "New Age" movement. Fred Turner calls it "digital utopianism" in his book "From Counterculture to Cyberculture" (2008).

Just like prophetic books mediated between science and religion back then, today it is science fiction that has mediated between religion and technology. Critical studies such as David Ketterer's "New Worlds for Old" (1974) showed that science fiction routinely borrows concepts from the Christian scriptures. Studies such as Thomas Disch's "The Dreams Our Stuff Is Made Of" (1998) and Jason Pontin's "On Science Fiction" (2007) documented how science fiction exerted a huge influence on A.I. scientists. Pontin once wrote "Science fiction is to technology as romance novels are to marriage: a form of propaganda" (MIT Technology Review, 2005). Many future A.I. scientists were inspired to enter the A.I. field precisely because they were fans of science fiction: Isaac Asimov's "I Robot" stories (originally written between 1940 and 1950) and "Multivac" stories (starting with "Franchise" of 1955); Osamu Tezuka's manga "Tetsuwan Atomu/ Astro Boy" (1951); Groff Conklin's "Science Fiction - Thinking Machines" (1954), an anthology of stories about robots, androids and computers; Arthur Clarke's short story "Dial F for Frankenstein" (1964); Frank Herbert's "Do I Wake or Dream/ Destination Void" (1965); Brian Aldiss' short story "Super-Toys Last All Summer Long" (1968); Philip Dick's "Do Androids Dream of Electric Sheep" (1968); Algis Budrys' "Michaelmas" (1977); Douglas Adams' "The Hitchhiker's Guide to the Galaxy" (1979); Vernon Vinge's novella "True Names" (1981); William Gibson's "Neuromancer" (1984), which was predated by his short story "Burning Chrome" (1982); Neal Stephenson's "Snow Crash" (1991); etc. After all, even the great theoretical physicist Freeman Dyson wrote in his visionary book "Imagined Worlds" (1998) that "science is my territory, but science fiction is the landscape of my dreams".

Coincidence or not, the first Artificial Intelligence conference took place in 1956, exactly at the peak of a frenzied boom of sci-fi movies: Christian Nyby's "The Thing From Another World" (1951), Edgar Ulmer's "The Man from Planet X" (1951), Robert Wise's "The Day The Earth Stood Still" (1951), Rudolph Mate's "When Worlds Collide" (1951), Felix Feist's "Donovan's Brain" (1953), Jack Arnold's "It Came From Outer Space" (1953), Phil Tucker's "Robot Monster" (1953), Gordon Douglas's "Them!" (1954), Herbert Strock's "Gog" (1954), Jack Arnold's "Creature From the Black Lagoon" (1954), Don Siegel's "Invasion of the Body Snatchers"

(1956), Fred Wilcox's "Forbidden Planet" (1956), Val Guest's "The Quatermass Xperiment" (1956), etc. At least three featured robots: "The Day The Earth Stood Still", "Gog" and "Forbidden Planet".

Science fiction inspired the "transhumanist" movement way before the Singularity became a popular concept. The Extropian movement believed in the power of science and technology to yield immortality. Its members practiced cryogenics to preserve their brain after death. The term "extropy" was coined by Tom Bell, juxtaposing it to "entropy". The Oxford philosopher Max More had helped set up the first cryonic service in Europe (later renamed Alcor). Relocating to Los Angeles, in 1988 More started the magazine Extropy, subtitled "journal of transhumanist thought" and founded the Extropy Institute, which in 1991 had its own online forum. The Extropian movement had strong anti-government libertarian/anarchic political views, predicting a technocratic society in which power would be wielded directly by the people. By the time Wired published the influential article "Meet The Extropians" in 1994, the extropian movement included members and sympathizers such as Hans Moravec, Ralph Merkle, Nick Szabo, Hal Finney, as well as co-founders Tom Bell (Tom Morrow) and Perry Metzger. Merkle would go on to become a leader in nanotechnology, Szabo and Finney would pioneer Bitcoin, Metzger would launch the cryptography mailing-list.

Nick Bostrom, a Swedish philosopher at Oxford University, has pursued more social and ethical concerns in the several organizations that he established: in 1998 Bostrom and fellow philosopher David Pearce founded the World Transhumanist Association that later changed name to Humanity+, and in the same year Bostrom published "How Long Before Superintelligence?" (1998). In 2004 Bostrom and James Hughes founded the Institute for Ethics and Emerging Technologies; and in 2005 Bostrom founded the Future of Humanity Institute at Oxford University.

Kevin Kelly explored the connection between information and God in the article "Nerd Theology" (1999).

The Belgian sociologist Armand Mattelart wrote in his book "History of the Information Society" (2001): "Each new generation of technology revived the discourse of salvation".

In 2006 the Italian physicist Giulio Prisco became an advocate in virtual reality for the transhumanist movement, initially through his avatar Giulio Perhaps in Second Life. In 2007 he published the article "Engineering Transcendence" predicting that in the future it will be possible to become immortal inside cyberspace and to create perfect simulations of the past that will revive all those who have ever been alive. In 2008 he founded the Order of Cosmic Engineers (in a virtual world) and in 2010 the Turing Church (in the real world). The latter was initially just a "mailing list about the intersection of transhumanism and spirituality", but in 2014 it evolved into a "minimalist, open, extensible" religion whose manifesto preaches: "We will go to the stars and find Gods, build Gods, become Gods, and resurrect the dead from the past with advanced science"

These "un-religions" (religions with neither a hierarchy of priests nor immutable dogmas) are reminiscent of the church of engineers envisioned by August Comte, the founder of positivism, in his book "Catechism of Positive Religion" (1852). Comte was hoping to replace all religious institutions (in his view outdated) with a scientistic religion.

There is a strand of the publishing business that warns against the power of the high-tech world to corrupt the soul of humanity. For example: Noam Cohen's "The Know-It-Alls - The Rise of Silicon Valley as a Political Powerhouse and Social Wrecking Ball" (2017), Franklin Foer's "World Without Mind - The Existential Threat of Big Tech Hardcover" (2017), Amy Webb's "The Big Nine - How the Tech Titans and Their Thinking Machines Could Warp Humanity" (2019). There are striking similarities between these books and medieval Christian treatises warning against the influence of Satan's angels. There are echoes of Paul's "Letter to the Romans" (in the New Testament), in which he describes Satan as "god of this world" and warns about the "secret power of wickedness", echoes of the medieval prayers for the recommendation of the soul (see the "Ordo Excommunicandi et Absolvendi") and of course of the Christian belief that humanity is preparing for a titanic fight between Jesus and Satan, which will precede the Second Coming and the Last Judgement, as predicted in the Book of Revelation. This prophecy can be summarized as: the Antichrist appears, sin becomes widespread, and Satan will demonstrate superhuman power. And so these books become the modern equivalent of the medieval manuals of exorcism such as the "Compendio dell'Arte Essorcistica" published in 1572 by the Franciscan friar Girolamo Menghi, a best-seller of its time.

Anthropologists may be interested in analyzing the different ways in which Europe, the USA and China perceived the studies on Artificial Intelligence. Europe's A.I. was largely immune to the mood swings of the East Coast and West Coast of the USA. Europe's A.I. was "secular": it came out of scientific laboratories and continued to produce papers and demos even during the "A.I. winter". Ditto for Canada and Japan. In fact, all the inventors of deep learning (today's most popular branch of A.I.) that i can name were born and raised outside the USA: Fukushima, Schmidhuber, Hinton, Lecun, Bengio (as well as the vast majority of their students) all the way to DeepMind's founders. Even those who did their PhD studies in US universities were typically born outside the USA (for example, Sebastian Thrun, father of Google's self-driving car, and Andrew

Ng). Kurzweil's "The Singularity Is Near" came out in 2005, predating the boom of deep learning by 7 years. Very few people were taking AI seriously in 2005, especially in the USA. Moravec's "Mind Children" came out even earlier in 1988, in the middle of the A.I. winter. A.I. was kept alive in the USA mainly by Hollywood movies: "Terminator" (1984), "Robocop" (1987), "Matrix" (1999), "A.I." (2001), "I Robot" (2004)... Looking at the timeline, one has to wonder whether the Singularists (Moravec, Kurzweil, etc) had an impact on the renewed academic and business investment in A.I. that led to Deep Learning and then to the new mythology of the AlphaGo era. China, on the other hand, missed 60 years of the history of A.I. and looked into it only after 2012. China is largely indifferent to the debate of what constitutes real A.I. China's "communist" A.I. is simply a synonym for mass automation to fulfill the Communist Party's program of efficient technological collectivism. Progress in Europe's secular A.I. (deep learning) lent credibility to the mystical A.I. of the USA and fueled the rise of China's communist A.I.

"There are things known and there are things unknown, and in between are the doors of perception" (Aldous Huxley).

Will truly intelligent machines become religious?

Historical Footnote: New Atheism

The first salvoes in the war between religion and science were shot at the end of the 19th century, during the post-Darwin euphoria. New York University's chemist John-William Draper, better known as a pioneer of photography, wrote a "History of the Conflict Between Religion and Science" (1874). Andrew Dickson-White, the historian who had founded Cornell University, in his influential "A History of the Warfare of Science with Theology in Christendom" (1896), recognized a "conflict between two epochs in the evolution of human thought - the theological and the scientific". A few decades later, during the post-Einstein euphoria, the British philosopher Bertrand Russell became the most celebrated critic of religion. He started with the lecture "Why I Am Not a Christian" (1927): "We can now begin a little to understand things, and a little to master them by help of science, which has forced its way step by step against the Christian religion, against the Churches... Science can teach us ... no longer to look round for imaginary supports, no longer to invent allies in the sky". The essay also contains the sentence "Science can help us to get over this craven fear in which mankind has lived for so many generations" which is rather bizarre: the exact opposite is likely to happen after you stopped believing in an afterlife.

In the 21st century the Oxford University biologist Richard Dawkins ("The God Delusion", 2006), the Cambridge University physicist Stephen Hawking ("Grand Design", 2010) and the UCLA neuroscientist Sam Harris ("Science Must Destroy Religion", 2006) led the charge against religion. It took one year for the radical right-wing news channel Fox News to pick up the news but, when it did, it panicked and posted an article titled "Physicist Stephen Hawking Says There Is No Heaven"! Hawking's view is actually humbler: "This doesn't prove that there is no God, only that God is not necessary" (interview published in German magazine Der Speigel in 1988). In 2002 the British philosopher Christopher Hitchens called the three monotheistic religions (Christianity, Judaism and Islam) the "axis of evil" and then published "God Is Not Great" (2007). The British novelist Philip Pullman excoriated religion in the fictionalized biography "The Good Man Jesus and the Scoundrel Christ" (2010). In 2006 Gary Wolfe wrote an article in Wired magazine about them titled "The Church of the Non-Believers" and his expression "New Atheism" stuck.

However, in 2011 political activist Noam Chomsky at MIT, hardly a defender of organized religions, and another child of secular Jews, called the New Atheists "religious fanatics" on his Myspace blog (which in 2017 shows "Page not Found", a reminder of how ephemeral all online debates will be). Two years later, in 2013, Chomsky, interviewed by futurist Nikola Danaylov, dismissed the singularity as "science fiction". Surprisingly, he missed the connection with the religious fanatics.

Historical Trivia: The Ancient Roots of A.I.

The Greeks indulged in quite a few legends of artificial beings before inventing science and mathematics. Homer in the "Iliad" (dated to about the 8th century BC) credits Hephaestus with building, among other devices, a group of golden female androids that "looked like real girls and could not only speak and use their limbs but were also endowed with intelligence". The legend of Talos, a giant killer robot made of bronze, may date back to the 7th century BC but is first recorded in coin and vase paintings of the 5th century BC. It probably inspired Talus, the iron man, in Edmund Spenser's lengthy poem "The Faerie Queene" (1590). In Hesiod's "Theogony" (7th century BC) Zeus commands Hephaestus to make the seductive android Pandora in order to punish men (Pandora is programmed to open the most famous box in history and thereby release the sufferings that have plagued humankind ever since). Actually, according to these Greek legends, humans are the first androids: they were built by Prometheus. Some suspect that the giants of Mont'e Prama erected by the

Nuragic civilization (9th century BC) in the Italian island of Sardinia may also represent mechanical robots (either that or the sculptors were not too good at portraying human faces). And, according to a Buddhist legend (the "Lokapannatti", known both in Sanskrit and in Mandarin), King Ajatashatru, who reportedly knew both Buddha and Mahavira (the founder of Jainism), had mechanical robots built to guard the corpse of Siddhartha Gautama (aka Buddha). These robots, modeled after Greek technology (so says the text), were eventually disarmed by king Ashoka who then proceeded to spread Buddha relics all over the world.

Analog vs Digital

Most machines in the history of human civilization were and are analog machines, from the waterwheel to your car's engine. A lot of the marvel about computers comes from the fact that they are digital devices: once you digitize texts, sounds, images, films and so forth the digital machine can perform, at incredible speed, operations that used to take armies of human workers or specialists armed with expensive specialized machines. Basically, digitizing something means reducing it to numbers, and, therefore, the mind-boggling speed at which computers perform calculations, gets automatically transferred to other fields, such as managing texts, audio and video. The "editing" feature is, in fact, one of the great revolutions that came with the digital world. Previously, it was difficult and time-consuming to edit anything (text, audio, photos, video). Filing, editing and transmitting are operations that have been dramatically revolutionized by progress in digital technology and by the parallel process of digitizing everything, one process fueling the other.

Now that television broadcasts, rented movies, songs and books are produced and distributed in digital formats, i wonder if people of the future will even know what "analog" means. Analog is any physical property whose measurable values vary in a continuous range. Everything in nature is analog: the weight of boulders, the distance between cities, the color of cherries, etc. (At microscopic levels nature is not so analog, hence Quantum Theory, but that's another story). Digital is a physical property whose measurable values are only a few. The digital devices of today can typically handle only two values: zero and one. Actually, i don't know any digital device that is not binary. Hence, de facto, in our age "digital" and "binary" mean the same thing. Numbers other than zero and one can be represented by sequences of zeroes and ones (e.g. a computer internally turns 5 into 101). Texts, sounds and images are represented according to specific codes (such as ASCII, MP3 and MP4) that turn texts, sounds and images into strings of zeroes and ones.

The easiest way to visualize the difference between analog and digital is to think of the century-old bell-tower clocks (with the two hands crawling between the 12 Roman numerals) and the digital clock (that simply displays the time in hours/minutes).

When we turn a property from analog to digital we enable computers to deal with it. Therefore you can now edit, copy and email a song (with simple commands) because it has been reduced to a music file (to a string of zeroes and ones).

Audiophiles still argue whether digital "sounds" the same as analog. I personally think that it does (at today's bit rates) but the stubborn audiophile has a point: whenever we digitize an item, something is lost. The digital clock that displays "12:45" does not possess the information of how many seconds are missing to 12:46. Yesterday's analog clock contained that information in the exact position of the minute hand. That piece of information may have been useless (and obtainable only by someone equipped with a magnifying glass and a pocket calculator) but nonetheless the device had it. The music file is not an exact replica of the song: when the musicians performed it, they were producing an analog object. Once that analog object is turned into a digital file, an infinite number of details have been lost. The human ear is limited and therefore won't notice (except the above said stubborn audiophiles). We don't mind because our senses can only experience a limited range of audio and visual frequencies. And we don't mind also because amazing features become available with digital files, for example, the ability to improve the colors of a photograph so we can pretend that it was a beautiful vacation when in fact it rained all the time.

When machines carry out human activities, they are "digitizing" those activities; and they are digitizing the "mental" processes that lie behind those activities. In fact, machines can manage those human activities only after humans digitized (turned into computer files) everything that those human activities require, for example maps of the territory.

Using digital electronic computers that employ Boole's binary logic (zeroes and ones) to mimic the brain is particularly tempting. After all, in 1943 Warren McCulloch and Walter Pitts had proven that a network of binary neurons is fully equivalent to a Universal Turing Machine. They had "not" proven that it is fully equivalent to the human brain, but that it is fully equivalent to a machine that can solve all possible logical problems.

There is, however, a catch: McCulloch's binary neurons integrate their input signals at discrete intervals of time, rather than continuously as our brain's neurons do. Every computer has a central clock that sets the pace for its logic, whereas the brain relies on asynchronous signaling because there is no synchronizing central clock. If you get into the finer details of how the brain works, there are more "analog" processes at work, and there are analog processes inside the neuron itself (which is not just an on/off switch).

One could argue that the brain is regulated by the body's internal clocks (that regulate every function, from your heart to your vision) and therefore the brain behaves like a digital machine; and that everything is made of discrete objects all the way down to quarks and leptons; hence nothing in nature is truly analog. Even if you want to be picky and invoke Quantum Theory, the fact remains that a brain uses a lot more than zeroes and ones; a computer can only deal with zeroes and ones. As tempting as it is to see the brain as a machine based on binary logic, the difference between the human brain and any computer system (no matter how complex the latter becomes) is that a computer is way more "digital" than a brain. We know so little about the brain that it is difficult to estimate how many of its processes involve a lot more than on/off switching, but a safe guess is that there are several hundreds. Despite the illusion created by the McCulloch-Pitts neuron, a computer is a binary machine which the brain is not.

There might be a reason if a brain operates at 10-100 Hz whereas today's common microprocessors need to operate at 2-3 Gigahertz (billions of Hz), hundreds of millions of times faster, to do a lot less; also human brains consume about 20 watts and can do a lot more things than a supercomputer that consumes millions of watts. Biological brains need to be low-power consumption machines or they would not survive. There are obviously principles at work in a brain that have eluded computer scientists.

Carver Mead's "neuromorphic" approach to machine intelligence is not feasible for the simple reason that we don't know how the brain works. Based upon the Human Genome Project (that successfully decoded the human genome in 2003), the USA launched the "Brain Initiative" in April 2013 to map every neuron and every synapse in the brain.

There are also government-funded projects to build an electronic model of the brain: Europe's Human Brain Project and the USA's Systems of Neuromorphic Adaptive Plastic Scalable Electronics (SYNAPSE), sponsored by the same agency, DARPA, that originally sponsored the Arpanet/Internet. Both Karlheinz Meier in Germany and Giacomo Indiveri in Switzerland are toying with analog machines. The signaling from one node to the others better mimics the "action potentials" that trigger the work of neurons in the human brain and requires much less power than the ones employed in digital computers. SYNAPSE (2008) spawned two projects in California, one run by Narayan Srinivasa at Hughes Research Laboratories (HRL) and the other run by Dharmendra Modha at IBM's Almaden Labs in Silicon Valley. The latter announced in 2012 that a supercomputer was able to simulate 100 trillion synapses from a monkey brain, and in 2013 unveiled its "neuromorphic" chip TrueNorth (not built according to the traditional John von Neumann architecture) that can simulate 1 million neurons and 256 million synapses. This represented the first building block to push computer science beyond the Von Neumann architecture that has ruled since the early days of electronic computation. Interestingly, this chip (consuming only 70 milliwatts of power) was also one of the most power-efficient chips in the history of computing... just like the human brain.

Analog Computation/ Reservoir Computing

In 1992 Hava Siegelmann of Bar-Ilan University in Israel and Eduardo Sontag of Rutgers University developed "analogic recurrent neural networks" ("Analog Computation via Neural Networks", a paper submitted in 1992 but published only in 1994).

For 60 years it had been assumed that no computing device could be more powerful than a Universal Turing Machine. Hava Siegelmann proved mathematically that analog RNNs can achieve super-Turing computing ("On the Computational Power of Neural Nets", 1992). Alan Turing himself had tried to imagine a way to extend the computational power of his universal machine ("Systems of Logic Based on Ordinals", 1938), but his idea cannot be implemented in practice. Siegelmann's system was not the first system to break the Turing limit using real numbers, and nobody has built a computer yet that can perform operations on real numbers in a single step.

Recurrent networks are harder to train with gradient descent methods than feed-forward networks. Luckily, two techniques introduced a new paradigm for training recurrent neural networks: "echo state networks", developed in 2001 by German chaos theorist Herbert Jaeger at University of Bremen for classifying and forecasting time series such as speech ("The Echo State Approach to Analysing and Training Recurrent Neural Networks", 2001), and "liquid state machines", developed in 2002 by the German mathematician Wolfgang Maass and the South-african neuroscientist Henry Markram at the Graz University of Technology in Austria as a biologically plausible model of spiking neurons ("Real-time Computing Without Stable States", 2002). They came from different disciplines (computer science and neuroscience) but arrived at the same trick: they trained only the final, non-recurrent output layer (the "readout layer"), while the other layers (the "reservoir") were randomly initialized. Therefore most of the weights in the network are assigned only once and at random. The difference between the two reservoir models is minimal, but, in a nutshell, liquid state machines are more general and therefore encompass echo state networks. The idea of random networks with a trained readout layer was mentioned by Frank Rosenblatt in his 1962 book "Principles of Neurodynamics", and the "context reverberation network" developed by Kevin Kirby at Wright State University in Ohio ("Context Dynamics in Neural Sequential Learning", 1991) and the neural network developed by Peter Dominey at the French National Institute of Health and Medical Research to model "complex sensory-motor sequences" in the brain such as speech recognition ("Complex Sensory-Motor Sequence Learning Based on Recurrent State Representation and Reinforcement Learning", 1995) were de facto already implementations of that idea, but it became accepted only three decades later, after Dean Buonomano in Michael Merzenich's laboratory at UC San Francisco, explained how the brain encodes time ("Temporal Information Transformed into a Spatial Code by a Neural Network with Realistic Properties", 1995). Reservoir computing greatly facilitated the practical application of recurrent neural networks. Reservoir computing provided a much needed alternative to gradient-descent methods for training recurrent neural networks. Echo state networks have also been implemented in hardware, e.g. by the team of Zambia-born physicist Serge Massar at University of Brussels in Belgium ("Brian-inspired Photonic Signal Processor For Generating Periodic Patterns And Emulating Chaotic Systems," 2017).

Reservoir computing wasn't just a cute trick to train neural networks. It stealthily represented a devastating critique of Artificial Intelligence. Turing machines are not well-suited for modeling the behavior of brain circuits, which are analog, not digital, and don't work in discrete time steps but continuously. Brains perform real-time computations on continuous input streams (or "time series"), whereas Turing machines perform off-line computations on discrete input values (basically, zeroes and ones). Brain states are "liquid" and that makes them well-suited for computing perturbations.

Reservoir computing provided a simpler way to train recurrent neural networks, but was soon made obsolete by the rise of deep learning. However, A.I. witnessed a resurgence of reservoir computing in 2017 when the team of physicist Edward Ott at University of Maryland showed that reservoir computing can closely simulate the evolution of a chaotic system ("Model-free Prediction of Large Spatiotemporally Chaotic System from Data", 2018). The term "butterfly effect", named after an Edward Lorenz lecture ("Does The Flap Of A Butterfly's Wings In Brazil Set Off A Tornado In Texas?", 1972), expresses the fact that a small change in the initial conditions can grow exponentially quickly, making long-term predictions of chaotic systems impossible. That's why weather predictions are still so unreliable: a mathematical model of the atmosphere is a chaotic model. Why reservoir computing is so good at learning the dynamics of chaotic systems is not yet well understood. To be fair, Sapsis Themistoklis and his student Zhong-yi Wan at MIT achieved similar results with an LSTM neural network ("Data-assisted Reduced-Order Modeling of Extreme Events In Complex Dynamical Systems", 2018).

Neural networks in computer science are sets of mathematical equations that represent stage changes with continuous variables. Real neurons, instead, interact primarily via discontinuous "spiking" (action potentials). Software simulations of real neurons were pioneered by Neuron, first published in 1984 and mainly developed by mathematician Michael Hines in the lab of Kacy Cole's student John Moore, now at Duke University; and by GENESIS (the GEneral NEural SImulation System), developed in 1988 by James Bower's team at Caltech. Then came NEST (Neural Simulation Tool), developed in 2002 by Markus Diesmann and Marc-Oliver Gewaltig at EPFL in Switzerland; and Brian, developed in 2008 by Romain Brette and Dan Goodman at the École Normale Supérieure in France. Hardware implementations include SpiNNaker (Spiking Neural Network Architecture), designed in 2005 by Steve Furber at University of Manchester, and Neurogrid, built in 2009 by Kwabena Boahen's team at Stanford University.

"The brain is an organ of minor importance" (Aristotle in "De Motu Animalium")

Teaser: Machine Ethics

If we ever create a machine that is a fully-functioning brain totally equivalent to a human brain, will it be ethical to experiment on it? Will it be ethical to program it? Will it be ethical to modify it, and to destroy it at the end?

The discussion about "machine ethics" is usually about ethics towards humans, not ethics towards machines (what is good behavior by a human towards a machine?) or between machines (what is good behavior by a machine towards another machine?). Morality between machines would entail a discussion about what are machine values, which would be a great topic of discussion but we are humans and don't really care what machines can do to each other.

Morality towards humans should be an easier discussion, except that, alas, humans are far from having reached consensus on what is ethically correct or not. For example, i personally don't consider religions to be very ethical (often, just the opposite), but apparently i am outnumbered by at least one billion Muslims and more than one billion Christians and Jews. And many of them, in turn, may not consider Darwin or Einstein as role models. Therefore we humans are a far cry from having reached consensus on what is good and what is bad.

Last but not least, we humans change our minds all the time about what is ethical and what is not. Not long ago it was really bad for an unmarried woman not to be a virgin (now it is almost the opposite) and not long ago it was really bad to say that gods don't exist (it will soon be the opposite). I am not even sure that our changing morality constitutes "moral progress". For example, the West has gone back and forth on homosexuality since Greek times. We cannot rationally prove that premarital sex is good or bad; only that right now and right here, to this generation of Westerners, it looks ok. It seems obvious to me that we shouldn't even think of teaching ethics to machines. We humans have killed, enslaved and oppressed way too many fellow humans in the name of our ethical principles.

In 2001 Eliezer Yudkowsky founded the field of "Friendly A.I." according to which it is imperative to design A.I. systems in such a way that they can never become dangerous to us. Steve Omohundro, however, warned that machines programmed to improve their skills, no matter how narrow that skill originally is, may develop an uncontrolled utilitarian "drive" that will make them the equivalent of human sociopaths ("The Basic A.I. Drives", 2008). Wendell Wallach, Colin Allen, and Iva Smit, introduced the term "artificial moral agents" ("Machine Morality", 2008). With all due respect for these thinkers (much smarter and more knowledgeable than me), i find their philosophical arguments to be weak and their mathematical "proofs" even weaker. I am not sure that i would want a machine that will never ever under any circumstances, kill a human being: my father would have died in a Nazi concentration camp if bombs and machine guns had refused to kill Nazis. Hence sometimes it is ok to harm humans. But we have no consensus on when that is the case. And the last people whom i would trust with making this decision are the software engineers.

Even creating an omnipotent supremely good god (if we could do it) isn't quite as "good" an idea as it sounds. In 2010 a user of the blog LessWrong, founded by the same Eliezer Yudkowsky one year earlier, imagined an artificial intelligence, the Basilisk, that has been created for only one purpose: to create the greatest good for the greatest number of people. One of the conclusions that it is likely to reach is that, in order to achieve its goal, it has to torture and possibly eliminate anybody who does not contribute to its goal, for example any software engineer who doesn't help improve the Basilisk itself, or, for example, any writer like me who merely wonders whether the Basilisk would be a good or bad idea. Would you want this omnipotent machine programmed to achieve the greatest good for the greatest number of people, at the cost of torturing and eliminating anybody who doubts or (anathema) opposes it? Trivia: Roko's Basilisk became an Internet meme because Yudkowsky banned discussion on it. Obviously Yudkowsky never studied the history of taboos.

Incidentally, we humans still have to agree on the source of morality. Several biologists, from Charles Darwin in person to Frans de Waal (who published "Primates and Philosophers" in 2006) think that morality evolved naturally, and not only in humans; whereas Thomas Hobbes thought that we are amoral animals kept in line by the brute force of the state (by the "leviathan"). That debate never ended.

How not to Build an Artificial General Intelligence – Part 1: The Many-task Mind

In April 2013 i saw a presentation at Stanford's Artificial Intelligence Lab by the team of Kenneth Salisbury in collaboration with Willow Garage about a robot that can take the elevator and walk upstairs to buy a cup of coffee. This implies operations that are trivial for humans: recognizing that a transparent glass door is a door (not just a hole in the wall and never mind the reflection of the robot itself in the glass), identifying the right type of door (revolving, sliding or automatic), finding the handle to open the door, realizing that it's a spring-loaded door so it doesn't open as easily as regular doors, finding the elevator door, pressing the button to call the elevator, entering the elevator, finding the buttons for the floors inside an elevator whose walls are reflective glass (therefore the robot keeps seeing reflections of itself), pressing the button to go upstairs, locating the counter where to place the order, paying, picking up the coffee, and all the time dealing with humans (people coming out of the door, sharing the space in the elevator, waiting in line) and avoiding unpredictable obstacles; if instructions are posted, read the instructions, understand what they mean (e.g. this elevator is out of order or the coffee shop is closed) and change plan accordingly. Eventually, the robot got it right. It took the robot 40 minutes to return with the cup of coffee. It is not impossible. It is certainly coming. I'll let the experts estimate how many years it will take to have a robot that can go and buy a cup of coffee upstairs in all circumstances (not just those programmed by the engineer) and do it in 5 minutes like humans do. The fundamental question, however, is whether this robot can be considered an intelligent being because it can go and buy a cup of coffee or it is simply another kind of appliance.

It will take time (probably much longer than the optimists claim) but some kind of "artificial intelligence" is indeed coming. How soon depends on your definition of artificial intelligence.

Nick Bostrom wrote that the reason A.I. scientists have failed so badly in predicting the future of their own field is that the technical difficulties have been greater than they expected. I don't think so. I think those scientists had a good understanding of what they were trying to build. The reason why "the expected arrival date [of artificial intelligence] has been receding at a rate of one year per year" (Bostrom's estimate) is that we keep changing the definition. There never was a proper definition of what we mean by "artificial intelligence" and there still isn't. No wonder that the original A.I. scientists were not concerned with safety or ethical concerns: of course, the machines that they had in mind were chess players and theorem provers. That's what "artificial intelligence" originally meant. Being poor philosophers and poor historians, they did not realize that they belonged to the centuries-old history of automation, leading to greater and greater automata. And they couldn't foresee that, within a few decades, all these automata would become millions of times faster, billions of times cheaper, and would be massively interconnected. The real progress has not been in A.I. but in miniaturization. Miniaturization has made it possible to use thousands of tiny cheap processors and to connect them massively. The resulting "intelligence" is still rather poor, but its consequences are much more intimidating.

To start with, it is wise to make a distinction between an artificial intelligence and an A.G.I. (artificial general intelligence). Artificial intelligence is coming very soon if you don't make a big deal of it, and it might already be here: we are just using a quasi-religious term for "automation", a process that started with the waterwheels of ancient Greece if not earlier. Search engines (using very old fashioned algorithms and a huge number of very modern computers housed in "server farms") will find an answer to any question you may have. Robots (thanks to progress in manufacturing and to rapidly declining prices) will become pervasive in all fields, and become household items, just like washing machines and toilets; and eventually some robots will become multifunctional (just like today's smartphones combine the functions of yesterday's watches, cameras, phones, etc; and, even before smartphones, cars acquired a radio and an air conditioning unit, and planes acquired all sorts of sophisticated instruments).

Millions of jobs will be created to take care of the infrastructure required to build robots, and to build robots that build robots, required to build robots, and to build robots that build robots, and ditto for search engines, websites and whatever comes next. Some robots will come sooner, some will take centuries. And miniaturization will make them smaller and smaller, cheaper and cheaper. At some point we will be surrounded for real by Neil Stephenson's "intelligent dust" (see his novel "Diamond Age"), i.e. by countless tiny robots each performing one function that used to be exclusive to humans. If you want to call these one-function programs "artificial intelligence", suit yourself.

We wouldn't call "intelligent" a human being whose brain can do only one thing.

An AGI, instead, would be more like us: maybe none of us does anything well, but we do many things and we are capable of doing many things that we will actually never do. An AGI would not be limited to one or two or twenty tasks: it would be able to perform ALL the tasks that human beings perform, although not necessarily excel at any of them.

Making predictions about the coming of an AGI without having a clear definition of what constitutes an AGI is as scientific as making predictions about the coming of Jesus. An AGI could be implemented as a collection of one-function programs, each one specialized in performing one specific task. In this case someone has to tell the A.I. specialist which tasks we expect from an AGI. Someone has to list whether AGI requires being able to ride a bus in Zambia and to exchange money in Haiti or whether it only requires the ability to sort out huge amounts of data at lightning speed or what else. Once we have that list, we can ask the world's specialists to make reasonable estimates and predictions on how long it will take to achieve each of the functions that constitutes the AGI.

This is an old debate. Many decades ago the founders of computational mathematics (Alan Turing, Claude Shannon, Norbert Wiener, John von Neumann and so forth) discussed which tasks can become "mechanic", i.e. be performed by a computing machine, which can and cannot be computed, i.e. can be outsourced to a machine and what kind of machine that has to be. Today's computers that perform today's deep-learning algorithms, such as playing go/weiqi, are still Universal Turing Machines, subject to the theorems proven for those classes of machines. Therefore, Alan Turing's original work still applies. The whole point of inventing (conceptually) the Turing Machine in 1936 was to prove whether a general algorithm to solve the "halting problem" for all possible program-input pairs exists, and the answer was a resounding "no": there is always at least one program that cannot be "decided", i.e. will never halt. And in 1951 Henry Gordon Rice generalized this conclusion with an even more formidable statement, "Rice's Theorem": any nontrivial property about the behavior of a Turing machine is undecidable, a much more general statement about the undecidability of Turing machines. In other words, it is proven that there is a limit to what machines can "understand", no matter how much progress is made, if they are Universal Turing Machines (as virtually all of today's computers are).

Nonetheless, by employing thousands of these machines, the "brute force" approach has achieved sensational feats such as machines that can beat go/weiqi champions and recognize cats. So you might be tempted to accept that an AGI will be created by sheer "brute force": creating a one-function program for each possible task and then somehow putting them all together in one machine that will then be able to carry out any human function.

Some of us doubt that the human mind works that way. We have seen no neurological evidence that the human brain is a collection of one-function programs. We have seen evidence of the opposite: that the human mind is capable of applying the skills of one function to a different function, and sometimes without even being told to do so. We are AGIs because our brain can approach new tasks and find a way to perform them even if nobody trained us to carry out such new tasks.

VanGogh and Nietzsche went mad in the same year, 1888, one year after Emile Berliner invented the gramophone (that records sounds) and in the same year in which Kodak introduced the first consumer camera (that records images). What about it? Your brain is already at work to find some connection, right? Nobody programmed your brain to find out why a painter and a philosopher went mad in the same year that an invention hit the market, but you can do it effortlessly. And, yes, it's probably a useless and pointless exercise. Nonetheless, that's what a multi-tasking intelligence can do, and does all the time.

How not to Build an Artificial General Intelligence - Part II: Smart, not Deep, Learning

Simply telling me that Artificial Intelligence and robotics research will keep producing better and smarter devices (that are fundamentally not "intelligent" the way humans are) does not tell me much about the chances of a breakthrough towards a different kind of machine that will match (general) human intelligence.

I don't know what such a breakthrough should look like, but i know what it doesn't look like. The machine that beat the world champion of go/weiqi was programmed with knowledge of virtually every major go/weiqi game ever played, and it was allowed to run millions of logical steps before making any move. That obviously put the human contender at a huge disadvantage. Even the greatest go/weiqi champion with the best memory can only remember so many games. The human player relies on intuition and creativity, whereas the machine relies on massive doses of knowledge and processing. Shrink the knowledge base that the machine is using to the knowledge base that we have and limit the number of logical steps it can perform to the number of logical steps that the human mind can perform before it is timed out, and then we'll test how often it wins against ordinary players, let alone world champions.

Having a computer (or, better, a huge knowledge base) play chess against a human being is like having a gorilla fight a boxing match with me: i'm not sure what conclusion you could draw from the result of the boxing match about our respective degrees of intelligence.

I wrote that little progress has been made in Natural Language Processing. The key word is "natural". Machines can actually speak quite well in unnatural language, a language that is grammatically correct but from which all creativity has been removed: "subject verb object - subject verb object - subject verb object - etc." The catch is that humans don't do that. If i ask you ten times to describe a scene, you will use different words each time.

Language is an art. That is the problem. How many machines do we have that can create art? How far are we from having a computer that switches itself on in the middle of the night and writes a poem or draws a picture just because the inspiration came? Human minds are unpredictable. And not only adult human minds: pets often surprise us, and children surprise us all the time. When is the last time that a machine surprised you? (Other than surprising you because they are still so dumb). Machines simply do their job, over and over again, with absolutely no imagination.

Here is what would constitute a real breakthrough: a machine that has only a limited knowledge of all the go/weiqi games ever played and is allowed to run only so many logical steps before making a move and that can still play well. That machine will have to use intuition and creativity. That's a machine that would probably wake up in the middle of the night and write a poem. That's a machine that would probably learn a human language in a few months just like even the most disadvantaged children do. That is a machine that would not translate "'Thou' is an ancient English word" into "'Tu' e` un'antica parola Inglese", and that will not stop at a red traffic light if it creates a dangerous situation.

I suspect that this will require some major redesigning of the very architecture of today's computers. For example, a breakthrough could be a transition from digital architectures to analog architectures. Another breakthrough could be a transition from silicon (never used by Nature to construct intelligent beings) to carbon (the stuff of which all natural brains are made). And another one, of course, could be the creation of an artificial being that is self-conscious.

Today it is commonplace to argue that in the 1970s A.I. scientists gave up too quickly on neural networks and connectionism. My gut feeling is that in the 2000s we gave up a bit too quickly on the symbolic processing (knowledge-based) program. Basically, we did to the logical approach what we had done before to the connectionist approach: in the 1970s neural networks fell into oblivion because knowledge-based systems were delivering practical results... only to find out that knowledge-based systems were very limited and that neural networks were capable of doing more.

My guess is that there was nothing wrong with the knowledge-based approach. Unfortunately, we never figured out an adequate way to represent human knowledge. Representation is one of the oldest problems in philosophy, and I don't think we got any closer to solving it now that we have powerful computers. The speed of the computer does little to fix a wrong theory of representation.

So we decided that the knowledge-based approach was wrong and we opted for neural networks (deep learning and the likes). And neural networks have proven very good at simulating specialized tasks: each neural network does one thing well, but doesn't do what every human, even the dumbest one, and even animals, do well: use the exact same brain to carry out thousands (potentially an infinite number) of different tasks.

"If a machine is expected to be infallible, it cannot also be intelligent" (Alan Turing, 1947).

The Timeframe of Artificial General Intelligence

If by "artificial intelligence" we simply mean a machine that can do something (not everything) that we can do (like recognizing cats or playing chess), but not "everything" that we can do (both the mouse and the chess player do a lot of other things), then all machines and certainly all appliances qualify. Some of them (radio, telephone, television) are even forms of superhuman intelligence because they can do things that human brains cannot do.

Definitions do matter: there is no single answer to the questions "when will machines become intelligent" and "when will superhuman intelligence appear". It depends on what we mean by those words. My answer can be "it's already here" or "never".

As it stands, predictions about the future of (really) intelligent machines (of AGI) are predictions about a technology that doesn't exist. You can ask a rocket scientist for a prediction for when a human being will travel to

Pluto: that technology exists and one can speculate what it will take to use that technology for that specific mission. On the contrary, my sense is that, using current technology, there is no way that we can create a machine that is even remotely capable of performing our routine cognitive tasks. The technology that is required does not yet exist. The machine that is supposed to become more intelligent than us and not only steal your job but even rule the world (and either kill us all or make us immortal) is pure imagination, just like angels and ghosts.

It is difficult to predict the future because we tend to predict one future instead of predicting all possible futures. Nobody (as far as i know) predicted that the idea of expert systems would become irrelevant in most fields because millions of volunteers would post knowledge for free on something called World-wide Web accessible by anybody equipped with a small computer-telephone. That was one possible future but there were so many possible futures that nobody predicted this one. By the same token, it is hard to predict what will make sense in ten years, let alone in 50 years.

What if 3D printing and some other technology makes it possible for ordinary people to create cheap gadgets that solve all sorts of problems. Why would we still need robots? What if synthetic biology starts creating alternative forms of life capable of all sorts of amazing functions. Why would we still need machines? There is one obvious future, the one based on what is around today, in which machines would continue to multiply and improve. There are many other futures in which computers and robots would become irrelevant because of something that does not exist today.

Anders Sandberg and Nick Bostrom, authors of "Whole Brain Emulation" (2008), conducted a "Machine Intelligence Survey" (2011) that starts with a definition of what an artificial intelligence should be: a system "that can substitute for humans in virtually all cognitive tasks, including those requiring scientific creativity, common sense, and social skills." My estimate for the advent of such a being is roughly 200,000 years: the timescale of natural evolution to produce a new species that will be at least as intelligent as us. If Artificial Intelligence has to be achieved by incremental engineering steps starting from the machines that we have today, my estimate about when a machine will be able to carry out a conversation like this one with you is: "Never". I am simply projecting the progress that i have witnessed in Artificial Intelligence (very little and very slow) and therefore i obtain an infinite time required for humans to invent such a machine.

But then, again, we'd probably have a lengthy discussion about what the expression "all cognitive tasks" really means. For example, leaving out consciousness from the category of cognitive tasks is like leaving out

Beethoven from the category of musicians simply because we can't explain his talent.

As i wrote, machines are making us somewhat dumber (or, better the environments we design for automation make us dumber), and there is an increasing number of fields (from arithmetic to navigation) in which machines are now "smarter" than humans not only because machines got smarter but also because humans have lost skills that they used to have. If i project this trend to the future, there is a serious chance that humans will get so much dumber that the bar for artificial general intelligence will be lower and therefore rendering artificial intelligence more feasible than it is today; and "superhuman" intelligence may then happen, but it should really be called "subhuman" intelligence.

How NOT to Find a Breakthrough

Don't ask me what the breakthrough will be in A.I. If i knew it, i wouldn't be wasting my time writing articles like this one. But i have a hunch it has to do with recursive mechanisms for endlessly remodeling internal states: not data storage, but real "memory".

For historians a more interesting question is what conditions may foster such a breakthrough. In my opinion, it is not the abundance of a resource (such as computing power or information) that triggers a major paradigm shift but the scarcity of a resource. For example, James Watt invented the modern steam engine when and because Britain was in the middle of a fuel crisis (caused by the utter deforestation of the country). For example, Edwin Drake discovered petroleum ("oil") in Pennsylvania when and because whale oil for lamps was becoming scarce. Both innovations caused an economic and social revolution (a kind of "exponential progress") that completely changed the face of the world. The steam engine created an economic boom, reshaped the landscape, revolutionized transportation, and dramatically improved living conditions. Petroleum went on to provide much more than lighting to the point that the contemporary world is (alas) addicted to it. I doubt that either revolution would have happened in a world with infinite amounts of wood and infinite amounts of whale oil.

The very fact that computational power is becoming an infinite inexpensive resource makes me doubt that it will lead to a breakthrough in Artificial Intelligence.

Water power was widely available to Romans and Chinese, and they had the scientific know-how to create machines propelled by water; but the industrial revolution had to wait more than one thousand years. One reason (not the only one but a key one) why the Romans and the Chinese never started an industrial revolution is simple: they had plentiful cheap labor (the Romans had slaves, the Chinese emperors specialized in mobilizing masses of subjects).

Abundance of a resource is the greatest deterrent to finding an alternative to that resource. If "necessity is the mother of invention", as Plato said, then abundance is the killer of invention.

We live in the age of plentiful computational power. To some observers this looks like evidence that super-human machine intelligence is around the corner; to me this looks like evidence that our age doesn't even have the motivation to try.

My fear, in other words, is that the current success in "brute-force A.I." is slowing down (not accelerating) research in higher-level intelligence (the real meaning of "human intelligence"). If a robot can fix a car without knowing anything about cars, why bother to teach the robot how cars work? The success in (occasionally) recognizing cats, beating go/weiqi champions and so forth is indirectly reducing the motivation to understand how the human mind (or, for that matter, the chimp's mind, or even a worm's mind) manages to recognize so many things in a split second and perform all sorts of actions. The success in building robots that perform this or that task with amazing dexterity is indirectly reducing the motivation to understand how the human mind can control the human body in such sophisticated ways in all sorts of situations and sometimes in completely novel ways.

Bill Joy wrote that "The future doesn't need us", but maybe it's the other way around: we won't need the future if the present starts giving us machines that can do everything we need.

The Real Breakthrough: Not Superintelligent Machines but Superintelligent Humans

On the other hand, i have seen astonishing (quasi exponential) progress in biotechnology, and therefore my estimate for when biotech will be able to create an "artificial intelligence" is very different: it could already happen in one year. And my estimate of when biotech might create a "superhuman" intelligence is also more optimistic: it could happen in a decade. I am simply basing my estimates on the progress that i have witnessed over the last 50 years; which might be misleading (again, most technologies eventually reach a plateau and then progress slows down), but at least this one has truly been "accelerating" progress. It would be interesting to discuss how biotech might achieve this feat: will it be a new being created in the laboratory, or the intended or the accidental evolution of a being, or a cell by cell replica of the human body? But that's for another book.

The 21st century will not be the age of the Singularity but it will be the age of "designer babies". We will design babies the same way we design buildings. The 21st century will be the age of "Disruptive Reproductive Technologies" (the title of a 2017 article by Glenn Cohen and others).

There are more than two million couples that are infertile just in the USA. Worldwide, there are tens of millions. The most common procedure for these couples to have a baby is "in-vitro fertilization" (IVF). The first "test-tube baby" was born in 1978 in Britain: Louise Brown. There are now thousands of "test-tube babies". But IVF is an unreliable procedure, successful for about 20% of couples, and it is a painful procedure for the woman. PGD (preimplantation genetic diagnosis) is a procedure that combines IVF and genetic screening. The procedure is IVF (eggs are taken from a woman's ovaries and fertilized in the laboratory with the man's sperm), but PGD includes the genomic testing: after three days the embryos already contain all the genetic information that the scientists need to determine the future health of that baby. PGD immediately tells the scientist if the embryo has any genetic defect. PGD was the result of research by Mark Hughes at Baylor College of Medicine in Texas, Robert Winston at Imperial College London, and Alan Handyside at University of London. Originally, PGD was conceived to help couples who carry genetic disorders: the risk that their children will have a horrible disease is high, and PGD is a way to make sure that their babies will be healthy. Another important use of this technique is to help parents who have a child with a disease that can be cured only with a transplant from a healthy donor. This procedure can be used to "design" a baby who will be a healthy sibling of the child: that's the best possible donor. Handyside is the one who in 1989 carried out the PGD procedure that led to the birth of the first PGD baby in 1990 (in that first case PGD was simply used to pick the sex of the baby because the parents needed a daughter to avoid a disease that only affects boys). In 2014 there were already 3,000 PGD babies, i.e. 3,000 humans who have been "designed" in the laboratory. Mark Hughes now heads the Genesis Genetics Institute in Detroit, the leading provider of PGD.

In 2013 Shoukhrat Mitalipov at the Oregon Health & Science University created human embryonic stem cells from cloned embryos ("Human embryonic stem cells derived by somatic cell nuclear transfer", 2013).

In 2015 Matthew Rabinowitz, the founder of Natera, and Jay Shendure of University of Washington sequenced in detail the genomes of two parents undergoing IVF and then inferred the genome sequence of the embryo ("Whole genome prediction for preimplantation genetic diagnosis", 2015).

The ethical problem with these procedures is that the laboratory creates a number of embryos and then only one is kept. In other words, all the other embryos are killed. The scientists pick the healthiest one. Of course, this means that the parents too could pick an embryo and kill all the others.

The 21st century will not be the age of the Singularity but it will be the age of "designer babies". We will design babies the same way we design buildings. The 21st century will be the age of "Disruptive Reproductive Technologies" (the title of a 2017 article by Glenn Cohen and others).

Using IVF and PGD the parents can create dozens of embryos and then pick the ones that (who?) will become tall, or blonde, or more similar to granpa; or someday, when we know more about the relationship between genes and intelligence, they could pick the ones who are more likely to become a scientist, or a painter, or a business man. CRISPR makes it easier and easier to edit genes out. So you can "design" the baby that you want. But it also means that you kill all the embryos that you don't want. All of this because we invented a way to create embryonic stem cells via in vitro fertilization.

In fact, soon it will be possible to do all of this without in vitro fertilization. At the end of 2015 a new technique called "in vitro gametogenesis" (IVG) has been tested in mice in Japan by Katsuhiko Hayashi of Kyushu University. This technique allows scientists to create both eggs and sperm in the laboratory out of skin cells. They took skin cells from the mice and created eggs and sperm, then fertilized the eggs to create hundreds of embryos and finally implanted the eggs into a female mouse, and several healthy pups were born. In 2012 Hayashi, working in the team of Mitinori Saitou at Kyoto University, had already discovered how to reprogram skin cells to behave like the cells that generate eggs, and earlier in 2016 Yayoi Obata's team at Tokyo University of Agriculture found out how to turn these cells into eggs without placing them into a female body. A few weeks later Hayashi combined the two procedures and obtained the embryos.

When in 2014 the Michigan State University physicist Stephen Hsu wrote the article "Super-Intelligent Humans Are Coming" in Nautilus magazine with the subtitle "genetic engineering will one day create the smartest humans who have ever lived", he wasn't dreaming: he was simply summarizing progress in genetics.

One can use either Shinya Yamanaka's technique or Shoukhrat Mitalipov's technique to reprogram adult cells, such as skin cells, to behave like embryonic stem cells. Then one can use Hayashi's procedure to program these stem cells so they become eggs or sperm. Then you can fertilize the egg and you get an embryo. In fact, usually the scientists produce many embryos, even hundreds. The goal is to select the "best" one.

This has been done only in mice, but in 2014 Jacob Hanna in Israel has already created the primordial human cells needed to generate human eggs. The next step (to generate the eggs, fertilize them and obtain the embryos) is not far away. Hanna is working with Azim Surani at Cambridge University to complete the cycle.

As Hank Greely has written in his book "The End of Sex and the Future of Human Reproduction" (2016), procedures like IVG will complete the process of separating sexual intercourse and reproduction. In the future a doctor will only need a few cells from a woman's skin and a few cells from a man's skin to create as many embryos as desired. Then these parents will be told the "features" of each embryo and pick their favorite. IVG can even create eggs from the skin cells of a man, and sperm from the skin cells of a woman. IVG will allow lesbians to have babies. It will even allow a woman to be both the father and the mother of a baby, a uniparent (although this presents the same genetic problems of incest).

Imagine a computer program that allows the parents to play with 100 different embryos: the parents can see a simulation of how each embryo will look like at the ages of 5, 10, 15 ,20... 80. The parents can simulate the life of each embryo and then decide which one they want to be. All the other embryos get thrown in the garbage.

It is just a matter of time before we will be flooded with apps to design our children. To start with, HumanCode, founded in Denver in 2017 by Chris Glode and Ryan Trunck, offers an app that runs on the online platform Helix and costs just $200 and tells future parents how tall their children will be. In 2017 a physicist at Michigan State University, Stephen Hsu (now the founder of Genomic Prediction), unveiled a genetic "predictor" that uses machine learning to estimate height from a person's DNA.

That day is not far away. In 2013 the first "designer baby" was born, a baby boy named Connor: his parents carefully selected one embryo out of seven grown in the laboratory of Dagan Wells at University of Oxford. All the other embryos were thrown in the garbage.

I personally don't blame the scientists for "playing God": every doctor plays God when s/he saves the life of someone who is dying of a disease. But society will be truly disrupted when it becomes possible to "design" babies, and it is not clear what the rights and the duties are. Should the parents have the right to choose which embryo will live and kill all the others? And nobody can ask a baby if she wants to be a genetic experiment. Someone could take a paper coffee cup that you casually tossed in the trash and turn you into a parent without your knowledge or consent. I am not sure that i would like to be the product of parents who wanted to program my beauty, my intelligence, my skills and maybe even my hobbies.

In 2015 Junjiu Huang's team at Sun Yat-sen University in Guangzhou used CRISPR to edit the genes of a human embryo but failed to produce a working embryo: some of the cells did not have the edited gene and some cells had damaged DNA ("CRISPR/Cas9-mediated Gene Editing in Human Tripronuclear Zygotes", 2015). In 2017 Shoukhrat Mitalipov's team succeeded: they removed a gene that causes a hereditary heart defect (hypertrophic cardiomyopathy) from a number of embryos without damaging the rest of the DNA ("Correction of a Pathogenic Gene Mutation in Human Embryo", 2017). The embryos were allowed to develop for only five days, but that is enough to prove the concept. We are not too far from the day when scientists will implant gene-edited embryos in women.

Greely argues that the "killer app" of synthetic biology will be the procedure to avoid rare diseases in babies (there are 6,000 rare diseases, with a chance of about 1% that your child will have one). He then argues that the same killer app will be used to treat infertility, and to give children to lesbian parents. Once it is approved, the social benefits of having healthy babies will become obvious: babies with fewer diseases mean lower costs for health care (especially later in life) and fewer epidemics. Greely thinks that states will provide something like IVG for free because of its public health benefits (exactly like vaccines are mandatory). Imagine a society in which there are no more disabilities. Harvard University's synthetic biology pioneer George Church, in his book "Regenesis" (2012), envisioned a factory-style process to produce humans who will be immune to all viruses.

But what willl happen to the others? Will these superhumans, equipped with a superpowerful immune system and with the perfect genes, be willing to live among regular humans who get sick and spread diseases, or with humans who are born physically handicapped and are a financial burden on the healthy ones? Will these perfect humans be willing to pay taxes that help cure the illnesses of regular humans?

The discussion about intelligent machines and the Singularity is a distraction when in fact a much bigger revolution (and a much bigger moral discussion) is about to shake humanity.

Humorous Intermezzo: Bayes and The End of the World

In 1983 physicist Brandon Carter introduced the "Doomsday Argument" later popularized by philosopher John Leslie in his book "The End of the

World: The Science and Ethics of Human Extinction" (1996). This was a simple mathematical theorem based on Bayes's theorem demonstrating that we can be 95% certain that we are among the last 95% of all the humans ever to be born. Leslie calculated that we will reach this point in about 10,000 years. It has been tweaked up and down by various dissenters but, unlike Singularity Science, it sits on solid mathematical foundations: in fact, on the exact same foundations (Bayesian reasoning) as those buttressing today's deep-learning neural-networks like AlphaGo.

The Future of Miniaturization: the Next big Breakthrough?

If i am right and the widely advertised progress in machine intelligence is mainly due to rapid progress in miniaturization and cost reduction, then it would be more interesting to focus on the future of miniaturization. Whatever miniaturization achieves next is likely to determine the "intelligence" of future machines.

While IBM's Watson was stealing the limelight with its ability to answer trivial questions, others at IBM were achieving impressive results in Nanotechnology. In 1989 Don Eigler's team at IBM's Almaden Research Center, using the scanning tunneling microscope built in 1981 by Gerd Binnig and Heinrich Rohrer, carried out a spectacular manipulation of atoms that resulted in the atoms forming the three letters "IBM". In 2012 Andreas Heinrich's team (in the same research center) stored one magnetic bit of data in 12 atoms of iron, and a byte of data in 96 atoms; and in 2013 that laboratory "released" a movie titled "A Boy and His Atom" made by moving individual atoms.

This ability to trap, move and position individual atoms using temperature, pressure and energy could potentially create a whole new genealogy of machines.

The Real Future of Computing

In 1988 Mark Weiser envisioned a future in which computers will be integrated into everyday objects ("ubiquitous computing") and these objects will be connected with each other. This became known as the "Internet of Things" after 1998 when two MIT experts in Radio-frequency identification (RFID), David Brock and Sanjay Sarma, figured out a way to track products through the supply chain with a "tag" linking to an online database.

The technology is already here: sensors and actuators have become so cheap that embedding them into ordinary objects will not significantly increase the price of the object. Secondly, there are enough wireless ways to pick up and broadcast data that it is just a matter of agreeing on some standards. Monitoring these data will represent the next big wave in software applications.

Facebook capitalized on the desire of people to keep track of their friends. People own many more "things" than friends, spend more time with things than with friends, and do a lot more with things than with friends. The equivalent of Facebook for "things" does not exist yet, but potentially it is an order of magnitude bigger.

At the same time, one has to be aware that the proliferation of digital control also means that we will live in a world under systematic and all-pervasive surveillance in which machines will keep a record of everything that has happened and that is happening. Machines indirectly become spies. In fact, your computer (desktop, laptop, notepad, smartphone) is already a sophisticated and highly accurate spy that records every move you make: what you read, what you buy, whom you talk to, where you travel, etc. All this information is on your hard disk and can easily be retrieved by forensic experts.

The focus of computer science is shifting towards collecting, channeling, analyzing and reacting to billions of data arriving from all directions. Luckily, the "Internet of Things" will be driven by highly structured data.

The data explosion is proceeding faster than the increase in processing speed: exploring data is becoming increasingly more difficult with traditional John von Neumann computer architectures that were designed for calculations.

If I am sceptical about the creation of an agent that will be an artificial general intelligence, i am very aware that we are rapidly creating a sort of global intelligence as we connect more and more software and this giant network produces all sorts of positive feedback loops. This network is already out of control and gets harder to control with each passing year.

A Brief History of Bionic Humans, Cyborgs and Neuroengineering

I suspect that the science fiction novels and movies about cyborgs are more realistic than the ones about robots and artificial intelligences.

The first electrical implant in an ear was the work of French surgeons Andre Djourno and Charles Eyries in 1957. Building upon their work, in 1961 William House invented the "cochlear implant", an electronic implant that sends signals from the ear directly to the auditory nerve (as opposed to hearing aids that simply amplify the sound in the ear).

Spanish-born neuroscientist Jose Delgado is credited with publishing the first paper on implanting electrodes into human brains: "Permanent

Implantation of Multi-lead Electrodes in the Brain" (1952). In 1965 he famously managed to control a bull via a remote device, injecting fear at will into the bull's brain. He then published his dystopian vision in the book "Physical Control of the Mind - Toward a Psychocivilized Society" (1969). In 1969 he created the first bidirectional brain-machine-brain interface when he implanted devices in the brain of a monkey and then sent signals in response to the brain's activity.

For a while the discipline of brain-machine interfaces lay dormant because the machines were just not up to the task. Nonetheless, Jacques Vidal at UCLA was implanting simple sensors within the brains of rats, mice, monkeys, and eventually humans. He published the first academic paper about brain-machine interfaces ("Toward direct brain-computer communication", 1973). A decade later, Emanuel Donchin and Larry Farwell at University of Illinois introduced the concept of "brain fingerprinting ("Talking off the top of your head: toward a mental prosthesis utilizing event-related brain potentials", 1988).

In 2000 William Dobelle in Portugal developed an implanted vision system that allowed blind people to see outlines of the scene. His patients Jens Naumann and Cheri Robertson became "bionic" celebrities as Dobelle continued to refine his artificial vision system.

The electrical interfacing of semiconductors and neurons is not trivial because neurons communicate using ions whereas semiconductors use electrons. In 1991 Peter Fromherz at the Max Planck Institute in Munich solved the problem of sensing the electrical field of a neuron on an electronic chip, and in 1995 he solved the problem of stimulating a neuron with an electronic chip (he used neurons of leeches). In 2001 he was therefore able to build a hybrid circuit of electronics and neurons (a snail's neurons).

In 2002 John Chapin and Sanjiv Talwar at the State University of New York debuted their "roborats", rats whose brains were fed electrical signals via a remote computer to guide their movements.

As for getting data out of the brain into a machine (output neuroprosthetics) in 1998 the Irish-born scientist Philip Kennedy at Georgia Tech developed a brain implant that could capture the "will" of a paralyzed man (Johnny Ray) to move an arm. In 1987 Kennedy had founded Neural Signals to develop a brain-computer interface, the first bionic startup. (Ray died in 2002 and in 2014 Kennedy himself almost died when he courageously chose to have electrodes surgically implanted in his own brain)

In 1998 Kevin Warwick at University of Reading in Britain implanted a transmitter in his arm to activate computer-controlled devices, a bionic precursor of the "Internet of Things". (In the same year Warwick also created an artificial intelligence to compose pop songs). In 2002 Warwick used a BrainGate device to connect his nervous system to the Internet.

In 2002 Brown University spun off Cyberkinetics, a startup charged with developing its BrainGate technology. In 2005 John Donoghue's team implanted a BrainGate device in the brain of a paralyzed woman, Cathy Hutchinson, which allowed her to operate a robotic arm.

In 2002 the Brazilian-born scientist Miguel Nicolelis at Duke University implanted a microchip into a monkey's brain that allowed the monkey to control a robotic arm.

In 2004 Theodore Berger at University of Southern California in Los Angeles demonstrated a hippocampal prosthesis conceived to replace the long-term-memory function lost by a damaged hippocampus. His lab would become another major hub of bionic research. In 2011 Berger developed "memory chips" that can turn memories on and off in a mouse's brain, and in 2015 Berger and Dong Song built a brain prosthesis to help people suffering from memory loss

In 2004 color-blind artist Neil Harbisson (born in Britain, raised in Spain, relocated to New York) became the first person in the world to have an antenna implanted in his skull, a device that transformed color into sound. (In 2010 Harbisson founded the Cyborg Foundation to defend "cyborg rights", the equivalent of "human rights" for cyborgs like him).

In 2004 PositiveID in Florida started selling VeriChip, an RFID chip implant for humans developed in Texas at Destron Fearing, a company that manufactures RFID tags for animal identification.

In 2003 the psychologist Marcel Just and the machine-learning guru Tom Mitchell at CMU started collaborating on a system to read minds: identify patterns of brain activity in fMRI associated with different objects ("Predicting Human Brain Activity Associated with the Meanings of Nouns", 2008).

This is when the government stepped in. In 2006 the Defense Advanced Research Projects Agency (DARPA) asked scientists to submit "innovative proposals to develop technology to create insect-cyborgs".

And this is also when the transhumanist movement adopted bionics. In 2006 Seattle-based transhumanist Amal Graafstra boasted a microchip in each hand, one for storing data (that could be uploaded and downloaded from/to a smartphone) and one for a code that unlocked his front door and logged him into his computer. In 2012 Graafstra implanted chips on attendees of the Toorcamp for $50 each, and in 2013 he started a website to sell home implants, dangerousthings.com.

In 2010 Epoc in Australia released a neuroheadset for videogames, Emotiv, to play videogames with your brain waves.

The laboratory of Finnish-born engineer Arto Nurmikko at Brown University that had inherited the BrainGate project from Cyberkinetics. By 2008 this device had become a wireless transmitter for paralyzed patients with a neural implant that bypassed the spinal cord. In 2011 Leigh Hochberg of that team used BrainGate to make a paralyzed woman operate a robotic arm simply by thinking about the movement.

Experiments on brains became more and more ambitious. In 2011 Matti Mintz in Israel replaced a rat's cerebellum with a computerized cerebellum. In 2012 the brain implant designed by Sam Deadwyler at Wake Forest University managed to improve the long-term memory of monkeys.

At the same time some independents began to view implants as the tattoos of the 21st century. In 2013 biohacker Rich Lee in Utah hired Steve Haworth in Arizona to implant headphones into his ears. Haworth had pioneered "body modification", a high-tech evolution of "body piercing" that implants devices (typically magnets) under the skin.

Two-way transmission was just a matter of combining existing technologies. In 2013 Nicolelis made two rats communicate (and they were located in two different countries) by capturing the "thoughts" of one rat's brain and sending them to the other rat's brain over the Internet and an electrode. In 2015 Nicolelis connected the brains of monkeys so that they could collaborate to perform a task.

In 2013 the Indian-born computer scientist Rajesh Rao and the Italian-born psychologist Andrea Stocco at University of Washington devised a way to send a brain signal from Rao's brain to Stocco's hand over the Internet, i.e. Rao made Stocco's hand move, probably the first time that a human was capable of controlling the body part of another human. In that year Rao (also a scholar of the ancient Indus script and of classical Indian painting) published "Brain-Computer Interfacing" (2013).

It was just a matter of time before someone thought of expanding the cyborgs beyond vision, sound and movement. In 2014 the team led by Italian-born electrical engineer Silvestro Micera at the Federal Institute of Technology (EPFL) in Switzerland designed an artificial hand for an amputee, Dennis Aabo-Soerensen. This hand sends electrical signals to the nervous system so as to create the sensation of touch.

In 2014 Chinese-born wireless scientist Ada Poon at Stanford invented a safe way to transfer energy to chips implanted in the body (to "electroceutical devices").

In 2015 Zoran Nenadic and An Do of University of California at Irvine attached an electroencephalograph device to the head of a paraplegyc man and made him walk a few steps.

In 2015 EPFL built a robotic wheelchair for paralyzed people. This chair combines brain control with artificial intelligence. In 2016 Gregoire Courtine at EPFL used BrainGate to restore movement to a monkey's paralyzed leg.

In 2016 Nick Ramsey's team in Holland (at University Medical Center Utrecht) inserted wireless electrodes into the skull of a paralyzed patient (unable to speak or move) so that she could control a computer mouse simply by thinking of moving her fingers.

Meanwhile, also in 2016, Niels Birbaumer at University of Tuebingen worked with patients affected by complete motor paralysis but perfectly lucid as far as mental processes go (the state called "complete locked-in"). Using functional near-infrared spectroscopy (fNIRS), they were able to answer yes/no questions with their thoughts.

It may soon be possible to do the same things without a brain implant. In 2016 Bin He at University of Minnesota demonstrated an EEG cap fitted with 64 electrodes that can convert the "thoughts" of a person into the movement of a robotic arm for grasping objects in a room.

In 2017 Bill Kochevar, a man with complete paralysis, was able to feed himself thanks to a brain-controlled arm designed by Bolu Ajiboye at Case Western Reserve University in Ohio.

The idea of developing neuroprostheses to enhance the human brain became popular with rich entrepreneurs who in fact started the two most hyped bionic ventures of the age: in 2016 Elon Musk launched Neuralink in San Francisco and Bryan Johnson founded Kernel in Los Angeles (Johnson had founded in 2007 the online and mobile payment platform Braintree, acquired by PayPal in 2013). But they were not the first startups in this field. Among the pioneers were Thomas Oxley's Synchron and Matt Angle's Paradromics. MIT graduates Robert McIntyre and Michael McCanna founded Nectome in 2015 to upload brains to computers, and in 2019 the media reported that billionaire Sam Altman had signed up for the service to digitally embalm his brain.

In 2017 Facebook announced a project (led by Mark Chevillet) to decode speech directly from the brain and Eberhard Fetz at the Washington University began adapting a chip by ARM to design chips for brains. In 2017 ARM, maker of the most popular chip for smartphones, opened a Center for Sensorimotor Neural Engineering.

In 2018 Stefan Harrer of IBM Australia announced GraspNet, a system that uses Deep Learning (running on an embedded Nvidia chip) to decode EEG signals and control a robotic arm. This had been done before but it was extremely difficult to decipher the very weak EEG signals. By using A.I., Harrer's team managed to get a clearer signal.

In May 2018 David Glanzman's team at UCLA removed memories from one sea snail and injected them into another sea snail in the form of RNA ("RNA from Trained Aplysia Can Induce an Epigenetic Engram for Long-Term Sensitization in Untrained Aplysia", 2018).

In 2017 bionic projects were carried out by surgeon Eric Leuthardt at Washington University, by Newton Howard at Oxford University (formerly the director of the MIT Mind Machine Project), whose neural implant used technology from Intel and Qualcomm, and by Dong Song at University of Southern California, whose brain implant boosted human memory.

In 2013 Michel Maharbiz and Jose Carmena at UC Berkeley presented "neural dust", by which they referred to a dust-sized, wireless, battery-less sensor that can be implanted in the nervous system as well as in muscles and in organs and can be activated with ultrasounds.

In 2016 DARPA's Neural Engineering Systems Design project, run by Phillip Alvelda, funded several bionic projects, including the project by Edward Chang at UC San Francisco to treat mental illnesses.

In 2018 Andrea Stocco's team at University of Washington, the pioneers of non-invasive direct brain-to-brain communication, demonstrated a network of brains interacting via electroencephalograms (EEGs), which records electrical activity in the brain, and transcranial magnetic stimulations (TMSs), which transmits information into the brain ("BrainNet - A Multi-Person Brain-to-Brain Interface for Direct Collaboration Between Brains", 2018).

Electroceuticals are implantable devices that deliver electrical impulses to the neural circuits of organs in order to treat ailments. The most famous "electroceutical" is the pacemaker. The first implantable pacemaker was invented by Rune Elmqvist in Sweden and placed into Arne Larsson by the surgeon Ake Senning in 1958 (and Larsson went on to outlive Senning by five years). In 1998 Kevin Tracey at Northwell Health near New York discovered that the nervous system and the immune system communicate via the vagus nerve: in particular, the vagus nerve emits chemicals that regulate the immune system. Tracey and others then began developing the technology of "vagal nerve stimulation" or VNS for therapeutic uses. In other words, VNS is a tool for "hacking" the nervous system.

Can A.I. visualize your thoughts? Yukiyasu Kamitani's lab at Kyoto University in Japan had experimented with "deep image reconstruction" ("Deep Image Reconstruction from Human Brain Activity", 2017). And what about dreams? Can A.I. visualize your dreams? Wouldn't be exciting if we could videotape our dreams and then project them on a screen? The same Kamitani lab decoded visual imagery during sleep ("Neural Decoding of Visual Imagery During Sleep", 2013) and Jack Gallant's lab at UC Berkeley captured the brain activity related to watching movies ("Decoding the Semantic Content of Natural Movies from Human Brain Activity", 2016).

Mind-reading to generate speech (to literally "listen" to what you are thinking) is not science-fiction anymore. For example, Edward Chang's team at UC San Francisco used a recurrent neural network to decode cortical signals into simulations of movements of the vocal tract, and then to transform these virtual movements into spoken sentences ("Speech synthesis from neural decoding of spoken sentences", 2019).

This is certainly an exciting field that can restore movement to paralyzed people, but there are concerns about how this will be used. Most people are concerned that governments could use this technology to control people, but maybe we should be more concerned about how ordinary people will use it. Here is an example from a very similar field. In 1946 a surgeon called Robert Heath at Tulane University invented a technique to implant electrodes into brains through small holes drilled into the skull in order to stimulate specific brain regions. There is a book about Heath's invention: Lone Frank's "The Pleasure Shock". Why "the pleasure shock"? Because this procedure tends to induce a feeling of pleasure in the patient. After it was approved in 2002 in the USA, this procedure is now called "deep brain stimulation" and has been performed on about 100,000 people with brain diseases such as Parkinson's disease. Deep brain stimulation costs about $50,000 so it is done only in emergency cases, but one can see a future in which it will become cheaper and people will start using it not to treat mental illness but simply to get pleasure. In other words, there is a risk of creating an epidemic of "pleasure addicts" who spend their spare time self-stimulating their brain with this procedure the same way that some people take drugs or watch pornographic videos.

We are moving closer to developing telepathy. The future of "man-machine interfaces" could be just thought. In that case thought will also become the natural form of communication between people. Telepathy will have become reality. Then maybe spammers will start flooding my brain with thoughts about useless products, and government agencies will start listening to my thoughts. History tends to repeat itself.

After Machine Inteligence: Machine Creativity – Can Machines do Art?

The question "Can machines think?" is rapidly becoming obsolete. I have no way of knowing whether you "think". We cannot enter someone else's brain and find out if that person has feelings, emotions, thoughts, etc. All we know about other people's inner lives is that it generates a behavior very similar to our own, and therefore we conclude that other people too must have the same kind of inner lives that we have (feelings, emotions, thoughts, etc). Since we cannot even determine with absolute certainty the consciousness of other people, it sounds a bit useless to discuss whether machines can be conscious. Can machines think? Maybe, but we'll never find out for sure, just like we'll never find out for sure if all humans think.

The question "Can machines be creative?" is much more interesting. Humans have always thought of themselves as creative beings, but always failed to explain what that really means. The humble spider can make a very beautiful spider-web. Some birds create spectacular nests. Bees perform intricate dances. Most humans don't think that the individual spider or the individual bird is a "creative being". Humans assume that something in its genes made it do what it did, no matter how complex and brilliant. But what exactly is the difference between the spider or the bird and Shakespeare, Michelangelo or Beethoven?

If you think that the history of human civilization is mainly the history of a uniquely creative species, it is surprising how little the human race has investigated the topic of creativity. The first major philosopher to use the term in the title of a book was probably John Dewey in "Creative Intelligence" (1917) but he was referring to the "being in the world" that, in his opinion, characterizes human intelligence. British social psychologist Graham Wallas in "The Art of Thought" (1926) outlined a four-stage model of the creative process, and others explored the topic in a superficial manner. The turning point came in 1950, when Joy-Paul Guilford of University of Southern California gave his presidential address to the American Psychological Association (later published as "Creativity"), defining creativity as the ability to generate novel ideas. In 1951 Guilford launched the Aptitudes Research Project at his university (the results of that project were later compiled in the book "The Nature of Human Intelligence", 1967). However, someone in California had already started a scholarly project on creativity.

In the 1930s Donald MacKinnon who had studied in Henry Murray's psychological clinic at Harvard University, and during World War II operated a secret laboratory at a remote Maryland farmhouse whose task was to select spies to infiltrate in Europe (on behalf of the Office of Strategic Services). In 1949 MacKinnon, now a professor at UC Berkeley, founded the Institute of Personality Assessment and Research (IPAR). IPAR interviewed and tested "creative" thinkers from various disciplines (such as writers, architects, scientists and mathematicians). MacKinnon concluded that engineering students were not creative ("Fostering Creativity in Students of Engineering", 1961) and outlined his own model of human intelligence ("The Nature and Nurture of Creative Talent", 1962). IPAR's most prominent researcher, Frank Barron, published the seminal book "Creativity and Psychological Health" (1963) and in 1969 moved to UC Santa Cruz, where he taught an influential course on creativity.

Meanwhile, Alex Osborn had been refining his "brainstorming" technique to stimulate creativity, first published in the book "Applied Imagination" (1953) but already applied successfully to both military and corporate organizations. In 1955 he founded the Creative Problem-Solving Institute at University of Buffalo that hosted a yearly conference. Also important where the Utah Creativity Research Conferences, inaugurated in 1955, and symposia at University of Michigan State in 1957 and 1958. In 1960 Paul Torrance at University of Minnesota launched the Minnesota Tests of Creative Thinking (MTCT), today better known as Torrance Tests of Creative Thinking (TTCT). The debate among these psychologists was about whether intelligence and creativity were the same mental process or two different processes. Within a few years "creativity" had become a popular buzzword. Creativity was discussed in popular books, from Arthur Koestler's "The Act of Creation" (1964) to Howard Gardner's "Frames of Mind" (1983), and, in more sensationalistic terms, Margaret Boden's "The Creative Mind" (1990), in which she updated Guilford's definition of creativity to "the ability to generate novel and valuable ideas" (but also fell for some quasi-scams of fake creative programs).

This research paralleled promises and progress in Artificial Intelligence. While Hubert Dreyfus maintained the impossibility of machines to be creative ("What computers still can't do", 1992), Douglas Hofstadter wrote a book about building such machines ("Fluid Concepts and Creative Analogies", 1995). Margaret Boden spoke about "Creativity and Computers" at Stanford's 1993 AAAI Spring Symposium on "Artificial Intelligence and Creativity" In the same year an International Symposium on Creativity and Cognition was held at Loughborough University. In the same year Gilles Fauconnier introduced "conceptual blending" at UC San Diego in California ("Conceptual Integration Networks", 1994).

Humans use tools to make art (with nothing else, but a pen). But the border between artist and tool has gotten blurred since Harold Cohen conceived AARON, a painting machine, in 1973. Cohen asked: "What are the minimum conditions under which a set of marks functions as an image?" I would rephrase it as "What are the minimum conditions under which a set of signs functions as art?" Even Marcel Duchamp's "Fountain" (1917), which is simply a urinal, is considered "art" by the majority of art critics. Abstract art is mostly about... abstract signs. Why are Piet Mondrian's or Wassily Kandinsky's simple lines considered art? Most paintings by Vincent Van Gogh and Pablo Picasso are just "wrong" representations of the subject: why are they art, and, in fact, great art?

Enter the machine. The Cybernetic Serendipity exhibition that ran in London from August to October 1968 featured computer-generated images (including one by Norbert Wiener), a live drawing computer, several computer-generated poems, Peter Zinovieff's music computer that could improvise a song based on a melody whistled by the user, and interactive robotic sculptures such as Bruce Lacey's ROSA Bosom (1965), that, incidentally, had been "best man" at his wedding (ROSA stood for "Radio Operated Simulated Actress"). The filmmaker Malcolm LeGrice created "Typo Drama" (1969), premiered in London at the Event One art exhibition in April 1969, a system that generated the text and the actions for the actors of a theatrical performance (the software was written by Alan Sutcliffe, founder of the Computer Arts Society).

It is not difficult to write a program that will write a book. Already in 1967 a program designed by the Fluxus artist Alison Knowles and the composer James Tenney to randomly assemble stanzas produced "The House of Dust" (1967), a computer-generated poem. It all depends on how random you want the sentences to be. In 1983 New York freelance writers and programmers William Chamberlain and Thomas Etter published "The Policeman's Beard Is Half Constructed", subtitled "the first book ever written by a computer", a collection of poems allegedly written by their program Racter, a program remarkably written in Basic on a personal computer with 64 kilobytes of memory. In 1993 Scott French used a program to compose "Just This Once", a romance novel in the style of Jacqueline Susann (and one of most stereotypical novels ever published), but French's manual contribution was probably massive. In 1992 the Polish artist Wojciech Bruszewski wrote a computer program that generated sonnets in a nonexistent (but pronounceable) language. These sonnets were published in eight volumes. (This is my favorite: if the machine has to be creative, let it invent its own language).

In 1996 Naoko Tosa at ATR in Japan built an art installation called "Interactive Poem" that consisted in a verbal collaboration between a person and a computer to write poems.

David Cope at UC Santa Cruz had experimented since 1981 with automatic music composition (his expert system EMI and its various descendants). Peter Todd at Stanford employed a recurrent neural network to compose melodies: his network was trained to predict the note following the current note ("A Connectionist Approach to Algorithmic Composition", 1989); but this note-by-note approach was clearly limited. LSTM networks were more appropriate to generate music: his inventor, Schmidhuber, collaborated with Douglas Eck at IDSIA to learn the characteristics of blues music ("Learning the Long-term Structure of the Blues", 2002) and then to compose music ("A First Look at Music Composition using LSTM Recurrent Neural Networks", 2002).

Hod Lipson and Jordan Pollack (now at Brandeis University) used Ken Stanley's NEAT algorithm to build EndlessForms, a program that keeps generating industrial designs ("Automatic Design and Manufacture of Robotic Lifeforms", 2000).

During the 1990s and 2000s several experiments further blurred the line between computer and human creativity: Ken Goldberg's painting machine "Power and Water" at University of South California (1992); Matthew Stein's PumaPaint at Wilkes University (1998), who came up with an online robot that allows Internet users to create original artwork; Jurg Lehni's graffiti-spraying machine Hektor in Switzerland (2002); David Cope's program Emily Howell for music composition, that was conceived in 2003 and went on to release the albums "From Darkness Light" (2009) and "Breathless" (2012); the painting robots developed since 2006 by Washington-based software engineer Pindar Van Arman; RAP (Robotic Action Painter), installed by the Portuguese artist Leonel Moura in 2007 at the American Museum of Natural History, a robot that created drawings (he also created the Robotarium, a zoo for robots and artificial life); and Vangobot (2008) (pronounced "Van Gogh bot"), a robot built by Nebraska-based artists Luke Kelly and Doug Marx renders images according to preprogrammed artistic styles.

The pundits, such as Juergen Schmidhuber at IDSIA in Switzerland ("Curious Model-building Control Systems", 1991, later expanded into a "Formal Theory of Creativity, Fun, and Intrinsic Motivation") and Geraint Wiggins at City University of London ("Towards a more Precise Characterisation of Creativity in A.I.", 2001), keep debating how to make machines creative.

After a Kickstarter campaign in 2010, Chicago-based artist Harvey Moon built drawing machines, set their "aesthetic" rules, and let them do the actual drawing. In 2013 Oliver Deussen's team at University of Konstanz in Germany demonstrated e-David (Drawing Apparatus for Vivid Interactive Display), a robot capable of painting with real colors on a real canvas. In 2013 the Galerie Oberkampf in Paris showed paintings produced over a number of years by a computer program called "The Painting Fool", which was designed by Simon Colton at Goldsmiths College in London. The Living Machines exhibition of 2013 at London's Natural History Museum and Science Museum featured "Paul", a creative robot capable of sketching a portrait, developed by French inventor Patrick Tresset in 2011, and BNJMN (pronounced "Benjamin"), a robot capable of generating images built for the occasion by Travis Purrington and Danilo Wanner from the Basel Academy of Art and Design.

Also in 2011, the computer Iamus at Malaga University in Spain premiered a piece in its own style (Cope's algorithm were imitating masters). Four of Iamus' compositions for full orchestra are performed by the London Symphony Orchestra in the album "Iamus" (2012).

In 2012 neuroevolutionary veteran Ken Stanley (of NEAT fame) and his students at University of Central Florida unveiled MaestroGenesis, a program that creates polyphonic music from simple monophonic melodies.

John Supko, a music scholar at Duke University, and digital media artist Bill Seaman created the software that composed the music released on the album "S_traits" (2014), voted by the critics of the New York Times as one of the best recordings of the year (it wouldn't make my top 1000 but that's personal taste).

While each of these systems caused headlines in the press, none was autonomous and the "trick" was easy to detect.

Then deep learning happened. Deep learning consists in a multi-layer network that is trained to recognize an object. The training consists in showing the network many instances of that object (say, many cats). Andrew Zisserman's team at Oxford University was probably the first to think of asking a neural network to show what it was learning during this training ("Deep Inside Convolutional Networks", 2014). Basically, they used the neural network to generate the image of the object being learned (say, what the neural network has learned a cat to be like).

In May 2015 a Russian engineer at Google's Swiss labs, Alexander Mordvintsev, used that idea to make a neural network produce psychedelic images. One month later he posted a paper titled "Inceptionism" (jointly with Christopher Olah, an intern at Jeff Dean's Google Brain team in Silicon Valley, and with Mike Tyka, an artist working for Google in Seattle) that sort of coined a new art movement. Neural nets trained to recognize images can be run in reverse so that they instead generate images. More importantly, the networks can be asked to identify objects that actually don't exist, like when you see a face in a cloud. By feeding back this "optical illusion" into the network over and over again, the network eventually displays a detailed image, which is basically the machine's equivalent of a human hallucination. For example, a neural network trained to recognize animals will identify inexistent animals in a cloudy sky.

In 2015 two students (Leon Gatys and Alexander Ecker) of Matthias Bethge's lab at University of Tubingen in Germany taught a neural network to capture an artistic style and then applied the artistic style to any picture ("A Neural Algorithm of Artistic Style", 2015). A neural network can imitate the style of any maestro. A neural network trained to recognize an object tends to separate content and style, and the "style" side of it can be applied to other objects, therefore obtaining a version of those objects in the style that the network previously learned. In other words, the neural network captures an artistic style and then applies the artistic style to any picture, turning it into a painting in that artistic style.

In September 2015, at the International Computer Music Conference, Donya Quick, a composer working at Paul Hudak's lab at Yale University, presented a computer program called Kulitta for automated music composition. In February 2016 she published on Soundcloud a playlist of Kulitta-made pieces.

In February 2016 Google staged an auction of 29 paintings made by its artificial intelligence at the Grand Theater in San Francisco in collaboration with the Gray Area Foundation for the Arts ("DeepDream: The Art of Neural Networks").

In March 2016 a 20-year-old Princeton University student, Ji-Sung Kim, and his friend Evan Chow created a neural network that can improvise like a jazz musician on Pat Metheny's "And Then I Knew" (1995).

In April 2016 a new Rembrandt portrait was unveiled in Amsterdam, 347 years after the painter's death: Joris Dik at Delft University of Technology created this 3D-printed fake Rembrandt consisting of more than 148 million pixels based on 168,263 fragments from 346 of Rembrandt's paintings. (To be fair, a similar feat had been achieved in 2014 by Jeroen van der Most whose computer program had generated a "lost Van Gogh" after analyzing statistically 129 real paintings of the master).

In May 2016 Daniel Rockmore at Dartmouth College organized the first Neukom Institute Prizes in Computational Arts (soon nicknamed the "Turing Tests in the Creative Arts"), that included three contests to build computer programs that can create respectively a short story, a sonnet, and a DJ set. Spanish students Jaume Parera and Pritish Chandna won the prize for the DJ set, while three students of Kevin Knight's lab at University of Southern California won the prize for the sonnet ("And from the other side of my apartment/ An empty room behind the inner wall/ A thousand pictures on the kitchen floor/ Talked about a hundred years or more").

Combining a convolutional neural network to learn a person's favorite style of fashion with a generative adversarial network, in 2017 Julian McAuley's student Wang-Cheng Kang at UC San Diego in collaboration with Adobe created a system that can generate personalized clothing.

In July 2016 a Bay Area software engineer, Karmel Allison, launched CuratedAI, an online magazine of poems and prose written by A.I. programs.

In September 2016 Google published a paper on WaveNet, a neural network that can generate music.

Mario Klingemann, a 2016 artist in residence at Google Cultural Institute in Paris, learned how to use generative adversarial networks and became perhaps the first professional artist to specialize in A.I.-based artworks. In 2017 (at the peak of the "fake news" debate) he became famous for "artworks" that consisted in artificially-generated audio and video of non-existing events that felt real.

In 2016 an LSTM created by Ross Goodwin of New York University and named Benjamin, and trained with sci-fi screenplays from the 1980s and 1990s, scripted the sci-fi movie "Sunspring" that was directed by Oscar Sharp and was presented at the Sci-Fi London annual film festival.

At the end of 2016 Maya Ackerman of San Jose State University and David Loker debuted Alysia, a computer program that generates a melody based on a text. A few months later Ackerman, also an opera singer, performed songs whose melodies had been composed by Alysia based on lyrics written by another computer program, Mable, based on Rafael Perez y Perez's Mexica. (Vanity: this performance took place at a Leonardo Art Science Evening Rendezvous, a series that i founded in 2008). The problem with these music-composing programs has been and still is that the music they produce is incredibly boring. If (like me) you think that pop music is mostly garbage, you're in for a real nightmare, because these machines can only make pop music that is even worse than your least favorite pop star's songs. Pure torture for my ears. I am not sure if this is what was meant in 2016 by Douglas Eck (now at Google) when he announced the Magenta project to make art with deep learning techniques when he mentioned "the completely, frankly, astonishing improvements in the state of the art".

A poetry book written by Microsoft's chatbot Xiaoice was published in China in May 2017.

A 2017 blog entry on the Magenta.as website by William Anderson of Huge Inc explained how to achieve pretty much the same kind of "creativity" as inceptionism by simply using old-fashioned Markov chains (the title of the online paper is unfortunately "Using Machine Learning to Make Art", 2017).

In 2017 Chris Donahue's team at UC San Diego trained a neural network with a dance videogame whose users have created dances for many popular songs. The neural network, named Dance Dance Convolution, can generate a dance for any new song.

In 2017 Ahmed Elgammal's group at Rutgers University collaborated with art historian Marian Mazzone of the College of Charleston in South Carolina to use a generative adversarial network (GAN) that created a system that learns about art styles and then "deviates" from norms to create its own style ("Creative Adversarial Networks", 2017). Perhaps more importantly in 2015 the same group had written an algorithm that could identify the artist, genre, and style of an artwork, and find correlations between styles, which is the job of the art historian.

In 2017 Ian Simon and Sageev Oore at Google's Magenta subsidiary developed Performance RNN, an LSTM-based recurrent neural network and trained it on the Yamaha e-Piano Competition dataset, which contains MIDI captures of 1,400 performances by skilled pianists, so that the network outputs polyphonic music.

In July 2017 San Francisco's McLoughlin Gallery hosted the exhibition "Artificial Intelligence: The End of Art as We Know It". It showed "portraits" by mural artist (and former Silicon Valley entrepreneur) Matty Mo, who since 2014 signs his artworks as "The Most Famous Artist", and who is mostly famous for stealing ideas from other artists. His portraits were jointly produced with an A.I. program created by a group of hackers.

In 2017 Stephen Thaler's DABUS (Device for the Autonomous Bootstrapping of Unified Sentience) at the University of Surrey generated artistic images that look like surrealistic hallucinations. DABUS was designed to generate novel ideas and in 2019 Thaler claimed that DABUS had even invented new devices and should be allowed to file its own patent applications for those inventions. (Then, again, in 2016 Thaler had claimed that "conscious computers have already been created in our labs", and all of his claims are based on childish theories of creativity and even more childish theories of consciousness, but the press loves them).

In October 2018 the New York auction house Christie's sold an A.I.-made print, "Portrait of Edmond de Belamy", for a significant sum that most human painters cannot fetch (although the winning bid came anonymously by telephone, which led many to suspect a scam to inflate its value). The money went to the "artist" who ran the algorithm (an open-source generative adversarial network that anyone could use) and who trained it using a dataset of old paintings, the "Old Masters" dataset (this too open-source). In other words, the "artist" (a French collective called Obvious) didn't create anything at all: not the algorithm, not the training dataset.

In March 2019 the HG Contemporary Gallery in New York's art district of Chelsea opened a solo show by Ahmed Elgammal, who now was founding director of the Art and Artificial Intelligence Lab at Rutgers University, an exhibition titled "Faceless Portraits Transcending Time" of

A.I.-made prints. It was billed as the first solo gallery exhibit devoted to an "A.I. artist". To generate the prints, Elgammal used a deep-learning algorithm called AICAN (a "creative adversarial network" or CAN) first introduced in 2017. AICAN-based works had already been exhibited in December 2018 at the Scope gallery in Miami Beach.

In May 2016 the TED crowd got to hear a talk by Blaise Aguera y Arcas, principal scientist at Google, titled "We're on the edge of a new frontier in art and creativity — and it's not human".

The standard objection to machine art is that the artwork was not produced by the machine: a human being designed the machine and programmed it to do what it did, hence the machine should get no credit for its "artwork". Because of their nonlinearity, neural networks distance the programmer from the working of the program, but ultimately the same objection holds.

However, if you are painting, it means that a complex architecture of neural processes in your brain made you paint, and those processes are a result of the joint work of a genetic program and of environmental forces. Why should you get credit for your artwork?

If what a human brain does is art, then what a machine does is also art.

A sceptical friend, who is a distinguished art scholar at UC Berkeley, told me: "I haven't seen anything I'd take seriously as art". But that's a weak argument: many people cannot take seriously as art the objects exhibited in museums of contemporary art, not to mention performance art, body art and dissonant music. How does humankind decide what qualifies as art?

The Turing Test of art is simple. We are biased when they tell us "this was done by a computer". But what if they show us the art piece and tell us it was done by an Indonesian artist named Namur Saldakan? I bet there will be at least one influential art critic ready to write a lengthy analysis of how Saldakan's art reflects the traditions of Indonesia in the context of globalization etc., etc.

In fact, the way that a neural network can be "hijacked" to do art may help us to understand the brain of the artist. It could lead to a conceptual breakthrough for neuroscientists. After all, nobody ever came up with a decent scientific theory of creativity. Maybe those who thought of playing the neural net in reverse told us something important about what "creativity" is.

This machine art poses other interesting questions for the art world.

What did the art collectors buy at the Google auction? The output of a neural network is a digital file, which can be copied in a split second: why would you pay for something, of which an unlimited number of copies can be made? In order to guarantee that no other copies will ever be made, we need to physically destroy the machine or... to re-train the neural network so it will never generate those images again.

What is missing to declare this art? Nothing: it is art. There is no doubt in my mind that machines can make art. Just like animals can make art: i have seen amazing spiderwebs and amazing bird nests. The Earth has created art that millions of tourists visit every year, from Iguazu Falls to the Namib desert.

What is missing is not the art, but, rather, the art critic. Art is a conversation between a producer and a consumer, and it's a conversation that often lasts a lifetime; in fact, it lasts centuries and millennia, from generation to generation. Art critics and art historians write books on how to appreciate art. The public visits museums to try and understand what the art critic saw in the art, and this becomes also a conversation between the art critic and the public. An algorithm can produce in a millisecond a positive/negative response based on some criteria, but art cultivates patience. There is no final answer. My reaction to a painting can (and in most cases will) change over my lifetime. I used to be moved by music that i now find tedious, and viceversa i have discovered unspeakable meanings (meanings that cannot be expressed in words) in music that i used to ignore. Different people have different reactions because they have different brains, different stories, different contexts. What is missing from machine art is not the art: what is missing is the ability to appreciate the art. Sure: a machine can suggest to me, based on my preferences, what music to listen to next (which is usually a really tedious suggestion) but that is precisely "not" what a music critic does: the music critic tells me to listen to music that i never even dreamed existed, and the music critic can explain to me how it relates to a virtually infinite number of cultural, social elements, including other music. A neural network can learn patterns (e.g., habits), but what a music, literary or art critic does is precisely to break those patterns and provide some kind of rationale for why it matters that the pattern is broken.

In a sense, i disagree with Charles Darwin who in "The Descent of Man" (1871) wrote: "The Imagination is one of the highest prerogatives of man. By this faculty he unites former images and ideas, independently of the will, and thus creates brilliant and novel results". In my opinion, all of this doesn't happen in the mind of the artist but rather in the mind of the critic or historian who deliberately and very rationally makes those "brilliant and novel results" brilliant and novel. It doesn't happen "independently of the will" but, on the contrary, very much deliberately. I disagree with Darwin on who is the "creator". All musicians can improvise but not all musicians will be considered great improvisers by jazz critics. The meme spreads not because of the musician but because of the critic/historian. The critic/historian may well be a fellow musician, who will endorse a previous musician and elevate him to a classic. Painting and singing probably preceded language but it is only when language emerged that we can talk of Art, because it is only then that we can... talk.

To paraphrase the philosopher Stephen Asma, author of "The Evolution of Imagination" (2017), humans were graphically literate before we were verbally literate. In fact, emotional communication (that now we normally classify as "art") was probably serving a very simple evolutionary function: to emphasize important facts about the environment to fellow humans and possibly trigger rapid reactions in those fellow humans.

Artists may not like to hear this, but art and music have always been overrated. The difficult part is to decide what is art and what is not, what deserves to be saved for future generations, and to explain why. Art is a meaning generator, but the generator is not quite the artist: the meaning if generated when the art is placed in a context and relations with that context are revealed. Both animals and machines can "make" things. Humans are uniquely equipped to criticize what is being made and to put it in a historical context.

I believe that some day human experts will place machine art in a historical and social context, but it will still be the humans deciding which machine art deserves to be remembered and why and in which context. The real breakthrough would be a machine that can do the same: place machine (and human) art in a historical and social context, judge it, analyze it, criticize it. Upon reading this statement, our beloved software engineer will rush to design a neural network that can say something meaningful about art (whether made by humans or machines). But, of course, by creating such a neural network, you've already let me know who the real critic was: the person who handpicked the dataset to train the neural network, and who crafted the architecture of the neural network, i.e., you. And then i will write a book in which i will mention the historical fact that art was made by machines and that you even created a machine to value the art. What is (still) uniquely human is the ability to write a history of what is being made (and how, why and by whom), not necessarily the ability to "make it". Art tells us a lot about the viewer, and music about the listener, while they tell us very little about the creators.

In 1961 the Italian artist Piero Manzoni displayed 90 tin cans labeled "Artist's Shit" in an art gallery with a price fixed to reflect the fluctuation in gold prices. This is what the Tate Gallery has to say in 2017 about Manzoni's 1961 shit: "Manzoni's critical and metaphorical reification of the artist's body, its processes and products, pointed the way towards an understanding of the persona of the artist and the product of the artist's body as a consumable object. The "Merda d'Artista", the artist's shit, dried naturally and canned 'with no added preservatives', was the perfect metaphor for the bodied and disembodied nature of artistic labour: the work of art as fully incorporated raw material, and its violent expulsion as commodity. Manzoni understood the creative act as part of the cycle of consumption: as a constant reprocessing, packaging, marketing, consuming, reprocessing, packaging, ad infinitum. I am certainly not an art expert, but it is obvious to me that Manzoni's "work of art" has indeed the value of shit (whether made by the artist or by someone else) and it becomes "art" only because the art critic has decided so and has written that elaborate interpretation that suddenly enlightens all of us savages to the exhilarating, life-changing meanings of Manzoni's shit. As game designer Ian Bogost of the Georgia Institute of Technology wrote in 2017: "Before art was culture it was ritual". I am sure that machines, and animals, can make art... but it becomes Art only when it becomes part of a human ritual. Without the human observer, it is not Art just like an electron is neither here nor there until the observer observes it.

In the age of AlphaGo (of machines that simply keep "playing" a game in eternity, producing all possible variations of it), we are tempted to equate quality with quantity. Of course, if one builds a simple computer program that tries all possible combinations of letters, just like in Jorge-Luis Borges' "The Library of Babel" (1941), eventually it will type all the masterpieces of world literature, including all of Shakespeare's tragedies and all of Dostoevsky's novels; and many more novels and poems that humans have never written. And a simple program that paints all possible paintings will eventually paint all Michelangelos and all Monets; and many more paintings that have never been painted by humans and maybe never will be. And a simple program that plays all possible combinations of sounds will eventually play all of Bach and Beethoven, and many more symphonies and sonatas and rock songs that humans have never composed. And a simple program that tries all possible equations in all possible combinations will eventually come up with Newton's Mechanics and Einstein's Relativity. And, by the way, a 3D printer capable of making all possible molecules in all possible combinations will eventually produce organs, limbs, animals, humans and even you.

"Art is what you can get away with" (Andy Warhol)

The Moral Issue: Who's Responsible for a Machine's Action?

During the 2000s, drones and robotic warfare stepped out of science-fiction movies and into reality. According to the Bureau of Investigative Journalism, an independent non-profit organization founded by David and

Elaine Potter in 2010, US drones have killed between 2500 and 4,000 people in at least seven countries (Afghanistan, Pakistan, Syria, Iraq, Yemen, Libya and Somalia). About 1,000 of them were civilians, about 200 were children.

These weapons represent the ultimate example of how machines can relieve us of the sense of guilt. If i accidentally kill three children, i will feel guilty for the rest of my life and perhaps commit suicide. But who feels guilty if the three children are killed by mistake by a drone that was programmed 5.000 kms away by a team using Google maps, Pakistani intelligence inputs and Artificial Intelligence software, a strike authorized by a general or by the president in person? The beauty of delegating tasks to machines is that we decouple the action from the perpetrator. We dilute the responsibility so much that it becomes easier to "pull the trigger" than not to pull it. What if the mistake was due to malfunctioning software? Will the software engineer feel guilty? She may not even learn that there was a "bug" in her piece of software; and, if she does, she may never realize that the bug caused the death of three children.

This process of divorcing the killing from the killer is not new. It started at least in first World War with the first aerial bombings (a practice later immortalized by Pablo Picasso, when it still sounded horrible, in his painting "Guernica") and that happened precisely because humans were using machines (the airplanes) to drop the bombs on invisible citizens instead of throwing grenades or shooting guns against visible enemies. The killer will never know nor see the people he killed.

What applies to warfare applies to everything else. The use of machines to carry out an action basically relieves the machine's designers and its operators of real responsibility for that action.

The same concept can be applied, for example, to surgery: if the operation performed by a machine fails and the patient dies, who is to blame? The team that controlled the machine? The company that built the machine? The doctor who prescribed the use of that specific machine? I suspect that none of these will feel particularly guilty. There will simply be a counter that will mechanically add one to a statistical number of failed procedures. "Oops: you are dead". That will be society's reaction to a terrible incident.

You don't need to think of armed drones to visualize the problem. Think of a fast-food chain. You order at a counter, then you move down the counter to pay at the cash register, and then you hang out by the pick-up area. Eventually some other kid will bring you the food that you ordered. If what you get is not what you ordered, it is natural to complain with the kid who delivered it; but he does not feel guilty (correctly so) and his main concern is to continue his job of serving the other customers who are waiting for their food. In theory, you could go back to the ordering counter, but that would imply either standing in line again or upsetting the people who are in line. You could summon the manager, who was not even present when the incident happened, and blame him for the lousy service. The manager would certainly apologize (it is his job), but even the manager would be unable to pinpoint who is responsible for the mistake (the kid who took the order? the chef? the pen that wasn't writing properly?)

In fact, many businesses and government agencies neatly separate you from the chain of responsibility so that you will not be able to have an argument with a specific person. When something goes wrong and you get upset, each person will reply "I just did my job". You can blame the system in its totality, but in most cases nobody within that system is guilty or gets punished. And, still, you feel that the system let you down, that you are the victim of an unfair treatment.

This manner of decoupling the service from the servers has become so pervasive that younger generations take it for granted that often you won't get what you ordered.

The decoupling of action and responsibility via a machine is becoming pervasive now that ordinary people use machines all the time. Increasingly, people shift responsibility for their failures to the machines that they are using. For example, people who are late for an appointment routinely blame their gadgets. Such as, "The navigator sent me to the wrong address" or "The online maps are confusing" or "My phone's batteries died". In all of these cases the implicit assumption is that you are not responsible, the machine is. The fact that you decided to use a navigator (instead of asking local people for directions) or that you decided to use those online maps (instead of the official government maps) or that you forgot to recharge your phone doesn't seem to matter anymore. It is taken for granted that your life depends on machines that are supposed to do the job for you and, if they don't, it is not your fault.

There are many other ethical issues that are not obvious. Being a writer who is bombarded with copyright issues all the time, here is one favorite. Let us imagine a future in which someone can create an exact replica of any person. The replica is just a machine, although it looks and feels and behaves exactly like the original person. You are a pretty girl and a man is obsessed with you. That man goes online and purchases a replica of you. The replica is delivered by mail. He opens the package, enters an activation code and the replica starts behaving exactly like you would. Nonetheless, the replica is, technically and legally speaking, just a toy. The manufacturer guarantees that this toy has no feelings/emotions, it simply simulates the behavior that your feelings/emotions would cause. Then this man proceeds to abuse that replica of you and later "kills" it. This is a toy bought from a toy store, so it is perfectly legal to do anything the buyer wants to do with it, even to rape it and kill it. I think you get the point: we have laws that protect this very sentence that you are reading from being plagiarized and my statements from being distorted, but no law protects a full replica of us.

Back to our robots capable of critical missions: since they are becoming easier and cheaper, they are likely to be used more and more often to carry out these mission-critical tasks. Easy, cheap and efficient: no moral doubts, no second thoughts, no double crossing. The temptation to use machines instead of humans whenever the ethical boundaries are fuzzy will be too strong to resist.

The program of neural networks is increasingly becoming a program for building a silicon copy of the human brain, which will then pilot a body to perform human-level tasks. Hidden behind this program is the unspoken goal of human nature: deprive other human beings of their rights and make them work for us. The neural network that will achieve full parity with the human brain will, ultimately, be a human being without human rights. We can do anything we want to a machine and to the software that the machine is running, whereas we have laws that limit what we can do to other humans. When we have a machine that is fully equivalent to a human being, we will be able to satisfy our secret desire to use human beings without having to worry about their rights.

I wonder if it is technology that drives the process of de-responsibilization or it is the desire to be relieved of moral responsibility that drives the adoption of new technology. I wonder whether society is aiming for the technology that minimizes our responsibilities rather than aiming for the technology that maximizes our effectiveness. What society should do instead is aim for the technology that maximizes our accountability.

The Dangers of Machine Intelligence: Machine Credibility

The world has indeed changed: these days humans have more faith in machines than in gods.

GPS mapping and navigation software is not completely reliable when you drive on secondary mountain roads. When my hiking group is heading to the mountains, we have to turn on the most popular "navigator" because some of my friends insist on using it even if there is someone in the car who knows the route very well. They will stop the car if the navigation system stops working. And they tend to defend the service even when faced with overwhelming evidence that it took us to the wrong place or via a ridiculous route.

In September 2013 i posted on Facebook that YouTube was returning an ad about (sic) pooping girls when i looked for "Gandhi videos". An incredible number of people wrote back that the ad was based on my search history. I replied that i was not logged into YouTube, Gmail or any other product. A friend (who has been in the software industry all his life) then wrote "It doesn't matter, Google knows". It was pointless to try and explain that if you are not logged in, the software (whether Google, Bing or anything else) does not know who is doing the search (it could be a guest of mine using my computer, or it could be someone who just moved into my house using the same IP address that i used to have). And it was pointless to swear that i had never searched for pooping girls! (For the last week or so i had been doing a research to compile a timeline of modern India). Anyway, the point is not that i was innocent, but that an incredible number of people were adamant that the software knows that i am the one doing that search. People believe that the software knows everything that you do. It reminded me of the Catholic priest in elementary school: "God knows!"

Maybe we're going down the same path. People will believe that software can perform miracles when in fact most software has bugs that make it incredibly stupid.

Maybe we are witnessing what happened in ancient times with the birth of religions. (Next they started burning at the stakes the heretics like me who refused to believe).

The faith that an ordinary user places in a digital gadget wildly exceeds the faith that its very creators place in it.

If i make a mistake just once giving directions, i lose credibility for a long time; if the navigation system makes a mistake, most users will simply assume it was an occasional glitch and will keep trusting it. The tolerance for mistakes seems to be a lot higher when it comes to machines.

People tend to believe machines more than they believe humans, and, surprisingly, seem to trust machine-mediated opinions better than first-hand opinions from an expert. For example, they will trust the opinions expressed on websites like Amazon or Yelp more than trusting the opinion of the world's experts on books and restaurants. They believe their navigation system more than they believe someone who has spent her entire life in the neighborhood.

The evidence (e.g. political elections) show that we are a lot less smart than we think, and we can easily be fooled by humans. When we use a computer, we seem to become even more gullible. Think of how successful "spam" is, or even of how successful the ads posted by your favorite search engine and social media are. If we were smarter, those search engines and social media would rapidly go out of business. They thrive because millions of people click on those links.

The more "intelligent" software becomes, the more likely that people trust it. Unfortunately, at the same time the more "intelligent" it becomes, the more capable of harming people it will get. It doesn't have to be "intentionally" evil: it can just be a software bug, one of the many that software engineers routinely leave behind as they roll out new software releases that most of us never asked for.

Imagine a machine that broadcasts false news, for example that an epidemic is spreading around New York killing people at every corner. No matter what the most reputable reporters write, people will start fleeing New York. Panic would rapidly spread, from city to city, amplified by the very behavior of the millions of panicking citizens (and, presumably, by all the other machines that analyze, process and broadcast the data fed by that one machine).

In June 2016 Baidu published the news that an Indian woman gave birth to 11 twins. Those of you who are old enough will remember this story. It was false the first time it came out in 2011 and it is still false today, but it keeps being repeated on websites throughout the world. The Baidu spider simply scours the web for interesting news and has no way to understand whether the news is correct or not. An investigative report, or for that matter any intelligent being with 20 minutes to spare, can easily find out that the news was fabricated in 2011 (in Zambia, apparently). The scary thing is not that the spiders are dumb enough to believe all sorts of scams; the scary thing is that this becomes news on Baidu, the main source of news in China. Millions of Chinese people are now convinced that (quote) "A Woman Gave Birth to 11 Babies at a Time in India".

Drone strikes seem to enjoy the tacit support of the majority of citizens in the USA. That tacit support arises not only from military calculations (that a drone strike reduces the need to deploy foot soldiers in dangerous places) but also from the belief that drone strikes are accurate and will mainly kill terrorists. However, the drone strikes that the USA routinely hails as having killed terrorists are often reported by local media and eyewitnesses in Pakistan, Afghanistan, Yemen and so on as having killed a lot of harmless civilians, including children. People who believe that machines are intelligent are more likely to support drone strikes. Those who believe that machines are still very dumb are very unlikely to support drone strikes. The latter (including me) believe that the odds of killing innocents are colossal because machines are so dumb and are likely to make awful mistakes (just like the odds that the next release of your favorite operating system has a bug are almost 100%). If everybody were fully aware of how inaccurate these machines are, i doubt that drone programs would exist for much longer.

In other words, i am not so much afraid of machine intelligence as of human gullibility.

The Dangers of Machine Intelligence: Speed Limits for Machines?

In our immediate future i don't see the danger that future machines will be conceptually difficult to understand (superhuman intelligence), but i do see the danger that future machines will be so fast that controlling them will be a major project in itself. We already cannot control a machine that computes millions of times faster than our brain, and this speed will keep increasing in the foreseeable future.

That's not to say that we cannot understand what the machine does: we perfectly understand the algorithm that is being computed. In fact, we wrote it and fed it into the machine. It is computed at a much higher speed than the smartest mathematician could. When that algorithm leads to some automatic action (say, buying stocks on the stock market), the human being is left out of the loop and has to accept the result. When thousands of these algorithms (each perfectly understandable by humans) are run at incredible speed by thousands of machines interacting with each other, humans have to trust the computation. It's the speed that creates the "superhuman" intelligence: not an intelligence that we cannot understand, but an intelligence vastly "inferior" to ours that computes very quickly. The danger is that nobody can make sure that the algorithm was designed correctly, especially when it interacts with a multitude of algorithms.

The only thing that could be so fast is another algorithm. I suspect that this problem will be solved by introducing the equivalent of speed limits: algorithms will be allowed to compute at only a certain speed, and only the "cops" (the algorithms that stop algorithms from causing problems) will be allowed to run faster.

The Dangers of Machine Intelligence: Criminalizing Common Sense

There is something disturbing about the machines that intelligent humans are building with the specific mandate to overcome the individuality of intelligent humans. Stupid machines in charge of making sure that human intelligence does not interfere with rules and regulations are becoming widespread in every aspect of life.

I'll take a simple example because i find it even more telling than the ones that control lives at higher and more sinister levels. I live in the San Francisco Bay Area, one of the most technologically advanced regions in the world. We hold evening events in one of the most prestigious universities of the Bay Area. Because of a world-famous fog, the weather is chilly (if not cold) in the summer, especially in the evening. Nonetheless, the computers have been programmed to provide air conditioning throughout the campus for as long as there are people at work, no matter how cold it is getting outside. People literally bring sweaters and even winter coats at these evening classes. Never mind the total waste of energy; the interesting point is that nobody knows anymore how to tell the machines to stop doing that. After months of trying different offices, we still are "not sure who else to contact about it" (quoting the head of a department in the School of Science) "apparently it is very difficult to reset the building's thermostat".

This is the real danger of having machines run the world. I don't think any of us would call a thermostat "intelligent" when it directs very cold air into a room during a very cold evening. In fact, we view it as utterly stupid. However, it is very difficult for the wildly more intelligent race that created it to control its behavior according to common sense. The reason is that this incredibly stupid machine was created to overcome the common sense with which more intelligent beings are equipped. Think about it and probably your computer-controlled car, some of your computer-controlled appliances and systems around you often prohibit you from performing actions that common sense and ordinary intelligence would demand, even when those cars and appliances work perfectly well, and, in fact, precisely because they work so well.

In 2018 and 2019 two Boeing 737 Max airplanes crashed, killing more than 300 people, because the pilots couldn't find a way to counter an "intelligent" flight-control system that repeatedly pushed the airplane' nose downward.

I am afraid of the millions of machines that will operate within human society with the specific goal of making sure that humans don't use common sense but simply follow the rules; in other words, with the specific goal of making us stupid.

The Dangers of Machine Intelligence: You Are a Budget

Another danger is that what will truly increase exponentially is the current trend to use computing power as a sales tool. The reason that people are willing to accept the terms and conditions of e-commerce websites is that these companies have been very good at concealing what they do with the information that they collect about you. The best minds of Hammerbacher's generation are thinking about how to make people click ads, and they do it by exploiting every tiny data that they can put their hands on. It's not only the best minds but also the best machines. Artificial Intelligence techniques are already being used to gather information on you (what used to be called "espionnage") for the purpose of targeting you with more effective sales strategies.

The original purpose of the World-wide Web was not to create a world in which smart software controls every move you make online and uses it to tailor your online experience; but that is precisely what it risks becoming. Computer science is becoming the discipline of turning your life into somebody else's business opportunity.

The Next Breakthrough: Understanding the Brain

Looking back, one can see that progress in neural networks came almost simultaneously from physics (Hopfield discovered recurrent neural networks studying the annealing process) and from neuroscience (Fukushima discovered convolutions studying the cat's visual system). These two insights led to a rediscovery and adaptation of the old mathematics of optimization and control methods. I suspect that this will continue to be the case in the future. But the physics is mostly old, whereas the neuroscience is new. So most of the progress in A.I. is likely to come from neuroscience.

There is much to be learned from what we already know about the brain. Whenever we discovered one of its tricks, it becomes obvious how to improve the performance of a machine. For example, how do you recognize that the people around you are scared? Paul Whalen's team at University of Wisconsin discovered that our brain can automatically recognize that someone is scared by simply looking at the eye whites: if they are enlarged, most likely the person is scared ("Human Amygdala Responsivity to Masked Fearful Eye Whites", 2004). Our brain doesn't need to scan the whole face and process a lot of information about the situation. Evolution has equipped our brain with a large repertory of tricks to reach quick approximate conclusions of all sorts.

National and international research programs have a bad reputation among people who don't like to pay taxes, but those programs have accounted for immense progress over the last century. Franklin Roosevelt's

Manhattan Project (1941) to build the atomic bomb, John Kennedy's Apollo Program (1963) to send a man to the Moon, Richard Nixon's Arpanet (1969) to create a national network of computers, and George Herbert Bush's Human Genome Project (1990) to decode the human genome have delivered byproducts for countless disciplines (and millions of well-paid jobs for the tax-payers). Barack Obama's BRAIN Initiative (2013), as well as the European Union's Human Brain Project (2013), have the same potential. (Believe it or not, BRAIN stands for Brain Research through Advancing Innovative Neurotechnologies). The BRAIN initiative contained the Machine Intelligence from Cortical Networks (MICrONS) project to reverse-engineer the brain, conceived by Jacob Vogelstein and led by David Cox at Harvard, by Tai Sing Lee at CMU and by Andreas Tolias at the Baylor College of Medicine. The bad news is that, according to the OpenWorm project, we have not even simulated the Caenorhabditis Elegans worm yet. Long way to go.

Understanding the brain, however, may or may not be equivalent to understanding "us", as philosophers never tire of discussing: when we find out everything about my brain, when we design the great neural network and write the great equations that fully describe my brain, will we find out everything about me? Is consciousness just a mathematical formula?

Having been trained as a mathematician in theoretical physics, i find an intriguing parallel with the dilemma faced by physicists. Richard Feynman said "If all mathematics disappeared today, physics would be set back exactly one week". But Eugene Wigner was puzzled by how well mathematics describes reality, i.e. by (quote) "the unreasonable effectiveness of mathematics in the natural sciences."

In the paper "Why Does Deep And Cheap Learning Work So Well?" (2016) Henry Lin, a physicist at Harvard University, and Max Tegmark, a mathematician at MIT, advanced the hypothesis that multilayer neural networks may have something profound in common with the nature of our universe.

Niels Bohr was fond of the principle of complementarity, that there exist dual formulations of reality, such as the the particle and wave aspects of physical systems, and that you have to choose one or the other but are never allowed to mix them (a fact that had just led Werner Heisenberg to his famous principle of indeterminacy). Later in life Bohr expanded the principle of complementarity to deal with philosophical topics: the mind-body problem, vitalism, yin and yang, etc. I suspect that Bohr would have happily accepted as both one and many the brain as purely computational mathematics and the brain as a conscious being.

Is neuroscience the final answer or just another question?

The Dangers of Machine Intelligence: Machine Stupidity

Perhaps the biggest danger of an artificial intelligence is its own stupidity. After the 2016 elections Google's answer to "who won the popular vote" was "Donald Trump", which was false at every stage of the vote counting process. The reason for that error is simple: a right-wing organization had easily found a way to become the top result on Google's search engine when that question was asked. You can easily fool an artificial intelligence, and this will remain true for many, many decades. Any service replaced by an artificial intelligence is (and will be) easily vulnerable to the machinations of ill-intentioned people.

The Dangers of Machine Intelligence: Scientific Socialism

The real danger is not that an artificial intelligence takes over the world, but that Artificial Intelligence scientists take over the world.

Plato, a philosopher, argued that only philosophers were smart enough to rule over kingdoms. Plato envisioned selective mating to produce the best of all leaders. Since then, countless experts have argued in favor of a rule by (unelected) experts (i.e. themselves), and in some cases (the old Soviet Union and today's China) this has been partially implemented (with mixed results). Karl Marx's buddy Friedrich Engels called it "scientific socialism". In 1925 Herbert Croly, a cofounder of the magazine New Republic, was looking forward to "the beneficent activities of expert social engineers." In the 1880s even the idea of improving the human population through breeding became popular, thanks to the writings of the British biologist Francis Galton. This led to the founding in 1907 of the British Eugenics Education Society and to the International Eugenics Conferences, held in 1912 in London. In 1906 the Race Betterment Foundation was created in Michigan, followed in 1911 by the Eugenics Record Office in New York, thanks to the advocacy of the US biologist Charles Davenport. The idea was to create a smarter society, run by smarter people.

In theory, the idea of a society ruled by impartial algorithms sounds really good: no more corruption among bureaucrats, no more lobbyists, no more incompetent presidents elected by dumb and ignorant voters and supported by dishonest media that spread fake news. But lurking behind the project of Artificial Intelligence there may be the secret desire to replace political institutions with omniscient (and omnipotent?) algorithms designed by... Artificial Intelligence experts.

The Dangers of Machine Intelligence: Who will get there first?

Traditionally, most of the investment for A.I. research has come (directly or indirectly) from military agencies such as DARPA whose purpose is to build war machines. Now those big corporations whose business plan is based on advertising are also investing massively in A.I.

Judging from where the money is (or, better, comes from), one is tempted to conclude that the first general-purpose A.I., the first A.G.I., will emerge from either the warmongering military-industrial complex or from the greedy corporate world. We have to face the realistic scenario that the Singularity is more likely to arise as the descendant of a military machine or of an advertising algorithm than as the evolution of humanitarian software.

If it looks like a duck, swims like a duck, and quacks like a duck, it is probably an advertisement for ducks.

The Dangers of Machine Intelligence: Amplifying Human Evil (or A.I. in The Economy of Hate)

Harvard economist Edward Glaeser wrote that "hatred is the result of an equilibrium where politicians supply stories of past atrocities in order to discredit the opposition and consumers listen to them" ("The Political Economy of Hatred", 2002). The real problem, however, is not the viciousness of the politicians: it is the willingness of the public to listen to stories of hatred. Increasingly, entrepreneurs are realizing that hatred "sells". The Economy of Hate is the entertainment industry that has mushroomed around the need for hatred in the age of mass media. Television and Internet platforms have amplified the phenomenon because they rely on advertising revenues and advertising revenues depend on number of viewers, and it turns out that hatred is a simple and effective generator of viewers; the more incendiary the better. Social-networking platforms (or, at least, their algorithms) encourage this economy of hate. They have a vested interest that you hate, and that you are hated back. It is nothing new. Jerry Springer made a fortune and became an icon of US television with a talk-show, debuted in 1991, in which his guests were insulting each other and even fist-fighting on stage. Nevermind that it was voted "worst show of all time" in a TV Guide viewers poll of 2002: it remained a hit throughout the 2000s and survived in the 2010s. When it declined in the mid 2010s, it was promptly replaced by a president of the country who transferred precisely Springer's hateful tone and tactics into politics. While theoretically more upscale, the contemporary Montel

Williams Show was not much better, with guests designed to be despised by the audience such as a teenage girl who boasted of having already had more than 100 sexual partners and a serial rapist of prostitutes committed to spreading AIDS in that profession. Their ratings depended on how vulgar and violent their guests were, and the rating translated into money for their TV channel, which therefore de facto encouraged the most vulgar and violent exchanges. Millions of people have been watching the matches of the World Wrestling Entertainment, the largest wrestling promotion in the world, founded in 1979 in Massachusetts as Titan Sports by Vince McMahon; except that its matches are fake, staged and scripted. The viewers cannot possibly be attracted by the sporting events themselves: they are attracted by the ritual around these matches, that consist in hate-filled insults by each wrestler against the other. Television simply broadcast hate to a wider audience, but the passion for displays of hate was already there. To this day the cruel entertainment of "la corrida" (bull-fighting) is legal in Spain, and "sabong" (cock-fighting) remains a top entertainment in the Philippines (the five-day World Slasher Cup is held in the Araneta Coliseum of Manila). Boxing has been a favorite Western sport since at least the 18th century. Throughout history, public executions have always attracted large crowds of cheering viewers.

Now that public executions are no longer public spectacles, the public can watch "virtual" executions on social media or television in which the victims are insulted and defamed in the most atrocious manner. The Communications Decency Act of 1996 holds that content platforms (such as a social-networking platform) are not responsible for the speech they host. It doesn't say that the platforms should make money out of hate speech but it turns out that is precisely what it allows them to do. Defamation laws protect free speech more than your reputation. Alas, your reputation can easily be monetized by anyone attacking it. For example, in 2017 a user of the YouNow platform, Charles Marlowe, launched a vicious campaign against another user, Anna Scanlon. Scanlon claims that Marlowe lied and defamed her, but YouNow's algorithms are not designed to decide what is true and what is false: they are designed to maximize YouNow's revenues. The more Marlowe insulted her, the larger his following became, and therefore the more profitable his account... for both him and for YouNow. YouNow's algorithm rewarded Marlowe for his growing number of followers. Most likely, a "recommendation algorithm" also helped spread the controversy to other users of the social network. Using sexist, racist and homophobic language is a plus. Fabricating conspiracy theories is another plus. The platform that financially rewards traffic as a generator of advertising money automatically rewards the hate-mongers. The problem is not bullying and harassment, but public bullying and harassment. Shifting to another field, imagine a political candidate who dodged the draft during the Vietnam War and still calls out a Vietnam war hero as a coward because this hero was taken prisoner while this political candidate was mainly busy having sex without wearing condoms (his own words): would you imagine that this political candidate becomes the president of the USA? And imagine if this same candidate had previously invented a bogus conspiracy theory that Barack Obama was not born in the USA. Still: 63 million voters chose this very hatemonger. Given that A.I.'s most successful application is still that it calculates how to make Internet viewers click on links in order to maximize advertising profits, we'll soon have A.I. programs that help create as much "hate" as possible. (Changing defamation laws might help: in Britain, where defamation laws are not so protective of free speech, blogger Jack Monroe won a libel suit against defaming journalist Katie Hopkins, coincidentally a former contestant in the TV show Apprentice, at the time run by abovesaid future president).

The fact however remains that "hatred" sells and any algorithm designed for profit will inevitably tend to increase the amount of hatred if that is what maximizes profits. The more intelligent the algorithm, the more hatred it will create.

Intermezzo: A Remedy to the Echo Chamber

Humans cannot be trusted with their most important decisions. While studying the psychology of judgment, the Israeli-born psychologists Daniel Kahneman at Princeton and Amos Tversky at Stanford demonstrated that we pretend (and convince ourselves) to be objective and rational when in fact we are not ("Subjective Probability", 1972). Emily Pronin introduced the expression "bias blind spot" to describe the fact that we excel at noticing the flaws of others, not our own ("The Bias Blind Spot", 2002). In a nutshell: i am too stupid to understand how stupid i am. If you think that you are too intelligent to fall into this category, think again: a study by Richard West of James Madison University and Keith Stanovich of University of Toronto showed that you are precisely the problem ("Cognitive Sophistication does not Attenuate the Bias Blind Spot", 2012) . In fact, the smarter you are, the more vulnerable you are to some silly mistakes. And there is nothing worse than someone who is not smart but thinks that he is smart: unfortunately, David Dunning and Justin Kruger of Cornell University discovered that this is a very common phenomenon ("Unskilled and Unaware of It", 1999). We assume to be "competent enough" for sound judgement when we are actually "incompetent enough" to infer terrible judgement. It is precisely my incompetence that makes me think that i am competent.

Possibly the most human of tendencies is the "confirmation bias" (as Peter Wason called it in 1960), already noticed 500 years ago by Francis Bacon in his "Novum Organum": we tend to pick the information that confirms what we already believe. We want to live inside an "echo chamber". That's precisely the opposite of what Karl Popper asked science to do in 1959: falsifiability is the criterion that distinguishes science from mere chatter. It turns out that we don't really like to be proven wrong. We tend to like people who like us. Subconsciously, i apply this simple rule to every person: the faster you realize that i am smarter than you, the smarter you must be; i rate your I.Q. as inversely proportional to how long it takes you to fully agree with me, i.e. to recognize that i am more intelligent than you.

Social media have compounded this tendency. As of 2017, Google filters the search results based on your previous clicks, and Facebook customizes your feeds based on your past behavior (what you clicked on and what you liked), so you mostly consume news and opinions that confirm your opinions, and you mostly find what you already found before. Farhad Manjoo, in her book "True Enough" (2008), argued that the new digital media are restructuring society into a set of isolated tribes whose members consume only that information which confirms their beliefs. The Internet is not the "global village" that Marshall McLuhan foresaw in "The Gutenberg Galaxy" (1962) but a federation of tribes that are drifting further apart each day because digital media makes us live inside a rapidly expanding echo chamber. Our echo chamber is the biggest it has ever been.

This long preamble is meant to show you that you are not to be trusted with any decision: you are biased, you always were, and now you are even more biased because you live inside the biggest echo chamber ever.

In 1933 the British philosopher Bertrand Russell wrote an essay titled "The Triumph of Stupidity" after the Nazis under Adolf Hitler won the German elections. Had only A.I. bots been allowed to vote in the 1933 elections, would Hitler have won? Had only A.I. bots voted in the 2016 election, would a vulgar real-estate tycoon with a sordid past and dubious links to Russia have won? Any person is biased, no matter how much they have been trained not to be. An A.I. bot designed not to be biased is not biased. For real: it is not racist, it is not sexist, it doesn't live inside an echo chamber. Artificial Intelligence has the potential to build a more fair society.

Hyper-employment

In 2017 the media were full of stories about how machines will automate most of today's jobs. Many conclude that machines will cause massive unemployment. I suspect that the truth is closer to the exact opposite conclusion: that technology will lead to hyper-employment (no unemployment and more than one job per person), and hyper-employment will cause a global catastrophe similar to and possibly including hyperinflation, with consequences that we can't imagine.

Many of the jobs that existed in 1900 had been automated by 1960. Today most of us have jobs that didn't exist in 1900. Therefore the fact that tasks will be automated by machines is nothing new. It would be a scary sign of technological decline if it didn't happen. Despite all the jobs that were automated between 1900 and my birth, i wasn't born into a world of massive unemployment. On the contrary, i was born in a world that offered much better and better paid jobs than the jobs that had been available to my grandfather.

So the real question is whether in the near future the percentage of jobs that will be automated will be roughly the same as the percentage of jobs that were automated in the past or less or more; and whether the new jobs created by this future automation will match the new jobs created by automation in the past. For example, the world will need a lot more engineers to build, program and maintain millions of software and hardware robots. A 2016 study by Robert Atkinson and John Wu, "False Alarmism: Technological Disruption and the U.S Labor Market, 1850-2015" doesn't show any major deviation from the past... so far.

I certainly wish that jobs like those of plumbers or electricians, etc were automated. Alas, i suspect that the world will need a lot more of them, not fewer, in the next 25 years.

Personally, i think that, as it is often the case, the media tend to look in the wrong direction. Machines that replace human jobs are nothing new. Machines to which you can outsource part of a job are, instead, something really new. These machines allow you to outsource not just one job but many, as many as you can and desire to, manage. I suspect that in the near future many people will have two or three or 25 or 2,000 jobs, thanks to tools that will allow us to massively multitask. In which case, unemployment will de facto become negative: more than one job per person, even for the very old and the very young. A 90-year-old man will be able to carry out the activities of a 20-year-old and in fact the activities of many young men. Energy and health will not be obstacles anymore. A 5-year-old will be able to generate money by using bots to make and market something.

My prediction is therefore exactly the opposite of the most popular "doom and gloom" predictions: the global threat to social stability of the 21st century will be hyper-employment.

We may already be living in that age. What has truly accelerated is not the intelligence of machines but our ability to produce and consume. I met a Lyft driver who was a stock trader when not driving around, an online tutor at night, a realtor during the weekend and a full-time landlord: technology allows him to work four jobs that would normally be done by four different people. Most of us work way more than 8 hours/day. We see studies about employment all the time, but those count only traditional work, the kind of work that will disappear. I'd like to see a study about the number of hours worked per capita. I suspect that number is going up. Automation is creating so much work that each person can work a lot more than in the past, potentially 16 hours a day, potentially 24 hours a day: you'll have a swarm of bots working for you even when you sleep. We forgot that, in medieval times, people worked only when they wanted, and some worked only six months a year. In the last century it was common in Italy to have a lengthy breakfast, a 3-hour lunch break and play bocce or cards in the evening before dinner. Juliet Schor in "The Overworked American" (1992), using a variety of historical studies, estimated that in the 13th century an adult male peasant in Britain worked 1620 hours compared with 1980 hours worked between 1400 and 1600 by an adult male farmer or miner and compared with 1949 hours worked in 1987 by the average worker in the USA. The Bureau of Labor Statistics report of 2000 showed that more than 25 million Americans (20.5% of the total workforce) worked at least 49 hours a week. OECD maintains statistics about the average annual hours actually worked per worker that show a 2% decline between 2000 and 2015 in the USA (from 1836 hours to 1790), but these statistics made in the age of the "net economy" can be misleading. They don't count all the independent additional income made by many of us on the Internet, and certainly don't count all the work done for free by the majority of Internet users when they supply content to social media. Ian Bogost in the Atlantic magazine argued that we're already hyperemployed, except that we mostly work for free, providing Internet companies the content that they monetize (for themselves). Until now we were working multiple jobs without even knowing that we were doing so, but we will soon become more and more aware that our leisure activities are sold as commercial goods by some corporation.

In 1930 the economist John Maynard Keynes wrote the essay "Economic Possibilities for our Grandchildren" in which he argued that future generations would soon be able to replace work with leisure thanks to widespread wealth and surplus. "I predict that both of the two opposed errors of pessimism which now make so much noise in the world will be proved wrong in our own time-the pessimism of the revolutionaries who think that things are so bad that nothing can save us but violent change, and the pessimism of the reactionaries who consider the balance of our economic and social life so precarious that we must risk no experiments." He was wrong on human psychology though: people's preferred form of entertainment is not poetry, nor painting, nor music, but making money. We, turned into hyper-employed workers by swarms of bots, will use our "leisure time" to make more and more money, selling our products to people who are making more and more money selling their things to people who are making more and more money selling their...

Economic booms tend to follow an increase in the workforce. The Golden Century followed the introduction of the reading glasses that enabled people to keep working even when eyesight declined. The post-war economic boom of the USA followed the mass entry of married women into the workforce. And viceversa: the rapidly shrinking working-age population of Japan and Italy has caused their economies to stall. The new technologies (call them "intelligent" or "automation") will enable older people to stay fully active. Today an elderly person who wants to work is unlikely to find a firm willing to hire her, but the new technologies will allow elderly people to carry out valuable (and lucrative) services. Technologies such as social media and search engines, invented by young people for young people, are more likely to benefit the elderly. Who benefits most from conversational user interfaces? For young people it is just a novelty, but for people with arthritis it opens new horizons. Who benefits more from the tools that allow you to run your business from home? Those who cannot drive anymore. Who can benefit most from wearables? Those who should check their health all the time. The smart home, the smart city and the on-demand economy will largely erase the advantages that younger people have over older people, starting with continuing education. The technological revolution of the Silicon Valleys of the world has so far been driven by young people like Steve Jobs and Mark Zuckerberg who could see what young people wanted. It is just a matter of time before we get an 80-year-old equivalent of Steve Jobs.

"I hope I die before I get old" (The Who, 1965)

Jobs - Utopia and Dystopia

As more and more jobs get automated, one can also envision a society in which not everybody has to work. The modern origins of this idea the "universal basic income", are in Britain, which actually didn't have much of a tradition in anarcho-socialist ideologies (way more prominent in France and Germany). Bertrand Russell had just finished publishing the "Principia Mathematica" when he wrote "Roads to Freedom" (1918) in which he proposed the universal basic income, and in the same year the British politician Dennis Milner and his wife Mabel published "Scheme for a State Bonus" (1918). This idea became a movement thanks to the amateur economist Clifford Davis' book "Social Credit" (1924). Davis was inspired by the same premise that inspires today's singularists: machines were automating so much of the production of goods that it wasn't really necessary to have every person add more work to the system, while it was still necessary to have every person consume those goods. This thread was picked up a few decades later in the USA by economists such as Milton Friedman in his book "Capitalism and Freedom" (1962), the book that introduced the concept of a "negative tax". Towards the end of the century the discussion moved to Europe again, where in 1986 the Basic Income European Network (BIEN) was established. In 2017 Finland became the first country to actually do it: the country launched a two-year trial of basic income for 2,000 unemployed people. As the concept spread to Silicon Valley entrepreneurs, the USA regained the leadership. In 2018 a joint team from MIT, Princeton and UC San Diego started providing universal basic income to 5,000 people in Kenya (via the organization Give Directly) within an experimental 12-year program. This is a scientific experiment to determine what humans do if they don't have to work.

Philippe Van Parijs and Yannick Vanderborght made a powerful argument in favor of universal basic income in their book "Basic Income" (2017). It would simplify the world, rectify some unjustice and, lo and behold, save money (no more need for welfare programs). The society of (software and hardware) robots may afford a chance to get rid of the idea that people without a job must starve or beg. Maybe people without a job can still lead a decent life and contribute what they can to society, instead of being forced to understand what society wants them to contribute even when they have neither the skills nor the education to fulfill those functions.

At the same time, a society of millions of humans with spare time is actually a scary thought. Many individuals don't do much with their spare time: watch movies, chat with friends, mostly sleep. But others spend their spare time building narratives, and most of them are not competent at all: the human brain is programmed to build narratives, whether true or fictional. Idle brains tend to reach the wrong conclusion on the wrong subject at the wrong time. Humans also have a very violent history (with a visceral passion for torture and genocide), and work is one of the activities that keep humans from devoting too much of their time to violence. I am not sure that we should feel comfortable with idle brains any more than we are with psychos.

"Remember that sometimes not getting what you want is a wonderful stroke of luck" (Tenzin Gyatso, 14th Dalai Lama).

Post-scriptum on Jobs - A.I. Brain Drain

As A.I. became more and more popular and corporations started hiring just about anybody whose PhD dissertation was remotely related to A.I., the other fear that began to mount was that this "brain drain" of university scientists would result in a slow-down in pure research. Google hired Sebastian Thrun, Geoff Hinton and most of his team, Ian Goodfellow, Oriol Vinyals, Christian Szegedy, Andrew Ng and Fei-fei Li, and acquired DeepMind which in turn hired a lot of talent like Daan Wierstra. Salesforce hired Richard Socher. Facebook hired Yann LeCun, Ross Girshick, Kaiming He, Tomas Mikolov, Rob Fergus, Jason Weston and Ronan Collobert. OpenAI hired Ilya Sutskever. Megvii hired Jian Sun, Tencent hired Dong Yu. In 2016, just when a presidential candidate was getting elected thanks to the blue-collar voters of the Midwest who were losing their manufacturing jobs, salaries for A.I. specialists were skyrocketing in Silicon Valley, attracting more and more graduates who would have normally stayed in academia most of their lives. The website Venture Scanner counted 2,000 A.I. startups at the end of 2017, many of them founded by promising talent in A.I.

Many viewed the scale of the exodus from the universities as unprecedented. Silicon Valley, however, should have known better. Silicon Valley got that nickname because William Shockley, one of the three inventors of the transistor, moved to Palo Alto and started hiring the best graduates that he could get. Out of his lab came Fairchild and out of Fairchild came dozens of "silicon" startups. That's how the legend of Silicon Valley came to be what it is. Given the spectacular success that those firms had in revolutionizing the electronic field, it is difficult to argue that Shockley and his descendants caused harm to research by "stealing" the best brains from universities. It is difficult to imagine that those same brains, left in their universities, would have achieved the same kind of technological revolution.

One can in fact conclude the opposite: without the transfusion of brains into the Silicon Valleys of the world we would know less (not more) about electronic engineering, and we probably wouldn't have the smartphone and all those other convenient gadgets.

"Silicon Valley is not so much a place as a state of mind" (an often repeated adage in Silicon Valley).

Dystopian Quartet 1. Andante: Lessons Learned from A.I.

The A.I. scientists of the 1950s have been mocked for making unreasonable predictions about the intelligence of computers. It turns out that they were mostly right. Those scientists lacked machines powerful enough to implement the theory, but, decades later, the theory can be implemented on powerful GPUs (especially when munificent benefactors, like Google, volunteer to buy thousands of them) and it works. Neural networks made of these powerful GPUs can indeed simulate a lot of "intelligent" tasks.

Therefore A.I. scientists have proven to neuroscientists that a lot of our "intelligence" is simply computational math, algorithms, formulas, very intricate systems of equations.

However, some tasks are too intelligent for us and too intelligent for the machine as well. For example, the original app: weather forecast. We still can't accurately predict the weather (except in sunny Silicon Valley). Weather forecast involves too many factors, and we don't have the dataset of past behavior that could help "recognize" the pattern for the future.

There is a second limitation to machine intelligence, and sometimes to human intelligence: lack of common sense. Common sense requires an almost infinite amount of knowledge about ordinary things. It turns out that the ordinary is much more difficult to map than the extraordinary. It is not difficult to describe mathematically how a master plays chess or weiqi/go. It is very difficult to describe mathematically how a waiter cleans a table at the restaurant.

Common sense is hard to replicate with algorithms. It has proven much easier to remove the need for common sense. To a large extent, the history of human civilization is the history of removing the need for common sense. We don't need to run away from tigers anymore and we don't even need common sense to cross a very busy street. We have structured the environment so that life can be easy, safe, and predictable. In the "developed" world, there are rules and regulations for just about everything. We train children from a very early age to abandon common sense and follow rules. We strive to remove every possible glitch from our factories, subways, malls, and streets. Removing the need for common sense enables even the dumbest people (like me) to easily survive. I can write, teach, travel and enjoy my hobbies regardless of whether today my mind is sharp or not: there is a vast system of rules and regulations that takes care of me. We turned human behavior and the human environment into a well-organized, highly predictable, dumb, machine-friendly system.

No wonder that machines can easily coexist with humans. The machines don't have to be very intelligent to deal with people because people have built an environment that doesn't require the one thing that machines don't have: common sense. And no wonder that we can easily replace humans with machines: humans have been trained all their life to behave like machines. So the issue now, as far as business is concerned, is not whether a machine can do the same job (it can, given a well-structured environment) but which one is overall cheaper to "operate".

We transferred "intelligence" from the human brain to the environment so that humans don't need intelligence anymore to survive in the environment; but that also means that machines can now do what only humans were capable of doing when the environment was chaotic and surviving in it required intelligence.

Dystopian Quartet 2. Allegro Forte: We are Surrounded

One day i walked into the subway station of a Chinese city and looked at the wall. It has machines that sell tickets, machines that sell drinks, machines that give you cash, machines that take passport-size photos of you and even machines that sell toys. I bought my ticket and then put my bag into a machine that detects metal and then inserted my ticket into a machine that validates the ticket. I walked into an escalator that took me to my train. And, of course, the train is a machine (controlled by machines).

Traffic control is entirely automated in most cities of the world. There are machines called "traffic lights" that decide which cars have to move and which cars have to stop. They are often connected to sensors or cameras: based on how many cars are waiting in each direction of the intersection, the traffic light decides which human drivers can go through the intersection. They have replaced thousands of traffic cops. We have already put behind us all the jobs that have disappeared. Think of the workers of an average sawmill one century ago: machinists, saw-setter, electricians, stokers, greasers, sawyers, edgers, sorters, stackers, ... What happened to them?

We are already surrounded by machines. These machines perform a lot of jobs that used to be done by people, but today most people have better-paid jobs building or selling or maintaining these machines. The Chinese are wealthier today than when they didn't have machines.

(On top of being surrounded by machines, machines are also beginning to populate and colonize our body: a number of people already live in symbiosis with a machine that has been implanted or is attached to their body, whether it be an artificial heart or a hearing aid).

We do not worry that these boxed machines may someday conquer the world. We, instead, react emotionally when the machine looks like a human being and walks. This has more to do with psychology than with anything else. The machines that walk and look like humans are actually the least useful. Most of them are tourist attractions: they serve to attract people into the store. Buy a Pepper robot and the whole neighborhood will come to take a picture with the robot. These walking humanoids are far less useful than the immobile machines of the subway station.

Dystopian Quartet 3. A Cappella: We are being Programmed

The problem is not that we are surrounded by machines. The problem is that we are asked to behave like (very stupid) machines in order to make these machines useful. My recent flight from Beijing to San Francisco is typical. From the moment i entered the Beijing airport to the moment i exited the San Francisco airport i was behaving like a machine. I had to stand in one line after the other, and there were precise rules to follow at each line. In Beijing i even had to stand in a specific position so that the security guard's camera could take a good picture of me. In San Francisco i used an incredibly stupid terminal to get my passport checked. The whole experience requires that you abandon any notion of being human. This is the future, which is increasingly the present: we have to follow rules and regulations so that we can be surrounded by machines.

We are being programmed as much as the machines that surround us.

There is a moment at the airport when you don't have to follow rules and regulations: during the long journey from the security checkpoint to your gate. Before you reach your gate (no matter how late you are) they want you to shop at countless shops. Why? The money that you spend on your way to the gate helps them to pay for all those machines. That's the world we live in. Increasingly, we are asked to behave like machines (i.e. like stupid humans) when we interact with machines, so that machines can perform some useful task. We are given freedom to spend our money. That money is actually used to fund machines. Anything you do in your spare time has a cost, and that cost, ultimately, is money for machines. When you make a phone call to a friend, you are actually paying to support a chain of machines extending from your phone to your friend's phone via countless communication machines. When you watch a movie, you are paying to support machines that create, store, organize, and deliver entertainment. And so forth. You have no freedom when you interact with machines: you must obey precise rules and regulations just as if you were a machine. And during the few moments when you have freedom, you are actually spending money to pay for those machines. If your only purpose in society is to pay for the machines that surround you and to behave like a machine when you interact with those machines, the question "What is the meaning of life?" acquires a whole new dimension. Isn't it easier to replace you with machines even in your "free" moments? Why can't we just have machines building and controlling machines, and just remove you and your money from the loop? Are we slowly but steadily making humans redundant? What is the real meaning of this increasingly automated consumer society? Consuming so that we can pay for automation that helps us consume? Is the ultimate goal the "consuming" or the "automating"? What is left at the end? So far it looks like machines are expanding and humanity is retreating. Extrapolate, and you can foresee a future in which humans will be redundant, or, worse, just an annoyance.

Note that this is due to the stupidity (not intelligence) of machines. If machines were intelligent, we wouldn't need to lower our intelligence, to behave like machines. We have to create a highly structured and regulated society (i.e. to enforce machine-like behavior on humans) because otherwise machines would not be able to deal with us humans. The process of turning us into machines is being triggered not by the intelligence of machines but by their utter and probably incurable stupidity.

(The astute reader has already been asking "Who are they, the people who want to turn us into machines?" Of course, it's us. We are de facto committing collective suicide on a massive scale).

Dystopian Quartet 4. Adagio: Humanity without Humanity

We live in a partially automated world: we travel using machines (cars, buses, trains, planes), kitchen appliances do most of the household chores, machines such as television sets and personal computers provide our entertainment.

Our interaction with other humans is increasingly limited because machines perform many of the functions that used to be performed by humans. Who gives you cash at the bank? An automatic teller machine. Who hands you the ticket at the parking garage? A machine.

We tend to look at the machines that replace humans purely in economic terms: the service is now available 24/7 and it is cheap or even free; a job is lost; we can create more jobs elsewhere because we saved money here; etc. But there is a more important story behind the multiplication of machines: if the people around me are replaced by machines, it means that i will interact less with humans. Every time a human is replaced by a machine, it decreases the interaction that i will have with other humans. We talk a lot about human-machine interaction, and tend to ignore the fact that a consequence of human-machine interaction is the decline of human-human interaction.

This trend has been going on for at least a century. There used to be armies of telephone operators to direct phone calls, there used to be armies of secretaries typing documents, there used to be armies of sales people serving the customers, etc. Today these human mediators have disappeared and we are increasingly alone in a world of machines.

This trend will continue into the age of Artificial Intelligence to the point that many individuals, especially the older ones, will only interact with machines. Machines will take care of our house, of our errands, of our health, of our entertainment. This will dramatically reduce our need to interact with other human beings; even with our own family, as family support will become less and less necessary.

Your co-workers will be robots. Your friends will be robots. Maybe your lovers will be robots. Your last friend, who will see you die, will be a robot. In many hospitals around the world the last one to take care of a patient on her dying bed is already a machine.

What happens to humanity when you don't interact with humans anymore?

You Are an Algorithm: The Vast Algorithmic Bureaucracy

The developed world is rapidly evolving towards a dystopia in which what is not mandatory is forbidden. (The Soviet Union was the future, not the past). There are rules and regulations for everything, and they are getting increasingly precise and pervasive. There are rules for driving, for paying at a supermarket, for sitting in a restaurant, for applying for a job, for applying for health care, for applying for retirement, for applying for citizenship, and even for hiking in the wilderness. If there is no rule for something, most likely it means that it is illegal. What other reason could there be for some actions not to be regulated by a bureaucracy?

It used to be that prohibitions prevailed: "Do not touch"; "Do not litter"; "Do not trespass"; etc. However, in vast algorithmic bureaucracies the balance will tilt towards what is mandatory. Algorithms will prescribe what you must do, and what you cannot do will become a consequence of what you must do. When you fill a form online, you often have no choice but to enter the data that the algorithm requires. Nobody says that you cannot enter different data, but there is just no way to do so. In fact, the vast algorithmic bureaucracy does more than define what is mandatory: it makes it the only option. You cannot break the law when you are inside an algorithm (just like a part on a conveyor belt cannot escape the assembly line). Other options are not forbidden: they are just not possible.

When someone working in a bureaucracy tells you "Don't worry", it means that you have entered an algorithm: you have become just a string of data that is being fed to an algorithm, and you shouldn't worry for the simple reason that now you don't have any choice anymore.

An increasing number of activities are turned into algorithms that humans must follow strictly, whether it's preparing a sandwich in a fast food joint or registering a patient in a clinic. On most flights the flight attendants give passengers a little bag of peanuts and a paper napkin: tell them that you don't want the peanuts, and they won't give you the paper napkin.

What terrifies me is not a future in which machines rule people, but a future in which vast heartless bureaucracies rule the world. They already rule airport security, insurance claims and even the emergency rooms of hospitals and the educational system. These are places in which humanity and compassion are increasingly replaced by a mechanical procedure, and it makes no difference whether the mechanical procedure is carried out by humans or machines. You often hear people excuse themselves with the expression "I am just doing my job" which really means "I work for a vast heartless bureaucracy and they pay me to enforce this rule and we don't care about your particular situation. The only thing that matters to us are the rules and regulations that define and protect this bureaucracy". Their job has little to do with your needs. They don't have the power to care for your needs, and they quickly lose any motivation to care.

The expression "I am sorry" is losing its original meaning: when you hear "I am sorry" from someone who is denying you a service, you know that in reality nobody is sorry. It is just a rule that they have to say "I am sorry" when they say "no". You probably got many letters that started with a "We are sorry" denying you something that you wanted. Either the message was sent by a machine with no human intervention or it was sent by a human who pressed a key on a keyboard. Nobody in that organization is sorry, and in fact probably nobody even knows what happened except the one who pressed a button. Not even my friends are sorry, because they know how this works: we're all in the same boat.

In fact, it may soon be that we will prefer the care of a machine, that hopefully has been programmed to show some civility and compassion, rather than the care of a security agent, of an insurance adjustor or of a hospital receptionist, people who are losing any motivation to treat us with civility and increasingly view us like annoyances if we don't comply with the rules and regulations that they are paid to enforce. As we turn human activities into mechanical procedures, we are making it easier for machines not only to perform them more accurately but even to perform them more humanely. It is hard for a human being to smile all day long when simply repeating the same procedure over and over again. It is easy to program a robot to smile all day long while doing the exact same thing. As more and more of our activities are turned into a set of rules and regulations devised and enforced by these vast heartless bureaucracies, we may actually turn to the robot for hope. Alone, helpless and adrift into the intricate network of vast heartless bureaucracies of the future, our only hope will be that robots will mediate for us, that robots will be programmed wisely to treat us like human beings, not objects, and thus compensate for the fact that those vast heartless bureaucracies treat us like objects. Robots will mitigate for us the arrogance (and sometimes the incompetence) of the bureaucrats.

Don't blame it on the machine: it is humans who have invented the vast algorithmic bureaucracies and who keep making them bigger and more algorithmic. It is equally unfair to blame governments and big corporations for the growth and expansion of the vast algorithmic bureaucracies: we want the algorithms. When there is no algorithm, the "user experience" is often annoying and upsetting. For example, when you land in a US airport, immigration officers are entitled to ask you questions about your trip. You have nothing to hide, but having nothing to hide is never a reason to be relaxed when you are confronted by someone wearing a uniform in a place where it is forbidden to make recordings and with no witnesses next to you. The unnerving factor is the unpredictability of what the officer may ask us. If we knew the questions, we would have the answers ready. Instead, each time it's a lottery. The truth is that we, the citizens, normally prefer to deal with algorithms, because they tend to be predictable; and therefore we, the citizens, contribute to the establishment of algorithmic bureaucracies.

We have created highly structured societies in which what is not mandatory is increasingly forbidden, societies in which behavior is determined by a body of rules and regulations. The original motivation was probably to make sure that even idiots can survive: all these rules and regulations minimize the intelligence required to survive and to thrive. All we have to do is to follow the rules and regulations. We have de facto turned persons into machines. By doing so, we have created societies in which machines can perform as well as humans; not because machines have become more like us, but because we have become a lot more like them. Automation in general, and Artificial Intelligence in particular, has little to do with "intelligence" and a lot to do with these societies of rules and regulations in which humans are forced to behave like machines and therefore machines can easily behave like humans. If all you long for is a life made of safe and unconscious routines, your dream is coming true. If, on the other hand, you like adventure, uncertainty and challenges, what we are building is hell on Earth. But some do ask: why would anyone want adventure, uncertainty and challenges? Why not behave like a machine and be treated like a machine?

A "smart city" is a city where everything has been turned into an efficient algorithm connected to all other algorithms. Smart cities are about maximizing efficiency. But what exactly gets optimized? Efficient for what? Cities are not just buildings, streets and cars. There are also people. The term "smart city" makes you think of an intelligent city working for its citizens when, in fact, a "smart city" could become a high-tech concentration camp where citizens are treated like numbers.

During the economic boom of the 1960s the Japanese government built giant tenements called "danchi". There are strict rules on what you can do and not do in these tenements. In 2000 the authorities discovered the skeleton (not even corpse) of a tenant who had died three years earlier. It took three years because for three years the dead man had paid on time all his bills: the amounts were automatically withdrawn from his bank account. The pension was automatically deposited there, and his bills were paid automatically from there. Algorithms were working perfectly fine even if he was long dead. Nobody noticed that he was missing until his bank account ran out of money. Then the government sent someone to make him pay and that's when his skeleton was discovered. (A moving book has been published by a humble woman who spent most of her life in one of these tenements, "Chieko's 53 Years in Tokiwadaira Danchi").

When Wystan Auden wrote his lengthy poem "The Age of Anxiety" (1947), he was sympathizing with the plight of humanity in an increasingly mechanized society. The next step in that process of "anxiety" is the plight of humans when they have to deal with these vast heartless bureaucracies that surreptitiously remove the human/humane element and replace it with algorithms. You can feel that anxiety, in particular, when you try to do something that is not included in the algorithm. If your case is an "exception", the bureaucracy still tries to fit you into its algorithm, and most likely this will cause you anger, frustration and anxiety.

One day something was wrong with my car and i took it to the mechanic. The first thing he asked me was to fill a form. A few days later a friend broke her wrist while skateboarding and i took her to the hospital. She was obviously in pain, and obviously couldn't write. Nevertheless, the first thing they asked her at the hospital was to fill a form. I filled the form for her and noticed that this form to take care of a wounded human being was quite similar to the form for repairing a malfunctioning car.

In the age of the Internet, many aspiring writers want to get their writing published on popular "platforms" such as Medium. The most important step is to read the "submission guidelines". You can't just write an article and hope that someone will publish it because it's a good article. The content is largely irrelevant. The platform mostly cares about the formatting of your text. You need to follow strict guidelines about how to format your article. Then you can submit it. Virtually nobody will read it to check how good it is. At best, a "bot" will check that your text doesn't contain banned words. Send them a badly formatted text that is the most creative article ever written and it won't get published. Send them an incredibly trivial text that is perfectly formatted and it will get published. It is not about the content, it is about the algorithm.

We scientists spend an incredible amount of time filling forms and drawing spreadsheets just to justify how we spend the money for doing important research. Whenever we have an idea for a great conference, the first concern is about who is going to do all the paperwork and "red tape". The vast algorithmic bureaucracy of the university will not allow you to realize your project if first you don't find out the exact procedure to get the budget, to book the rooms, to order microphones and so forth, and who will write the cheques for honoraria, all the way down to the humble employee who must enter the tax information of each participant. One has to wonder how many great conferences, workshops and symposia never happened because no scientist had either the time or the patience to fill all those forms and send all those emails; how much intellectual capital is wasted by society to make sure that its algorithms are properly activated and obeyed.

It is not only that you live in an increasingly mechanized world, but your own life has been mechanized: your life is being reduced to an algorithm so that a vast bureaucracy of computers can easily and quickly process it (or, better, "process you").

Get over it: you are an algorithm.

The Importance of the Human Mind

That's the real question, because, if the human mind is not important, then let's just flatten our world in such a way that human minds are no longer required (nor tolerated). Algorithms can certainly run a perfectly smooth society with no need for human minds. In fact, with no need for humans.

The Dystopia of Vast Algorithmic Bureaucracies - Post-truth Algorithms

People are both accomplices and victims of this standardization of life. Not long ago, the tradition was that, at year's end, you would write (handwrite!) personalized letters to your relatives and friends summarizing the main happenings of the year. Then it became a routine to buy standard "happy holidays" cards and simply (hand)write two or three personalized wishes for each person. Now most people send a mass email to all their acquaintances, or a message on a social-networking platform. There is a smartphone app that sends out automatically birthday wishes to your friends on their birthday: you don't need to remember, the app remembers for you, and even customizes a message for you. Some day someone will develop an app that automatically replies to birthday wishes, so you won't have to reply to all the people who send you birthday wishes. The two apps will exchange messages such as "Happy birthday, old man!" and "Thank you, i hope you are doing well" without either person knowing that this is going on. I had lots of uncles and aunts and i remember each one being a little eccentric. The new generations are increasingly uniform in their life routines and hobbies. I suppose this will create a standardized human being. It won't make much of a difference whether you hunt mushrooms with this uncle or tend the garden with that uncle or watch aunt Teresa cook. They will all make the same jokes, wear the same clothes, shop at the same mall, and tell you to brush your teeth and do your homework.

The beauty of algorithmic bureaucracies, as any lawyer knows, is also that it doesn't even matter whether your answers are truthful or not, only whether they match what the algorithm is programmed to accept.

When you are looking for a job, the "headhunter" knows exactly what you should write in the resume in order to get an interview. What your skills really are is largely irrelevant. An idiot with a well-designed resume is more likely to get a job than a genius with a horrible resume.

Tutors can help the dumbest student pass the most difficult examinations without studying too much: an examination is just a game of rules, and they know which are the rules. (They don't necessarily know much about the subject matter).

If you want certainty to obtain a tourist visa to visit a foreign country, apply for the visa through a travel agency specialized in visas for that country: this agency knows exactly the rules to obtain a visa, they may invent a fake itinerary with fake hotel reservations and even fake flights, and you will obtain your visa by express procedure. If you apply by yourself and tell the truth, chances are that you will write something that will cause your application to be denied. Machines have nothing to do with this: the visa is denied and approved by human beings. Those human beings simply follow rules and regulations. Their job is to check if your story fits the rules and regulations. Note that their job is not to find out whether you are telling the truth or lying. The rules are more important than the truth.

When you file a claim for a car accident, you have a choice that will determine what happens to your case: if you hire an attorney, you enter an algorithm in which your attorney will speak with their attorney, whose goal is to close your case as soon as possible; if you don't hire an attorney, you enter an algorithm in which you will have to deal with an adjustor who is paid to minimize the amount that the insurance has to reimbourse you.

The future is a bleak algorithmic society. Don't blame it on the machines. Blame it on the obsession that humans have for structuring their environment.

Common sense is increasingly uncommon.

The Dystopia of Vast Algorithmic Bureaucracies - The Age of Apathy

The algorithmic society is also redefining morality. You are "good" if you obey the rules and regulations, and you are bad if you disobey; and you are "smart" proportional to how well you can play the rules and regulations to your benefit. In 2017 the USA inaugurated a president who didn't seem to have any moral feelings. To him, it was all about playing the rules and then winning or losing. The rules are not right or wrong: they are the rules. Life is a game of using your intelligence to play with those rules and try to win. The people who voted for him were precisely the people who in the old days voted for the most moral candidate. When in May 2017 that president wanted to condemn a terrorist attack in Britain, he said: "I won't call them monsters, because they would like that term. I will call them from now on losers." To him being a "loser" is much worse than being a "monster". Being evil is not such a bad thing, it is in fact a meaningless term; but being a loser... that is easier to define and, in his world, it really hurts.

A perfect example of vast algorithmic bureaucracy is the court trial in the USA. Nobody is truly interested in whether the defendant is guilty or innocent. A court trial is just a performance of rules, as stated and enforced by the judge (who is not there to "judge" the defendant but only to judge whether the attorneys, witnesses and jurors follow the rules). The jury is not allowed to interrogate the witnesses, let alone the defendant. The attorneys follow rules on how to question the witnesses and address the jury. And so on. In some states it is already mandatory for judges to use "computerized risk assessment tools" when deciding the sentence. One such program, Compas (developed by a firm called Northpointe that seems to have changed name to Equivant in 2017), was responsible for a Wisconsin judge sentencing a man (Eric Loomis) to eleven years in prison for minor crimes.

This is the age of the "plea bargain" in which the criminal plays a game with the prosecution. If they find a point of equilibrium, that saves money on both sides and doesn't cost the criminal too much time in jail, there will be no trial. The whole justice system is based on rules and regulations, not on the truth. Whether you have committed the crime counts far less than how good your attorney is at maneuvering within the maze of laws. If you defend yourself, you'll probably lose, even if you are absolutely innocent.

Morality is rapidly being reduced to defining "good" as the set of rules and regulations. If a rule does not exist, then you are free to behave as you wish. It is not surprising therefore that in July 2017 a group of Florida teenagers watched a man drown in the river and only thought of making a video for YouTube of the scene: there is no law that mandates you should help a fellow human who is drowning.

There is also something fundamentally unfair in the fact that the algorithmic society treats everybody the same way. It makes no difference whether you are a polite and gentle soul or an aggressive obnoxious jerk: the rules and regulations are the same, the punishment is the same. The motivation to be nice to other human beings disappears as we are all treated the same. Civility and good manners are useless, and a bit old-fashioned, within vast algorithmic bureaucracies.

The likely consequence of an increasingly algorithmic society is a humankind that, instead of searching for meaning, will be reduced to intellectual, political, spiritual and emotional apathy. Then humankind may become a mere footnote in the history of technology.

The Ultimate Dystopia: a Perfect World

In the 2010s the trend in A.I. is about systems that learn by themselves to operate in environments where the rules are finite and very clear, such as games. These systems work wonders as long as we can perfectly simulate the world for them. My fear is that, instead of making machines as intelligent as humans (a feat that seems virtually impossible with today's technology), we will make the human world increasingly "perfect" (devoid of exceptions and creativity) so that machines can perform as well as humans without any need for human-level intelligence. The amazing skill of human beings is to be able to survive and thrive in a world that is far from perfect. It would be much easier to live in a perfect world. In fact, a perfect world would probably be very boring for humans. Perfection does not require intelligence. Perfection is not super-human: it is sub-human.

The Dystopia of Vast Algorithmic Bureaucracies - Appendix: the Self-driving car

Self-driving cars are a mirage in real-world traffic. But the vast algorithmic bureaucracy could simply introduce rules and regulations to turn real-world traffic into highly structured traffic, i.e. traffic where anybody could drive a car, even someone who never took driving lessons.

In my opinion, adoption of the self-driving car will be driven not by Artificial Intelligence but instead by economic interests that eventually will impose new limitations on our lives in order to make the self-driving car a reality. As of May 2017, Uber boasted one billion passengers and 1.5 million drivers worldwide. More than 50% of the revenues were pocketed by the drivers. Hence Uber's motivation to get rid of the drivers. In 2015 Uber hired more than 40 scientists from CMU's prestigious robotics laboratory. This is the laboratory led by Red Whittaker that won the first DARPA Grand Challenge for autonomous vehicles in 2004 in the Mojave desert and that came second in 2005 to a Stanford vehicle built by their former scientist Sebastian Thrun, and then won again in 2007 DARPA's Urban Challenge. Red Whittaker had been building robots since 1979 when CMU helped clean up the mess of the nearby Three Mile Island nuclear reactor accident. Unfortunately the self-driving car doesn't work, and it won't work any time soon, so firms like Uber will inevitably come up with a scheme to pressure cities to limit what people can do. If we turn streets into railways with no rails, then the self-driving car becomes a reality.

The big car manufacturers would be perfectly happy with this outcome. Ride-sharing firms like Uber don't make cars and someone will have to make Uber's self-driving cars. The advantage is that firms like Uber will standardize what a self-driving car will look like along with its features. If the cars of the future are purchased by Uber's standardizing bureaucrats instead than by 100 million idiosyncratic consumers, the car manufacturers will be able to save a lot of money and earn a lot more of money.

Last but not least, car manufacturers may have smelled another source of revenues: providing remote driving assistance. A startup called Phantom provides a remote driver who can see what the "self-driving car" is doing and can take over the wheel remotely via a fast wireless telephone line.

Economic interest is, to me, the real driver of the self-driving car (sorry for the pun).

"Knowing what I know about computer vision, I wouldn't take my hands off the steering wheel" (computer vision pioneer Jitendra Malik in 2016).

Corollary to the Dystopia of Vast Algorithmic Bureaucracies: Leviathan, Panopticon, Biopower

I view the history of human civilization as divided in three chronological stages. In the first stage it was mostly religion that enforced rules of conduct, i.e. social behavior. Religious dogmas prescribe how to live, how to behave towards each other, duties and rights. The effect of religious dogmas was measured in generations. Then states began enacting secular laws to enforce social behavior, and the laws, again, prescribed duties and guaranteed rights. The impact of laws is usually measured in years, if not decades.

Now human civilization is entering a stage in which algorithms enforce social behavior, what you can do and what you must do. The impact is measured in months, if not days.

The first stage, the religious stage, relied on sacred texts that left a lot to be interpreted: "do not steal" is a very general commandment that does not quite define "theft". In the second stage, the secular stage, the state came up with much stricter definitions of social behavior, and a definition of "theft" is now embedded in every code of law. There is still some latitude in the interpretation, as proven by the fact that there are millions of lawyers and they are generally very rich (Note that the definition of "theft" carefully avoids calling "theft" what lawyers do). The third stage removes any ambiguity in the interpretation: the algorithm unequivocally determines what you must do and what you cannot do. What you "can" do is a subset of what you "must" do.

We need to update the British philosopher Francis Bacon's "knowledge is power" (a paraphrase of what he wrote in 1597 in "Meditationes Sacrae"). In 1651 his former assistant Thomas Hobbes published a book titled "The Leviathan" in which he compared the state to that Biblical monster. Hobbes argued that humans are chaotic and selfish animals that need a strong central government else they would live in permanent civil war. Hence, the ideal situation is a "social contract" between a government that provides benefits to the citizen as long as the citizen obeys the laws, and viceversa, a sort of Faustian pact between the individual and the monster for the sake of order and safety. (The 1668 edition of this book added the sentence "knowledge is power" and popularized it). The Leviathan is being implemented as a vast library of algorithms that will eventually rule every aspect of our lives. In 1785 the British philosopher Jeremy Bentham drew a diagram of an ideal prison called the Panopticon in which control towers observed the convicts without the convicts realizing that they are being constantly watched.

The French philosopher Michel Foucault, in his book "Discipline and Punish" (1975), saw the modern state as a Panopticon of sorts, one that relies on technology to constantly surveille its population. But we need to update that notion too. We are not only watched by algorithms: we are physically forced to live organized lives. Foucault's "disciplinary" society is a hierarchy of guards, each layer observing unseen the lower layer with the aim to discipline them. This disciplinary society "normalizes" each and every individual. It doesn't need to take recourse to torture anymore because technology has endowed it with "biopower" that consists in "techniques for achieving the subjugations of bodies and the control of populations" ("History of Sexuality", 1976). Today the user of algorithms lives in such a Panopticon that is augmented with the biopower of pervasive and omnipotent technologies of digitization. These algorithms objectify individuals as mathematical quantities in order to manage and control them as efficiently as possible. Algorithms are more than Foucault's technologies of surveillance: they are technologies of enforcement. Biopower turns the individual into a passive program, a normalized mathematical quantity.

Religious and state laws always had a secondary role, besides enforcing behavior: they also "trained" people to think in specific ways. They didn't simply forbid you to steal, they "trained" you to believe that theft is evil, to prevent theft and to disparage thieves. Likewise, algorithms do more than simply force you to behave in a specific way: they "train" you that way. Religion and state created the knowledge that people consumed and that shaped their way of thinking. Biopower creates the knowledge that individuals consume. People's minds are being trained as they are being forced to adopt a specific rule of conduct.

The totalitarian regimes of the 20th century depended not only on repression but on convincing the masses that the repression was justified and, in fact, in the interest of the very masses that were being repressed. The masses actively collaborated in keeping those totalitarian regimes in power. The two most famous dystopic novels of the 20th century, Aldous Huxley's "Brave New World" (1932) and George Orwell's "1984" (1949), can be read as psychological studies, not just parables. The vast algorithmic bureaucracy does something similar but without psychological subtlety: it provides people with algorithms that are justified as improving everybody's efficiency, and then uses those algorithms to enforce the only approved code of conduct. Clifford Nass at Stanford wrote a book called "Media Equation" (1996), in which he discussed computers as social actors, and his student Brian Fogg coined the term "persuasive technology": people can be trained by algorithms ("Mass Interpersonal Persuasion", 2007).

Niccolò Machiavelli in his book "The Prince", published posthumously in 1532, wrote: "The chief foundation of every state... are good laws and good arms... Where there are good arms, good laws inevitably follow". That form of repression too has been superseded by the algorithms. The state or, better, the vast algorithmic bureaucracy, is rarely challenged by armed citizens: it protects itself with algorithms that limit what citizens can do, and mostly forces them to do precisely what makes the bureaucracy so powerful.

"Maybe the goal nowadays is not to discover what we are but to refuse what we are" (Michel Foucault).

The Ethics of Algorithms

The vast algorithmic bureaucracy can also program ethics. Of course, we first need to choose among the many variants of ethics. John Stuart Mill's "utilitarianism" states that humankind should strive for the greatest possible happiness for the greatest number of people. Utilitarianism cannot, per se, decide whether war or peace is better, whether lying or telling the truth is better, etc. It all depends on the consequences.

At the other extreme, Immanuel Kant believed in universal laws, regardless of the consequences, and therefore he considered lying always morally wrong.

A few centuries earlier, Thomas Aquinas worked out a hierarchy of lies, all of them wrong but to different degrees: malicious lies, joking lies, helpful lies. Today's algorithmic bureaucracies have no concept of helpful or joking lies: they don't admit a lie as a legitimate answer. In the future, as they become more "intelligent", they will probably lean towards Mill's utilitarianism of maximizing benefits for the maximum number of people, maybe with the provision that the happiness of rich and powerful people (and advertisers?) should count more. The genetic consequences of medicine, for example, could be a weakening of the human genetic pool because medicine allows the "unfit" to reproduce. Saving one life today could harm many lives in the future. If medicine is going to cause a catastrophic genetic decline, is it morally good?

The vast algorithmic bureaucracy will likely adopt an extreme form of rationalism and turn everything into numbers and formulas.

The Australian philosopher Peter Singer, and Princeton University's first professor of bioethics, advocated a strictly rational approach to action in his book "Practical Ethics" (1979), even to the extent of viewing euthanasia and infanticide as necessities. Singer calculated the ethical value of an action simply based on the outcome: if killing a child will make, in the long run, a family happier, why not? That's where Nazi supremacism and enlightened liberalism meet.

But this approach has indeed a rational advantage. For example, would you risk your life to benefit the people who will live in the year 3000? Most likely not: your life is important to support your family now, and you don't really care for the people of the year 3000. You have no emotional attachment to them, even though some of them will be your direct descendants. This is irrational behavior. What is the point of caring for your children if you don't care for their children's children's children's children's ... children? You presumably want your children to be happy with their children, and their children to be happy with their children, and so forth: don't you? Then why stop before the year 3000? The vast algorithmic bureaucracy, instead, can be programmed to reward actions that maximize the benefits to humankind no matter how far in the future. So, if survival of the species is the ultimate goal, why not use an algorithm to calculate the moral value of your actions and remove all emotional attachments? Why not calculate the benefits and costs to the future of the human race before deciding to save you from a disease?

Peter Singer's ethics is reminiscent of Adolf Hitler's minus the racism. Hitler was inspired by evolutionary ethics to pursue the utopian project of biologically improving the human race. Richard Weikart's "Hitler's Ethic" (2009) makes the case that, rationally speaking, Hitler wasn't a monster, but rather a very moral person for whom the ultimate goal (improving the human race) justified scientific genocide. Of course, the specific genocide that he picked was not really going to improve the race: blame it on the algorithm!

I consider Singer one of the greatest philosophers of our age so don't take this as criticism against him personally. In a later book, "The Expanding Circle" (1981), Singer studied the parallel development of reasoning and ethics, and argued that the human species inevitably tends towards a more and more universal morality. The circle of altruism expands from the family to the tribe to the nation to the race to even other species (as in today's animal-rights movement).

Whatever its evolutionary advantage, evolution has equipped humans with the faculty of reasoning, and reasoning now plays against our self-interest in making us develop a more and more universal moral code. Leon Festinger in "A Theory of Cognitive Dissonance" (1957) introduced the notion that we, mathematical beings, instinctively try to remove inconsistencies and therefore keep moving towards a more objective point of view. We strive for an objective viewpoint based on logic.

From this point of view one could conceive a different kind of "ethical" algorithm, one designed to maximize not the benefits for the most but equal treatment for the most. This algorithm would continue the human race's evolution towards a universal ethics with the big advantage that such an algorithm would not be biased towards siblings, friends or neighbors.

After all, technology is the way of dealing with the world so that we don't have to deal with it.

Bureaucracy Compounds A.I.'s Stupidity

One thing that is important to realize is that the A.I. programs or "bots" employed by the vast bureaucratic bureaucracies can be extremely stupid, in fact beyond human stupidity. I didn't know it was possible to go beyond human stupidity until we started building artificial ones.

Unfortunately, each new bot replaces a human being, which means that there is nobody to which you can explain the stupidity of the bot. For example, my Internet provider Comcast has emailed me a couple of times about a "Notification of Claimed Copyright Infringement Made under the DMCA". This is not the result of a forensic investigation by a team of human sleuths but simply an automatic message sent out by a bot. Most bots are triggered automatically when some conditions arise. We had downloaded nothing illegally but it was impossible to find out what had awakened the copyright-infringement. After fighting for ten or twenty minutes with Comcast's automated customer support, we finally managed to speak with a human being. The Comcast representative graciously accepted to remove this incident from my record, but he didn't even ask me for the name of the website from which we downloaded the file. I guess they don't have the (human) staff to investigate each and every incident, so they just trust the angry customer who is willing to make a phone call and press 1-4-3-2-6-2 etc to get to speak to someone!

Having removed the physical staff that used to collect the bridge toll, and having forced all drivers to pay by phone or online, the Bay Area public-transportation system (ironally called "Fast Track") then had to face the fact that people driving a rental car would not receive the bill at home. Therefore this is what they emailed me when i asked for a clarification (9 September 2017): "A One-Time Payment for a rental vehicle must be created within 48 hours of the crossing to clear the toll through the One-Time Payment. If you receive notifications from the rental company that means the payment was posted to the rental companies' account and you will need to contact them to resolve any subsequent issues regarding the transaction. When contacting the rental company, please have your One-Time Payment confirmation available so they will know you attempted to pay for the toll before it posted to the rental account. Remember that if you add a rental vehicle to your FastTrack account, you must promptly remove the vehicle when you have returned it because rental vehicles may have additional tolls or violations for other drivers. Thank you, Bay Area FastTrack Customer Service Center Team. Please do not reply to this message. This is an automatically generated notification." Simple, painless, error-proof and... user-friendly, isn't it?

The problem with these incredibly stupid Artificial Intelligence programs is, of course, that there is no way to talk to them and explain how stupid they are.

"If you obey all the rules, you miss all the fun" (Katharine Hepburn)

Transcendental Intermezzo: You are a Robot - The Demise of Free Will

I suspect that Artificial Intelligence will have (is having) a powerful effect on the human condition. The shocking revelation coming from A.I. is not that machines can do everything that we can do (we could easily live with this one) but in the specular realization: that we can only do what a machine can do; i.e. that we are just a machine. Robots don't have free will. You can always find the electromechanical cause of a robot's action. You can backtrack and find out the exact sequence of events that triggered the action of a machine. It only depends on how far back in time you want to go, but there is a clear path of causes that have had effects that eventually resulted in that machine lifting its arm or saying, "This life has no meaning". Once these machines start doing everything that we do, it becomes obvious that our actions too are simply the effects of events outside our control.

Think of memory. Humans knew that their memory was fallible, but machines show us on a daily basis how pathetic our memory is. We are literally helpless without a computer memory to remind us of our appointments, our important data, our documents, our favorite videos, etc. We are being diminished by the constant reminder that comes from using a machine to do the things that humans do.

Neuroscience already knew that our actions are caused by electrochemical reactions in the brain, but most of us have always ignored the literal meaning of this finding. Free will is an illusion: even my thoughts about free will are due to neural events in my brain that are beyond my control.

Robots are machines guided by an operating system, by some training, by a program and by some inputs. We are machines guided by genes, by an upbringing, by the ideas of our times, and by the events of our times. Neurological diseases and nutritional deficiencies can alter our behavior just like a software bug or a power outage can affect a robot's behavior.

Evil is, ultimately, due to a combination of factors (genes, upbringing, ideologies, life events) over which we don't have any control just like a robot has no control over how it was built and programmed. Good and evil are meaningless: we are simply machines that are programmed by external forces. A serial killer is no more guilty than an earthquake. The very feelings of morality, revenge, justice and so forth arise from neural processes in the brain that are due to genes, life events and external influences.

Robots will remind us every single second that we are just machines, made of flesh and blood; ultimately, just machines like the machines made of plastic and silicon.

I suspect that, far from leading us to more exciting levels of understanding, the rise of intelligent machines will be a humbling experience for humankind. Copernicus showed us that we are not at the center of the universe. Newton and Einstein showed us that the future is predetermined by the past. Artificial Intelligence will show us that we are not even in control of our actions: we are robots, just like "them".

Transcendental Intermezzo: Neural Variant

Neuroscience has pinpointed specific areas of the brain that are responsible for the loss of specific skills. If a certain region of the brain is impaired, then the person loses the ability to do something. While neuroscientists warn that the functioning of the brain is not so simple, and that many regions can interact during an apparently linear process, the temptation to view the brain as a set of neural networks is hard to resist. If artificial general intelligence (A.G.I.) is simply the sum of its parts, then it is just a matter of devising all the neural networks required to perform the functions of an average person. Today's neural networks tend to be highly specialized, and their performance is highly dependent on the dataset used to train them, but it is entirely plausible to believe that progress in their design will lead to more robust architectures, and that eventually someone will find a way to integrate a myriad of neural networks into an artificial brain that will match the human brain in every single activity. If you believe that A.G.I. is merely the set of all artificial neural networks required to perform all the tasks that we perform, there is an important consequence but not the one that most people discuss. It is not that the machine will some day be free willing, but that... we don't have free will. If you believe that our brain is shaped by genes (just like the brain of an artificial neural network is shaped by its designer) and by "datasets" (experience), then we, just like the artificial neural networks, simply execute what is programmed in our brain. Where is my free will? I can blame all my sins on the genes that initially shaped my brain and on the "datasets" that trained my brain to produce the behavior that it does. The physiologist Benjamin Libet at UC San Francisco discovered that we become aware of our actions only "after the fact", after the brain has already sent the signal to the limbs to act ("Cortical Activation in Conscious and Unconscious Experience", 1965). Michael Gazzaniga at Cornell University popularized this notion in his book "Social Brain" (1985): an "interpreter" in the brain makes sense of the decisions that the other brain modules have made, and that interpreter is your consciousness. I am not in control of my actions, hence don't blame me if i commit a crime. The law specifically relieves people of responsibility and liability when it can be proven that they were not in control of their actions. Where is my free will? St Augustine grappled with the same problem in the context of an omnipotent and infinitely good God: if God created the sinner and God decides everything that happens in the universe, why should we blame the sinners for their sins? It is not trivial to prove free will once you assume that external forces determine our behavior. For those interested in the debate whether the universe is fully deterministic or not, check out the various philosophical theories about the implications of the randomness of quantum mechanics (e.g. in my book "Thinking about

Thought"), and don't forget that randomness is often just a measure of our ignorance.

"You are not controlling the storm, and you are not lost in it. You are the storm" (Sam Harris)

Transcendental Intermezzo Revised: You are a...?

Science explained too much, even things that we didn't know had to be explained. Science found out that there is a multitude of equations at work that regulate every event in the universe. Science introduced determinism in our thinking: maybe everything is already planned and inevitably proceeds from the initial conditions according to some formulas, and we are just cogs in this giant clockwork.

But we could live with that. Say what you will of me being driven by physical equations, i still feel otherwise.

However, when machines get to the point that they can do everything that we can do... well, it will be harder to feel that there is something special about us. It will be proven that we are just machines.

Underlying the project of Artificial Intelligent is precisely the tacit belief that humans are machines. If we think that we are not machines, we can have a philosophical discussion about the possibility that machines can become human like us. If we think that we are just machines, the discussion is not philosophical: it is technological. It becomes a problem of reverse engineering.

Therefore, when machines finally match all our capabilities, technology will have proven conclusively what Science only hinted at. All those equations failed to convince us that we are mere machines, but technology may prove it beyond any reasonable doubt by showing us a machine that can do everything we can and even better. Then what?

Maybe we do need a new religion to give meaning to this bundle of pulsating organs wrapped in skin that is me.

Why the Singularity is a Waste of Time and Why we Need A.I. – A Call to Action

The first and immediate reason why obsessive discussions about the coming of machine super-intelligence and human immortality are harmful is that they completely miss the point.

We live in an age of declining innovation. Fewer and fewer people have the means or the will to become the next Edison or Einstein. The great success stories in Silicon Valley (Google, Facebook, Apple) are of companies, started by individuals with very limited visions, that introduced small improvements over existing technologies. Entire nations

(China and India, to name the obvious ones) are focusing on copying, not inventing.

Scholars from all sorts of disciplines are discussing the stagnation of innovation. A short recent bibliography: Tyler Cowen's e-book "The Great Stagnation" (2010) by an economist; Neal Stephenson's article "Innovation Starvation" (2011) by a sci-fi writer; Peter Thiel's article "The End of the Future" (2011) by a Silicon Valley venture capitalist; Max Marmer's "Reversing The Decline In Big Ideas" (2012) by another Silicon Valley entrepreneur; Jason Pontin's "Why We Can't Solve Big Problems" (2012) by a technology magazine editor; Rick Searle's article "How Science and Technology Slammed into a Wall and What We Should Do About It" (2013) by a political scientist; "The Innovation Illusion" (2016) by economist Fredrik Erixon and investor Björn Weigel; Robert Gordon's book "The Rise and Fall of American Growth" (2016) shows that the pace of innovation has slowed since 1970 after a rapid rise between 1800 and 1970.

Sadly, the ages of real "accelerating progress" were the ages of total warface. Technological progress over the centuries was often (largely?) driven by warfare. World War I was the reason to accelerate progress in radio transmissions and in airplanes. World War II was the reason to accelerate progress in nuclear physics, the radar and electronics, and the first electronic computer was built for military purposes. The Cold War gave us the Internet, the Global Positioning System (GPS) and Artificial Intelligence (almost entirely funded by DARPA until the 1970s).

A study by Ashish Arora and Sharon Belenzon at Duke University "Killing the Golden Goose?" (2015) shows that in the 27 years after 1980 the USA has experienced a shift from long-term invention towards short-term innovation. In 2016 Nicholas Bloom and Charles Jones of Stanford University and John Van Reenen of MIT wrote a paper titled "Are Ideas Getting Harder to Find?" in which they calculate the productivity of scientists and conclude that "research productivity is declining sharply".

Then there is the fundamental issue of priorities. The hypothetical world of the Singularity distracts us from the real world. The irrational exuberance about the coming Singularity distracts a lot of people from realizing the dangers of unsustainable growth, dangers that may actually wipe out all forms of intelligence from this planet.

Let's assume for a second that climate scientists like Paul Ehrlich and Chris Field (i met both in person at Stanford) are right about the coming environmental apocalypse. Their climate science is ultimately based on the same science that happens to be right about what that bomb would do to Hiroshima (as unlikely as Einstein's formula may look), that is right about what happens when you speak in that rectangular device called smartphone (as unlikely as it may seem that someone far away will hear your voice), that is right about what happens when someone broadcasts a signal in that frequency range to a box sitting in your living room called TV set (as unlikely as it may seem that the box will then display the image of someone located far away), that is right about what happens when you turn on that switch (as unlikely as it is that turning on a switch will light up a room); and it's the same science that got it right on the polio vaccine (as unlikely as it may look that invisible organisms cause diseases) and many other incredible affairs.

The claims about the Singularity, on the other hand, rely on a science (Artificial Intelligence) whose main achievement has been to win board games. One would expect that whoever believes wholeheartedly in the coming of the Singularity would believe tenfold stronger that the human race is in peril.

Let's assume for a second that the same science that has been right about almost everything that it predicted is also right about the consequences of rapid climate change and therefore the situation is exactly the opposite of the optimistic one based mostly on speculation depicted by A.I. science: the human race may actually go extinct before it even produces a single decent artificial intelligence.

In about one century the Earth's mean surface temperature has increased by about 0.8 degrees. Since it is increasing faster today than it was back then, the next 0.8 degrees will come even faster, and there is widespread agreement that 2 degrees above what we have today will be a significant tipping point. Recall that a simple heat wave in summer 2003 led to 15,000 deaths in France alone. Noah Diffenbaugh and Filippo Giorgi (authors of "Heat Stress Intensification in the Mediterranean Climate Change Hotspot ", 2007) have created simulations of what will happen to the Earth with a mean temperature 3.8 degrees above today's temperature: it would be unrecognizable. That temperature, as things stand, is coming for sure, and coming quickly, whereas super-intelligence is just a theoretical hypothesis and, in my humble opinion, is not coming any time soon.

Climate scientists fear that we may be rapidly approaching a "collapse" of civilization as we know it. There are, not one, but several environmental crises. Some are well known: extinction of species (with unpredictable biological consequences: one such is the declining population of bees which may pose a threat to fruit farms), pollution of air and water, epidemics, and, of course, anthropogenic (human-made) climate change. See the "Red List of Threatened Species" published periodically by the International Union for Conservation of Nature (IUCN). See University of North Carolina's study "Global Premature Mortality Due To Anthropogenic Outdoor Air Pollution and the Contribution of Past Climate

Change" (2013) that estimated that air pollution is likely to cause the deaths of over two million people annually. A Cornell University study led by David Pimentel, "Ecology of Increasing Diseases" (2007), estimated that water, air and soil pollution account for 40% of worldwide deaths. A 2004 study by the Population Resource Center found that 2.2 million children die each year from diarrhea caused by contaminated water and food. And, lest we think that epidemics are a thing of the past, it is worth reminding ourselves that AIDS (according to the World Health Organization) has killed about 35 million people between 1981 and 2012, and in 2012 about 34 million people were infected with HIV (Human Immunodeficiency Virus, the cause of AIDS), which makes it the fourth worst epidemic of all time. Cholera, tuberculosis and malaria are still killing millions every year; and "new" viruses routinely pop up in the most unexpected places (Ebola, West Nile virus, Hantavirus, Avian influenza, Zika virus, etc).

Some environmental crises are less advertised but no less terrifying. For example, global toxification: we filled the planet with toxic substances, and now the odds that some of them interact/combine in some deadly runaway chemical experiment never tried before are increasing exponentially every year. Many scientists point out the various ways in which humans are hurting our ecosystem, but few single out the fact that some of these ways may combine and become something that is more lethal than the sum of its parts. There is a "non-linear" aspect to what we are doing to the planet that makes it impossible to predict the consequences.

The next addition of one billion people to the population of the planet will have a much bigger impact than the previous one billion. The reason is that human civilizations have already used up all the cheap, rich and ubiquitous resources. Naturally enough, humans started with the cheap, rich and ubiquitous ones, whether forests or oil wells. A huge amount of resources is still left, but those will be much more difficult to harness. For example, oil wells have to be much deeper than they used to. Therefore one liter of gasoline today does not equal one liter of gasoline a century from now: a century from now they will have to do a lot more work to get that liter of gasoline. It is not only that some resources are being depleted, but even the resources that will be left are, by definition, those that are difficult to extract and use (a classic case of "diminishing rate of return").

The United Nations' "World Population Prospects" (2013) estimated that the current population of 7.2 billion will reach 9.6 billion by 2050, and population growth will mainly come from developing countries, particularly in Africa: the world's 49 least developed countries may double in size from around 900 million people in 2013 to 1.8 billion in 2050.

A catastrophic event is not only coming, but the combination of different kinds of environmental problems makes it likely that it is coming even sooner than the pessimists predict and in a fashion that we cannot quite predict.

The result of exposure to lead, mercury and pesticides on fetal brain development has been known for a while In 2006 Philippe Grandjean at Harvard and Philip Landrigan at Mount Sinai in New York wrote of a "silent pandemic" of "neurotoxins" that are damaging the brains of unborn children ("Developmental Neurotoxicity of Industrial Chemicals," 2006). In 2012 the Harvard neurologist David Bellinger showed that intelligence quotients among children whose mothers had been exposed to "neurotoxins" while pregnant were lower than the IQs of children whose mothers were not exposed to those toxins ("A Strategy for Comparing the Contributions of Environmental Chemicals and Other Risk Factors to Neurodevelopment of Children", 2012). But now the CDC (Center for Disease Control) warns that 212 industrial chemicals are widely disseminated in the natural environment and can routinely be found inside the bodies of people ("The Fourth National Report on Human Exposure to Environmental Chemicals", 2009). In 1937 Lowell Thomas made a film on the story of "bakelite", the first synthetic plastic invented in 1907 by Leo Baekeland using fossil-fuel derivatives. Thomas talked of a fourth kingdom (besides animals, plants and fungi: the kingdom of synthetics). Within a century someone could make a film to show how the kingdom of synthetics infiltrated all other kingdoms. (In what was basically a preview of the Singularity movement, Baekeland chose the infinity symbol as the logo of his plastic-manufacturing firm). The word "petro-topia", popularized by the book "Petrochemical America" (2012), composed by the photographer Richard Misrach and the landscape architect Kate Orff , and by the book "Living Oil - Petroleum Culture in the American Century" (2014), written by an environmental scientist at University of Oregon, Stephanie LeMenager. Petro-topia could well turn into "petro-calypse".

For the record, the environmentalists are joined by an increasingly diversified chorus of experts in all sorts of disciplines. For example, Jeremy Grantham who is an economist (managing 100 billion dollars of investments). His main point (see, for example, his 2013 interview on Charlie Rose's television program) is that the "accelerated progress" that the Singularity crowd likes to emphasize started 250 years ago with the exploitation of coal and then truly accelerated with the exploitation of oil. The availability of cheap and plentiful energy made it possible to defy, in a sense, the laws of Physics. Without fossil fuels the human race would not have experienced such dramatic progress in merely 250 years. Now the planet is rapidly reaching a point of saturation: there aren't enough resources for all these people. Keeping what we have now is a major project in itself, and those who hail the coming super-intelligence miss the point the way a worker planning to buy a bigger house misses the point that he's about to get fired.

We are rapidly running out of cheap resources, which means that the age of steadily falling natural resource costs is coming to an end. In fact, the price of natural resources declined for a century until about 2002 and then in just 5 or 6 years that price regained everything that it had lost in the previous century (i am still quoting Grantham). This means that we may return to the world of 250 years ago, before the advent of the coal (and later oil) economy, when political and economic collapses were the norm; a return to, literally, the ages of starvation.

It is not only oil that is a finite resource: phosphates are a finite resource too, and the world's agriculture depends on them.

Population growth is actually a misleading parameter, because "overpopulation" is measured more in terms of material resources than in number of people: most developed countries are not overcrowded, not even crowded Singapore, because they are rich enough to provide a good life to their population; most underdeveloped countries are overcrowded because they can't sustain their population. In this sense, overpopulation will increase even in countries where population growth is declining: one billion Indians who ride bicycles is not the same as one billion Indians who drive cars, run A/C units and wrap everything in plastic. If you do it, why shouldn't they?

The very technologies that should improve people's lives (including your smartphone and the robots of the future) are likely to demand more energy, which for now comes mainly from the very fossil fuels that are leading us towards a catastrophe.

All those digital devices will require more "rare earths", more coltan, more lithium and many other materials that are becoming scarcer.

We also live in the age of Fukushima, when the largest economies are planning to get rid of nuclear power, which is the only form of clean alternative energy as effective as fossil fuels. Does anyone really think that we can power all those coming millions of robots with wind turbines and solar panels?

Chris Field has a nice diagram (expanded in the 2012 special report of the Intergovernmental Panel on Climate Change titled "Managing the Risks of Extreme Events and Disasters to Advance Climate Change Adaptation") that shows "Disaster Risk" as a function of "Climate Change" and "Vulnerability" (shown, for example, at a seminar at the Energy Biosciences Institute in 2013). It is worth pondering the effects of robots, A.I. and the likes on that equation. Manufacturing millions of machines will have an impact on anthropogenic climate change; economic development comes at the cost of exploitation of finite resources; and, if high technology truly succeeds in increasing the longevity of the human race, the population will keep expanding. In conclusion, the race to create intelligent machines might exacerbate the risk of disasters before these super-intelligent machines can find a way to reduce it.

Economists such as Robin Hanson ("Economics Of The Singularity", 2008) have studied the effects of the agricultural, industrial and digital revolutions. Each caused an acceleration in economic productivity. The world's GDP may double every 15 years on average in this century. That's an impressive feat, but it's nothing compared with what would happen if machines could replace people in every single task. Productivity could then double even before we can measure it. The problem with that scenario is that the resources of the Earth are finite, and most wars have been caused by scarcity of resources. Natural resources are already strained by today's economic growth. Imagine if that growth increased ten fold, and, worse, if those machines were able to mine ten or 100 times faster than human miners. It could literally lead to the end of the Earth as a livable planet. Imagine a world full of machines that rapidly multiply and improve, and basically use all of the Earth's resources within a few years.

Ehrlich calls it "growthmania": the belief that there can be exponential growth on a finite planet.

The optimists counter that digital technology can be "cleaner" than the old technology. For example, the advent of email has dramatically reduced the amount of paper that is consumed, which has reduced the number of trees that we need to fell. It is also reducing the amount of mail trucks that drive around cities to deliver letters and postcards. Unfortunately, in order to check email and text messages you need devices like laptops, notepads and smartphones. The demand for materials such as lithium and coltan has risen exponentially.

Technological progress in the internal combustion engine (i.e., in fuel-efficient vehicles), in hybrid cars, in electric cars and in public transportation is credited for the reduction in oil consumption since 2007 in developing countries. But Asia Pacific as a whole has posted a 46% increase in oil consumption in the first decade of the 21st century. In 2000 oil consumption in China was 4.8 million bpd (barrels per day), or 1.4 barrels per person per year. In 2010 China's consumption had grown to 9.1 million bpd. China and India together have about 37% of the world's population. The rate of cars per person in China (0.09%) is almost 1/10th the one in the USA (0.8%) and in India is one of the lowest in the world (0.02%). Hence analysts such as Ken Koyama, chief economist at the Institute of Energy Economics Japan, predict that global petroleum demand will grow 15% over the next two decades ("Growing Oil Demand and SPR Development in Asia", 2013).

Services such as Zipcar and Uber were hailed as technology that empowers people, but the side-effect has been to shift thousands of people from public transportation towards cars. Many of the Zipcar and Uber customers are people who used to take the bus. These services have caused an increase in road traffic. Ironically, they may have also caused longer (not shorter) times to reach your destination because of increased traffic jams. If self-driving cars ever do happen, they will further compound the problem of traffic congestions: it should be obvious even to the dumbest Silicon Valley engineer that 200 people riding a bus occupy less asphalt than 200 people riding 200 self-driving cars.

There are loops that are obviously spinning out of control. I am writing this in China. One can hear the constant buzz of the giant air conditioner of China's large train stations, blowing cold air nonstop in order to lower the indoor temperature. However, those giant air conditioners contribute to global warming, i.e. to higher temperatures; which will require even bigger air conditioners; which will cause even more global warming; etc. China has just relaxed the "one-child policy" because its aging society will soon need more youngsters to take care of the elderly (and to pay taxes to support their health care). Europeans and Japanese are encouraging their people to make children for the same reason. We were scared of the population explosion, but we have come to realize that a declining population is no less dangerous. We are doomed either way: if population keeps increasing and gets wealthier, we will run out of resources; if it doesn't increase, our economies will collapse.

George Mitchell pioneered fracking in 1998, releasing huge amounts of natural gas that were previously thought inaccessible. Natural gas may soon replace oil in power stations, petrochemical factories, domestic heaters and perhaps motor vehicles. The fact that there might be plenty of this resource in the near future proves that technology can extend the life expectancy of natural resources, but it does not change the fact that those resources are finite, and it might reduce the motivation to face the inevitable.

Technology is also creating a whole new biological ecosystem around us, a huge laboratory experiment never tried before. Humans have already experienced annihilation of populations by viruses. Interestingly, the three most famous ones took hold at a time of intense global trade: the plague of 1348 (the "black death") was probably brought to Europe by Italian traders who picked it up in Mongol-controlled regions at a time when travel between Europe and Asia was relatively common and safe; and the flu pandemic of 1918, that infected about 30% of the world's population and killed 50 million people, took hold thanks to the globalized world of the British and French empires and to World War I. The HIV came out in the 1980s when the Western economies had become so entangled and it spread to the whole world during the globalization decade of the 1990s. By the end of 2012 AIDS had killed 35 million people worldwide.

We now live in the fourth experiment of that kind: the most globalized world of all time, in which many people travel to many places; and they do so very quickly. There is one kind of virus that could be worse than the previous ones: a coronavirus, whose genes are written in RNA instead of DNA. The most famous epidemics caused by a coronavirus was the Severe Acute Respiratory Syndrome (SARS): in February 2003 it traveled in the body of a passenger from Hong Kong to Toronto, and within a few weeks it had spread all over East Asia. Luckily both Canada and China were equipped to deal with it and all the governments involved did the right thing; but we may not be as lucky next time. In 2012 a new coronavirus appeared in Saudi Arabia, the Middle East Respiratory Syndrome (MERS).

All of these race-threatening problems are unsolved because we don't have good models for them. One would hope that the hi-tech industry invest as much into creating good computational models that can be used to save the human race as into creating ever more lucrative machines. Otherwise, way before the technological singularity happens, we may enter an "ecological singularity".

Discussing super-human intelligence is a way to avoid discussing the environmental collapse that might lead to the disappearance of human intelligence. We may finally find the consensus to act on environmental problems only when the catastrophe starts happening. Meanwhile, the high-tech world will keep manufacturing, marketing and spreading the very items that make the problem worse (more vehicles, more electronic gadgets, and, soon, more robots); and my friends in Silicon Valley, firmly believing that we are living in an era of accelerating progress, will keep boasting about the latest gadgets…the things that environmental scientists call "unnecessarily environmentally damaging technologies".

Fans of high technology fill their blogs with news of ever more ingenious devices to help doctors, not realizing that the proliferation of such devices will require even more energy and cause even more pollution (of one sort or another). They might be planning a world in which we will have fantastic health care tools but we will all be dead.

I haven't seen a single roadmap that shows how technology will evolve in the next decades, leading up to the Singularity (to super-human intelligence). I have, instead, seen many roadmaps that show in detail what will happen to our planet under current trends.

In 2005 Norman Myers of Oxford University delivered a speech at the Economic Forum in which he predicted that climate change alone would cause the displacement of 250 million people by 2050. The number sounded hard to believe. However, a 2015 report by the Internal Displacement Monitoring Centre estimated 157.8 million people had been forced to flee their homes between 2008 and 2014 because of natural disasters. In 2015 The Economics of Land Degradation Initiative estimated that land affected by serious drought doubled between the 1970s and the early 2000s. A 2015 study by the UN Environment Program (UNEP) found that 70% of Africa's migrants left their countries because of poverty or unemployment. In 2015 a surge in illegal immigrants from Central America into the USA followed catastrophic crop failures. The biggest exodus of the 2010s took place in Syria, devastated by a civil war.: it might be a coincidence, but Syria suffered a crippling drought in the five years preceding the civil war, so even the civil war itself may be due to some extent to the climate. Millions of people move because changing climate has destroyed their traditional livelihood. Their job has not been stolen by robots but by climate change.

There is also plenty to worry about the Internet. As Ted Koppel wonderfully explained in his book "Lights Out" (2015), the chances of a massive cyber-attack, that would leave the USA without electricity, communications and even water for weeks, are very high. There are dozens of hacking incidents every day. Banks, retail chains, government agencies, even the smartphone of the director of the CIA and even Mark Zuckerberg's Facebook account have been hacked. And they are usually hacked by amateurs in search of publicity. Spy agencies can cause a lot more damage than amateurs. They are probably monitoring the system right now, and they will strike only when it is worth their while. Companies that boasted about being invulnerable to hacking attacks have frequently been subjected to humiliating hacking attacks. The fact is that the Internet cannot be defended. It was probably a strategic mistake to make so much of the economy and of the infrastructure depend on a computer network (any computer network). Computers are vulnerable in a way that humans are not. You need to capture me and torture me in order to extract information from me that would harm my friends, relatives and fellow citizens; but you don't need to capture and torture a computer. It is much easier than that. Computer networks can be easily fooled into providing access and information. The more intelligent you make the network of computers, the bigger the damage it can cause to the humans who use it.

A.I.'s promises of dramatic economic and social change have been very effective in obtaining public and private funding, but that has come at the expense of other disciplines. Steven Weinberg's book "Dreams of a Final Theory" (1993) failed miserably to convince the political establishment to fund a new expensive project, the Superconducting Super-Collider. He failed because he narrated the reality of scientific research. Ray Kurzweil's "The Age of Spiritual Machines" (1999), a provocative and enthusiastic (and wildly self-congratulatory) reaction to IBM's Deep Blue beating the world chess champion in 1997, was totally out of touch with reality but impressed the political establishment enough such that many A.I. scientists obtained funding for their research. Research in A.I. in the USA has always relied on funding from the government (mainly through its "defense" arm called DARPA, which is really a designer of weapons). It was true of the original A.I. labs at MIT and Stanford, it was true of the A.I. research at SRI that yielded the autonomous robot Shakey and eventually the conversational agent Siri, and it was true of Nicholas Negroponte's Media Lab at MIT. Capturing the imagination of the political and military establishment is imperative for the progress of a scientific program (in Europe a similar phenomenon is at work, although it is the social impact rather than the military one to be more valued). The media's passion for A.I. may end up draining legitimate disciplines of the funding they need to improve the lives of millions of people. Imagine if the enthusiasm for early A.I. had diverted funds that were spent on the Interstate Highway System or Social Security; or if today's enthusiasm ends up diverting some of the $7 billion that the government pays to the Center for Disease Control (CDC), our front line in fighting infectious diseases. I for one think that, in the grand scheme of things, the Superconducting Super-Collider would have been more useful than Siri.

Last but not least, we seem to have forgotten that a nuclear war (even if contained between two minor powers) would shorten the life expectancy of everybody on the planet, and possibly even make the planet uninhabitable. Last time I checked, the number of nuclear powers had increased, not decreased, and, thanks to rapid technological progress and to the electronic spread of knowledge, there are now many more entities capable of producing nuclear weapons.

The unbridled optimism of the Artificial Intelligence community, and of the media that propagate it, is not justified because A.I. is not helping to solve any of these impelling problems. We desperately need machines that will help us solve these problems. Unbridled optimism is not a replacement for practical solutions.

The enthusiastic faith that Rome was the "eternal city" and the firm belief that Venice was the "most serene republic" did not keep those empires from collapsing. Unbridled optimism can be the most lethal weapon of mass destruction.

The Future of Human Creativity

Maybe we should focus on what can make us (current Homo Sapiens people) more intelligent, instead of focusing on how to build more intelligent machines that will make our intelligence obsolete. Creativity is what truly sets Homo Sapiens apart from other species.

There are two myths here that i never bought. The first one is that adults are more intelligent than children, and therefore children have to learn from adults, not vice versa.

Children perform an impressive feat in just a few years, acquiring an incredible amount of knowledge and learning an incredible portfolio of skills. They are also fantastically creative in the way they deal with objects and people. Teenagers are still capable of quick learning (for example, foreign languages) and can be very creative (often upsetting parents and society that expect a more orthodox behavior, i.e. compliance with rules). Adults, on the other hand, tend to live routine lives and follow whatever rules they are told to obey.

When i look at the evidence, it seems to be that creativity, and therefore what is unique about human intelligence, declines with age. We get dumber and less creative, not smarter and more creative; and, once we become dumb adults, we do our best to make sure that children too become as dumb as us.

Secondly, the people of the rich developed high-tech world implicitly assume that they are more intelligent and creative than the people of the poor undeveloped low-tech world. In my opinion, nothing could be farther from the truth. The top of creativity is encountered in the slums and villages of the world. It is in the very poor neighborhoods that humans have to use their brain every single minute of their lives to come up with creative and unorthodox solutions, solutions that nobody taught them, to problems that nobody studied before. People manage to run businesses in places where there is no infrastructure, where at any time something unpredictable can (and will) happen. They manage to sell food without a store. They manage to trade without transportation. When they obtain a tool, they often use it not for the purpose for which it was originally designed but for some other purpose. They devise ever new ways to steal water, electricity, cable television and cellular phone service from public and private networks. They find ways to multiply and overlap the functions of the infrastructure (for example, a railway track also doubles as a farmer's market, and a police road-block becomes a snack stop). They help each other with informal safety networks that rival state bureaucracies (not in size or budget, but in effectiveness). The slums are veritable laboratories where almost every single individual (of a population of millions) is a living experiment (in finding new ways of surviving and prospering). There is no mercy for those who fail to "create" a new life for themselves every day: they stand no chance of "surviving".

If one could "measure" creativity, i think the slums of the world would easily outperform Silicon Valley.

Robots will replace Silicon Valley engineers way before they can replace the humble seller of pillows at the bus station who walks around barefoot trying to locate the most likely customer among the thousands of frantic long-distance passengers.

These highly creative people yearn for jobs in the "white-collar" economy, the economy of the elite that lives outside the slums. For that "white-collar" economy they may perform trivial repetitive jobs (chauffeur, cashier, window washer); which means that they have to leave their creativity at home. The "white-collar" economy has organized daily life in such a way (around "routines") that everybody is guaranteed to at least survive. The people of the slums use their brains only when they live and work in the slums. When they live or work outside the slums, they are required to stop being creative and merely follow procedures, procedures that were devised by vastly less creative people who would probably not survive one day in the slums. Routines maximize productivity precisely by reducing human creativity. Someone else "creates", and the worker only has to perform, a series of predefined steps. The slum dweller cannot be replaced by a machine, but the "routinized" worker can be.

The routine, however, is useful for businesses because it can "amplify" the effect of innovation. The innovation may be very small and very infrequent, but the effect of the routine performed by many workers (e.g., by many Silicon Valley engineers) is to make even the simplest innovation relevant for millions of individuals.

The creativity of slums and villages, on the other hand, is constant, but, lacking the infrastructure to turn it into routine, ends up solving only a small problem for a few individuals. The slums are a colossal reservoir of creative energies that the world is wasting, and, in fact, suppressing.

In our age we are speeding up the process by which (rule-breaking) children become (rule-obeying) adults and, at the same time, we are striving to turn the creativity of the slums into the routine of factories and offices. It seems to me that these two processes are more likely to lead to a state of lower rather than higher intelligence for the human race.

I suspect that removing the unpredictable from life means removing the very essence of the human experience and removing the very enabler of human intelligence. On the other hand, removing the unpredictable from life is the enabler of machine intelligence.

Why I am not Afraid of A.I.

Between 2014 and 2015 Silicon Valley's serial entrepreneur Elon Musk and British physicist Stephen Hawking, as well as the richest man in the world, Bill Gates, all sounded alarm bells about the danger posed to humankind by Artificial Intelligence. They were descendants of Bill Joy's "The Future doesn't need us". In 2016 Elon Musk and Peter Thiel founded OpenAI, a non-profit organization with the mission "to advance digital intelligence in the way that is most likely to benefit humanity as a whole". They hired Ilya Sutskever, formerly at Google and in Hinton's group, to lead the research, and hired advisors such as Pieter Abbeel of UC Berkeley, Yoshua Bengio, and personal-computer pioneer Alan Kay. Even former secretary of state Henry Kissinger weighed in, writing an article in the Atlantic titled "How the Enlightenment Ends" (June 2018).

Gates, Hawking and Musk had a predecessor in the visionary art historian Jack Burnham. In the final pages of his book "Beyond Modern Sculpture" (1968) Burnham predates today's apocalyptic transhumanism, hinting at the possibility that humans may be engineering the very life form that will succeed them on this planet.

I, on the other hand, am not afraid of A.I. because we are not even remotely close to having truly intelligent machines.

I am not afraid of A.I.: I am afraid that it will not arrive soon enough. Machines are essential to our well-being today, and will increasingly determine our well-being in the future, and more intelligent machines are probably indispensable to solve many of the gravest problems of our era.

A world without robots is a world in which humans have to work for very low wages in order to produce goods that ordinary families can afford to buy. It is a world in which only the rich could afford a car or even a TV set. A world without robots is a world in which humans would have to perform all sorts of dangerous and unhealthy jobs, such as cleaning up Fukushima's nuclear disaster, and would have to work in horrible conditions inside mines and steelworks. Robots are being used to disarm suicide bombers and to remove landmines. Without robots, these tasks would be carried out by human beings. A world without robots would be a terrible world.

Robots suffer from poor marketing. Robots are mostly presented as big, scary beasts. We should instead publicize the fact that someday the hardware store next door will offer tiny robots capable of crawling inside the plumbing of our house and of unclogging the pipes. Robotic "exosuits" will allow us to lift and carry heavy weights in the backyard. And so forth: robots will help us by solving practical problems around the house.

Do we need service robots? Do they steal jobs? Have you ever desperately looked for a human being in a large store, a shopping mall, a hotel lobby, a train station, a public office, or even in a street? How many times does it happen that you have a simple question and there's nobody to talk to? Or maybe there are people but they don't speak your language, or they are tourists like you, or they are wearing headsets and listening to music? Wouldn't it be nice if a robot could answer your question and, if it is about a location, even take you there? It could even be that humankind finally gets rid of the "hours of operation". Service robots can someday keep a store open nonstop even when all the human staff is asleep. That would be a much more dramatic achievement than super-intelligent machines.

These service robots do not steal jobs. Those jobs today don't exist. Maybe they existed a century ago, and maybe they still exist in poor countries. I have been in developing countries where a clerk welcomes you when you enter the store and gladly helps you to find the product that you are looking for. But this has become a rarity in developed countries.

Instead of worrying about the jobs that may be "stolen" by machines, we should worry about the jobs that we will soon need and for which we are not prepared. Taking care of elderly people is a prime example. There is virtually no country where population growth is accelerating. In most countries of the world, population growth is decelerating. In some countries it is turning negative. In many countries the population has peaked and will soon start decreasing while at the same time aging thanks to improvements in medicine. In other words, many countries need to prepare for a future with a lot of older people with fewer younger people who can take care of them. In the Western world the 1950s and 1960s were the age of the "baby boom". The big social revolution of the 21st century will be the boom of elderly people. The rich world is entering the age of the "elderly boomers". Who is going to take care of that aging population? Most of these aging people will not be able to afford full-time human care. The solution is robots: robots that can go shopping, that can clean the house, that can remind people to take their medicines that can check their blood pressure, etc. Robots could do all of these things day and night, every day of the year, and at an affordable price. I am afraid that A.I. will not come soon enough and we will face the aging apocalypse.

We profess that we want all the people in the world to become rich like the rich Western countries, but the truth is that any "rich" society needs poor people to perform all the vital jobs that the "richer" people refuse. Poor people take care of most of the chores that keep society working and that keep us alive. Those are humble and low-paid jobs such as collecting garbage and making sandwiches. We profess that we want all eight billion people of the planet to have the same standard of living that the rich world has, but what happens when all eight billion people become rich enough that nobody wants to take those humble and low-paid jobs? Who is going to collect the garbage once a week, who is going to make sandwiches at the lunch cafeteria, who is going to clean the public bathrooms, who is going to wash the windows of the office buildings? We don't want to admit it, but today we rely on the existence of millions of poor people who are willing to do those jobs that we don't want to do. I hope that we will indeed solve the problem of poverty in 50 years or even less; what that means is that we only have 50 years to invent robots that can do all the jobs that people will not want to do 50 years from now. I am not scared of robots, i am scared of what will happen in 50 years if we don't have intelligent robots to collect garbage, make sandwiches, clean bathrooms, etc.

A world without robots is a dysfunctional world, a world of very poor people working and living in horrible conditions, a world of societies that cannot care for the elderly and that cannot help people with permanent disabilities.

A world without robots is a scary place.

A Conclusion: Religion in the Age of A.I.

Humans have been expecting a supernatural event of some kind or another since prehistory. Human brains seem to be programmed to believe in the supernatural and to strive for immortality.

As mentioned at the beginning of this book, we are witnessing the birth of a new religion, a religion that believes in a supernatural world that exists not in this universe, not in the heavens, but in the dataverse.

According to this new "religion", A.I. will generate a kind of supernatural intelligence that will rule over the human world.

In retrospect, ancient religions were realistic: they admitted that we all have to die and looked for hope in the afterlife. This was an empirical and rational approach. The narrative of the Singularity denies the obvious: that everything has an ending. This is neither empirical (there is no evidence of eternal immortal beings) nor rational (there is no science that would justify something living longer than the lifetime of the universe, or, for that matter, just the lifetime of the Sun).

Rationality is under attack from both the right and the left, the "right" being modern spirituality and the "left" being the Singularity camp.

The modern spirituality (the "right") largely rejects the superstitions of Judaism, Christianity and Islam in favor of Daoism and Buddhism, the two philosophies (not quite religions) that seem to best match what we know about the universe; and the two philosophies that have been prominent in the San Francisco Bay Area since the 1960s, i.e. the two that Silicon Valley naturally encountered as it abstracted technology into a philosophy of life. The "right" was born approximately with Fritjof Capra's "The Tao of Physics" (1975), written by a physicist, and Michael Singer's bestseller "The Untethered Soul" (2007) or William Broad's "The Science of Yoga" (2012) are typical examples of its mature stage, leaving behind attempts to merge spirituality and physics such as Deepak Chopra's "Quantum Healing" (1989) and Danah Zohar's "The Quantum Self" (1990), co-written with a physicist.

The "left" is part of a more general movement that thinks of the universe as ruled by data. Israeli historian Yuval Harari calls it a new religion, Dataism, in his book "Sapiens - A Brief History of Humankind" (2011). This "left" has faith in science, and notably in computer science. According to the Pew Research Center, in 2015 a whopping 89% of US adults believed in some kind of god (in Europe we estimate that number to be about 77%), but very few of them are looking forward to the afterlife. Most of them are terrorized by the idea of dying. Their real religion is medicine, and, indirectly, science. They tolerate scientific research in chemistry, biology, physics and the like because they hope that it will contribute to progress in medicine. They were happy that in 1993 the government killed an expensive project to build the world's fastest particle accelerator (the "Superconducting Supercollider") and that in 2011 the government shut down the most powerful particle accelerator in the country (the Tevatron); but tell them that accelerating hadrons will prolong their lives and they will gladly pay taxes for the most expensive particle accelerator ever built. Tell them that space exploration will prolong their lives and they will gladly pay taxes for a mission to Saturn. Tell them that Artificial Intelligence will grant them immortality, and that a super-human intelligence (the Singularity) is truly coming soon, and they will react the way Jews, Christians and Muslims used to react to the news that the Messiah is coming: fear and hope; fear that the Messiah will send us to hell (a well-deserved hell, judging from my favorite history books) and hope that the Messiah will be so merciful as to grant us immortality.

I also submit that much of the "exponential progress" that we witness today is due to the retreat of religious institutions. Religious institutions, whether in the Catholic world or in the Islamic world, mostly resisted governments. They would not support and sometimes not even protect scientists, engineers, philosophers and physicians who hinted in any way that the soul does not exist or that it is the mere manifestation of electrochemical brain processes. Religion is naturally hostile to technological and scientific progress because progress distracts from the foundation of religious morality, the soul, besides clearly upsetting the traditional social life upon which priests rely for their power. Therefore, we live inside a positive-feedback loop: religions decline, their decline fosters scientific and technological progress, progress causes religion to decline, etc. No wonder that belief in spiritual superbeings is rapidly being replaced by belief in a technological superbeing.

Traditional religions worked well when there was no hope for a remedy to death. Hence the only remedy was faith in a god's mercy. The new technological religion offers a terrestrial remedy to death: terrestrial longevity if not immortality. Whether this new religion is any more realistic than the ancient Western religions is debatable; whether the Singularity or the Messiah or nobody comes to our rescue, remains to be seen.

The Singularity may become the new religion for the largely atheistic crowd of the high-tech world. Just like with Christianity and Islam, the eschatological mission then becomes how to save oneself from damnation when the Singularity comes, balanced by the faith in some kind of resurrection.

We've seen this movie before, haven't we?

P.S. What i did not Say

I want to emphasize what i did not say in this book.

I did not claim that an artificial general intelligence is impossible, (only that it requires a major revolution in the field); and i certainly did not argue that superhuman intelligence is not possible (in fact, i explained that it is already all around us); and i did not rail against technological progress (i lamented that its achievements and benefits are wildly exaggerated).

I did not write that technology makes you stupid. I wrote that rules and regulations make you stupid; and technology is used to establish, enforce and multiply those rules and regulations (often giving machines an unfair advantage over humans).

I did not write that there has been no progress in Artificial Intelligence: there has been a lot of progress, but mainly because of cheaper, smaller and faster processors, and because of better structured environments in which it is easier to operate (both for humans and for machines).

I did not write that humans will never create an artificial intelligence. We have already created Artificial Intelligence programs that do many useful things, as well as some truly obnoxious ones (like displaying ads on everything you do online). The definition of "intelligence" is so vague that the very first computer (or the very first clock) can be considered an artificial intelligence. In fact, early computers were called "electronic brains", not "electronic objects".

I did not write that Artificial Intelligence is useless. On the contrary, i think that it has helped neuroscience. Maybe the "enumeration" problem (the problem of enumerating all the intelligent tasks that are needed in order to achieve artificial general intelligence) is a clue that our own brain might not be "one" but a confederation of many brains, each specialized in one intelligent task.

I did not write that i am afraid of intelligent machines. Quite the opposite: we need more intelligent machines. Technological progress has had many downsides but, overall, it has helped humans live better lives. We do need the progress in machine intelligence that technology has promised and not delivered.

I am not afraid of intelligent machines. I am afraid that they will not come soon enough.

Appendix: The Myth of Longevity

The new cult of digital immortality goes hand in hand with the widely publicized increases in life expectancy.

For centuries life expectancy (at older ages) rose very little and very slowly. What truly changed was infant mortality which used to be very high. But towards the end of the 20th century life expectancy posted an impressive increase: according to the Human Mortality Database, in developed countries life expectancy at age 85 increased by only about one year between 1900 and 1960, but then increased by almost two years between 1960 and 1999. I call it the "100 curve": for citizens of developed countries the chance to live to 100 is now about 100 times higher than it was 100 years ago. In fact, if one projects the current trends according to the Human Mortality Database, most babies born since 2000 in developed countries will live to be 100.

James Vaupel, the founding director of the Max Planck Institute for Demographic Research, showed that the rate of increase in life expectancy is about 2.5 years per 10 years ("Demography", 2002). It means that every day our race's life expectancy increases by six hours. And Vaupel argues that life expectancy is likely to keep increasing.

These studies, however, often neglect facts of ordinary life. Since 1960 the conditions in which people live (especially urban people) have improved dramatically. For centuries people used to live with (and die because of) poor sanitation. The water supply of cities was chronically contaminated with sewage, garbage and carrion. Typhoid, dysentery and diarrhea were common. Outbreaks of smallpox, measles, polio, cholera, yellow fever, assorted plagues and even the flue killed millions before the invention of vaccines and the mandatory immunization programs of the last century. Before the 1960s polio would paralyze or kill over half a million people worldwide every year. Smallpox used to kill hundreds of thousands of Europeans annually (it was eradicated in 1979) and killed millions in the Americas after colonization. The World Health Organization estimates that measles has killed about 200 million people worldwide over the last 150 years (but almost nobody in developed countries in recent decades). Cholera killed 200,000 people in the Philippines in 1902-04, 110,000 in the Ukraine in 1910 and millions in India in the century before World War I. The flu killed at least 25 million people worldwide in 1918, four million in 1957 and 750,000 in 1968. These causes of death virtually disappeared from the statistics of developed countries in the last half century. After 1960 diseases are generally replaced by earthquakes, floods and hurricanes (and man-made famines in communist countries) in the list of the mass killers. The big exceptions, namely tuberculosis (more than one million deaths a year), AIDS (almost two million deaths a year) and malaria (more than 700,000 deaths a year), are now mostly confined to the under-developed countries that are not included in the studies on life expectancy (the World Bank estimates that 99% of deaths due to these three diseases occur in underdeveloped countries).

Another major factor that contributed to extending life expectancy is affordable professional health care. Health care used to be the responsibility of the family before it shifted towards the state. The state can provide more scientific health care, but it is expensive. Professional health care became affordable after World War II thanks to universal health care programs: France (1945), Britain (1948), Sweden (1955), Japan (1961), Canada (1972), Australia (1974), Italy (1978), Spain (1986), South Korea (1989), etc. Among major developed countries Germany (1889) is the only one that offered universal health care before World War II (and the USA is the only one that still does not have one in place).

After improved health care and reduced infectious disease rates, the economist Dora Costa's "Causes of Improving Health and Longevity at Older Ages" (2005) lists "reduced occupational stress" and "improved nutritional intake" as the other major factors that determine longevity. However, work stress is increasing for women, as they ascend the corporate ladder, and data on diets (e.g., obesity) seem to point in the opposite direction: people quit smoking, but now eat junk, and too much of it (and insert here your favorite rant against pesticides, hormone-raised meat and industrialized food in general).

Violent deaths have also decreased dramatically throughout the developed world: fewer and less bloody wars, and less violent crime. The rate of homicide deaths per 100,000 citizens is widely discussed in Steven Pinker's "The Better Angels of Our Nature" (2011). (Even in the USA where guns are widely available, and therefore violent crime kills exponentially more people than in Europe or Asia, the gun homicide rate decreased 49% between 1993 and 2013).

These factors certainly helped extend life expectancy in the developed world, but there is little improvement that they can still contribute going forward. In some cases one can even fear a regression. For example, no new classes of antibiotics have been introduced since 1987 whereas new pathogens are emerging every year, and existing bugs are developing resistance to current antibiotics. On the same day of March 2013 that a symposium in Australia predicted drugs to slow down the ageing process within a decade so that people can live to 150 years, the Chief Medical Officer for England, Dame Sally Davies, raised the alarm that antibiotics resistance may become a major killer in the near future. The Lancet, the British medical journal, estimated that in 2013 more than 58,000 babies died in India because they were born with bacterial infections that are resistant to known antibodies.

Drug-resistant tuberculosis killed an estimated 170,000 people in 2012. A 2016 report by the British government and the Wellcome Trust estimated that 700,000 people die every year from infections caused by drug-resistant pathogens. Instead of machine super-intelligence we should worry about biological super-bacteria.

The American Cancer Society calculated 1.6 million new cases of cancer and nearly 600,000 deaths in the USA in 2012, which means that the number of cancer deaths in the USA has increased by 74% since 1970. The World Health Organization's "World Cancer Report 2014" estimated that cancer cases will increase by 70 percent over the next 20 years.

The future promises more biomedical progress, and particularly therapies that may repair and reverse the causes of aging. This leads many to believe that human life can and will be extended dramatically, and maybe indefinitely.

However, health care has become too expensive for governments to continue footing the bill for the general population. Virtually every society in the developed world has been moving towards a larger base of elderly people and a smaller base of younger people who are supposed to pay for their health care. This equation is simply not sustainable. The professional health care that the average citizen receives may already have started to decline, and may continue to decline for a long time. It is just too expensive to keep the sick elderly alive forever for all the healthy youth who have to chip in. To compound the problem, statistics indicate that the number of people on disability programs is skyrocketing (14 million people in the USA in 2013, almost double the number of 15 years earlier). At the same time the tradition of domestic health care has largely been lost. You are on your own. This parallel development (unaffordable professional health care combined with the disappearance of domestic health care) is likely to reverse the longevity trend and lead to a diminished (not better) chance of living a long life.

There have already been several times in the history of the Earth when intelligent life almost went extinct; and natural events such as plagues, volcano eruptions and meteorite crashes still constitute a threat to the fragile bodies that host our minds; not to mention the possibility of self-destruction by a nuclear war or by some kind of collective religious martyrdom in the name of some medieval superstition; i.e. by the stupidity of our so-called "intelligent" minds.

Furthermore, the rate of suicide has been increasing steadily in most developed societies, and, for whatever reason, it usually goes hand in hand with a decline in birth rates. Hence this might be an accelerating loop. The country with the oldest people is Japan. That is also one of the countries with the highest suicide rates of all, and most of the suicides are committed by elderly people. Getting very old does not make you very happy. In 2013 the Center for Disease Control (CDC) found that the suicide rate among middle-aged people in the USA had increased 28% in a decade (40% for white people) and that since 2009 suicide had become the 10th leading cause of death in the country, overtaking car accidents.

As all countries reach the point of shrinking health care and accelerating suicide rates, life expectancy will start to decline for the first time in centuries.

I watched the billboards in San Francisco announcing "the first person to live to 150 has already been born" the day after reading the article titled "Have Smartphones Destroyed a Generation?" in The Atlantic (September 2017), written by psychologist Jean Twenge of San Diego State University, that basically describes how smartphones are raising a suicidal generation. A study published in 2017 by Judith Weissman's team at New York University found that mental illness and suicide are on the rise, as are stress, anxiety and depression. Meanwhile, the numbers from the Centers for Disease Control and Prevention show that suicide has become the second leading cause of deaths after car accidents among young people aged 10 to 24.

I am not sure how to count this, but the opioid epidemics that are ravaging the USA is hardly a promising phenomenon for the future of the species. In 2016, about 60,000 people died of a drug overdose in the USA alone (more victims than during the AIDS epidemics at its peak and five times more than the victims of the ebola outbreak in West Africa of 2014-15). In 2017 drug overdoses became officially the leading cause of death for people under the age of 50. In the same year a study by the National Institute on Alcohol Abuse and Alcoholism in New York (published in JAMA Psychiatry by Bridget Grant) found remarkable increases in alcohol abuse (and related mortality rates) over ten years. The number of people at risk because of alcohol abuse was about 30 million, which is 15 times more than the ones affected by drug addiction.

According to the National Center for Health Statistics of the US government, in 2015 the death rate rose for the first time in a decade. The CDC speculated that more people were dying from drug overdoses, suicide and Alzheimer's disease, but the death rate from heart disease, long in decline, was also slightly higher.

Jeanne Louise Calment died at the age of 122 in 1997. Since then no person in the developed world (where you can verify the age) has died at an older age. Even if you believed the claims from various supercentenarians in developing countries (countries in which no document can prove the age of very old people), you could hardly credit their achievement on technological or medical progress since those supercentenarians lived all their lives with virtually no help from technology or medicine. In other words, the real numbers tell us that in almost 20 years nobody has reached the age that someone reached in 1997 with the possible exception of people who lived in underdeveloped countries. It takes a lot of imagination to infer from this fact that we are witnessing a trend towards longer life-spans.

There is also a shift in value perception at work. The idea that the only measure of a life is the number of years it lasted, that dying of old age is "better" than, say, dying in a car accident at a young age, is very much grounded in an old society driven by the survival instinct: survive at all costs for as long as possible. As the (unconscious) survival instinct is progressively replaced by (conscious) philosophical meditation in modern societies, more and more people will decide that dying at 86 is not necessarily better than dying at 85. In the near future people may care more about other factors than the sheer number of years they lived. The attachment to life and the desire to live as long as possible is largely instinctive and irrational. As generations become more and more rational about life, longevity may not sound so attractive if one has to die anyway and be dead forever and be forgotten for the rest of eternity.

Then there are also many new habits that may contribute to creating a sicker species that will be more likely (not less likely) to die of diseases.

Most children in the developed world are now born to mothers (and fathers) aged 30 and older. As more parents delay childbearing, and the biological clock remains the same, the new generations are a veritable experiment (and sometimes literally a laboratory experiment). Fertility rates begin to decline gradually at age 30 and then decline exponentially, and to me that is nature's way of telling us when children should be born. In fact, babies born to older parents (as it is becoming common) have a higher risk of chromosome problems, as shown, for example, in a study led by Andrew Wyrobek and Brenda Eskenazi, "Advancing Age has Differential Effects on DNA Damage, Chromatin Integrity, Gene Mutations, and Chromosome Abnormalities in Sperm" (2006), and by a study led by Bronte Stone at the Reproductive Technology Laboratories, "Age thresholds for changes in semen parameters in men" (2013). Autism rates have risen 600 percent in the past two decades. While the age of the parents may not be the only cause, it is probably one significant cause. In Janie Shelton's and Irva Hertz-Picciotto's study "Independent and Dependent Contributions of Advanced Maternal and Paternal Ages to Autism Risk" (2010) the odds that a child would later be diagnosed with autism was 50% greater for a 40-year-old woman than for a woman between 25 and 29. To be fair, a study led by Mikko Myrskyla at the Max Planck Institute for Demographic Research in Germany, "Maternal Age and Offspring Adult Health" (2012), reassured many older mothers that education is the main factor determining the future health of their babies.

Nobody knows its causes and it is difficult to speculate on the effects, but several studies from European nations seem to show that the quality of sperm has deteriorated during the last half of the previous century. I doubt that this bodes well for the physical and mental health of our offspring. For example, a study led by the epidemiologist Joelle LeMoal ("Decline in Semen Concentration and Morphology in a Sample of 26609 Men Close to General Population Between 1989 and 2005 in France", 2012) found that sperm concentration of young men decreased by nearly 30% in 17 years. In 2017 an international team of researchers led by Shanna Swan of Mount Sinai Health System in New York published the result of a study that involved tens of thousands of men in Western countries between 1973 and 2011 and concluded that sperm counts have declined 52%. If similar numbers showed up for environmental problems in a certain territory, we would immediately evacuate the place.

Last but not least, antibiotics, filtered water, cesarean-section childbirths and other environmental and behavioral aspects of modern life in developed countries greatly weaken the beneficial bacteria that constitute the physical majority of the cells of the human body. Vaccinations are useful to prevent children from contracting horrible diseases, and i have nothing against them, but the immune system is largely shaped by the environment and, indirectly, vaccines fool the immune system into thinking that the environment is what it is not: as the number of vaccinations multiply even in adult life, we are de facto engineering an artificial immune system and we just hope that it will still work well in any real environment. Therefore health care itself is tampering with an immune system that evolution designed to be capable of adapting to unpredictable attackers. Whether it will work better than the unprotected immune system of our grandparents when attacked by new unknown viruses is debatable.

Personal mobility has greatly increased the chances that a deadly epidemic spreads worldwide killing millions of people.

Dana King's study "The Status of Baby Boomers' Health in the United States" (2013) seems to show that the "baby boomer" generation is less healthy than the previous generation. For example, twice as many baby boomers use a cane as did people of the same age in the previous generation. I personally have the feeling that the young people around me are less healthy than my generation was. Too many young people seem to have all sorts of physical problems and seem to get sick with every passing germ. They get food poisoning the moment they cross a border, and they start taking all sorts of pills in their 30s. I don't see how this bodes well for our race's longevity.

We now routinely eat genetically-manufactured food whose effects over the long term are yet to be determined.

One fears that most of the gains in life expectancy may have already occurred, and now the challenge will be to preserve them. I can't shake off the feeling that we are building a weaker and weaker species while creating a more and more dangerous world.

In December 2017 the US National Center for Health Statistics reported that the USA experienced a decline in life expectancy for two consecutive years. The last time that it fell for two consecutive years was in 1962-63 (source: Center for Disease Control and Prevention). If you think that this must be related to the crazy habits of US citizens like guns and drugs, think again: in 2016 life expectancy also declined in Italy, one of the countries that traditionally tops the lists of longevity (as i am typing, the oldest person in the world is an Italian woman).

If you think that this was caused by the opioid epidemics, a study by Steven Woolf's team at Virginia Commonwealth University that looked at data between 1999 and 2016 showed that all cause mortality in midlife increased among all racial groups of the USA ("Changes in Midlife Death Rates Across Racial and Ethnic Groups in the United States", 2018). The USA, where studies on longevity tend to be concentrated, has the lowest life expectancy among high-income countries, and life expectancy is even decreasing, obviously a sign that, so far, researching longevity is not the right way to achieve longevity.

Ironically, life expectancy is higher in Hiroshima (where people had to live with nuclear radiations after 1945) than in the USA (sources: Japanese Ministry of Health, Labour and Welfare and US Center for Disease Control and Prevention) which to me means that all the modern scientific progress in medicine is irrelevant compared with genetic, cultural and/or dietary factors.

I was fascinated in 2016 when i saw at the same time two recent statistics, one (by the Pew Research Center) about the exponential growth of social media and the other (by the Center for Disease Control) about suicide rates. The number of suicides in the USA has been rising since 1999 in every age group and for both sexes. The rate of suicide has increased from 10.5 per 100,000 in 1999 to 13 per 100,000 in 2014. A study by David Lester, one of the world's experts in suicide statistics, published in 2002 in the Journal of the American Medical Association, noted a decline in the overall suicide rate between 1987 and 1997. So the suicide rate in the USA declined until 1997, then it started rising again. The coincidence is interesting: 1997 is the year that the first social network, Six Degrees, was launched.

Appendix: The Cloud of Invisible Robots

We now thrive in the "data economy". A credit card company is no longer a financial company: it has become a data company (it knows what you buy). Social media are obviously data companies: they know what your interests are, whom you are connected with, and what you like. What do these technologies have in common: Internet of Things, biotech, wearables, social media...? They produce data. Billions of data. According to IDC, there will be 44 zettabytes of data by 2020. Just about every new technology is producing huge amounts of data, and deep learning is one technique to analyze those data. It turns out that "big data" is also what Artificial Intelligence needs to train neural networks. Deep learning needs large datasets; large datasets need deep learning. Traditionally, neural networks were about pattern recognition (such as image or speech recognition). As data become available on such a large scale, a new natural application (the "killer application") becomes increasingly tempting neural: profile recognition. If we put together what you buy, what you do, whom you befriend, etc., we obtain your digital behavior, and that's pretty much your mind (desires, beliefs, needs).

Invisible robots can analyze all the data that we leave behind on the Web and learn a lot about us. These invisible robots can deduct intimate facts about your life. Not only does the invisible robot know your digital behavior: it also knows the digital behavior of everybody else. It knows the correlation between circumstances and human behavior; and it can guess which behavior will follow from your circumstances. The next step is to affect your circumstances, to "persuade" you to adopt a certain behavior.

The invisible robot can "persuade" you because it knows your mind. It can customize a message for your mind that will generate the desired behavior. Information can be crafted to "design" your behavior. This is nothing new: marketing specialists have always done this. The difference is that now a) it can be customized for each individual; and b) the "bots" can find out a lot more about you than any marketing study could dream of.

The machine knows more about you than your best friends, and has more data about everybody else than you do; so the machine can use knowledge about humans to understand how humans act, and then use your data to understand how to manipulate you.

Sounds scary? It gets worse: thousands of invisible bots will collaborate on this project.

The bot doesn't understand your conversations and doesn't really "learn" anything; but it can do magic with the digital footprint that you leave behind when you buy something from an online store and when you engage with people on social media.

Invisible robots are the future not only of marketing. A bot trained with a dataset of voter behavior can customize a political candidate's message to maximize the number of people who will vote for that candidate. A bot trained with a dataset of performance appraisals can be more efficient than a human recruiter in finding the best person for a job. A bot, trained with the historical record of repeat offenders, can even advise a judge on the optimal sentence because it can guess the probability that the convicted defendant will commit the same crime again once released from prison.

Bruce Tognazzini, the influential designer of early Apple, has written that in our age an invisible computer works for an invisible user. I would rephrase it as: "an infinite crowd of invisible robots work nonstop for an infinite crowd of invisible users".

The engineer of the future is the "behavior designer". The engineer of the future will need to also understand the language of machines, not just the language of humans; and the language of machines is computational mathematics.

You want to be the one who shapes Artificial Intelligence, not the one who is shaped by Artificial Intelligence.

Appendix: The Era of Objects

As we speculate on what will be the next stage of life's evolution, we underestimate the real innovation that has happened since the invention of intelligent life: objects. Life started building objects.

Life has not evolved much in the last ten thousand years, but objects have: there has been an explosive proliferation of objects.

We tend to focus on objects that mimic life (robots and the like) as candidates for replacing life as we know it, and in the long term that might well be correct, but the objects that have truly multiplied and evolved at an astonishing rate are the ordinary static objects that populate our homes, our streets and our workplaces. There are objects virtually for anything.

When we look at life's evolution, we tend to look at how sophisticated the brain has become, but we tend to underestimate what that "sophisticated" brain is built to do: make more objects. The chimp's brain and the human brain are not all that different, and the behavior of the two species (eating, sleeping, sex, and perhaps even consciousness) are not all that different from the viewpoint of a (non-chimp and non-human) external observer, but the difference in terms of objects that they make is colossal. The real evolution of the brain is in terms of the objects it can build.

What the human race has truly accomplished is to turn a vast portion of the planet into objects: paved streets and sidewalks, buildings, cars, trains, appliances, clothes, furniture, kitchenware, etc.

Our lives revolve around objects. We work to buy a car or a home, and our work mostly consists in building or selling objects. We live to use them (usually in conjunction with other objects), to place them somewhere (usually inside other objects), to clean them (using other objects), etc.

The fundamental property of life is that it dies. Everything that was alive is now dead, except for those that are dying now. For us the planet is just a vast cemetery. For objects, instead, this planet is a vast factory because, before dying, each of us builds or buys thousands of objects that will survive us and that will motivate future generations to build and buy more objects.

It is neither living beings nor genes nor memes (ideas) that evolve and drive evolution on this planet: it is objects. Objects have evolved far faster than life or ideas. The explosive proliferation of objects is the one thing that would be visible to anyone playing the last ten thousand years of history on Earth. Everything else (politics, economics, natural disasters, etc) pales in comparison to the evolution and proliferation of objects. The human body has not changed much in 200,000 years. Ideas have changed but slowly. Objects, instead, have completely changed and keep changing rapidly.

For example, what caused the collapse of the Soviet Union (and of communism in general) was neither the Pope nor Afghanistan: it was consumerism. Soviet citizens wanted goods in their stores, lots of goods. They actually liked many features of the communist society (they still do) but they wanted the proliferation of goods that democratic/capitalist societies offer. It was all about objects. Hence the Soviet Union collapsed because it dared to challenge the domination of objects. For the same reason religions are declining everywhere: they are being replaced by philosophies of life that are more materialistic, i.e. that increase the evolution of objects.

Any system that challenges the absolute power of objects, or that doesn't contribute enough to the survival, proliferation and evolution of objects tends to lose. What benefits objects tends to succeed. Objects rule. Perhaps we are merely supposed to follow their orders, and that's the only meaning of life. We get annihilated if we dare contradict objects.

You may think that you are changing a car because you want a new car, but you can also see it the other way around: it is cars that want you to spend money that will go into making more and better cars.

In a sense, the consumer society is one stage in the evolution of objects, invented by objects in order to speed up their own evolution. Consumers are just the vehicle for objects to carry out their strategy of proliferation and domination.

Eventually objects will evolve into space stations and extraterrestrial colonies in order to expand outside this planet and begin the colonization of other parts of the universe in their quest to dominate all the matter that exists, until all matter in the universe will have been turned into objects by object-creating beings like us (in clear defiance of the second law of Thermodynamics).

We are even turning our food into objects as we increasingly eat packaged food. The food system has changed more over the last 40 years than in the previous 40,000 years.

Shoes, refrigerators, watches and underwear are the real protagonists of history. Everything else is just a footnote in their odyssey.

(By the same token, i think that videogames play people, not the other way around).

A Timeline of Artificial Intelligence

1935: Alonzo Church proves the undecidability of first order logic
1936: Alan Turing's Universal Machine ("On computable numbers, with an application to the Entscheidungsproblem")
1936: Alonzo Church's Lambda calculus
1941: Konrad Zuse's programmable electronic computer
1943: "Behavior, Purpose and Teleology" co-written by mathematician Norbert Wiener, physiologist Arturo Rosenblueth and engineer Julian Bigelow
1943: Kenneth Craik's "The Nature of Explanation"
1943: Warren McCulloch's and Walter Pitts' binary neuron ("A Logical Calculus of the Ideas Immanent in Nervous Activity")
1945: John Von Neumann publicizes the notion of a computer that holds its own instructions, the "stored-program architecture"
1946: The ENIAC computer
1946: The first Macy Conference on Cybernetics
1947: John Von Neumann's self-reproducing automata
1948: Alan Turing's "Intelligent Machinery"
1948: Norbert Wiener's "Cybernetics"
1949: Leon Dostert founds Georgetown University's Institute of Languages and Linguistics
1949: William Grey-Walter's Elmer and Elsie robots
1949: Warren Weaver's "Translation" memorandum
1949: The Ratio club
1950: Claude Shannon's "Programming a Computer for Playing Chess"
1950: Alan Turing's "Computing Machinery and Intelligence" (the "Turing Test")
1951: Claude Shannon's maze-solving robots ("electronic rats")
1951: Karl Lashley's "The problem of serial order in behavior"
1952: First International Conference on Machine Translation organized by Yehoshua Bar-Hillel
1952: Ross Ashby's "Design for a Brain"
1953: Harvey Chapman's "Garco" robot
1953: Marshall Rosenbluth invents the "Metropolis algorithm", implemented by his wife Arianna, the first Markov Chain Monte Carlo method
1954: Wesley Clark and Belmont Farley build the first computer simulation of a neural network
1954: Marvin Minsky's thesis on reinforcement learning
1954: Demonstration of a machine-translation system by Leon Dostert's team at Georgetown University and Cuthbert Hurd's team at IBM, possibly the first non-numerical application of a digital computer
1955: The Western Joint Computer Conference with papers by Newell, Selfridge, Clark, etc
1955: Arthur Samuel's Checkers, the world's first self-learning program, and the first implementation of the alpha-beta algorithm
1956: Allen Newell and Herbert Simon demonstrate the "Logic Theorist"

1956: Dartmouth conference on Artificial Intelligence
1956: Ray Solomonoff's inductive inference engine
1956: Gordon Pask's special-purpose electro-mechanical automata SAKI and Eucrates
1957: Richard Bellman's "Dynamic Programming"
1957: Frank Rosenblatt's Perceptron
1957: Newell & Simon's "General Problem Solver"
1957: Noam Chomsky's "Syntactic Structures" (transformational grammar)
1958: John McCarthy's LISP programming language
1958: Oliver Selfridge's Pandemonium
1958: John McCarthy's "Programs with Common Sense" focuses on knowledge representation
1958: Yehoshua Bar-Hillel's "proof" that machine translation is impossible
1959: John McCarthy and Marvin Minsky found the Artificial Intelligence Lab at MIT
1959: Noam Chomsky's review of a book by Skinner ends the domination of behaviorism and resurrects cognitivism
1959: The industrial robot Unimate is deployed at General Motors
1959: Bernard Widrow's and Ted Hoff's Adaline ((Adaptive Linear Neuron or later Adaptive Linear Element) that uses the Delta Rule for neural networks
1959: Zellig Harris' team writes the first natural-language parser
1960: Henry Kelley and Arthur Bryson invent backpropagation
1960: Donald Michie's reinforcement-learning system MENACE
1960: Hilary Putnam's Computational Functionalism
1961: Melvin Maron's "Automatic Indexing"
1961: Leonard Scheer's and John Chubbuck's Mod I (1962) and Mod II (1964)
1963 Irving John Good (Isidore Jacob Gudak) speculates about "ultraintelligent machines" (the "singularity")
1963 John McCarthy moves to Stanford and founds the Stanford Artificial Intelligence Laboratory (SAIL)
1963: Edward Feigenbaum's and Julian Feldman's "Computers and Thought"
1963: Vladimir Vapnik and Alexey Chervonenkis invent the support vector machine (SVM)
1964: IBM's "Shoebox" for speech recognition
1965: Alexey Ivakhnenko publishes the first learning algorithms for multilayer networks
1965: Ed Feigenbaum's Dendral expert system
1965: Lotfi Zadeh's Fuzzy Logic
1965: Bruce Lacey's robot Rosa Bosom at the Cybernetic Serendipity exhibition of computer art
1966: Leonard Baum's Hidden Markov Model
1966: Joe Weizenbaum's Eliza
1966: Ross Quillian's semantic networks
1967: Charles Fillmore's Case Frame Grammar
1968: Glenn Shafer and Stuart Dempster's "Theory of Evidence"
1968: Peter Toma founds Systran to commercialize machine-translation systems

1969: First International Joint Conference on Artificial Intelligence (IJCAI) at Stanford

1969: Marvin Minsky & Samuel Papert's "Perceptrons" kill neural networks

1969: Roger Schank's Conceptual Dependency Theory for natural language processing

1969: Cordell Green's automatic synthesis of programs

1969: Stanford Research Institute's Shakey the Robot

1970: Albert Uttley's Informon for adaptive pattern recognition

1970: William Woods' Augmented Transition Network (ATN) for natural language processing

1971: Richard Fikes' and Nils Nilsson's STRIPS planner

1971: Ingo Rechenberg's "Evolution Strategies"

1972: Alain Colmerauer's PROLOG programming language

1972: The first chatbot to chatbot conversation ever takes place over the Arpanet between Kenneth Colby's chatbot Parry at Stanford and Eliza at MIT

1972: Harry Klopf's "Brain Function and Adaptive Systems"

1972: Bruce Buchanan's MYCIN

1972: Hubert Dreyfus's "What Computers Can't Do"

1972: Terry Winograd's Shrdlu

1973: "Artificial Intelligence: A General Survey" by James Lighthill criticizes Artificial Intelligence for over-promising

1973: Ichiro Kato's Wabot, the first anthropomorphic walking robot

1973: Jim Baker applies the Hidden Markov Model to speech recognition

1974: Marvin Minsky's Frame

1974: Paul Werbos' Backpropagation algorithm for neural networks

1975: The first Artificial Intelligence in Medicine workshop at Rutgers University

1975: John Holland's genetic algorithms

1975: Roger Schank's Script

1975: Raj Reddy's team at CMU develops three speech-recognition systems (Bruce Lowerre's "Harpy", Hearsay-II and Jim Baker's Dragon)

1976: Stephen Grossberg's Adaptive Resonance Theory (ART) for unsupervised learning

1976: Fred Jelinek's "Continuous Speech Recognition by Statistical Methods"

1976: Richard Laing's paradigm of self-replication by self-inspection

1978: Ryszard Michalski builds the first practical system that learns from examples, AQ11

1978: David Marr's theory of vision

1978: John McDermott's expert system R1/XCON

1979: Johan DeKleer's qualitative reasoning

1979: Drew McDermott's non-monotonic logic

1979: William Clancey's Guidon

1980: Intellicorp, the first major start-up for Artificial Intelligence

1980: John Searle's "Minds, Brains, and Programs" on the "Chinese Room"

1980: Kunihiko Fukushima's Convolutional Neural Networks

1980: McCarthy's Circumscription

1981: Danny Hillis' Connection Machine

1981: Hans Kamp's Discourse Representation Theory
1981: The Parallel Distributed Processing research group at UC San Diego
1981: Automatix introduces the first commercial robot with a vision system
1982: Teuvo Kohonen's Self Organising Map (SOM)
1982: Japan's Fifth Generation Computer Systems project
1982: John Hopfield describes a new generation of neural networks, based on a simulation of annealing
1982: Judea Pearl's "Bayesian networks"
1982: Teuvo Kohonen's Self-Organized Maps (SOM) for unsupervised learning
1982: The Canadian Institute for Advanced Research (CIFAR) establishes Artificial Intelligence and Robotics as its very first program
1983: Scott Kirkpatrick's simulated annealing
1983: Yurii Nesterov's accelerated version of gradient descent (Nesterov momentum)
1983: Geoffrey Hinton's and Terry Sejnowski's Boltzmann machine
1983: Gerard Salton's "Introduction to Modern Information Retrieval"
1983: John Laird and Paul Rosenbloom's SOAR
1984: Valentino Braitenberg's "Vehicles"
1985: Ross Quinlan's ID3 for decision trees analysis
1985: Rodney Brooks' subsumption architecture for robots
1985: The first international conference on genetic algorithms
1986: Terrence Sejnowski's and Charles Rosenberg's NETtalk
1986: David Zipser's "autoencoder"
1986: Jeanny Herault's and Christian Jutten's independent component analysis
1986: David Rumelhart's "Parallel Distributed Processing" rediscovers Werbos' backpropagation algorithm
1986: Paul Smolensky's Restricted Boltzmann machine
1986: Barbara Grosz's "Attention, Intentions, and the Structure of Discourse"
1987: Dana Ballard uses unsupervised learning to build representations layer by layer
1987: Chris Langton coins the term "Artificial Life"
1987: Hinton moves to the Canadian Institute for Advanced Research (CIFAR)
1987: Marvin Minsky's "Society of Mind"
1988: Toshio Fukuda's self-reconfiguring robot CEBOT
1988: Dean Pomerleau's self-driving vehicle
1988: Fred Jelinek's team at IBM publishes "A statistical approach to language translation"
1988: Hilary Putnam: "Has artificial intelligence taught us anything of importance about the mind?"
1988: Philip Agre builds the first "Heideggerian AI", Pengi, a system that plays the arcade videogame Pengo
1988: Fred Jelinek's team at IBM publishes "A Statistical Approach to Language Translation"
1989: Yann LeCun's "Backpropagation Applied to Handwritten Zip Code Recognition"

1989: George Cybenko proves that neural networks can approximate continuous functions

1989: Chris Watkins' Q-learning

1989: George Cybenko proves that neural networks can approximate continuous functions

1989: Alex Waibel's "time-delay" neural network

1989: Yann LeCun's convolutional neural network for handwritten-digit recognition (LeNet-1)

1990: Robert Jacobs' "mixture-of-experts" architecture

1990: Carver Mead describes a neuromorphic processor

1990: Peter Brown at IBM implements a statistical machine translation system

1990: Ray Kurzweil's book "Age of Intelligent Machines"

1991: Isabelle Guyon applies Vapnik's "support vector machine" to pattern classification

1992: Thomas Ray develops "Tierra", a virtual world

1992: Hava Siegelmann and Eduardo Sontag prove that recurrent neural networks are equivalent to Turing machines

1992: Ron Williams' REINFORCE algorithm

1992: Long-ji Lin's "experience replay" algorithm for reinforcement learning

1993: Masayuki Inaba's remote-brained robots

1993: Tom Mitchell's Xavier robot

1994: Ernst Dickmanns' self-driving car drives more than 1,000 kms

1994: The first "Toward a Science of Consciousness" conference in Tucson, Arizona

1995: Geoffrey Hinton's Helmholtz machine

1995: Vladimir Vapnik's "Support-Vector Networks"

1995: First "No Free Lunch" theorem by David Wolpert

1996: David Field & Bruno Olshausen's sparse coding

1997: Sepp Hochreiter's and Jeurgen Schmidhuber's LSTM model

1997: IBM's "Deep Blue" chess machine beats world champion, Garry Kasparov

1997: NASA's Mars Pathfinder lands on Mars and deploys the first roving robot, Sojourner

1998: Yann LeCun's LeNet-5

1998: Sebastian Thrun's Minerva and Pearl robots

1998: Two Stanford students, Larry Page and Russian-born Sergey Brin, launch the search engine Google

1998: Thorsten Joachims' "Text Categorization With Support Vector Machines"

2000: Cynthia Breazeal's emotional robot, "Kismet"

2000: Honda's humanoid robot "Asimo"

2000: Seth Lloyd's "Ultimate physical limits to computation"

2000: Hirochika Inoue's humanoid robot H6

2001: Yoshua Bengio's "Neural Probabilistic Language Model"

2001: Juyang Weng's "Autonomous mental development by robots and animals"

2001: Nikolaus Hansen introduces the evolution strategy called "Covariance Matrix Adaptation" (CMA) for numerical optimization of non-linear problems

2001: Herbert Jaeger's echo state networks

2002: Wolfgang Maass and Henry Markram's liquid state machines
2002: iRobot's Roomba
2003: Hiroshi Ishiguro's Actroid, a robot that looks like a young woman
2003: Jackrit Suthakorn and Gregory Chirikjian at Johns Hopkins University build an autonomous self-replicating robot
2003: Yoshua Bengio's "Neural Probabilistic Language Model"
2003: Tai-Sing Lee's "Hierarchical Bayesian inference in the visual cortex"
2003: Klaus Loeffler's humanoid robot Johnnie
2004: Mark Tilden's biomorphic robot Robosapien
2005: Andrew Ng at Stanford launches the STAIR project (Stanford Artificial Intelligence Robot)
2005: Boston Dynamics' quadruped robot "BigDog"
2005: Jun-ho Oh's humanoid robot Hubo
2005: Hod Lipson's "self-assembling machine" at Cornell University
2005: Pietro Perona's and Fei-fei's "A Bayesian Hierarchical Model for Learning Natural Scene Categories"
2005: Sebastian Thrun's driverless car Stanley wins DARPA's Grand Challenge
2006: Geoffrey Hinton's Deep Belief Networks (a fast learning algorithm for restricted Boltzmann machines)
2006: Osamu Hasegawa's Self-Organising Incremental Neural Network (SOINN), a self-replicating neural network for unsupervised learning
2006: Robot startup Willow Garage is founded
2006: Alex Graves' connectionist temporal classification (CTC)
2007: Yoshua Bengio's Stacked Auto-Encoders
2007: Stanford unveils the Robot Operating System (ROS)
2008: Adrian Bowyer's 3D Printer builds a copy of itself
2008: Cynthia Breazeal's team at MIT's Media Lab unveils Nexi, the first mobile-dexterous-social (MDS) robot
2008: Dharmendra Modha at IBM launches a project to build a neuromorphic processor
2009: Fei-fei Li's ImageNet database of human-tagged images
2010: Daniela Rus' "Programmable Matter by Folding"
2010: Lola Canamero's Nao, a robot that can show its emotions
2010: The New York stock market is shut down after algorithmic trading wiped out a trillion dollars within a few seconds.
2010: James Kuffner coins the term "cloud robotics"
2011: Dong Yu's speech recognition using deep learning
2011: IBM's Watson debuts on a tv show
2011: Nick D'Aloisio releases the summarizing tool Trimit (later Summly) for smartphones
2011: Osamu Hasegawa's SOINN-based robot that learns functions it was not programmed to.
2012: Rodney Brooks' hand programmable robot "Baxter"
2012: Alex Krizhevsky and Ilya Sutskever demonstrate that deep learning outperforms traditional approaches to computer vision processing 200 billion images during training

2012: Andrew Ng's team demonstrates an unsupervised neural network that recognizes cats in videos

2012: Alex Krizhevsky and Ilya Sutskever from University of Toronto demonstrate that deep learning outperforms traditional approaches to computer vision processing 200 billion images during training (AlexNet)

2013: Ross Girshick's Region-based Convolutional Neural Networks (R-CNN)

2013: Max Welling's and Diederik Kingma's variational autoencoders

2013: Tomas Mikolov's Word2vec

2013: Volodymyr Mnih's Deep Q-Networks

2013: Yangqing Jia develops the deep-learning platform Caffe

2013: Nal Kalchbrenner's and Phil Blunsom's "sequence to sequence" learning

2014: Vladimir Veselov's and Eugene Demchenko's program Eugene Goostman, which simulates a 13-year-old Ukrainian boy

2014: Fei-fei Li's computer vision algorithm that can describe photos

2014: Alex Graves, Greg Wayne and Ivo Danihelka publish a paper on "Neural Turing Machines"

2014: Jason Weston, Sumit Chopra and Antoine Bordes publish a paper on "Memory Networks"

2014: Microsoft demonstrates a real-time spoken language translation system

2014: Ilya Sutskever and Oriol Vinyals use a recurrent neural network to improve machine translation at Google ("Sequence to Sequence Learning with Neural Networks")

2014: Ian Goodfellow's generative adversarial networks

2014: Karen Simonyan and Andrew Zisserman's VGG-16

2014: Kyunghyun Cho's encoder-decoder model and gated recurrent units (GRUs)

2014: Volodymyr Mnih's recurrent attention model (RAM)

2014: Alex Graves' LSTM without Hidden Markov Models for speech recognition

2014: Christian Szegedy's GoogLeNet

2014: Microsoft's Skype demonstrates a real-time spoken language translation system

2015: Microsoft's 152-layer Residual Net

2015: Alec Radford's deep convolutional generative adversarial networks

2015: Over 1,000 high-profile Artificial Intelligence scientists sign an open letter calling for a ban on "offensive autonomous weapons"

2015: Baidu's Deep Speech 2 that uses a GRU instead of a LSTM and no HMM

2015: Francois Chollet develops the deep-learning platform Keras

2016: Kaiming He's ResNet with identity mappings of 1001 layers

2016: AlphaGo, developed by Aja Huang's team at DeepMind, beats Go master Lee Se-dol

2016: Jianpeng Cheng's and Mirella Lapata's self-attention

2017: Hinton's Capsule Nets

2017: DeepMind's AlphaZero

2017: Alexei Efros' team generates images from sketches with Pix2pix

2017: More than 100 variants of GANs are introduced in 2017

2017: Google's "transformer" model for sentence analysis (Ashish Vaswani, Noam Shazeer, Jakob Uszkoreit)

2017: John Schulman's proximal policy optimization for reinforcement learning
2018: Ali Eslami's and Danilo Rezende's Generative Query Network – GQN
2018: OpenAI's OpenAI Five
2018: Xiaolong Wang's nonlocal neural networks
2018: Jeremy Howard's and Sebastian Ruder's ULMFiT
2018: David Duvenaud's Neural ODEs
2018: Jacob Devlin's BERT for reading comprehension
2019: OpenAI's GPT2 creates convincing articles
2019: DeepMind's AlphaStar wins against 99.8% of its human opponents at the videogame StarCraft II

Readings on the Singularity

Anderson, James & Rosenfeld, Edward: "Talking Nets" (1998)

Barricelli, Nils: "Suggestions for the Starting of Numeric Evolution Processes Intended to Evolve..." (1987)

Brynjolfsson, Erik & McAfee, Andrew: "Race Against the Machine" (2012)

Bostrom, Nick: "Superintelligence" (2014)

Carr, Nicholas: "Utopia Is Creepy and other Provocations" (2016)

Chalmers, David: "The Singularity - A Philosophical Analysis" (2010)

Cooke, Conrad: "Automata Old and New" (1893)

Domingos, Pedro: "The Master Algorithm" (2015)

Eden, Amnon & others: "The Singularity Hypotheses" (2013)

Geraci, Robert: "Apocalyptic A.I." (Oxford, 2010)

Gerstner, Wulfram: "Neuronal Dynamics" (2014)

Good, Irving John: "Speculations Concerning the First Ultraintelligent Machine" (1964)

Goodfellow, Ian & Bengio, Yoshua: "Deep Learning" (2016)

Kaku, Michio: The Future of Intelligence (2014)

Khan, Nora: Towards a Poetics of Artificial Superintelligence (2015)

Kurzweil, Ray: "The Age of Intelligent Machines" (1990)

Lanier, Jaron: "The Singularity Is Just a Religion for Digital Geeks" (2010)

Moravec, Hans: "Mind Children " (1988)

Mori, Masahiro: "The Buddha in the Robot" (1974)

Muehlhauser, Luke: "When Will AI Be Created?" (2013)

Nilsson, Nils: "The Quest for Artificial Intelligence" (2009)

Noble, David: "The Religion of Technology" (1997)

Norvig, Peter & Russell, Stuart: "Artificial Intelligence - A Modern Approach" (2018)

Scaruffi, Piero: "Thinking about Thought" (2015)

Theodoridis, Sergios: "Machine Learning" (2015)

Vinge, Vernon: "The Singularity" (1993)

IEEE Spectrum's "Special Report: The Singularity" (2008)

The Journal of Consciousness Studies's Volume 19/7-8 (2012):
    Corabi & Schneider: "The metaphysics of uploading"
    Dainton, Barry: "On singularities and simulations"
    Hutter, Marcus: "Can intelligence explode?"
    Steinhart, Eric: "The singularity: Beyond philosophy of mind"

Alphabetical Index

Piero Scaruffi is an amateur human being and professional free thinker. He graduated in Mathematics (summa cum laude) in 1982 from University of Turin, where he did work in General Theory of Relativity. For a number of years he was the founding director of the Artificial Intelligence Center at Olivetti, based in Cupertino, California, and later joined IntelliCorp, one of the earliest companies specializing in Artificial Intelligence. He has also conducted interdisciplinary research at Harvard University and Stanford University, has lectured in three continents (most recently at U.C. Berkeley), and has published a number of books, including *History of Silicon Valley* and *Thinking about Thought*, but also *A History of Rock Music* and *A Brief History of Knowledge*. He is the fonder of the Leonardo Art Science Evening Rendezvous (L.A.S.E.R.) series and of the Life Art Science Tech (L.A.S.T.) festival.

Praise for "Intelligence is not Artificial":

"When it comes to any technology it is crucial to consider its reality, and not simply the hype put out by the media, or by the individuals and industries that are poised to profit by those techonologies. 'Intelligence is Not Artificial' does a great job cutting through the mountains of propaganda to present sobering and refreshing realities of its subject."

(Mitch Altman, virtual-reality pioneer and founder of one of the earliest hackerspaces, Noisebridge)

"Creative thinking, outside the box. "Intelligence is not Artificial" is a well thought-out counterpoint to the contemporary stream of artificial intelligence prophecies and assertions."

(Eric Gordon, Consulting Professor at Stanford University, Senior Advisor at Skyline Ventures, President of Palantir Consulting)

"I loved already the premise: like so many past ages, we inflate our importance in the overall arch of history. And i love this book's wet blanket reality thrown over the arrogant self importance of the AI scientist/hipster class."

(John Law, cofounder of the Burning Man Festival)

www.ingramcontent.com/pod-product-compliance
Lightning Source LLC
Chambersburg PA
CBHW071245220526
45468CB00001B/6